Essentials of Biophysics

Essentials of Biophysics

(Second Edition)

P. Narayanan

ANSHAN LTD
11a Little Mount Sion,
Tunbridge Wells, Kent,
TN1 1YS

ESSENTIALS OF BIOPHYSICS
SECOND EDITION

Co-Published in the UK in 2010 by:

Anshan Ltd.
11a Little Mount Sion
Tunbridge Well
Kent TN1 1YS

Tel: +44 (0) 1892 557767
Fax:+44 (0) 1892 530358

e-mail: info@anshan.co.uk
Web site: www.anshan.co.uk

ISBN 978 1 848290 34 1

British Library Cataloguing in Publication Data
A catalogue record for this book is available from the British Library

Preface to the Second Edition

Biophysics is a multidisciplinary subject—a "bridge" between physical sciences (Physics, Chemistry and Mathematics) with biological, medical and allied sciences. There is a direct correlation and equivalence between physical and biological/medical sciences. X-ray crystallography, NMR and other spectroscopies are increasingly used in structure-function elucidation of biomolecular complexes at atomic and molecular resolutions. While imaging techniques (*e.g.,* microscopies and radiography) have been in use in medicine and biology, advances in microscopic and topographic technologies have ushered in high-resolution and better quality imaging, which have become regular features in modern-day medical diagnostics and therapy protocols.

Recent advances in physics and allied areas– scanning-, and surface-probe imaging methods, Bose-Einstein condensate matter and the possibility of single-atom lasers, laser-traps ("optical- tweezers"), Bioinformatics and nanotechnology, use of biomolecules in computer chips, memory devices (neuronal networks)—, and advances in molecular biology (*e.g.*, genome projects) have brought Biophysics to the centre stage as via medium between the physical and biological/ medical sciences.

As biology and medical sciences are becoming more data-rich, it is becoming increasingly apparent that analytical and computational methods, characteristics of physical sciences, are required to streamline and interpret vast amount of complex data and, therefore, there has been a growing awareness to link physical and biological/medical sciences more closely in research and teaching.

The main aim of this book, therefore, is to present practically all the topics that can be classified as essential to Biophysics. Such topics include chemical bonding (essentials of Quantum physics), Cellular, Molecular and Membrane biophysics, Biophysical chemistry of biomolecules, physiochemical techniques (spectroscopies, microscopies, NMR, X-ray diffraction etc.,), Bioenergetics, Biomechanics, Photo-, Chemo-, and Radiation biophysics, and Bioinformatics.

The contents of the book have been arranged under sections and subsections. An overview is given for each section that comprises several chapters with a common theme. In addition, synopsis is included for each chapter to provide readers a gist of the chapter in compact form. Further mathematical formulas and formalisms, Biostatistics and useful information are provided in appendix to make the book self-sufficient. A list of terms, concepts and phrases in the glossary and index would facilitate the readers a quick access to the technical terms and subject matter in the text. With these efforts, it is hoped that this book would serve the purpose that it is intended for and would catalyze in some way in bringing the subject of Biophysics to play a central role in the advancement of biological, biomedical and related areas.

In this book, a conscious effort has been made to avoid over-emphasis on overtly Physics or overtly Biology topics. Therefore, there is less attention given to Electronics, Computers, and Instrumentation and Genetics, Molecular biology, Biochemistry, Immunology, and Microbiology etc. They are separate subjects in their own right and there are number of books available on these subjects. This book is intended for all students and researchers with Physics, Chemistry, Biology and Medicine background. It is to serve people with biological and medical background to make them knowledgeable about physical principles and techniques, which have become an integral part of biological and medical sciences. It is also intended to serve as an introductory source to make the physicists and chemists and other physical scientists aware of the essential aspects of biological sciences and the trends and progresses in *natural sciences.*

P. Narayanan

Acknowledgements

It is with pleasure to acknowledge the help rendered by Professor M. Vijayan, Indian Institute of Science, Bangalore, for providing photographs and diagrams and allowing the use of Computer Graphics facilities at the Institute. Acknowledgements are also due to Prof. M.R.N. Murthy of the same institute for the X-ray diffraction photographs. All the members of the Bioinformatics Centre of the institute have been generous in extending their help and the author warmly acknowledges their cooperation and camaraderie. Acknowledgments are also due to Dr. D.N. Patkar, Director and colleagues of the Western Regional Instrumentation Centre, University of Mumbai, for their help in some of the artwork, and to Swarna Murthy, and Mr. T.K.S. Murthy for their help and encouragement.

Contents

SECTION II
Physical Techniques in Structure Determination

SECTION III
Bioenergetics

SECTION IV
Biological Systems

Introduction

The subject of Biophysics has always been and is going to be even more, multidisciplinary, bridging the areas of physical (Physics, Chemistry and Mathematics) and natural sciences (Biology, Pharmacology and Medicine). It is a discipline concerned with the application of the principles of physics and the other physical sciences to the solution of biological/ medical problems. Its main emphasis is on the quantitative analysis of the physical and chemical aspects of the functions of biological molecules, organisms and entities, *vis a vis* their structures at the cellular, molecular and atomic levels. The techniques and methodologies that Biophysics relies on are closer to Physics and Chemistry, but the arena of applications are in the biological, medical and related sciences. There is a direct correlation and equivalence between physical and biological/medical sciences (Table I.1).

Table I.1 Correlation between physics and biology/medicine.

Physics	*Biology/Medicine*
Statics (kinematics)	Skeletal function; orthopedics; internal body mechanics.
Dynamics (Motion)	Spatial movement; flagella, heart and lung function.
Mass transport	Fluid flow; metabolic functions.
Charge transport	Ion transport; nerve impulse transduction; molecular signaling.
Surface tension	Capillary action; fluid transport in plants and cells.
Osmosis	Fluid transport in cells.
Molecular Physics	Structure-function elucidation.
Thermodynamics	Bioenergetics; metabolic and reaction kinetics.
Optics	Spectrosocpics; microscopies; lasers; optics of vision.
Acoustics	Mechano-transduction.
Radiation Physics	Radioactivity; radioassays; diagnostics; imaging and therapy.
Tomographies	Diagnostics and therapy.

Biophysics had a good beginning, due to the efforts of the early stalwarts who had a progressive and wide-ranging outlook on science and technology. Many of the early discoveries in physical sciences were made by such scientists who had diverse scientific backgrounds and interests, and were not constrained by the narrow over-compartmentalisation of knowledge. In fact, quite a number of discoveries in the 19th century in physical sciences were from scientists who were trained in biological and medical sciences. They applied physicochemical methods of analysis to Biology and, being from the biological sciences, they were in a better position to understand the problems and complexities of extending physicochemical methods to the natural sciences.

To cite a few cases: **Luigi Galvani** was a man of medicine who was instrumental in the development of electrochemistry and neurophysiology. **Thomas Young** was a pioneer in optics and vision, in addition to his keen interest in deciphering the Egyptian hieroglyphs. **Hermann von Helmoltz** was a trained physician, but his contributions to Physics and Biology are vast and wide-ranging— from electromagnetism, thermodynamics (conservation of energy), physiological optics (he invented the ophthalmoscope), colour theory and colour vision to physiological acoustics. To such group can be added the names of pioneers like **von Humboldt**. With such a stimulating and holistic approach, the flow of knowledge, from physical to natural sciences and *vice versa*, was on equal footing.

Developments in science in the later part of the 19th century and the early part of 20th century were lopsided, with a narrow compartmentalisation of subjects. The historical development of science, the practical necessity for divisions and the rigid adherence to their narrow areas by the practicing scientists were mostly responsible to such a state of affairs. This situation led to Biophysics, an interdisciplinary subject, relegated and forgotten.

By and large, Biology as practiced till the mid 20th century (till the late 1940s) was descriptive and speculative. Scientists from Biology and Medicine were trained with little or no knowledge of the fundamentals Physics and Chemistry and also the quantitative aspect that physical sciences inculcate. Being devoid of such training, it is not surprising that there was no major cross-fertilisation of scientific ideas from the biological to the physical sciences. While the physical sciences were making epoch-making discoveries from the early part of the 20th century, a large body of biological scientists concentrated their best efforts on the classification and morphological studies, and qualitative trial-and-error and hit-and-miss methods for solution. Such a development in the natural sciences had a tremendous negative impact on Biophysics and its neglected status.

There has been a discernible change, from the mid 20th century on wards in the approach and attitude to tackling the problems in biological, medical and allied sciences with more emphasis on the quantitative approach. This has been partly due to the realisation that—

1. Traditional departments (compartmentalisation) can be physical barriers to cross-fertilisation of ideas.
2. A deeper understanding of complex biological systems will need a more quantitative type of biology that is closely integrated with the physical sciences.
3. Active participation of physical scientists in biomedical-related sciences is essential.
4. Advances in physical and physicochemical techniques and instrumentation. Many of today's high-profile biological/biomedical advances have their roots in physics-related principles and technologies. Biomedical instrumentation is widely used in medical fields.

Some of the physical scientists who provided impetus for the progress of Biophysics in the 20th century include **George Gamow, Max Delbrück, Erwin Schrödinger, Lawrence Bragg, Linus**

Pauling, J.D. Bernal, Dorothy Hodgkin, Max Pertuz, Francis Crick, G.N. Ramachandran, Hodgkin and **Huxley, Claude Shanon, Alan Turing** and many others. The recent change in the mindset towards biology has also been due to the realisation that—

(i) Understanding of the complex arrangement of living matter cannot be achieved by classification, speculation and insights, but by quantitative physicochemical methods.

(ii) Life processes, in spite of varied structural organisation and complexity, do not violate the basic laws of Physics.

(iii) All biological problems can be addressed at the cellular, molecular and atomic levels, due to recent advances in physicochemical techniques and computer-based processing and modeling methods applied to biological systems.

Modern quantitative biology (Genetics, Molecular biology, Biochemistry, Microbiology etc.) relies on the physicochemical methods to understand life processes at the cellular, molecular and atomic levels. Some of these physical techniques are electron microscopies, spectroscopies, hydrodynamical methods, magnetic resonance, and X-ray diffraction. In the early half of the 20th century there was a greater emphasis on Radiation biophysics. This was because of the preoccupation of a large school of physical scientists in nuclear-related fields and due to generous availability of funds, facilities and governmental and industrial patronage. Radiation physics, applied to biological systems, was considered synonymous with Biophysics. The shift towards Molecular Biophysics has been primarily due to feasibility of obtaining complete three-dimensional architectures of biological macromolecules by X-ray diffraction at molecular/atomic resolutions. Starting from **Max Perutz** and **John Kendrew**, who were instrumental in solving the three-dimensional structures of proteins, hemoglobin and myoglobin, respectively, X-ray crystallographers played (and are playing) a very important role in extending the quantitative approach in biological and medical sciences. The **Watson-Crick** model of the DNA molecule brought together the structural and biochemical approaches to Genetics and Molecular biology, and has been instrumental to bringing the Molecular biology to the forefront. Recent advances in solution studies by multidimensional nuclear magnetic resonance (NMR) spectroscopic techniques have brought forth an additional quantitative physical modality for elucidating the three-dimensional structure and functional features of smaller molecular size proteins, nucleic acids and other biomolecuar entities of importance.

Progress made in the chemical foundation of life, that is Biochemistry, has been due to important developments in Molecular chemistry by hydrodynamical methods, sequence determination by conventional and advanced methods (e.g. mass spectrometry), and tracer and enzyme-labelled techniques. Contributions by various groups to Bioenergetics (**Katalsky, Prigogione** and others of the Brussels's group); and in computer technology— the development of mathematical logic and formalisms for developing artificial intelligence and robotics (Cybernetics)—, initiated by **von Neumann**, **Norbert Wiener**, **Alan Turing**, **Claude Shanon** and others brought about the influx of computer era in biology and medicine, and emergence of Bioinformatics to the centre stage in biological/biomedical sciences.

Spectacular advances in Molecular biology in recent years have also contributed greatly in bringing Biophysics to the centre stage. For example: application of innovative physical techniques in Molecular biology— fluorescence methods, tagged with auto-fluorescent proteins (such as green fluorescence protein), enables the imaging of molecular interactions and dynamics in the cell; gene selection and amplification techniques (by polymerase chain reaction and other methods) and selective methods/mutation methods (such as site-directed mutagenesis) allow greater sophistication and precision to biologists to modify or manipulate biological macromolecules in a more rational and planned way. The feasibility of obtaining three-dimensional structures of biological molecular species by X-ray diffraction and NMR spectroscopic methods in a reasonable time, and computer-modelling (*computational biology/molecular tinkering)* methods offer solutions to rational design protocols to obtain new types of molecular species (*virtual mutants*). Molecular engineering, which is a combination of molecular biological methods, physical techniques (primarily X-ray diffraction and NMR spectroscopy), and knowledge-based computer modeling procedures (*Bioinformatics*) will play a central role, in the progress of biological and medical sciences, in our understanding of the functions of biological systems *vis a vis* their structure, and in designing new molecular species, tailor-made to improve or suit specific requirements.

Imaging techniques— radiological and optical, electron and other microscopies have been in use in medicine and biology. But, recent advances in imaging techniques— radiological (CT), nuclear (SPECT and PET), magnetic resonance (MRI and fMRI) and ultrasonography— have greatly enhanced the clarity and resolution of images, very useful in diagnostic procedures and organ function as well as therapy. Further, developments in scanning-probe microscopy (SPM) technology— laser scanning confocal (LSM), scanning-tunneling (STM), atomic force (AFM), aperture-less and chemical microscopy— augur fundamental changes in developing new methods to examine physical and chemical properties of single macromolecules.

Application of recent developments in physical sciences— *monoatomic* condensed matter (Bose-Einstein condensate), quantum dots, single-atom lasers, laser-traps ("*optical tweezers*"), Bioinformatics and nanotechnology will be the basis for future developments, for exquisitely sensitive measurements on biomolecules at atomic resolution, use of biomolecules in computer chips, memory deices (neuronal networks) and solar-energy cells etc., thus bring Biophysics to the centre stage, as *via* medium between physical and natural (biological/ biomedical) sciences.

As biological and medical sciences are becoming data-rich (e.g. with the advent of genome projects), it is becoming increasingly apparent that analytical and computational methods, characteristics of physical sciences, are required to streamline and interpret vast quantity of complex of data. With this growing awareness, there have been conscientious efforts to link physical and natural sciences more closely in research and teaching, to impart biologists the ability to separate biological problems into segments that are amenable to exact physical interpretation and to formulate hypotheses that can be tested by experiment. It is hoped that by imparting biological streams with more Physics, Chemistry and Mathematics— that is, orienting them to Biophysics (*Integrative biology*)— they can be made to expand their horizons and adapt quantitative approaches in research and teaching, for example, the ability to use complex physical theories (e.g. X-ray diffraction techniques) to analyse natural entities (cells; molecules etc.).

But, imparting physical sciences courses to biological sciences should not be done as a separate exercise or as an additional material. The approach should be integral and holistic, with concepts and principles of physics and chemistry and necessary mathematical formalisms and quantitative tools of analytical analysis forming an integral part of such an approach in biological and medical sciences. In the absence of such a holistic approach, even subjects like Biochemistry and Molecular Biology, which have closer affinity to physical sciences, would become "soft" subjects. Therein lies the importance and centrality of Biophysics, in the emerging areas in biological and medical sciences.

Abbreviations

AAS	:	Atomic Absorption Spectroscopy
AES	:	Atomic Emission Spectroscopy
AFM	:	Atomic Force Microscopy
αtα	:	Helix-Turn-Helix
CARS	:	Coherent AntiStokes Raman Spectroscopy
CD	:	Circular Dichorism
CIA	:	Chemiluminescence ImmunoAssay
DIC	:	Differential Interference Contrast
ENDOR	:	Electron Nuclear Double Resonance
ESR	:	Electron Spin Resonance
FID	:	Free Induction Decay
FRET	:	Fluorescence Resonance Energy Transfer
FT	:	Fourier Transform
GC	:	Gas Chromatography
HPLC	:	High-Performance Liquid Chromatography
IEP	:	IsoElectric Point
IMRA	:	ImmunoMetricRadioAssay
IPG	:	Immobilised pH Gradient
LCAO	:	Linear Combination of Atomic Orbitals
MRI	:	Magnetic Resonance Imaging
MS	:	Mass Spectrometry
NMR	:	Nuclear Magnetic Resonance
PAGE	:	PolyAcrylamide Gel Electrophoresis

PET	:	Positron Emission Tomography
PMT	:	PhotoMultiplier Tube
RIA	:	RadioImmunoAssay
SDS	:	SodiumDodecyl Gel Electrophoresis
SEM	:	Scanning Electron Microscopy
STM	:	Scanning-Tunnelling Microscopy
SPECT	:	Single-Photon Emission Computed Tomography
STEM	:	Scanning Transmission Electron Microscopy
TEM	:	Transmission Electron Microscopy

SECTION I

Biomolecular Structures

OVERVIEW

The nature of chemical bonding and practically all molecular interactions at the atomic level can be rationally explained only by invoking the quantum nature of such interactions. **Pauli's '*Aufbau*'** principle, the mode of filling up of orbitals by electrons, gives a rational picture to the periodic table of elements. The formation of molecules by covalent bonding is due to the combination of atomic orbitals — **s-**, **p-**, and **d**-orbitals.

The most important biomacromolecules are: proteins, nucleic acids, polysaccharides, membranes and their complexes. Of these, proteins/enzymes, nucleic acids and polysaccharides are biopolymers whereas membranes are not. A basic knowledge of these structures and conformation flexibilities of macromolecules is very necessary to an understanding of the structure *vis-a-vis* function and reaction mechanisms of various biochemical and biological processes. These are: protein-protein associations (e.g. immunoglobulin defence mechanisms), protein-nucleic acid interactions (e.g. translation of genetic information and hereditary characteristics), protein-carbohydrate interactions (e.g. cell recognition, motility and transport of mass and charge across cell membranes).

Proteins are linear polymers of twenty L-amino acids, each characterised by its side-chain functional group(s). Amino acids are in the 'zwitterionic' state at physiological pH and some side chains are acidic, some are basic, some are neutral, while others are hydrophobic. The functional diversity of proteins is due to the amino acid sequence and tertiary folding of the protein molecule as well as due to environmental influences. Higher level structural organisation of macromolecules (tertiary and quaternary) are due to weak non-covalent interactions. Various semiempirical methods of structure prediction of proteins from their amino acid sequence data attempt to reduce the astronomical number of possible conformations to probable ones, by incorporating experimental data and model-building procedures. These might suggest possible folding pathways of protein complexes (core of Bioinformatics).

Nucleic acids are also linear polymers of nucleotide units, comprising a sugar-phosphate backbone and planar bases: adenine, guanine, thymine (uracil) and cytosine. The bases are in keto/amino tautomeric forms and the sugar ring has either 2'- or 3'-endo conformation. Nucleic acids exist in right-handed A-, B-, and left-handed Z-families, depending on the humidity and salt concentrations. Microlevel structural differences lead to macrolevel functional features.

Carbohydrates are used by organisms as sources of energy and as structural material. They occur in nature predominantly as monosaccharides (glucose), disaccharides (sucrose, maltose and cellobiose) and linear as well as branched-chain polysaccharides (cellulose, starch and glycogen). Protein-membrane complexes form fluid-mosaic bilayers and act as highly selective permeability barriers (as gates, pumps and channels) to various ions and molecules. They serve as conduits for transport of mass, charge and stimuli. Phospholipids, sphingolipids, glycolipids and cholesterol are the major classes of membrane lipids. Membrane fluidity is controlled, at a given temperature, by fatty acid composition and cholesterol content.

The problem of prediction of secondary and tertiary structures from the amino acid sequence data of proteins is a complex one due to the cooperativity of stabilising interactions. There are many empirical methods of structure prediction; procedures relying on the matching of amino sequence with a given topology, taxonomic classification of structures into structural motifs and modules and dynamic methods of model-building, based on these structural elements, seem to have 'better' success rate of prediction. Protein-folding is related to evolutionary characteristics as tertiary folding is better conserved in evolution than amino acid sequence homologies. Comparison of tertiary structures of proteins is one of the ways of studying evolutionary processes at the molecular level, such as convergent and divergent evolutions, survival and adaptability to changing environments. Evolutions of cytochrome c and globin families are well-characterised examples.

1 Atoms, Bonds and Molecules

Synopsis: The laws of classical physics, governing the motions of macroscopic bodies, rest on the premises of continuum of energy and space and the possibility of the precise determination of trajectories of particles. These assumptions are erroneous and, therefore, cannot explain the interactions that take place at molecular, atomic and sub-atomic levels.

Classical physics cannot correctly explain the features of blackbody radiation, specific heats of atoms, line spectra of atoms and atomic structure. The correct explanation of these by quantum principles is the basis of modern physics and chemistry.

Max Planck put forward the idea of energy occurring in discrete packets (quanta) in integral multiples of $h\nu$, to explain the phenomenon of blackbody radiation. Planck's quantum hypothesis has been extended to explain various other phenomena—specific heats of atoms, the photoelectric effect, atomic and molecular spectra and the electronic configuration of atoms. **Schrödinger's** wave mechanics is the representation of a system by wave functions. Atomic orbitals represent regions in space in which a particle of particular energy is most likely to be found. The state of an atom can be determined by four quantum numbers: principal, angular, magnetic and spin quantum numbers. **Pauli's** *'Aufbau'* principle enables a systematic approach to filling up various orbitals by electrons and gives a rational explanation for the chemical nature of atomic structure and to chemical periodicity. This 'Aufbau' principle rests on the fundamental law governing electrons, which are **fermions**, and related to Pauli's exclusion principle.

Of atomic orbitals, **s**-orbitals are spherically symmetric, **p**-orbitals are distributed orthogonally in space and **d**-orbitals comprise three $\mathbf{2t_g}$ and two $\mathbf{e_g}$ orbitals. Covalent bonds between atoms are formed by sharing electron pairs. Combination of atomic orbitals (addition and subtraction) leads to molecular orbitals and chemical bonds. Single bonds are due to the combination of **s**- and **p**-orbitals ($\mathbf{sp^3}$-hybridisation), forming cylindrically symmetric σ-bonds. A combination of σ- and π-bonds ($\mathbf{sp^2}$-hybridisation) leads to the formation of double bonds. The aromatic nature of planar molecules, like benzene, is due to the π-bonding in these molecules, with π-electron clouds distributed above and below the plane of the molecule.

1.1 CONCEPTS OF CLASSICAL PHYSICS AND LIMITATIONS

Concepts of classical physics are based on two premises:

(i) continuum of energy and space where the particle can take up any arbitrary value and (ii) the motion of particles and their trajectories can be precisely determined. These assumptions, fundamental to classical physics, are actually approximations to the actual description of events and are normally applicable at the macroscopic level.

General laws of classical mechanics are primarily based on Newtonian kinematics. These are—

velocity, $v = x/t = dx/dt$

acceleration, $a = v/t = d^2x/dt^2$ ...(1.1a)

momentum, $p = mv$

angular momentum, $J = I\omega$

force, $F = ma = m\dfrac{d^2x}{dt^2} = -K.x$ (for harmonic oscillator)

(K is force constant)

energy, $E = \dfrac{1}{2}mv^2 + V$

K.E. P.E.

In addition, the kinetic theory of gases leads to the equipartition theorem—whereby the average value of energy in n-dimensions is

$$E = \frac{1}{2}nkT \quad \text{...(1.1b)}$$

where k is the **Boltzmann** constant and T is the absolute temperature.

But classical physics, which gives a fair description of the kinematics of macroscopic events, which are perceivable, cannot explain the microscopic (atomic and subatomic) nature of the events. The inadequacies of classical physics are apparent from the failure to explain, (i) wavelength-dependence of blackbody radiation, (ii) occurrence of discrete spectral lines, (iii) atomic structure and the nature of chemical bonds and other features (interactions) in the microscopic realm.

1.1.1 Blackbody Radiation

All hot substances emit electromagnetic radiation and a blackbody is one which absorbs and emits all frequencies of radiation uniformly. The plot of energy output as a function of temperature dependence (Fig. 1.1) indicates

(i) The peaks of curves shift towards shorter wavelengths with increase in temperature. This shift towards the blue-region is summarized under **Wien's** displacement law

$$T\lambda_{\max} = \frac{1}{5}C^2 \quad \text{...(1.2a)}$$

(ii) The energy density is proportional to the fourth power of the temperature, which is **Stefan's** law

$$U \propto \sigma T^4 \qquad \text{...(1.2b)}$$

where C^2 and σ are the second radiation constant and **Stefan-Boltzmann** constant, respectively, and λ is wavelength.

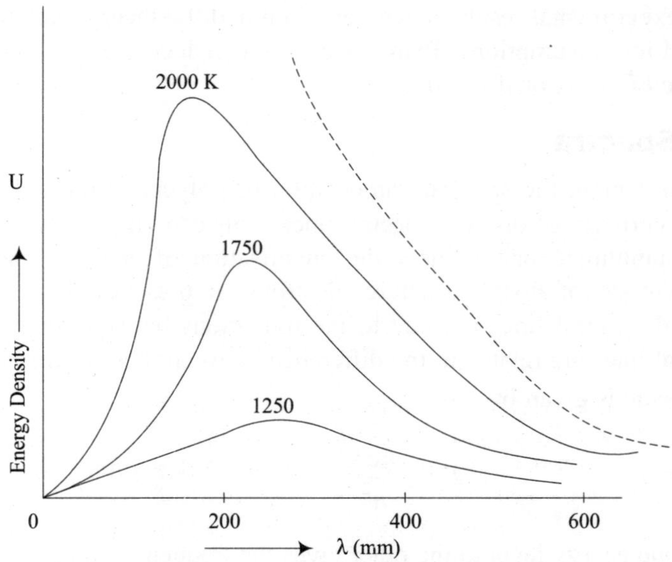

Fig. 1.1 Energy distribution of blackbody radiation as a function of wavelength, λ.... Rayleigh-Jeans curve.

Lord **Rayleigh** considered this problem from the standpoint of classical physics, treating electromagnetic radiation as an ensemble of harmonic oscillators, with an energy continuum. Applying the equipartion theorem for the average energy of the oscillators, the energy per unit volume per unit wavelength is given by the **Rayleigh-Jeans** equation

$$\rho = \frac{8\pi kT}{\lambda^4} \qquad \text{...(1.2c)}$$

The theory does not, however, fit the experimental results. The theory expects a large amount of energy to accumulate in the high-frequency region, which is called the ***'ultraviolet catastrophe'***. The objects should glow in the dark and there should be no darkness, according to the classical theory!

Planck's quantum theory. **Max Planck** resolved this problem by stating that the energy space is not continuous, but discrete. Therefore, the energy of each oscillator has a discrete value. The quantum of energy should be in integral multiples of $h\nu$.

$$E = nh\nu \qquad \textbf{Planck's law} \qquad \text{...(1.2d)}$$

where $n = 0, 1, 2, 3, \ldots$ and h is Planck's constant (6.626×10^{-34} J.Hz^{-1} and ν = frequency). Not all oscillators, but only those which acquire the necessary quanta of energy take part in the process and high frequency oscillators are not generally excited. The energy density is given by Planck's equation

$$\rho = \frac{8\pi h\mathbf{c}}{\lambda^5}\left\{\frac{1}{e^{hc/kT} - 1}\right\} \quad \text{Planck's equation} \qquad (1.2e)$$

The theory and experimental results agree very well and the theory explains Wien and Stefan's laws, without any ad-hoc assumptions. Planck's equation reduces to Rayleigh-Jeans equation for larger wavelengths $hc/kT \ll 1$; or if $h = 0$.

1.1.2 Atomic Spectra

Classical physics can explain the spectral composition of polychromatic light (white light), but it cannot explain the occurrence of discrete spectral lines. This can only be explained by invoking the basic principle of quantum theory, namely the quantisation of energy (discrete energy levels). According to **Bohr's** model of atomic structure, electrons are present only in certain allowed orbits (quantised states) and spectral lines are due to the transitions between these allowed orbits. The frequencies of spectral lines are related to the difference between these energy orbits (Fig. 1.2).

The energy of a state is given by

$$E_n = \frac{h\mathbf{c}R}{n^2} \qquad \text{...(1.3a)}$$

Transition from one energy level to the other gives the frequency, $\bar{\nu}$

$$\bar{\nu} = \frac{\nu}{\mathbf{c}} = R\left\{\frac{1}{n_1^2} - \frac{1}{n_2^2}\right\} \qquad (1.3b)$$

where n is the principal quantum number and R is the **Rydberg** constant.

$n_1 = 1, 2, 3, 4$ (for **Lyman, Balmer, Paschen** and **Brackett** series)

$n_2 = (n_1 + 1), (n_1 + 2), (n_1 + 3), (n_1 + 4) \ldots$

1.1.3 Atomic Structure

Classical physics cannot account at all for electronic configuration, that is, the spatial arrangement of electrons around the nucleus of an atom. According to the concepts of classical physics, electrons orbiting around a nucleus are unstable, because electrons in motion emit radiation and, therefore, their trajectories should be spiral, which should lead to their collapse into the nucleus. In order to avoid this serious pitfall **Niels Bohr** suggested, while proposing the atomic model for the line spectra of hydrogen atoms, that electrons in the allowed orbits (states) do not radiate energy, but the emission (or absorption) of radiation occurs only when there is an electronic transition from one energy state to another. Though this model could accurately explain the details of the hydrogen line spectrum, the model was based on ad-hoc assumptions to explain the experimental results. The electrons were

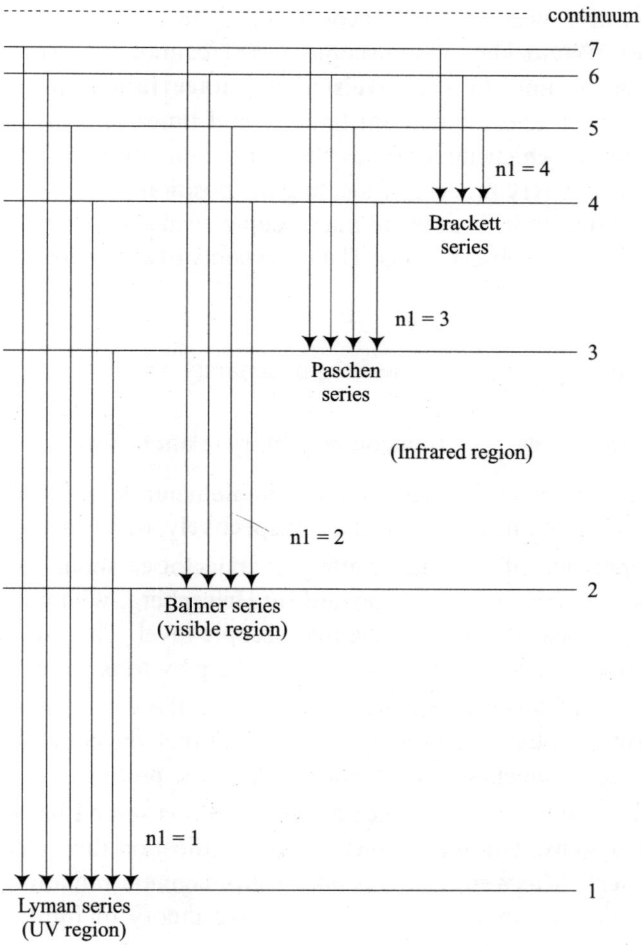

Fig. 1.2 Line spectrum of the hydrogen atom.

$$\bar{\nu} = R\left(\frac{1}{n_1^2} - \frac{1}{n_2^2}\right)$$

n_1 = 1 Lyman series = 2 Balmer series
= 3 Paschen series = 4 Brackett series

visualised as particles moving in defined orbits, like planetary motions in a solar system, and the model did not contradict the fundamental concepts of classical physics, except for energy quantisation. Extension of Bohr's model, incorporating elliptical orbits of electrons by **Sommerfeld** did improve the numerical accuracy, but did not alter the basic concepts. The assumption of radiationless electron orbits still smacked of arbitrariness.

The need to explain atomic structure in more rational terms led to two different approaches: matrix mechanics by **Werner Heisenberg** and wave mechanics by **Erwin Schrödinger**.

Heisenberg's suggestion was—not to accept in any type of model those terms which are not verifiable (observable). Day-to-day macroscopic terms cannot be extrapolated to explain the microscopic events. According to the **'Heisenberg uncertainty principle'**, complementary observables such as position and momentum (energy and time) of a particle cannot be precisely determined simultaneously, which undermines the very foundations of trajectory-based classical mechanics. For example, the very process of locating the position of an electron by a hypothetical γ-ray microscope would dislocate it. That is, all microscopic events are such that the observer is a part of the observable and affects the observations. The uncertainty of complementary observables is given by the relationship

$$\Delta\delta \; \Delta\varepsilon \geq \frac{\hbar}{2} \qquad \text{Heisenberg uncertainty principle} \qquad ...(1.4a)$$

$$(pq - qp)\,\psi = ih\,\psi \qquad \text{(noncommutability relationship)} \qquad ...(1.4b)$$

where $\Delta\delta$ and $\Delta\varepsilon$ are the uncertainties in the complementary observables and p and q are complementary operators of momentum and space, respectively, of a function ψ.

So, the classical approach of particles moving in trajectories should be abandoned to explain atomic structure. These concepts were put forward by Heisenberg, with considerable help from his professor, **Max Born,** to describe events at the microscopic level. The use of matrix theory to treat noncommutability relationships was suggested to Heisenberg by Max Born.

Schrödinger approached this problem with emphasis on the wave nature of electrons, based on the ideas of **Louis de Broglie**. Later, Schrödinger also established the equivalence of matrix and wave mechanics; the two are complementary approaches to the same problem.

The nature of light, whether it is corpuscular or wave, was settled in favour of wave theory of light by the work of **Huygens, Young, Fresnel** and the unification theory of electricity, magnetism and light by **James Clerk Maxwell**. Light is an electromagnetic radiation (wave). However, the photoelectric effect could be explained, not by the wave theory of light, but by the corpuscular (particle) theory by **Einstein**; quantising the energy and invoking light as particles which are called **photons**. For observations involving energy and momentum, light can be treated as photons and for those experiments involving the spatial distribution of light, it should be treated as a wave (as its wave properties come to the fore).

De Broglie surmised, from the equations of energy, E, and momentum, p, of particles, that particles (matter) should exhibit wave nature (Eqn. 1.5)

$$E = h\nu = mC^2$$

$$p = mC = \frac{E}{C} = \frac{h\nu}{C}$$

$$C = \lambda\nu$$

$$\therefore \qquad p = h/\lambda \qquad ...(1.5)$$

The wave nature of electrons was experimentally proved by **Davisson** and **Germer**.

According to Schrödinger, the wave function of a system can be represented by the wave equation

$$\Delta^2\psi + \frac{2m}{\hbar^2}(E - V)\psi = 0 \qquad ...(1.6a)$$

where m = mass of the particle; V = potential energy; $\hbar = h/2\pi$

Equation (1.6a) is the famous Schrödinger wave equation. It is an equation of the general type

(operator) (wave function) = (numerical value) (wave function)

$$H \quad \psi \quad E \quad \psi \qquad ...(1.6b)$$

An operator is a mathematical operation that has to be performed on a function, ψ, to extract information about an observable. For example, the gradient of a wave function (multiplied) by $\hbar$/i) at a point gives the momentum and the curvature of the wave function, the kinetic energy. The operator must be such that they yield real observables. That is, they are **hermitian operators** and should exhibit the noncommutability relationship ($pq - qp \neq 0$).

An operator (**Hamiltonian**), H, operating on a Schrödinger wave function ψ, would lead to the same wave function multiplied by observable(s) E_s (energy values). That is, Schrödinger wave functions are *eigenfunctions* and the energy observables are *eigenvalues*. Born's interpretation of the wave function puts further restrictions on wave function types. It is a probability function; the square of the wave function ($\psi\psi^*$) gives the spatial distribution probability of a given particle.

1.2 QUANTUM PRINCIPLES OF ATOMIC STRUCTURE

Wave functions of hydrogenic atoms (atoms and ions with a single electron) are called orbitals. Orbitals represent regions in space in which a particle of particular energy is most likely to be found. The state of an electron can be determined by wave functions represented by four quantum numbers—principal quantum number, **n**, angular (azimuthal) quantum number, **l**, magnetic quantum number, $\mathbf{m_l}$, and spin quantum number, **s**. The orbitals belonging to a principal quantum number, **n**, constitute a 'shell' and the energy of a state is determined by the principal quantum number. The 'shells' are denoted by capital letters of the alphabet.

n = 1, 2, 3, 4, … (K, L, M, N, … shells).

Orbitals of the same shell but different azimuthal values are called 'subshells'. These are denoted by lowerscript letters of the alphabet.

l = 0, 1, 2, 3, 4, … (**s, p, d, f, g** … subshells). An nth shell contains

$$\sum_{n=0}^{n-1}(2l+1) = \mathbf{n}^2 \text{ orbitals.} \qquad ...(1.7a)$$

The magnetic quantum number, $\mathbf{m_l}$ (space quantisation) will have

$$\mathbf{m_l} = 0, \pm 1, \pm 2, \pm 3, \ldots \pm l \text{ values} \qquad ...(1.7b)$$

Its significance is realized when a particle is under a magnetic field (**Zeemann** effect) (Fig. 1.3a).

The spin quantum number, **s**, can have only two values, $\mathbf{m_s} = + ½$ (↑) (called α-electron), or $\mathbf{m_s} = - ½$ (↓) (called β-electron) (Fig. 1.3b).

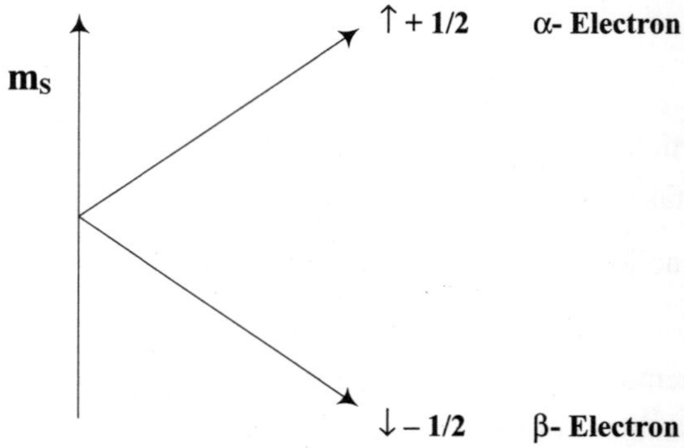

Fig. 1.3a Magnetic quantum number $\mathbf{m_l}$ (Space quantisation).

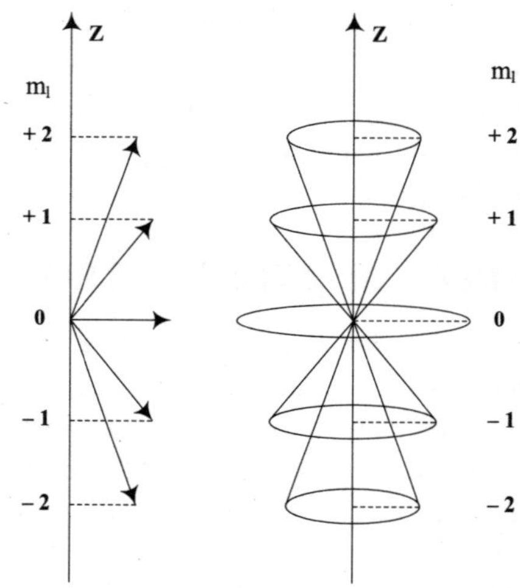

Fig. 1.3b Spin quantisation (m_s = +½ (↑) and −½ (↓)).

1.2.1 Pauli's Aufbau Principle of the Periodic Table

Pauli's Aufbau (building up) principle enables a rational way of determining the electronic configuration of an atom and thereby its essential chemical properties and chemical periodicity (the periodic table of elements).

An important and fundamental condition to the Aufbau principle is Pauli's 'exclusion principle' that applies to all **fermions** (all particles with half-integral spin, such as electrons, protons, neutrons, ^{13}C, ^{35}Cl etc.). This exclusion principle does not apply to **bosons** (particles with zero or integral spin such as photons, ^{12}C etc.). The **Aufbau** rules of filling up orbitals are:

(1) Electrons occupy the lowest energy orbitals first.

(2) A maximum of only two electrons per orbital is allowed (Pauli's 'exclusion principle').

(3) Electrons in the same energy level occupy separate orbitals with parallel spins first; only then are antiparallel spin electrons paired in each orbital (**Hund's** rule).

The electronic distributions of atomic structure are as in Table 1.1.

Table 1.1 Electronic distribution of atomic structure.

	s	p	d	f	g	h	Total	Total
l ⇒	*0*	*1*	*2*	*3*	*4*	*5*	*orbitals*	*electrons*
⇓ n								
1 (K)	1						1	2
2 (L)	1	3					4	8
3 (M)	1	3	5				9	18
4 (N)	1	3	5	7			16	32
5 (O)	1	3	5	7	9		25	50
6 (P)	1	3	5	7	9	11	36	72

The order of electronic configuration of atoms is given in Figs. 1.4a & 1.4b.

The order of filling electronic orbitals of an atom of atomic number, *Z* is **1s . 2s2p . 3s3p . 4s3d4p . 5s4d5p . 6s4f5d6p . 7s6d5f.**

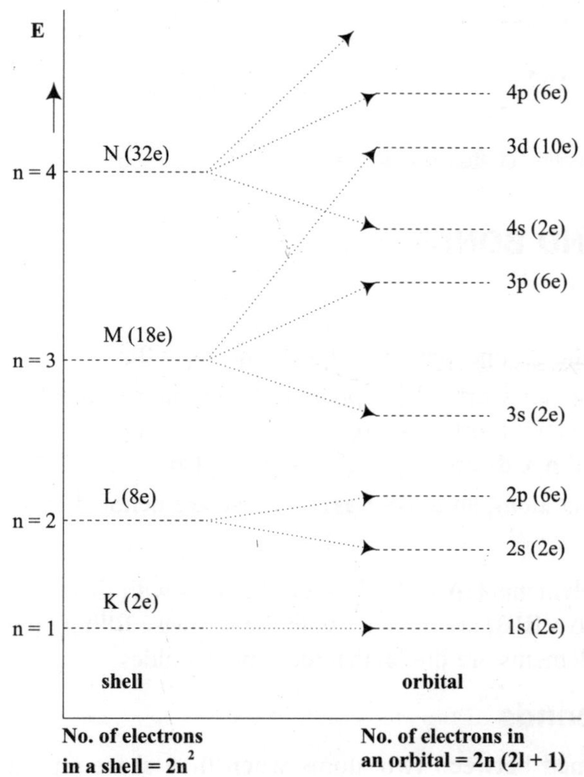

Fig. 1.4a Energy level diagram of atomic orbitals.

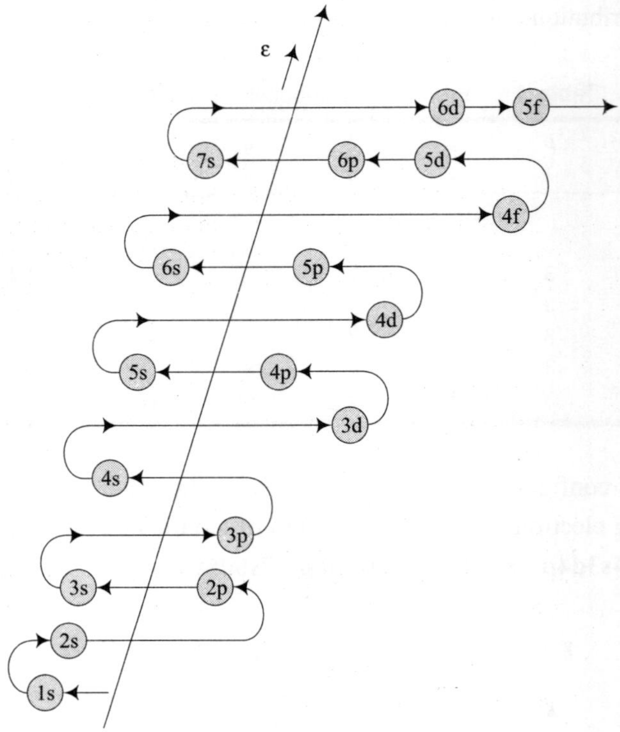

Fig. 1.4b Electronic configuration of atomic structure.

1.3 ORBITALS AND BONDS

1.3.1 Orbitals

s-orbitals correspond to the ground state with **l** = 0 (and $\mathbf{m_l}$ = 0).

The wave function of an **s**-orbital is spherically symmetric, **p**-orbital electrons have angular momentum, **l** = 1 and the three orbitals ($\mathbf{m_l}$ = –1, 0, 1) are directed orthogonally along the three Cartesian axes ($\mathbf{p_x}$, $\mathbf{p_y}$, and $\mathbf{p_z}$). **d**-orbitals are five — **l** = 2; (**m** = –2, –1, 0, 1, 2), corresponding to five different angular momenta along an axis (z-axis). These are denoted by d_{xy}, d_{yz}, d_{zx}, $d_{x^2-y^2}$ and d_{z^2} (Fig. 1.5).

The main group of elements (IA to VIIIA) are the ones with their **s**- and **p**-orbitals being filled. Transition elements (IB to VIIIB) are the ones with the stepwise filling up of their **d**-orbitals (**d**-block elements). The **f**-block elements are the lanthanides and actinides.

1.3.2 Chemical bonds

A covalent bond is formed between two atoms when they share an electron pair between them (**Lewis**). That is, chemical bond formation is due to the combination of several atomic orbitals leading to molecular orbitals. A molecular orbital is like an atomic orbital, but spreads throughout the molecule. Therefore, the *linear combination of atomic-orbitals* (LCAO) approach can be employed to

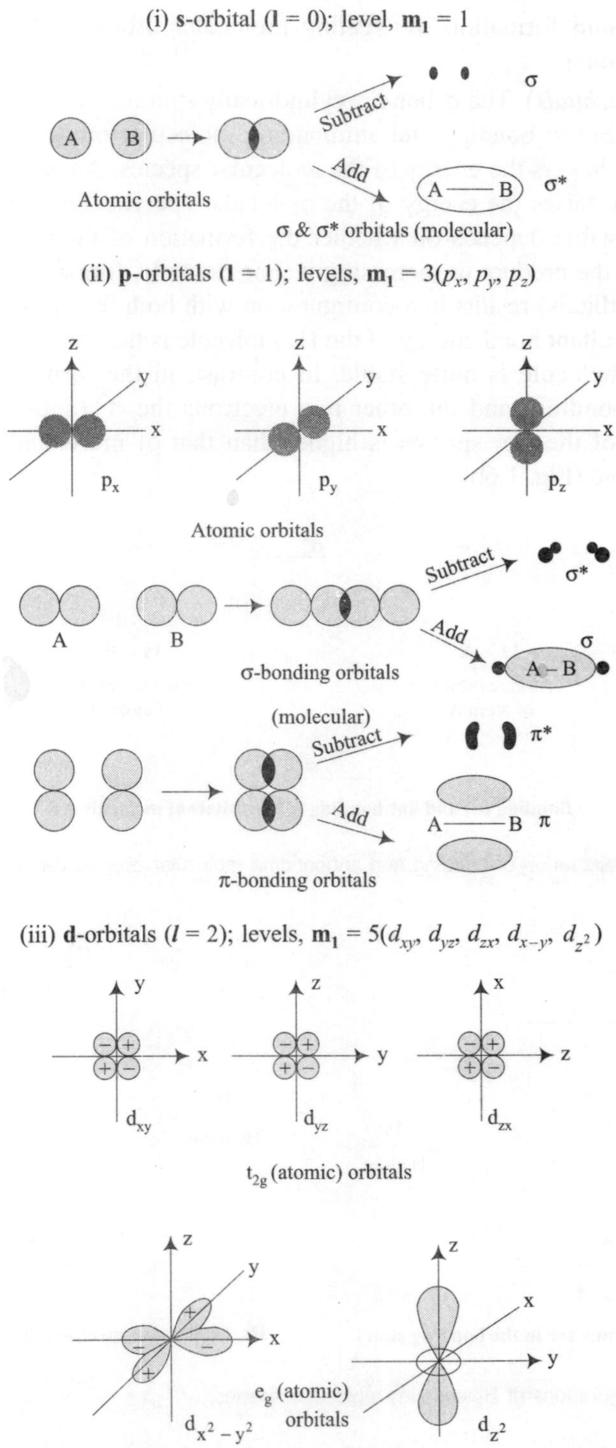

Fig. 1.5 Formation of molecular orbitals.

understand covalent bond formation by treating molecular orbital as a sum of atomic orbitals (additions and subtractions).

(i) *Single bond* (*σ-orbitals*). The σ-bond is cylindrically symmetrical. Addition and subtraction of two atomic **s**-orbitals lead to bonding and antibonding molecular orbitals, respectively. A bonding orbital (σ), if occupied, lowers the energy of the molecular species. An antibonding orbital σ^* on the other hand, if occupied, raises the energy of the molecular species (Fig. 1.6a). Whether a molecule formed is stable or unstable depends on whether the formation of the molecular species results in lowering the energy of the product or increasing it. For example, formation of a hydrogen molecule from two H atoms (**s**-orbitals) results in a configuration with both the electrons in the lowest orbital (bonding state). The resultant bond energy of the H_2 molecule is therefore less than that of individual H-atoms. So, the H_2 molecule is quite stable. In contrast, in the formation of He_2 species, two electrons saturate σ (bonding) and the other two electrons the σ^* (antibonding) orbitals and the resultant bond energy of the He_2 species is higher than that of individual He atoms. So, the He_2 species is highly unstable (Fig. 1.6b).

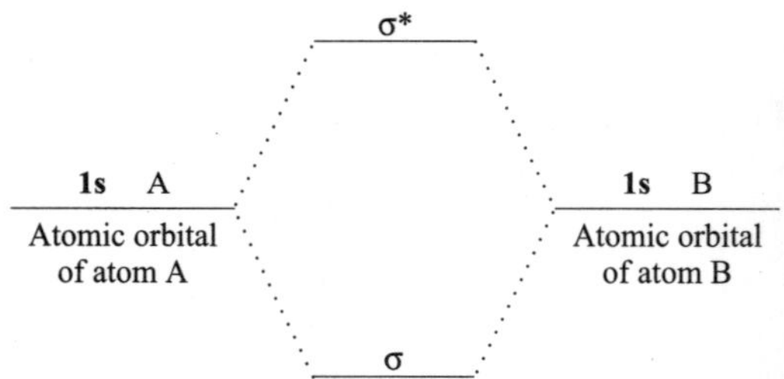

Bonding (σ) and antibonding (σ*) orbitals of molecule AB

Fig. 1.6a Schematic representation of bonding and antibonding molecular orbitals using the LCAO concept.

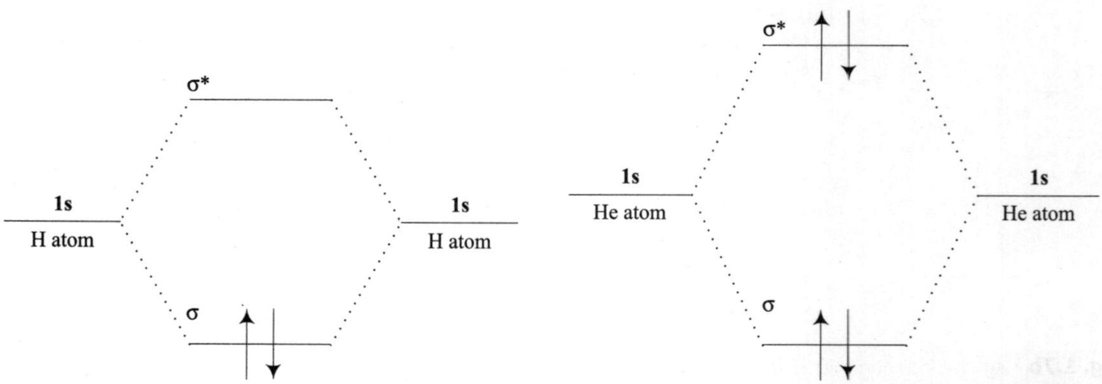

H_2 molecule (both electrons are in the bonding state) **He_2 (molecular species) (electron in both σ & σ^* states)**

Fig. 1.6b Electronic configurations of H_2 and He_2 molecular species.

The atoms of the first row of the periodic table can never be bound to more than four neighbouring atoms. Atoms C, N and O bind preferentially in tetrahedral (single bonding; **sp**^{3}-hybridisation) and trigonal (double bonding; **sp**^{2}-hybridisation) configurations (Fig. 1.7a). C and N atoms can have linear (triple bonding; **sp**-hybridisation) configurations also.

Fig. 1.7a **sp**^{3}-hybridisation (single bonds). (Combination of σ-orbitals) C = Carbon; N = Nitrogen; O = Oxygen; l = Lobe

(ii) *Double bond* (σ & π-*orbitals*). A σ and a π bond, with one electron pair each, can exist between two atoms at the same time. This is called a double bond. Electrons in π-molecular orbitals reside only above and below the bond axis (Fig. 1.7b).

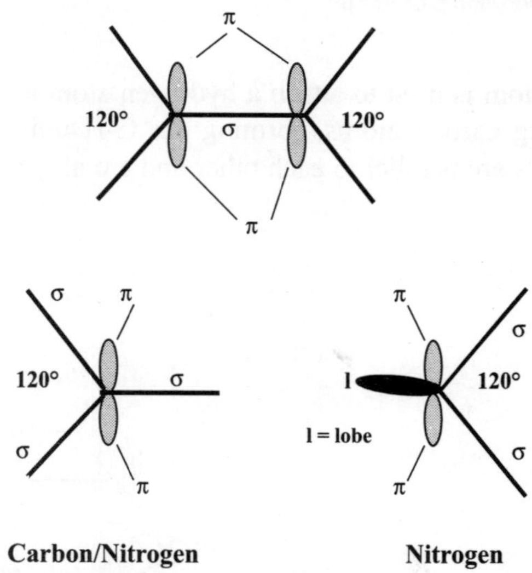

Fig. 1.7b **sp**^{2}-hybridisation (double bonds) (Combination of σ and π orbitals).

Phosphorus can form a singly negatively-charged link of a chain or a doubly negatively-charged terminal link (Fig. 1.7c). In the chemistry of phosphorus the lack of overlap of **p**π–σ orbitals and considerable overlap of **p**π–**d**π orbitals play a significant role in its chemical properties.

(iii) *Conjugated double-bonds*. The electronic model of benzene ring-type aromatic molecules involves the formation of double bonds by six (carbon) atoms by **sp**^{2}-hybridisation. One of the three

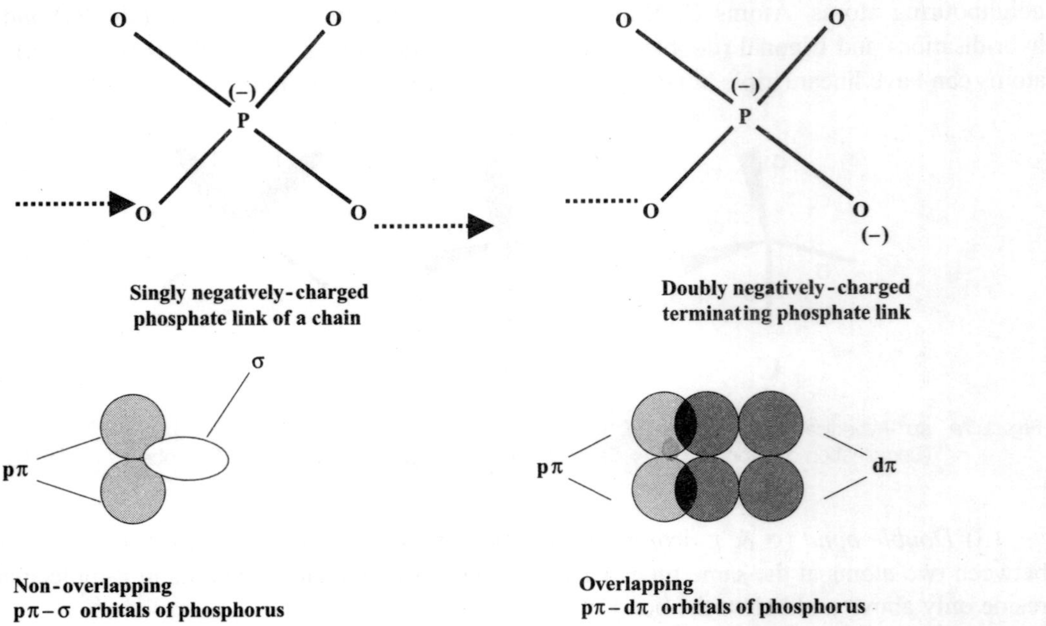

Fig. 1.7c Different orbitals in phosphorus chemistry.

sp^2 orbitals of each carbon atom is used to attach a hydrogen atom and the other orbital (**p_y**) makes bonds with the neighbouring carbon atoms, forming six C-H units, forming a planar molecule (Fig. 1.8a). The six **p_z** orbitals are parallel to each other and are all perpendicular to the planar C_6H_6

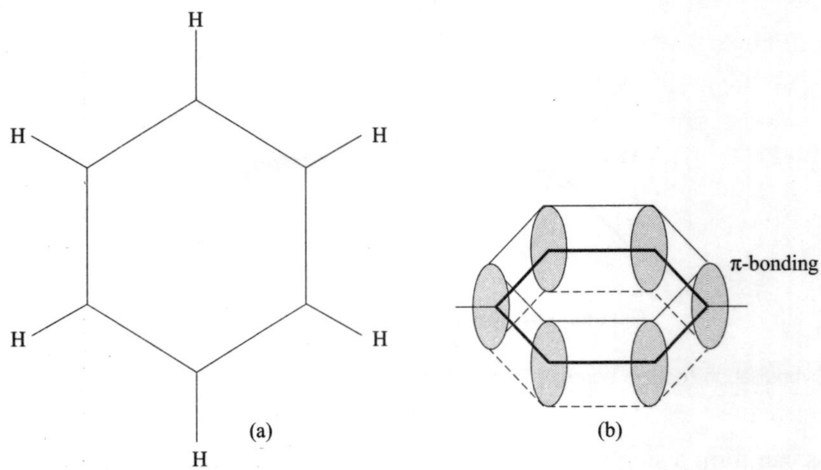

(a) The σ-skeleton of aromatic planar molecule formed by the overlap of **sp^2** orbitals.

(b) π-bonding orbitals with electron cloud above and below the planar molecule.

Fig. 1.8 Conjugated (double) bonds.

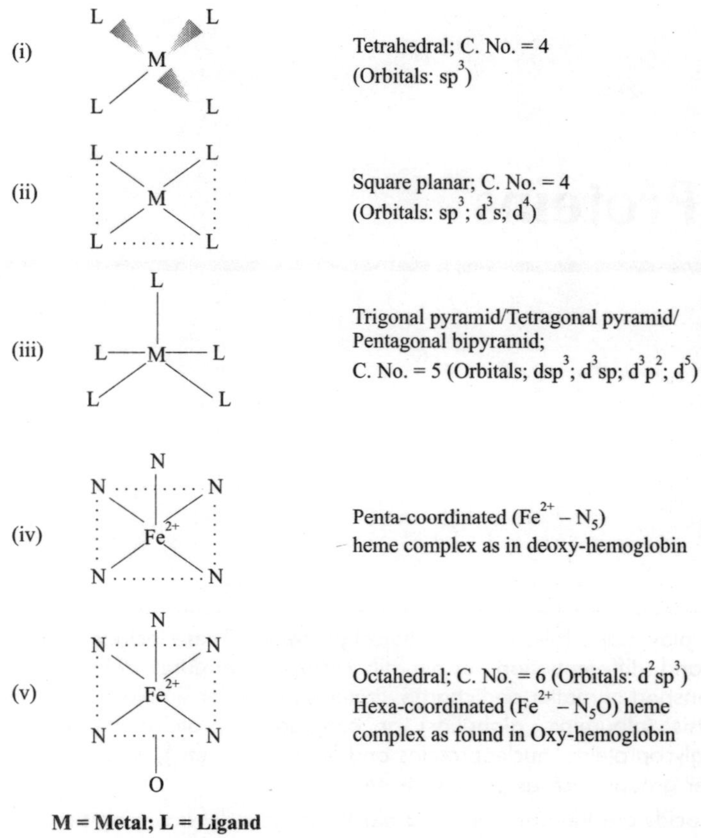

Fig. 1.9 Metal coordination complexes.

ring. These $\mathbf{p_z}$ orbitals overlap sideways to form π-bonding orbitals and the electrons in these π-orbitals are above and below the plane of the molecule (Fig. 1.8b).

(iv) *Coordination complexes*. In general, metal-coordinated complexes that are of biological interest are those with four, five and six-coordinated complexes and they occur in tetrahedral, square planar, trigonal pyramidal and octahedral forms (Fig. 1.9).

Further details on metal coordination are given in Chapters 8 and 9.

Suggested Reading

Atkins, P.W. *Physical Chemistry*. 6th Edn. Oxford University Press/Oxford. (1998).

Pauling, L. *The Nature of the Chemical Bond*. Cornell University Press/New York. (1960).

Rae, A.I.M. *Quantum Mechanics*. McGraw Hill/London. (1981).

(Many other introductory Modern Physics (Quantum physics) and Physical Chemistry books deal with this topic).

2 Proteins

Synopsis: Proteins play major roles in all biological processes. These include structural roles, mechanical support, growth and differentiation, enzymatic catalysis, immune defence, contractile processes, thrombosis and transport of matter and charge. Proteins are either simple proteins that produce α-amino acids on hydrolysis (albumins, globulins) or conjugated proteins (hemoproteins, flavoproteins, phosphoproteins, glycoproteins, nucleoproteins and lipoproteins etc.), which produce not only α-amino acids but also other groups such as porphyrins etc.

Twenty L-amino acids are the basic building blocks of naturally occurring proteins. Amino acids differ from each other structurally and functionally owing to the structure and chemical nature of their side chains. Under physiological conditions amino acids exist in zwitter-ionic form and they can be categorised as aliphalic, aromatic, basic, acidic and neutral species, based on the chemical nature of their side chains. The functional diversity and versatility of proteins are due to the structural and chemical diversity of these side chains. Proteins are linear polymers of amino acids, covalently linked through peptide (C-N) bonds. The structural and the functional features of a protein are addressed at four hierarchical levels of organisation. The primary structure of a protein is its amino acid sequence. The peptide unit $(-\underset{\underset{\text{O}}{\|}}{\text{C}}-\text{NH}-)$ is planar due to resonance and the *trans* configuration is the preferred configuration for the peptide unit side chains. Steric hindrance further restricts the conformations of the polypeptide backbone to three regions in the **Ramachandran** conformational map. The secondary structure deals with aspects of local organisation of the polypeptide, namely, ordered segments, α-helix, β-sheet, turns and loops etc. The α-helix is right-handed, with intramolecular hydrogen bonds parallel to the helix axis between an N-H group and C=O group of the fourth residue. β-pleated sheets have intermolecular hydrogen bonding (between strands), normal to the strand axis. The β-sheet exhibits a right-handed twist.

The tertiary structure of a polypeptide is its unique three-dimensional folding. The forces involved in stabilising the proper folding of macromolecules are nonbonded (noncovalent) interactions—ionic, van der Waals, hydrophobic and hydrogen bonds acting cooperatively. Van der Waals, hydrophobic and hydrogen bonds play very important roles in the structure and function of macromolecules. Quaternary structure is the subunit organisation and assembly of multimeric polypeptides of proteins containing, more than one polypeptide chains (Hemoglobins, ATPases etc.). Structurally proteins can be classified as globular, where the folding of the polypeptide chain renders a compact structure (myoglobins, immunoglobulins), and fibrous (keratin and collagen classes), which are stranded. Proteins such as, α-keratin, myosin, epidermin and fibrinogen belong to the keratin group. Many of the fibrous proteins are helical and frequently several helical monomers intertwine together to form coiled-coil structures. For example, keratin forms such a coiled-coil structure. Silk fibroin has an antiparallel β-sheet structure in which layers of glycine side chains (—H) alternate with layers of alanine and serine side chains ($—CH_3$ and —OH). Collagen is a rod-shaped fibrous protein formed by the intertwining of three polypeptide chains. Each monomer, due to a unique repeat sequence, is a left-handed poly-L-proline type helix. Internal hydrogen bonds render structural stability. The functional properties of fibrous proteins can be correlated with their structural features.

2.1 AMINO ACIDS

Amino acids are the basic structural constituents of naturally occurring proteins. They all consist of an amino group (NH^+_3), a carboxylate group (COO^-), a hydrogen atom and a substituent group, *R*, called a side chain, bonded to a central carbon atom (C_α atom) (Fig. 2.1). A simple way of remembering the nomenclature of the L- and D-forms of the amino acids is— assign the amino group to the thumb, H-atom to the little finger, COO^- group to the middle finger, and the R-group to the arm. Then, facing the palm up, the left hand, with the thumb to the left and the little finger to the right, will represent the *L*-configuration of the amino acid, and the right hand, the D-configuration, respectively.

There are twenty different types of L-amino acids, as a result of which twenty different side chains occur in proteins. The L-amino acids that occur in proteins are listed with their single and triple alphabet symbols in Table 2.1.

2.1.1 Dissociation Characteristics of Amino Acids

Compounds that undergo complete dissociation in water are termed strong electrolytes (e.g., NaCl, KCl) and weak electrolytes are those exhibiting partial dissociation in aqueous solution (e.g., acetic

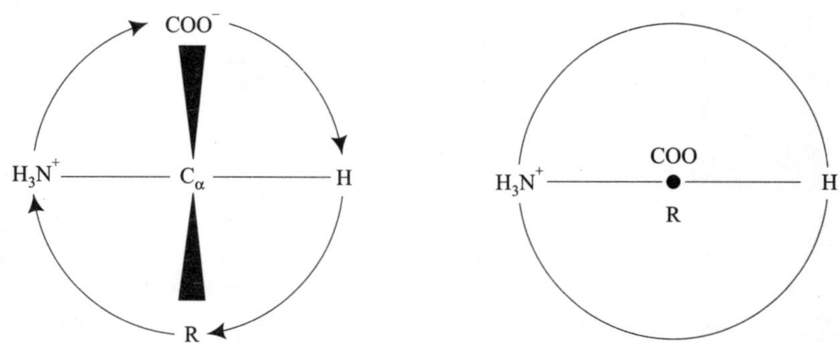

Fig. 2.1a Representation of L-amino acids. (R = side chain).
(i) **Fisher** Representation
(In CORN notation COO^-RN^+ has clockwise sense in L-amino acids)
(ii) **Neumann** Representation

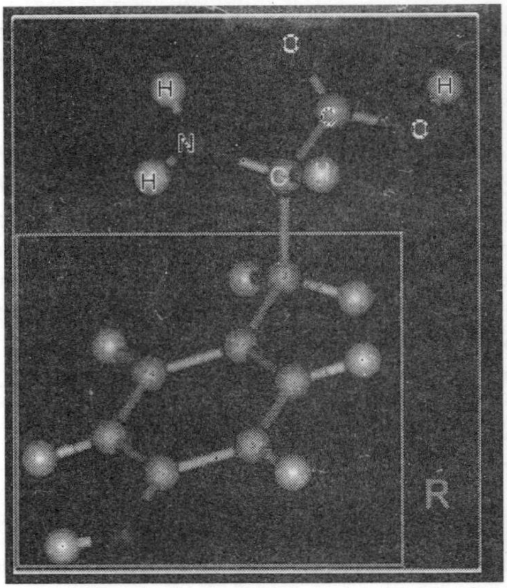

Fig. 2.1b The structure of an amino acid.

Table 2.1 Single and triple alphabet symbols of amino acids that occur in proteins.

	Single alphabet	*Triple alphabet*	*Amino acid*
(1)	A	Ala	Alanine
	B	Asn/Asp	Aspargine/Aspartic acid
(2)	C	Cys	Cysteine
(3)	D	Asp	Aspartic acid
(4)	E	Glu	Glutamic acid
(5)	F	Phe	Phenylalanine
(6)	G	Gly	Glycine
(7)	H	His	Histidine
(8)	I	Ile	Isoleucine
(9)	K	Lys	Lysine
(10)	L	Leu	Leucine
(11)	M	Met	Methionine
(12)	N	Asn	Asparagine
(13)	P	Pro	Proline
(14)	Q	Gln	Glutamine
(15)	R	Arg	Arginine
(16)	S	Ser	Serine
(17)	T	Thr	Threonine
(18)	V	Val	Valine
(19)	W	Trp	Tryptophan
(20)	Y	Tyr	Tyrosine
	Z	Gln/Glu	Glutamine/Glutamic acid
	X		Any of the above amino acids

acid). The polarisation of the covalent bonds involved in the dissociation will determine the extent to which ionic species are produced. Amino acids in solution exhibit the characteristics of weak acids.

Reiterating the acid/base concept of **Brønsted-Lowry**, an acid is a *proton donor* and a base is a *proton acceptor.*

$$\text{Acid} \rightleftharpoons H^+ + \text{Base}$$

$$HA \xrightarrow{H_2O} H^+ + A^- \qquad ...(2.1)$$

where HA and A^- are referred to as conjugate acid (proton donor) and conjugate base (proton acceptor), respectively. The dissociation constant, K_a, is

$$K_a = \frac{[H^+][A^-]}{[HA]}$$

for water,

$$HA \equiv H_2O;\ A^- \equiv OH^- \qquad ...(2.2a)$$

$$K_w = [H^+]\,[OH^-] \qquad ...(2.2b)$$

(as the concentration of $[H_2O]$ does not change)

$$\therefore \quad \log K_w = -\log [H^+] - \log [OH^-] \qquad ...(2.2c)$$

pH is defined as $-\log_{10} [H^+]$ ...(2.3a)

$$\therefore \quad pK_w = pH + p\,OH \qquad ...(2.3b)$$

The pH of a solution is a measure of its hydrogen ion concentration, $[H^+]$; that is, its dissociation. In general, for a weak acid

$$-\log [H^+] = -\log K + \log \frac{[A^-]}{[HA]} \qquad ...(2.4a)$$

$$\therefore \quad pH = pK_a + \log \frac{[A^-]}{[HA]}$$

$$pH = pK_a + \log \frac{[\text{Conjugate base}]}{[\text{Conjugate acid}]}$$ **Henderson-Hasselbalch** equation ...(2.4b)

Henderson-Hasselbalch equation establishes the relationship between pH and pK_a of acids and bases.

At the half-point dissociation of ionisable groups

$$pH = pK. \qquad ...(2.4c)$$

Isoelectric point, *pI*, of an ionisable species is the pH at which there is *no net charge.*

$$pI \sim \frac{1}{2}(pK_1, + pK_2) \qquad ...(2.5a)$$

The pK_a value of a compound with a single ionisable group, such as acetic acid, can be obtained from the titration curve of pH versus equivalent of alkali. pK_a is the mid point of such a titration curve, the pH at which the buffer displays its greatest ability to resist pH change.

The buffer value of a system, which is the slope of the titration curve, is useful in determining the maximum buffer value of a system.

$$\text{Buffer value} = \frac{d\,(\text{equivalent}\,[A^-])}{d[pH]} \qquad \text{...(2.5b)}$$

Maximum buffer value = 0.576 C

where, *C* is the initial concentration of the acid.

Useful pH range of a buffer is $\approx pH \pm 1.5$.

Amino acids are amphoteric, that is, they are capable of donating protons at low pH and accepting protons at high pH and thus they behave either as acids or bases, depending on the pH of the solution (Fig. 2.1c).

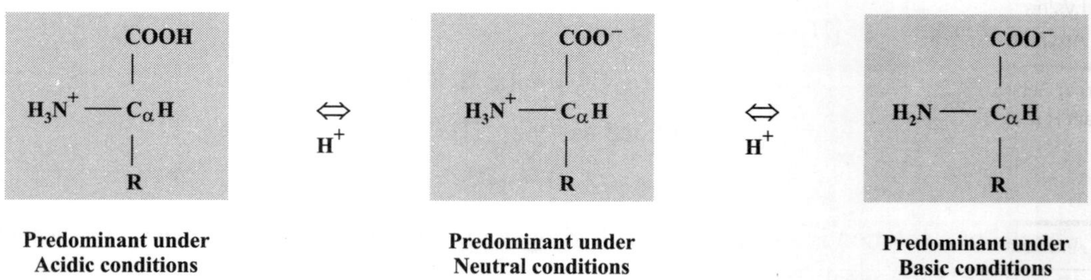

Fig. 2.1c Ionisation states of an amino acid as a function of pH

All amino acids have at least two dissociation constants (pK_1 and pK_2) — one for the α-carboxyl and one for the α-amino group (e.g. **gly, ser, ala**). Other amino acids such as **asp, his, lys,** with ionisable side chain groups have more than two pK_a values. pK_a values of some amino acids are given in Table 2.2a.

The pK_a values of α-COOH groups of the amino acids are in the range of ~2.2, so that above pH ~ 3.5 these groups are in the carboxylate form (COO^-). The α-amino groups have pK_a values at ~ 9.5 and are, therefore, in the protonated form ($NH^+{}_3$) below pH = 8.0.

Therefore, at neutral pH range ~ 7.0, amino acids exist in zwitterionic (double ionic) form. pI is the pH at which an amino acid is electrically neutral (zwitterionic form). Isoelectric pHs (pIs) of some important amino acids are given in Table 2.2b.

The pI values of amino acids such as **Gly, Ala** are pI $\approx \frac{1}{2}(pK_1 + pK_2)$. The pI values of amino acids with ionisable side chains can be obtained from the steep regions of the titration curves as shown in Figs. 2.2 and 2.3 for aspartic acid and lysine, respectively.

Table 2.2a Dissociation constants (pK_a) of some amino acids.

Amino acid	*pK'_1 (α-carboxyl)*	*pK'_2*		*pK'_3*	
Alanine	2.3	9.7	α-amino group	–	
Glycine	2.4	9.8		–	
Leucine	2.4	9.6		–	
Phenylalanine	1.8	9.1		–	
Proline	2.0	10.6		–	
Serine	2.2	9.2		–	
Threonine	2.1	9.1		~ 13	
Tryptophan	2.4	9.4		–	
Valine	2.3	9.6		–	
Aspartic acid	2.0	3.9	side chain (COOH)	10.0	α-amino group
Glutamic acid	2.2	4.3		9.4	
Histidine	1.8	6.0	(amino)	9.2	
Cysteine	1.8	8.4		10.8	
Methionine	2.1	9.3		–	
Tyrosine	2.2	9.0	α-amino group	11.0	(side chain OH)
Lysine	2.2	9.2		10.5	NH_3^+
Arginine	1.8	9.0		12.5	(guanidine)

(**Source:** Sober, H.A. (Ed.) *Handbook of Biochemistry and Molecular Biology*. 2nd Ed^n (1970).
Neurath H (Ed). *The Proteins.* Vol. II. 2nd Ed^n. (1964).

Table 2.2b Isoelectric pHs (pIs) of some amino acids representative of their class.

Amino acid	*Characteristic*	*pI*
An amino acid	Acidic	2.8 to 3.2
	Neutral	4.8 to 6.3
	Basic	7.6 to 10.8
Aspartic acid	Acidic	2.8
Glutamic acid	Acidic	3.2
Alanine	Nonpolar	6.0
Valine	Nonpolar	6.0
Serine	Neutral & Polar	5.7
Threonine	Neutral & Polar	5.6
Lysine	Basic	9.7
Arginine	Basic	10.8

2.1.2 Characteristics of Amino Acids

Amino acids can he categorised based on the shape, size and the chemical nature of their side chains. From their chemical nature the side chain can be i) charged, ii) neutral and iii) nonpolar R-groups.

(i) There are five amino acids with charged R-groups under physiological conditions. These are: basic amino acids lysine, arginine and histidine, and acidic amino acids aspartic and glutamic acids.

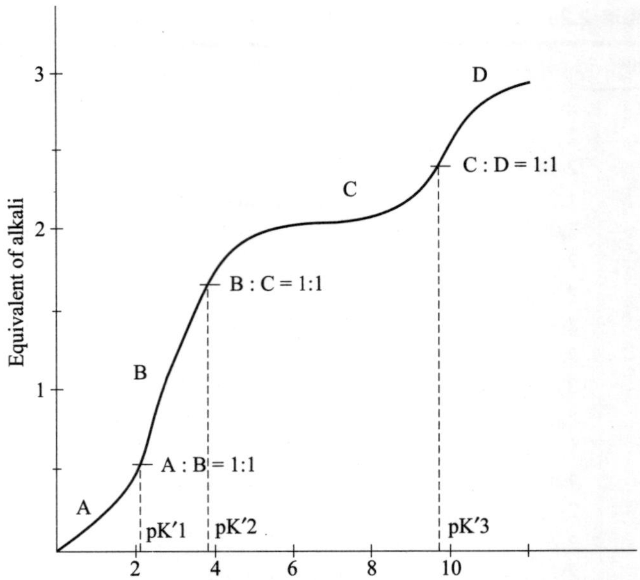

Fig 2.2a Ionisation profile of aspartic acid of species A, B, C & D. B is the isoelectric species.

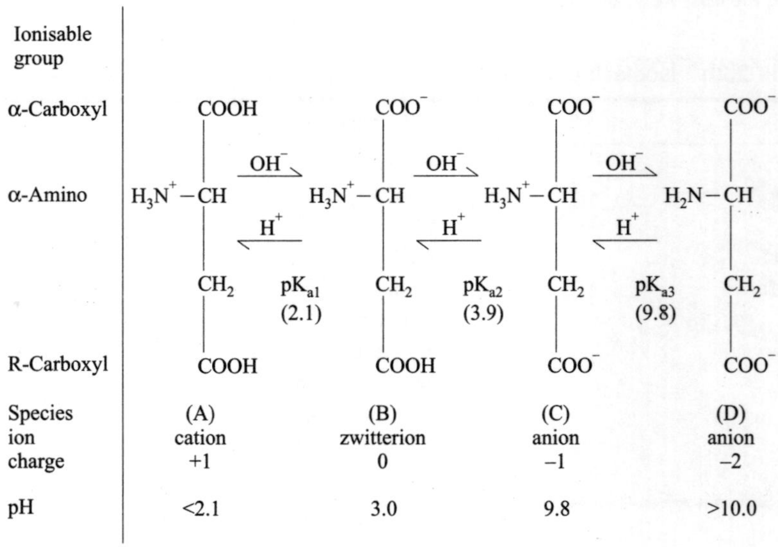

Fig 2.2b Chemical formulae of aspartic acid species A, B, C & D as a function of pH.

(ii) Neutral amino acids are: asparagine, glutamine, cysteine, serine and threonine.

(iii) Nonpolar amino acids are: glycine, alanine, leucine, isoleucine, valine, methionine, proline, phenylalanine, tryptophan and tyrosine (uncharged).

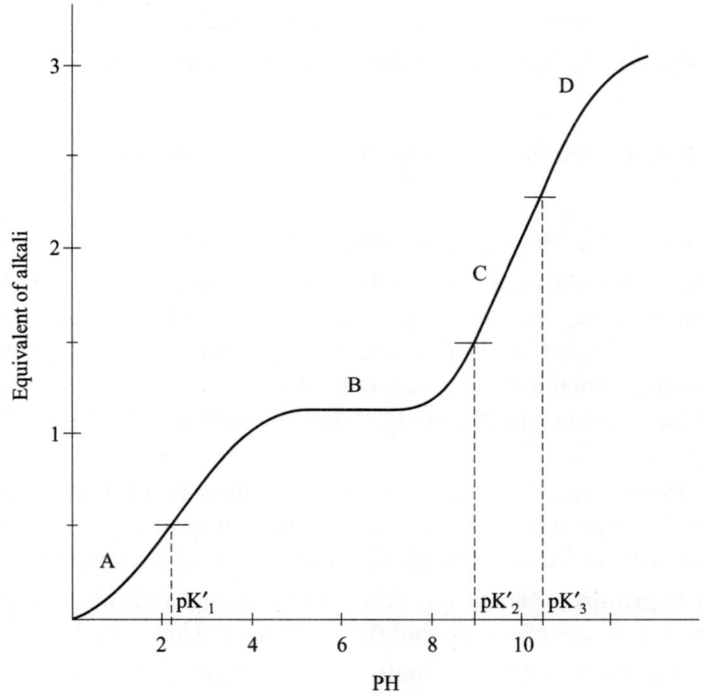

Fig 2.3a Ionisation profile of lysine of species A, B, C and D. C is the isoelectric species.

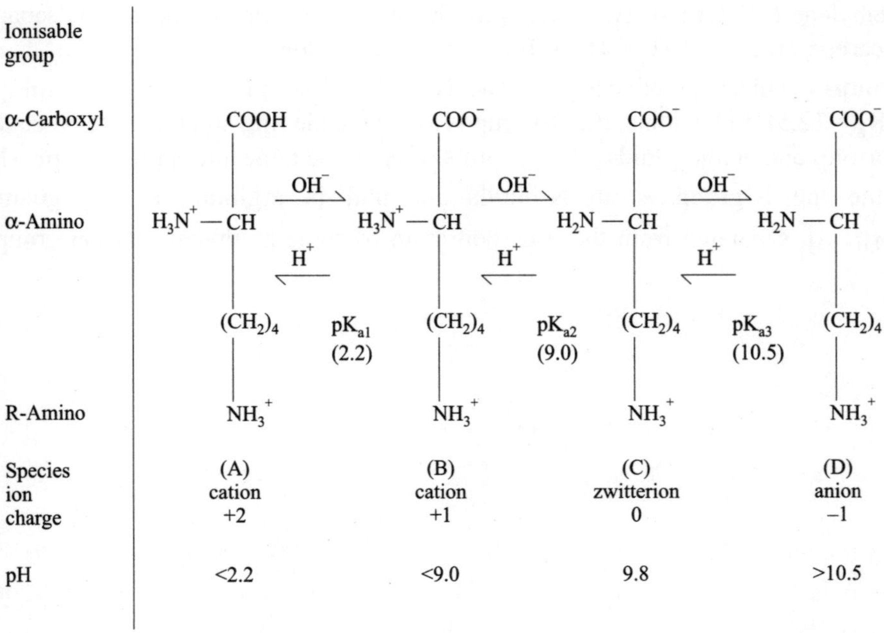

Fig. 2.3b Chemical formulae of the lysine species *A, B, C* and *D* as a function of pH.

Various R-groups of the amino acids with their characteristics are:

Alkyl R-groups. Glycine, alanine, leucine, isoleucine and valine. Valine has an isopropyl R-group

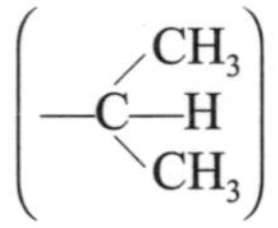

and in **Leu** and **Ile** the R-groups are butyl alkyl chains. In **Leu**, the branching methyl group is on the γ-carbon and in **Ile**, the butyl side chain is branched at the β-carbon.

Aromatic R-groups. Phenylalanine, tryptophan and tyrosine (pK = 10.0 for H). Tryptophan contains an indole, and tyrosine a phenol moiety. In all these cases the aromatic groups are attached to the C_α-atom through a —CH_2-group. Proline also has a hydrophobic side chain, but it differs from other amino acids in that it contains a secondary rather than a primary amino group. That is, proline is an α-imino acid and has a cyclic structure (Fig. 2.4a). All aromatic residues contain one methylene group (–CH_2) as a spacer.

Sulphur-containing R-groups. These are cysteine with thiomethyl (–$HSCH_2$) and methionine with thioether ($CH_3SCH_2CH_2$) R-groups. Two cysteine residues can form a cystine unit through a covalent S-S (disulphide) bond. This is the only branched covalent bond in proteins.

Hydroxyl (alcohol) R-groups. Serine (weakly acidic, pK = 13.6), has a hydroxymethyl moiety ($HOCH_2$) and threonine is a secondary alcohol (CH_3CHOH–). Threonine has a second chiral centre.

Asparagine and glutamine contain an amide moiety in their R-group, cannot be protonated, and are uncharged at physiological pH.

Acidic R-groups. Dicarboxylic, monoamino acids are aspartate and glutamate and they contain a charged carboxylate group at pH ~ 7. In aspartate the side chain carboxylic acid group is separated by a single methylene (–CH_2) moiety, whereas in glutamate the γ-carboxylic acid is separated by two methylene carbon groups (–CH_2–CH_2–) from the α-carbon atom.

Basic R-groups. Dibasic, monocarboxylic acids are histidine (pK_a = 6.0), lysine (pK_a = 10.5) and arginine (pK_a = 12.5). In histidine, the R-group is an imidazole ring and there are an equal number of basic (imidazole) and acidic (imidazolium) forms of histidine molecules at neutral pH (Fig. 2.4b).

In lysine the R-group is an N-butylamine and in arginine it is a guanidine group (NH_2—C(=NH)—HN—), separated from the α-carbon atom by three methylene carbon groups (–CH_2–)$_3$.

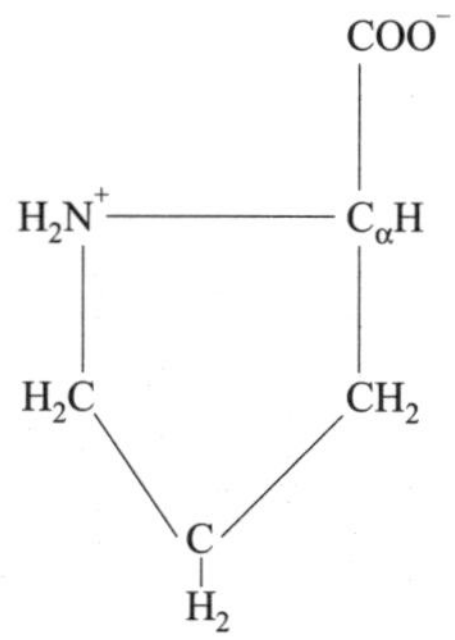

Fig. 2.4a Cyclic structure of proline.

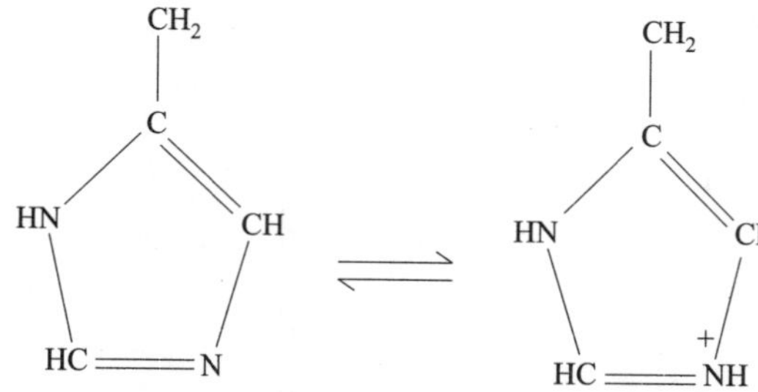

Imidazole Imidazolium

Fig. 2.4b Basic and acidic forms of the imidazole ring of histidine.

Both the amino groups of lysine and arginine are protonated at pH ~ 7.0 and are in their charged forms.

Amino acids can be grouped not only according to their charge but also according to their size.

Small and hydrophobic:	**Gly** and **Ala**.
Medium and hydrophobic:	**Cys, Leu, Ile, Met, Pro** and **Val**.
Large and hydrophobic:	**Phe, Tyr** and **Trp**.
Small and hydrophilic:	**Asp$^-$, Asn, Glu$^-$, Gln, Ser** and **Thr**.
Large and hydrophilic:	**His, Lys$^+$** and **Arg$^+$**.

The chemical structures of amino acids and their side chains are listed in Fig. 2.5 (a & b).

2.2 STRUCTURAL ORGANISATION OF PROTEINS

Proteins are linear polymers (except for the covalent S-S bond in cystine) of amino acids, synthesised from a library of twenty amino acids, covalently linked through peptide (C-N) bonds by the hydrolysis between α-amino and α-carboxyl of successive amino acids (Fig. 2.6).

A single diamino acid is termed a peptide, and a polymer of several peptides an oligopeptide. A polymer in the range 20–50 amino acid residues is termed either a peptide or a small protein and a polypeptide above 50 amino acid residues is called a protein. A polypeptide consists of repeating peptide units (main chain) with side chains (R_1, R_2, R_3 etc). The repeating unit along the main chain is

$$\mathrm{HN{-}\overset{\displaystyle H}{\overset{|}{C_\alpha}}{-}CO}.$$

The structural and functional features of proteins and protein complexes are addressed at four levels of hierarchal organisation. These are: 1) Primary structure (1°-Structure), 2) Secondary structure (2°-Structure), 3) Tertiary structure (3°-Structure) and 4) Quaternary structure (4°-Structure).

2.2.1 The Primary Structure (1°-structure) of Proteins

The amino acid sequence of a polypeptide chain(s) constitutes the primary structure of proteins. The convention for the designation of the order of amino acids is that the N-terminal end (free α-amino group) is to the left (and the number 1 amino acid) and the C-terminal end (end with the residue

Amino acids		*Chemical formula*	*Mol. Mass M*	*Characteristics*
1. Acidic amino acids				
Aspartic acid	Asp (D)	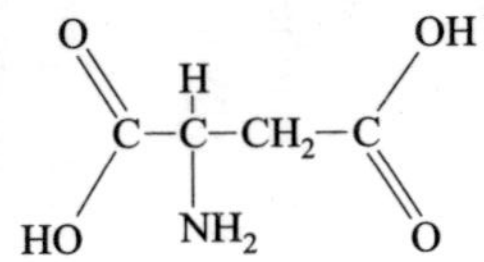	133	Acidic carboxyl group
Glutamic acid	Glu (E)	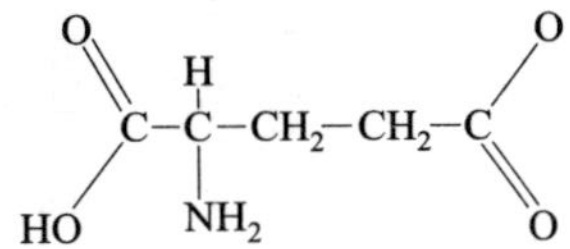	147	Acidic carboxyl group
2. Basic amino acids				
Lysine	Lys (K)	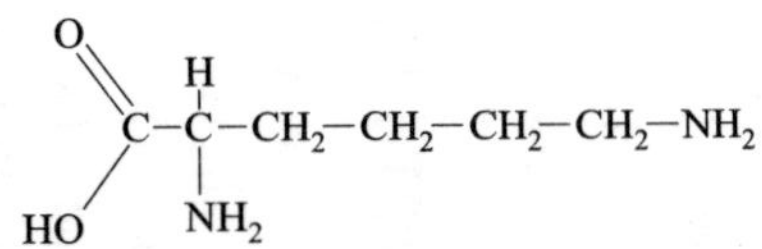	146	Basic Amino group
Arginine	Arg (R)	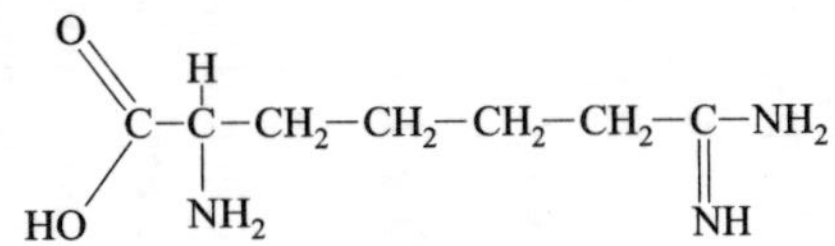	174	Basic guanidyl group (—N(H)—C(=NH)—NH_2) makes arginine almost as strong a base as NaOH.
Histidine	His (H)	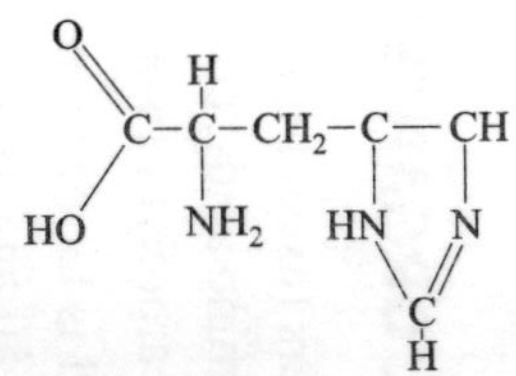	154	Contains the weak heterocyclic base imidazole.
3. Neutral & Polar amino acids				
Serine	Ser (S)	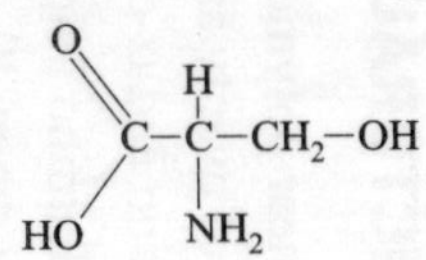	105	The hydroxyl group may take part in the usual alcoholic reactions such as ester formation. It occurs at the active site of many enzymes.
Threonine	Thr (T)	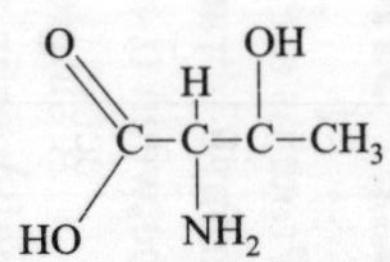	119	

(Contd.)

Amino acids		Chemical formula	Mol. Mass M	Characteristics
Cysteine	Cys (C)	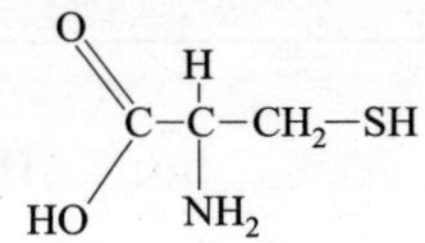	121	The S—H group is quite reactive and is easily oxidised and two molecules join to form cystine.
Methionine	Met (M)	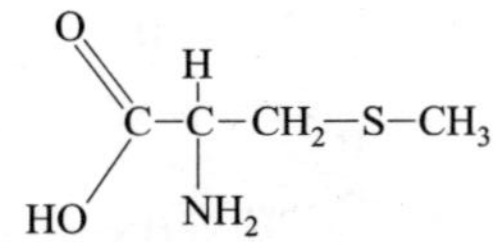	149	It is important as a donor of methyl groups.
Asparagine	Asn (N)	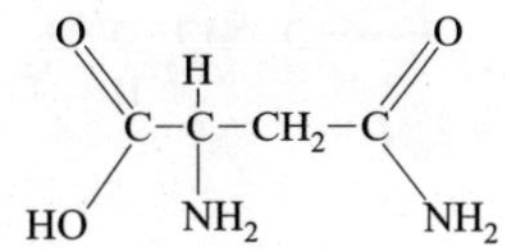	133	
Glutamine	Gln (Q)	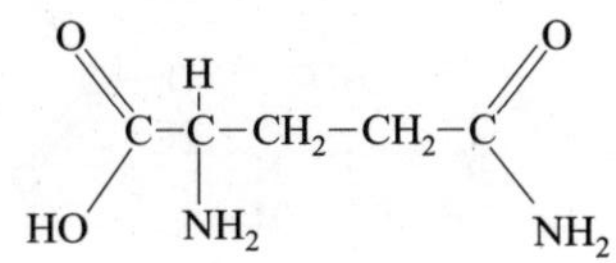	147	
4. Non-Polar amino acids Glycine	Gly (G)	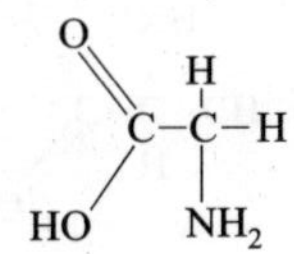	75	The only amino acid lacking an asymmetric carbon.
Alanine	Ala (A)	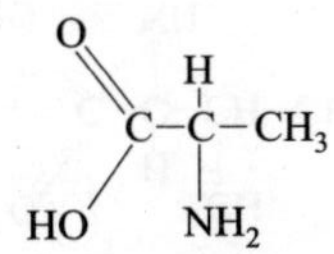	89	Simple nonpolar amino acid.
Valine	Val (V)	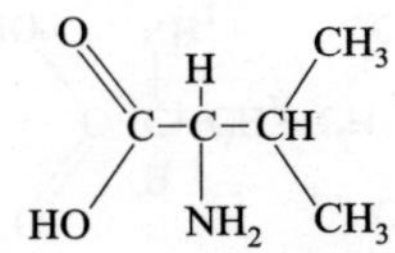	117	

(Contd.)

Amino acids		*Chemical formula*	*Mol. Mass M*	*Characteristics*
Leucine	Leu (L)	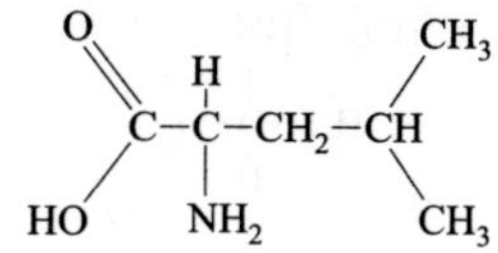	131	
Isoleucine	Ile (I)	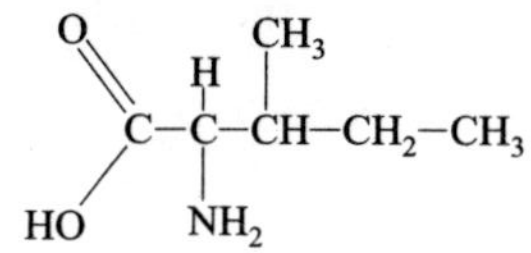	131	
Phenylalanine	Phe (F)	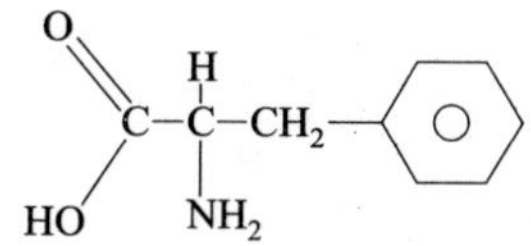	165	Highly nonpolar aromatic amino acid.
Proline	Pro (P)	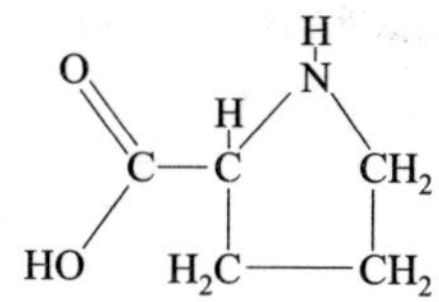	115	It contains a secondary amino group rather than a primary amino group. It is an imino acid.
5. Non-polar amino acids (with residual polarity) Tryptophan	Trp (W)	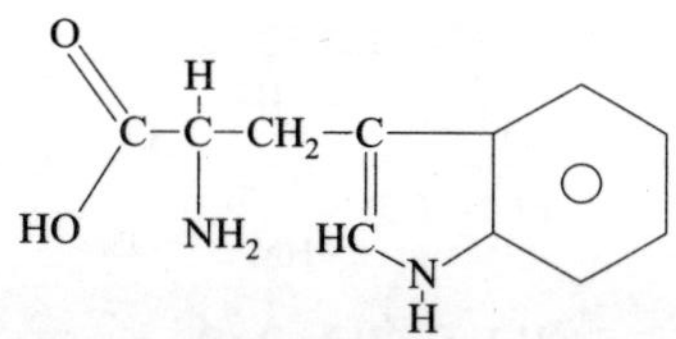	204	A heterocyclic imino acid (a derivative of indole).
Tyrosine	Tyr (Y)	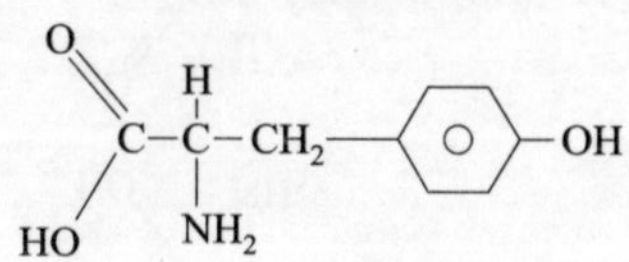	181	The phenolic group is weakly acidic. It loses its proton at pH values above 9.

Fig. 2.5a Amino acids with their side chains and their characteristics. Single alphabet nomenclature is given in brackets.

Side Chain	**Amino acid**
$—H$	Glycine
$—CH_3$	Alanine
$—CH_2—OH$	Serine
$—CH_2—SH$	Cysteine
$—C(H)(OH)—CH_3$	Threonine
$—CH—(CH_3)_2$	Valine
$—CH_2—\varphi$	Phenylalanine
$—CH_2—\varphi . OH$	Tyrosine
$—CH_2—$ Indole	Tryptophan
$—CH_2—$ Imidazole	Histidine
$—CH_2—COO^-$	Aspartic acid
$—CH_2—C(=O)—NH_2$	Asparagine
$—CH_2—CH_2—COO^-$	Glutamic acid
$—CH_2—CH_2—C(=O)—NH_2$	Glutamine
$—CH_2—CH—(CH_3)_2$	Leucine
$—C(H)(CH_3)—CH_2—CH_3$	Isoleucine
$—CH_2—CH_2—S—CH_3$	Methionine
$—CH_2—CH_2—CH_2—CH_2—NH_3^+$	Lysine
$—CH_2—CH_2—CH_2—NH—C(NH_2)=NH_2^+$	Arginine

Fig. 2.5b Side chains of amino acids.

containing a free α-carboxyl group) is to the right. Single alphabet nomenclature of amino acids is used in the amino acid sequence data. It is essential to obtain the primary structure of a protein to understand its molecular mechanism of action, to compare with other homologous proteins, to trace evolutionary paths, and to predict secondary and tertiary structures from the sequence homologies with related proteins (**structure prediction**). Primary structure data of a protein is generally a prerequisite for its structure determination by X-ray diffraction and NMR spectroscopic techniques.

The sequence determination of the insulin molecule by **Sanger** in 1953 was a landmark in protein chemistry. Since then progress in instrumentation has enabled automation in protein sequencing and larger proteins can be sequenced in a straightforward way. The speed and simplicity of gene sequencing methods, from which protein sequence can be inferred, has made the primary structure

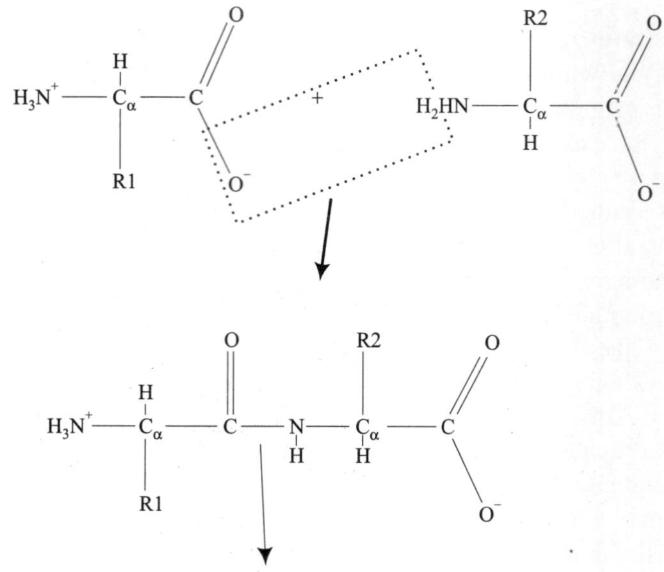

Fig. 2.6 Formation of a dipeptide from two monopeptides. (R_1 and R_2 are side chains).

determination of proteins an achievable task. The standard procedure for amino acid sequencing is i) end-group analysis, ii) cleavage of S-S bridges, iii) determination of amino acid composition and iv) specific cleavage of the polypeptide into smaller fragments and sequence determination of these fragments, and comparison of fragment-overlaps for determining the final sequence of the protein. The sequence of a protein can be inferred from the sequence of its nucleic acid fragment whenever it is feasible.

End-group analysis. The objective of end-group analysis is to establish whether the protein is a single polypeptide or composed of subunits. N-terminal identification is carried out either by the dansyl chloride (1-Dimethyl-aminonaphthalene-5-sulphonyl chloride) method or more elegantly by the **Edman** degradation method (Phenyl isothiocyanate). Aminopeptidases that cleave amino acids sequentially from the N-terminus of the polypeptide are used to determine N-terminal sequences. C-terminus identification is carried out by exopeptidases, such as carboxypeptidases, A, B, C and Y.

Cleavage of S-S bonds. For disulphide-containing proteins, cleavage of S-S bonds prior to any amino acid sequence analysis is necessary to permit the separation of polypeptide chains (if they are linked by S-S bridges) and to undo the native conformation that is retained if the S-S bonds are not cleaved. S-S bonds can be cleaved by 2-mercaptoethanol (reducing agent) or by **Clealand's** reagents (dithiothritol and dithioerythritol) under denaturing conditions and the resultant free-SH groups are alkylated by iodoacetic acid treatment to prevent reoxidation.

Separation and purification. Individual polypeptides are separated and purified by various fractionation methods such as ion exchange, gel filtration and affinity chromatography, and dialysis and reverse-phase HPLC. Various methods of protein separation and purification are explained in detail in any standard biochemistry textbook. (See chromatographic and electrophoretic methods in Chapter 6).

Amino acid composition and sequencing. The amino acid composition of a polypeptide is determined by hydrolysis with various proteases, some general and others more specific (papain, pepsin, thermolysin, trypsin etc.) and analysis of the amino acids. Automated amino acid sequencing is carried out either by ion exchange chromatography or by reverse-phase HPLC employing **Edman's** degradation method. Any protein with more than 50 amino acids has to be cleaved to obtain smaller fragments. This is done either enzymatically or chemically or by both. Trypsin is highly specific and cleaves peptide bonds after positively charged residues (**Lys** and **Arg**) if the next residue is not proline. Similarly, cyanogen bromide (CNBr) specifically cleaves peptide bonds after methionine residues. Individual peptide fragments thus cleaved are separated and purified by standard biochemical methods. The development of reverse-phase HPLC techniques has greatly improved purification procedures and has also reduced the amount of material needed for amino acid sequencing ($\leq 0.1\ \mu g$). After the completion of sequencing of individual fragments, the amino acid sequences are compared to identify the overlaps of specific cleavage sites between various fragments. This puts the reassembling of amino acid sequences of several fragments into a single frame. The final step in amino acid sequencing is to determine the positions of S-S bonds in disulphide-containing proteins. This is achieved by cleaving the native protein with the S-S bonds intact then these fragments are identified through diagonal electrophoresis.

Gene sequencing methods. The amino acid sequences of proteins can be inferred from the base sequences of corresponding nucleic acids. As the sequencing of nucleic acids is faster and easier than protein sequencing, this method is very attractive and followed wherever it is feasible. However, there are certain limitations and ambiguities in gene sequencing which should be taken note of while inferring the amino acid sequences of a protein from its nucleic acid sequence. These are: a) deletion and insertion of nucleotide(s) which can lead to an erroneous reading frame for the amino acids, b) the genetic code is not universal (the mitochondrial genetic code is different), c) many proteins are modified after biosynthesis and these modified amino acids can be detected only by direct protein sequencing and d) S-S bonds can be located only by protein sequencing and, therefore, direct protein sequencing is necessary for disulphide-containing proteins/enzymes.

2.2.2 The Secondary Structure (2°-Structure) of Proteins

The secondary structure of a protein deals with the ordered segments (helices, sheets), reverse turns and loops, and local hydrogen bonding of the polypeptide backbone. The secondary structure elements constitute the building blocks of the folding units in the globular proteins. The secondary structure of a polypeptide (protein) is determined largely by local sequence information.

Analysis of structural parameters (bond lengths and bond angles) of mono- and dipeptides, obtained from X-ray diffraction studies, led **Linus Pauling** to propose the essential features of the peptide unit and polypeptide backbone. The postulations are:

(i) The amide group is rigid and planar as a result of resonance.

The peptide bond, C–N, has about 30–40% double bond character. Therefore, rotation about the C–N bond is not feasible. This leads to only two rotational degrees of freedom per peptide. These are about the single bonds N–C_α (φ) and C_α–C (ψ) (Fig. 2.7). The conformation angles φ, ψ are referred to as **Ramachandran angles**.

(ii) The '*trans*' configuration is more stable than the *cis* configuration for the R-groups at the C_α-atom, because of steric hindrance between the substituents in the *cis* isomer.

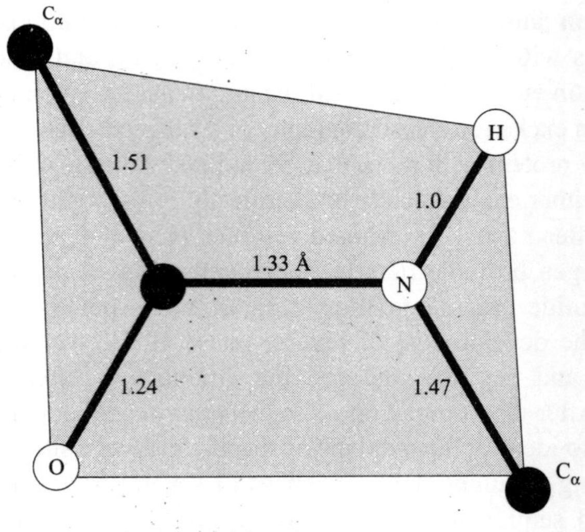

Fig. 2.7a Average bond distances (in Å) of the peptide unit. Angles around C and N atoms are ~120°, indicative of the planarity of the O = C – NH group. Planar peptide unit is shaded.

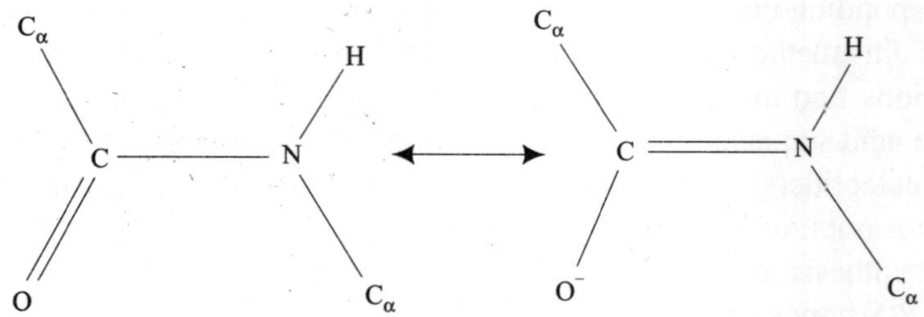

Fig. 2.7b Resonance structure of peptide group.

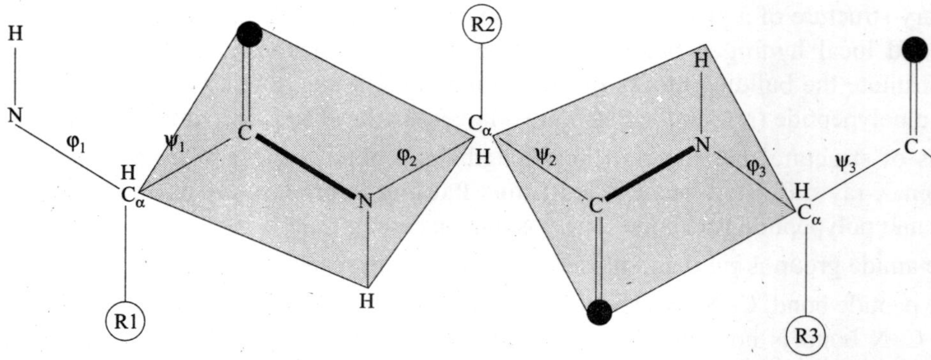

Fig. 2.7c Conformational freedom in peptide backbone. Rigid, planar peptide units are shaded. Peptide bond (C-N) is bold line.
• = Oxygen; R = Side chain residue

(iii) Hydrogen bonds play a crucial role in stabilising the polypeptide chain conformation.

From these conclusions Pauling postulated two ordered structures that should occur in the polypeptide backbone, namely the α-helix and β-sheet. The existence of such ordered segments in proteins was confirmed long before the first three-dimensional structure of a protein was determined by X-ray diffraction.

α-helix. The α-helix is a stable right-handed helical structure and is also called a 3.6_{13}-helix. It is a rod-like structure. The helix parameters are: 3.6 residues per turn (n = 3.6), rise per residue, d = 1.5 Å and the pitch, P = 5.4 Å. The conformation angles are φ = 122 (–58°) and ψ = 133 (–47°). The helix is stabilised by intra-molecular hydrogen bonds parallel to the helix axis occurring between an N-H of a residue and the C=O group of a 4th residue (Fig. 2.8a). α-helical segments are found in many globular proteins (myoglobins, troponin-C, calmodulin) (Fig. 2.8b & Fig. 2.8c). Two or more α-helices can entwine around each other to form a coiled-coil cable. Such α-helical coiled-coil structures are found in keratin type proteins.

The β-sheet. The β-sheet structure is a pleated sheet and results from inter-molecular hydrogen bonds (between N-H and C=O groups of neighbouring strands) perpendicular to the strand axis. The polypeptide chain is extended (d = 3.5 and P = 6.95 Å as against 1.5 and 5.4 Å in the α-helix) (Fig. 2.9a).

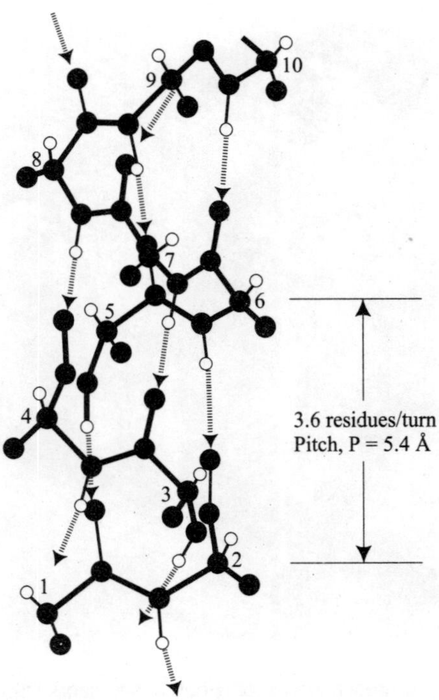

Fig. 2.8a Model of right-handed α-helix (α_R).
Thick lines show the polypeptide backbone. α-carbons are numbered
O (open circle) = Hydrogen
dashed lines = Hydrogen bonds (intra-molecular)
Intra-molecular hydrogen bonds have same polarity

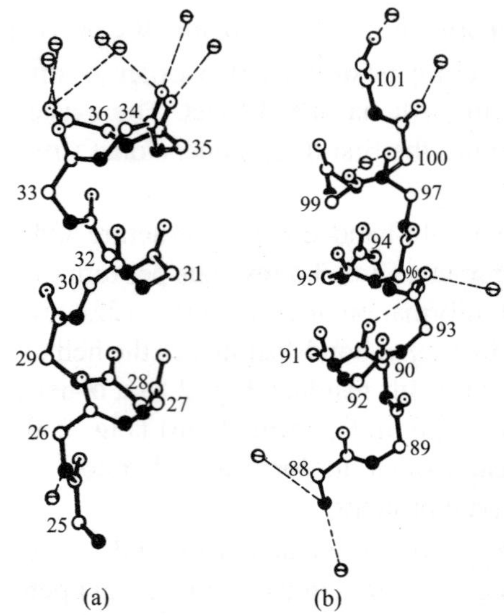

Fig. 2.8b α-helical regions (residues 25 → 36 and 88 → 101) stabilised by water of hydration in monoclinic lysozyme†.

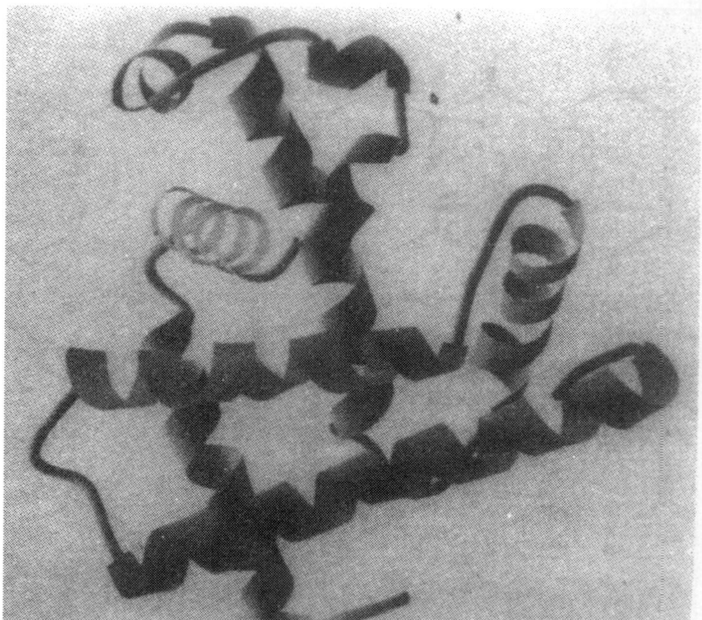

Fig. 2.8c Structure of myoglobin with predominance of α-helical segments (*).
(**Source:** Watson, S.C. & Kendrew, J.C. 1 MBN; 4/73).
(Ref: Carson. M. (1987) J. Mol. Graphics. 5; 103).:
The ribbon diagrams in the book indicated with (*) have been generated from the Computer Graphics System at the Bioinformatics Centre, Indian Institute of Science, Bangalore, using the 'Ribbons' Programme of Carson, M.
The coordinates for the generation of the diagrams were from the Brookhaven National Laboratory Protein Data Bank (PDB). The *source* indicates the author(s) followed by PDB code and the month/year of deposition of coordinates to the PDB.

β-sheets can occur either as parallel strands where the direction of the strands is the same (↑↑) or as antiparallel strands, in which case the direction of adjacent strands is opposite (⇄) (Fig. 2.9a). Also the β-sheet exhibits a slight right-handed twist (Fig. 2.9b). A β-sheet structure (antiparallel) is found in *Bombyx mori* silk fibroin. Both parallel and antiparallel β-sheet structures are found in globular proteins (e.g. structural motifs like immunoglobulin fold, β-meander and β-barrel etc.) (Figs. 2.9c & 2.9d). Often the β-secondary structure regions are stabilised by water molecules surrounding the protein molecules (water of hydration) (Fig. 2.9e). Structure of a globular protein with predominance of β-sheets is shown in Fig. 2.9f.

Turns and loops. While helices and sheets are termed 'ordered' structures because their residues have repeating backbone torsion angles φ, ψ and their N-H and C=O groups have periodic patterns of hydrogen bonding, turns and loops do not exhibit such regular secondary structural features.

Reversible turns, or β-turns are those regions in proteins where the backbone folds back on itself at ~180°. They, in effect, change the overall direction of the polypeptide chain. It is these chain reversals which give a protein globularity rather than linearity. A β-turn consists of a tetrapeptide unit with the C_α-atoms of the first and the fourth residues ~4.8 Å apart and a hydrogen bond between the C=O group of the first and the N-H group of the fourth residue (according to the Venkatachalam classification) (Fig. 2.10a). The hydrogen bond between the 1st and 4th residues is not a necessary condition for the β-turn. There are 11 types of β-turns, type I being most common. The β-turn is polar due to the presence of N-H and C=O groups. Its polarity is further enhanced because turn formation is promoted at chain sites where the average linear hydrophobicity is minimal.

A loop is a continuous segment that resembles Ω. A loop is defined by i) segment length (6 - 16 residues), (ii) distance between segment termini (< 10 Å) and iii) absence of regular secondary structure. The backbone of the loop surrounds an internal cavity that is packed with side-chain atoms from loop residues. A reverse turn, unlike a loop, has backbone groups that pack together closely, forcing the side chains to project outwards. Proteins without loops are generally small and could also be nonglobular (e.g. glucagon, insulin, mellitin etc.). Reverse turns and loops exhibit structural flexibility, and in proteins the unique folding in these regions is stabilised by water molecules (water of hydration) (Figs. 2.10b, c and d). Structure of a globular protein with predominance of turns and loops is shown in Fig. 2.10(e).

Turns, hairpin bends and loops are situated at the molecular surface, and they play an important role in determining conformation, specificity of enzyme active sites and antibody-antigen associations. Reverse turns are good candidates for protein engineering, because they have few consecutive residues (≤ 4) and are also restricted by packing considerations.

Polypeptide chain conformation. As the peptide bond, C-N, is essentially rigid, the polypeptide chain has only two degrees of rotational freedom, about $N\text{-}C_\alpha$ (φ) and $C_\alpha\text{-}C$ (ψ), per amino acid residue. Therefore, the conformation of the polypeptide main chain (backbone) is fully specified when the torsion angles, φ and ψ, for each amino acid residue of the protein are known. The torsion angle, Θ, about a single bond B-C is defined as the angle between the projected bonds A-B and C-D, viewed along the bond $B \rightarrow C$ (Fig. 2.11). The torsion angle, Θ, is zero if A-B and C-D are eclipsed (*cis* and coplanar) and the sign is taken as positive if the rear bond is rotated clockwise with respect to the near bond. It can also be defined as the angle between the two planes containing the atoms ABC and BCD.

φ–ψ Diagram (Ramachandran Map). Conformational flexibility of a polypeptide is restricted to certain regions because of the steric hindrances arising out of minimum contact distances permissible

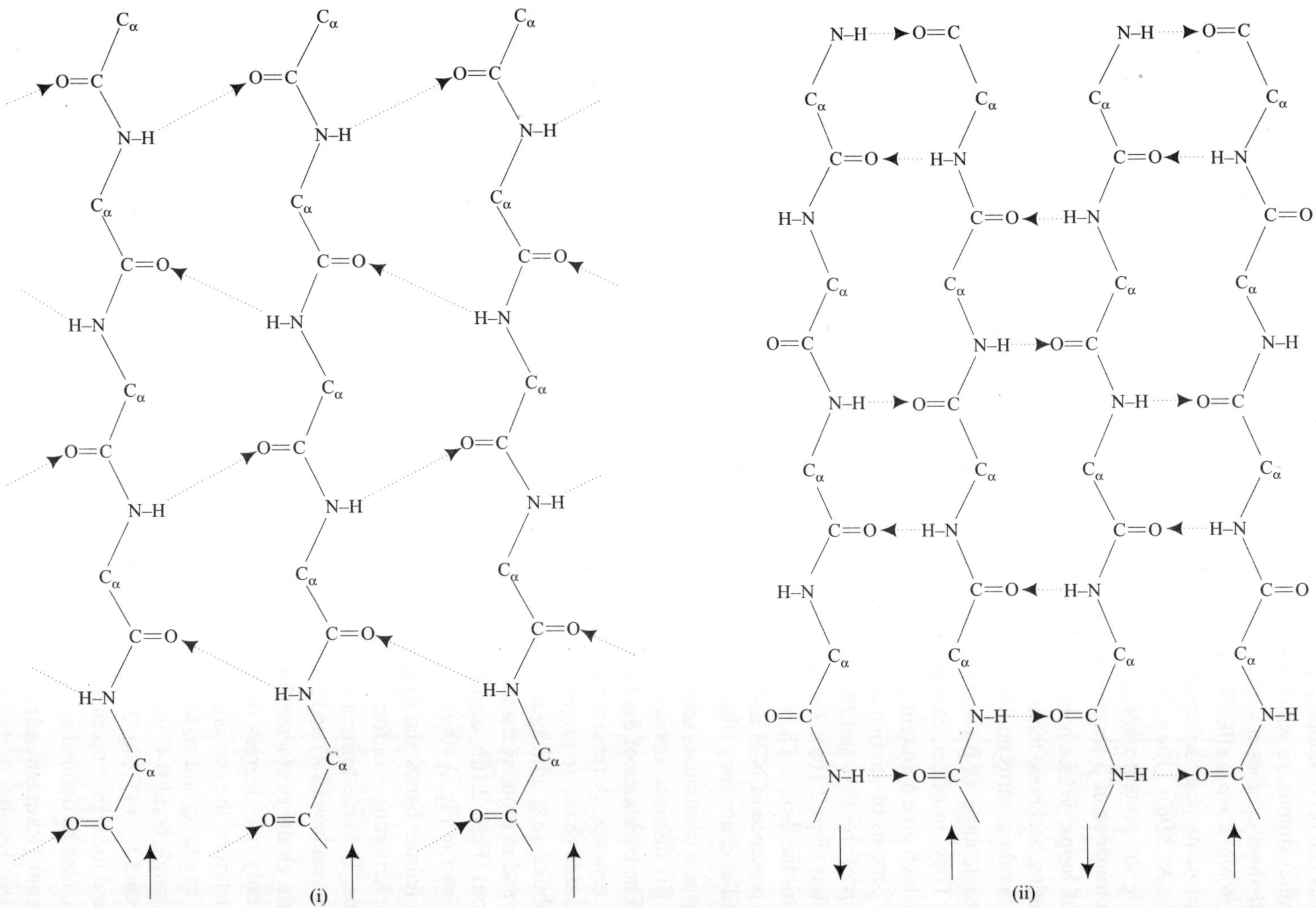

Fig. 2.9a Inter-molecular (Inter-strand) hydrogen bonding in β-pleated structures.
(i) Parallel β-sheet (ii) Antiparallel β-sheet
Hydrogen bonds are perpendicular to the strand axis and they are more symmetric in antiparallel β-sheet

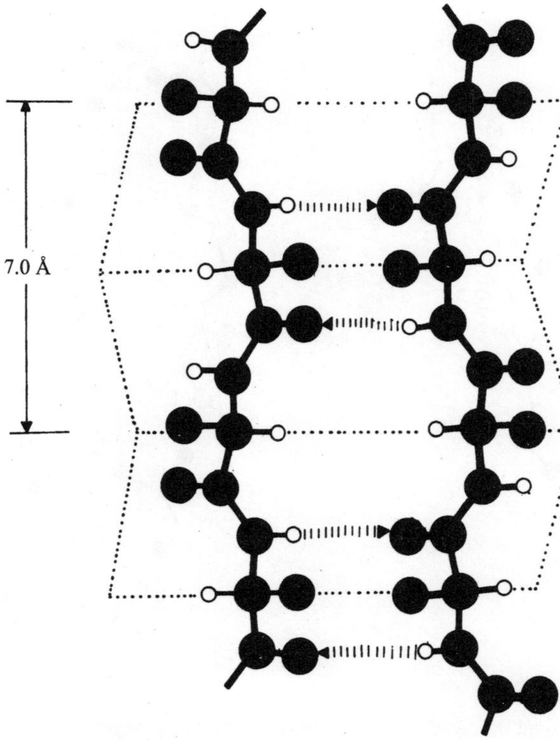

Fig. 2.9b Schematic diagram of β-sheet showing the pleated-sheet nature and extended conformation.

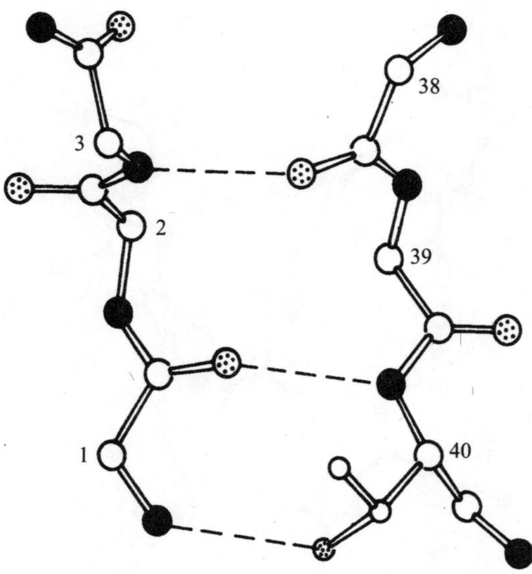

Fig. 2.9c Antiparallel β-strand in tetrahedral lysozyme structure†.

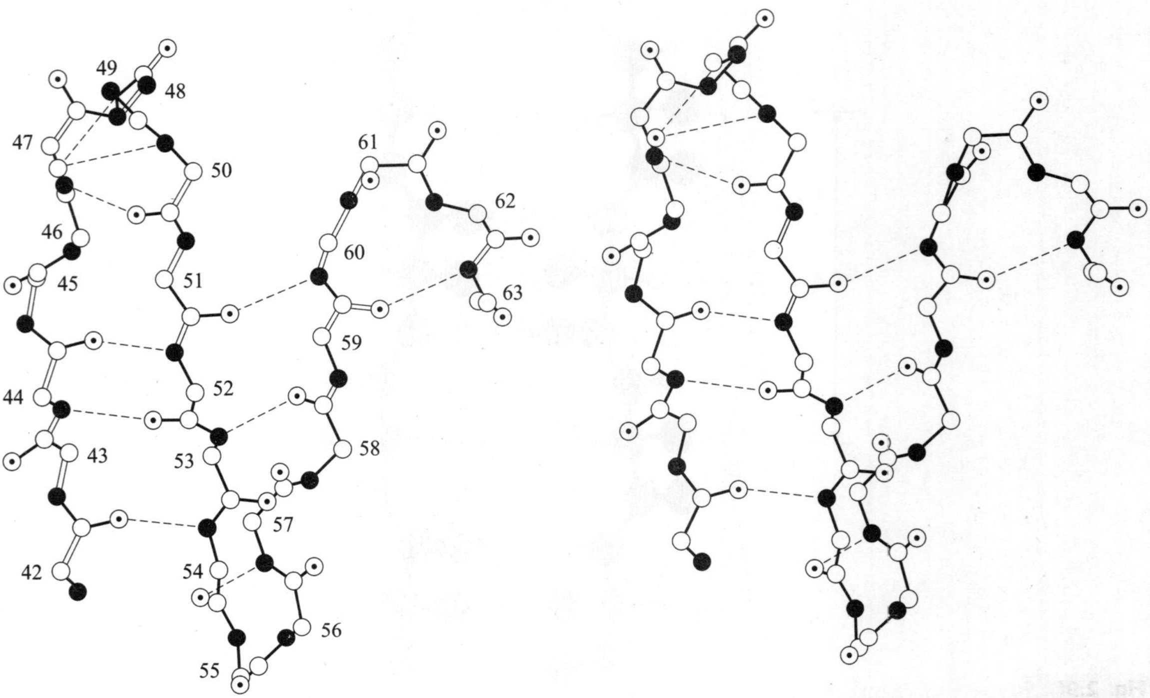

Fig. 2.9d Stereo view of the peptide backbone in the major β-structure region in low humidity tetrahedral lysozyme structure[†].

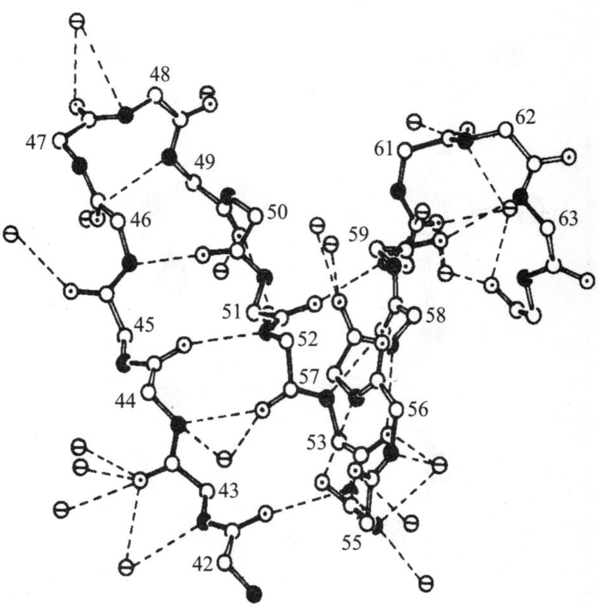

Fig. 2.9e Hydration of the main-chain atoms in the β-structure region in low-humidity monoclinic lysozyme structure[†].

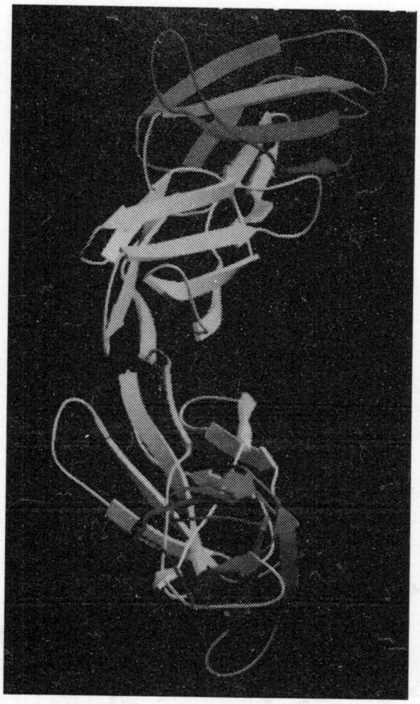

Fig. 2.9f Structure of Fibronectin with Predominance of β-sheets.
(Ref: Learhy, D.J., Aukhil, I. & Erickson, H.P. (1998). Cell, 84; 155)
(Source: Protein Data Bank: 1FNF.pdb)

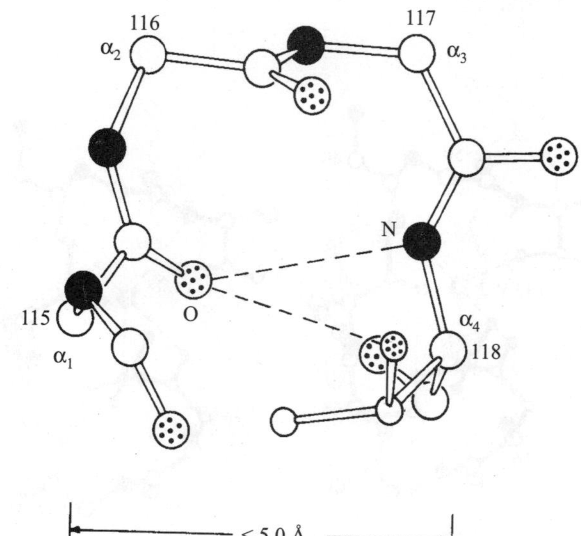

Fig. 2.10a Reverse turn (β-turn) stabilised by main-chain hydrogen bonding†.

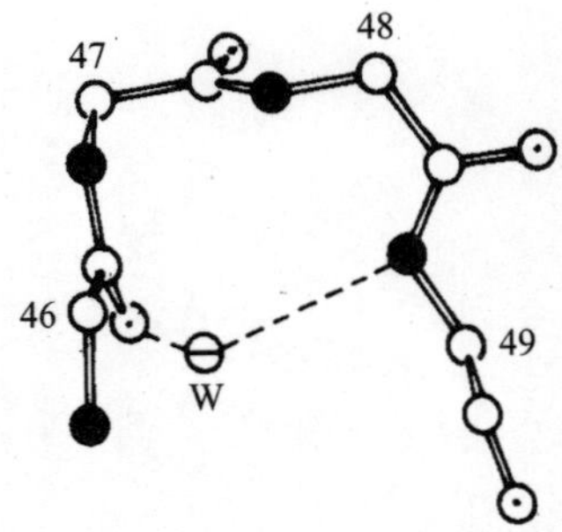

Fig. 2.10b Water-mediated reverse turn in low-humidity monoclinic lysozyme structure[†].

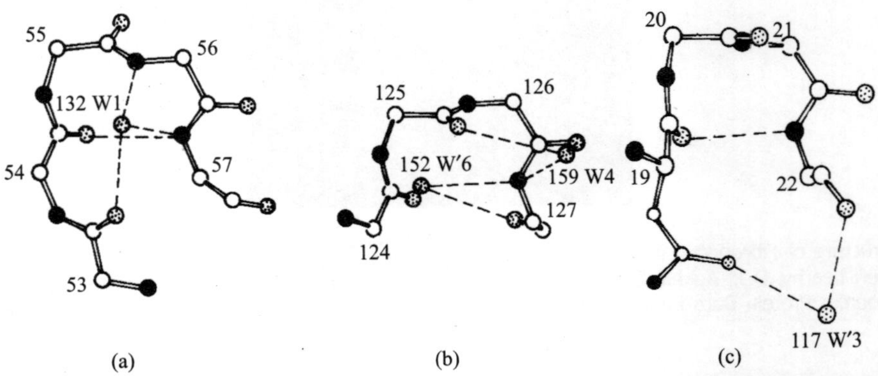

Fig. 2.10c Examples of reverse turns stabilized by water molecules in low-humidity tetrahedral lysozyme structure[†].

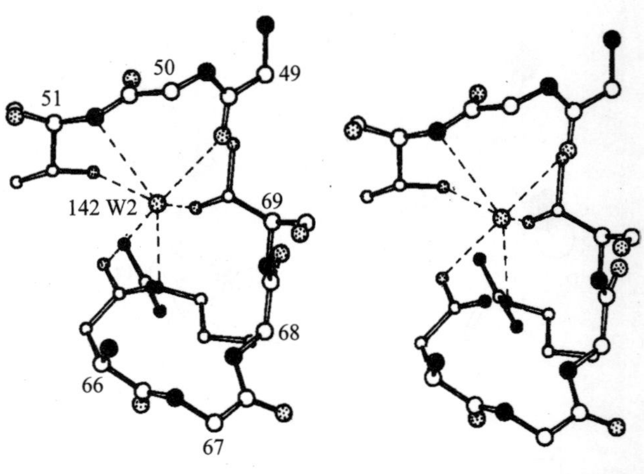

Fig. 2.10d Stereo view of the β-strand and loop regions stabilised by water of hydration in low-humidity tetragonal lysozyme structure[†].

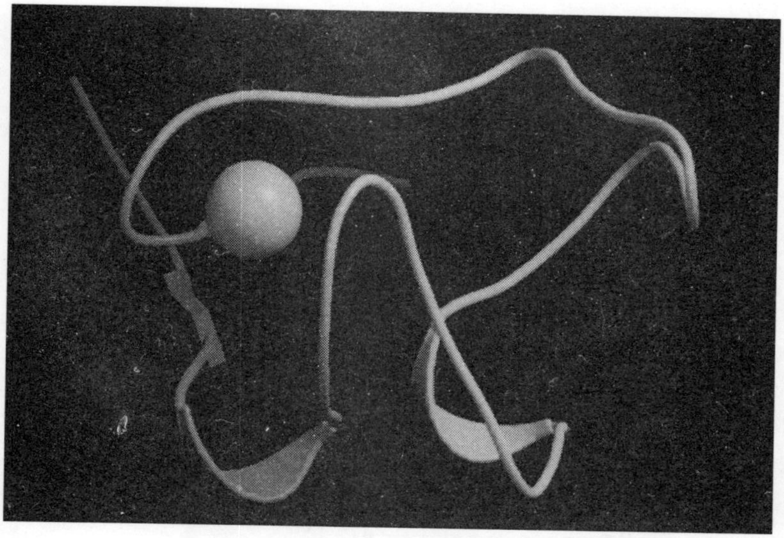

Fig. 2.10e Structure of Rubredoxin Mutant G10A with predominance of turns and loops. (Ref: Maher, M.J. et al. (1999). Acta Cryst D, Biol Crystallogr., 55; 962) (Source: Protein Data Bank: 1B13.pdb)

between various atoms (for C……C = 3.0, C……O = 2.8, O……O = 2.8 Å etc.). The sterically allowed and disallowed regions of φ and ψ can be determined by plotting φ, ψ angles of each residue of the polypeptide, as has been done by **Ramachandran**. The φ, ψ diagram, also known as a Ramachandran map, gives information on the allowed and not-allowed conformations of the polypeptide backbone (Fig. 2.12). For example, conformations like $\varphi = 0$ and $\psi = 180°$ or $\psi = 180$ and $\varphi = 0°$ are not allowed due to steric hindrance of atoms (Fig. 2.13). The Ramachandran map demonstrates that most φ, ψ values are conformationally not feasible for a non-glycine polypeptide backbone. Only three conformational regions are in the **Ramachandran map** for regular structures (α-helix, β-sheet and L-helix).

There are secondary structures, such as the 3_{10}-helix, the π-helix that do occur in proteins and also the L-polyproline type helix (which is the conformation of collagen monomers) that are also found in these conformational regions of the Ramachandran map (Fig. 2.14 and Table 2.3).

A Ramachandran map does not provide information on protein folding; that is, information on the nearest neighbours of amino acid residues in a polypeptide chain. This is because Ramachandran

---

†References to figures: 2.8b. 2.9c, 2.9d. 2.9e, 2.10a, 2.10b, 2.10c and 2.10d.

Figures 2.8b, 2.9e and 2.10b.

Madhusudan, Kodandapani, R. & Vijayan, M. *Acta Cryst.* D49; 234 (1993).

Figures 2.9c, 2.9d and 2.10d.

Vijayan, M. & Kodandapani, R. in "Molecular Conformation and Biological Interactions". p. 149. (Eds. Balaram, P. & Ramaseshan, S). Indian Academy of Sciences, Bangalore (1991).

Figures 2.10a and 2.10c.

Kodandapani, R., Suresh, C.G. & Vijayan, M. *J. Biol. Chem.* 27; 16126. (1990) (Diagrams: courtesy of Prof. M. Vijayan, Indian Institute of Science, Bangalore).

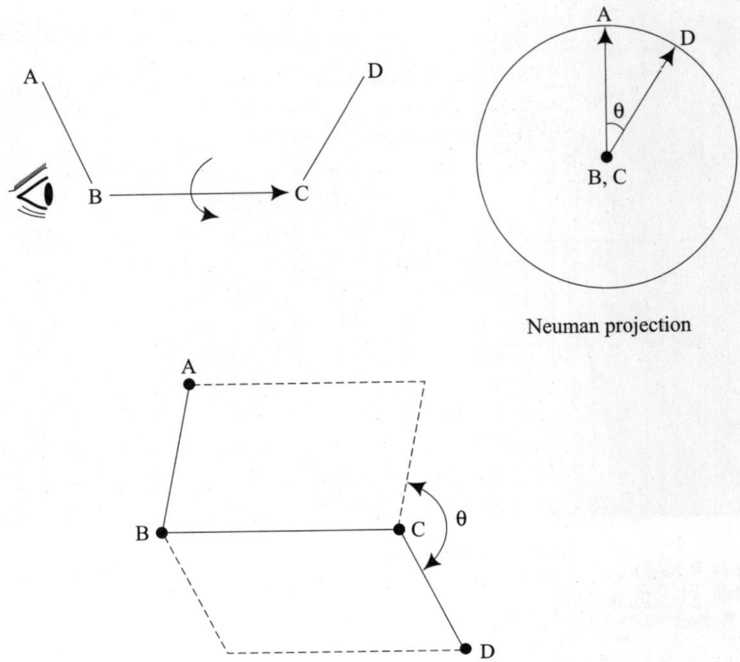

Fig. 2.11 Definition of torsion angle, Θ, about the bond $B \rightarrow C$.

analysis is a plot of conformational angles φ, ψ of amino acid residues of a polypeptide in a sequence. Such information is obtainable from a 'distance plot'. A **'distance plot'** or **'diagonal plot'** is a plot of distances between each amino acid residue and all other residues of a polypeptide (distances between all C_α-atoms of a polypeptide), which provides the folding topology of the polypeptide. Such a map provides information on amino acid residues which are sequentially distant but are nearest neighbours. Inspection of distance maps helps in discerning tertiary structural features of proteins and can provide correlation of structural units of a protein with its DNA-exonic regions, even in protein structures which do not show apparent domain structure.

2.2.3 The Tertiary Structure (3°-Structure) of Proteins

The three-dimensional structure (spatial folding) of a protein, consisting of only one polypeptide chain, is referred to as the tertiary structure. The tertiary structure (protein folding) of a protein is unique and specific to each protein (Fig. 2.15). The structural and functional features of proteins—binding site of ligands (protein-drug interactions), the active sites of an enzymes, or the binding sites for other proteins (protein-protein associations)— depend on their tertiary structures, and therefore, knowledge of the spatial folding of any protein is prerequisite to understand its structural and biochemical functions. This knowledge can also aid the understanding of how particular mutations or variations in the gene that encodes a particular protein lead to changes in the protein's behaviour that can result in disease or in differences in drug interactions among different individuals.

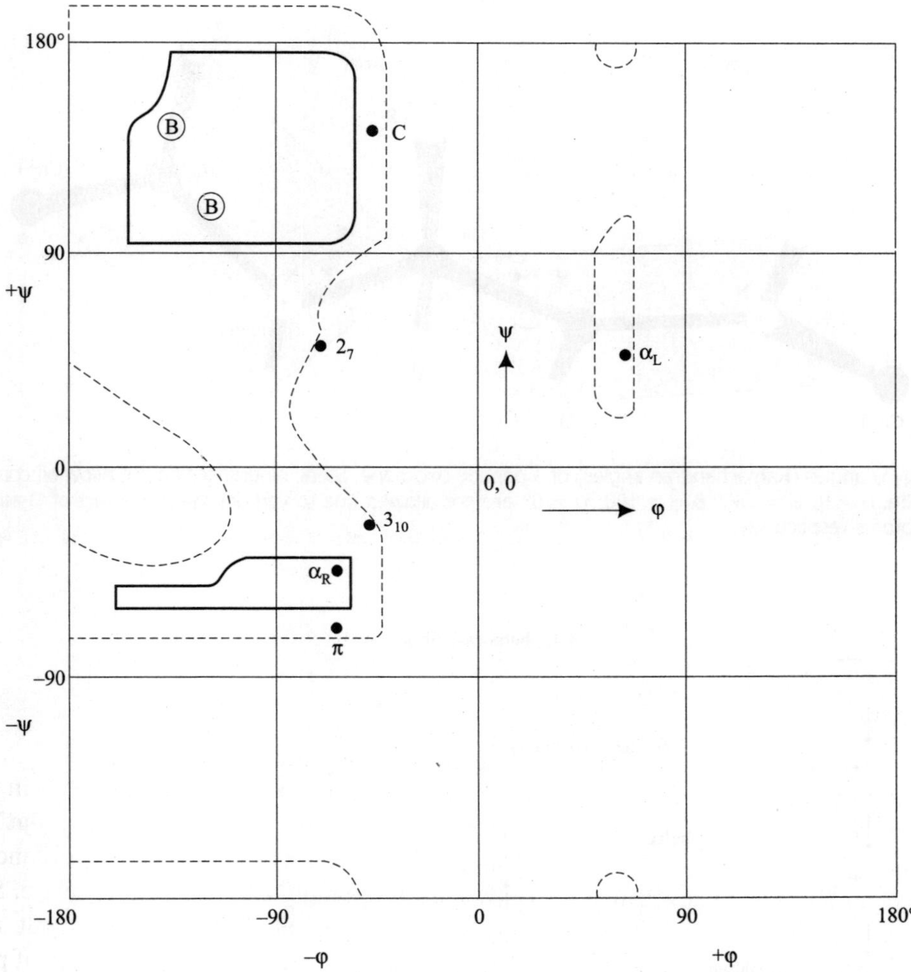

Fig. 2.12 Ramachandran conformational diagram of a polypeptide chain. The solid line indicates fully allowed regions, and broken lines the outer limit regions.
α_R = α-helix; α_L = α-helix (left); 3_{10} = 3_{10}-helix; π = π-helix ; 2_7 = 2.2_7 ribbon;
C = Collagen; β_p and β_a Parallel and antiparallel β-sheet, respectively.

The unique folding of a protein is the result of non-bonded interactions between the polypeptide backbone and the side chains, acting cooperatively. The non-bonded interactions that stabilise the proper folding of macromolecules are—ionic (salt bridges), dipole-dipole; van der Waals and hydrophobic interactions and hydrogen bonds. Ions, dipoles and induced-dipoles have electrical effects on one another by categories of electrostatic interactions: ion/ion, ion/dipole, dipole/induced-dipole and induced-dipole/induced-dipole. Types of interactions and their potentials are summarised in Table 2.4.

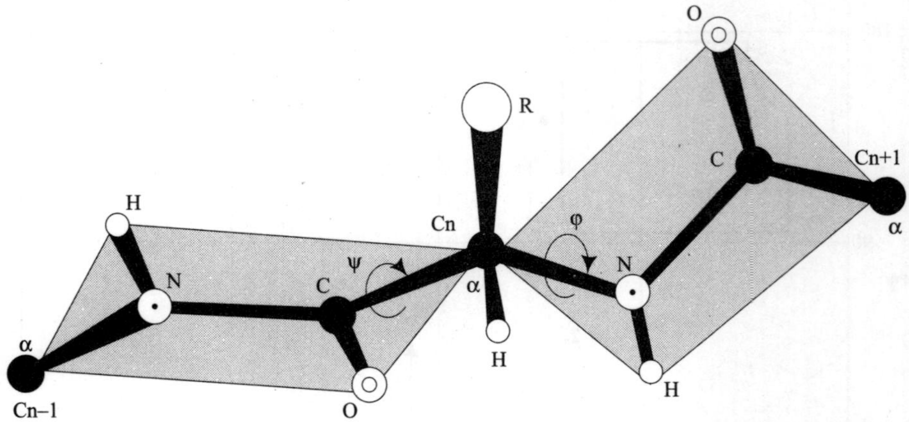

Fig. 2.13 φ, ψ angles (Ramachandran angles) of a peptide backbone. Steric hindrances do not allow all conformations. Eg. φ = 0; ψ = 180° & φ = 180; ψ = 0° are not allowed due to van der Waals contact of O-atoms and H-atoms respectively.

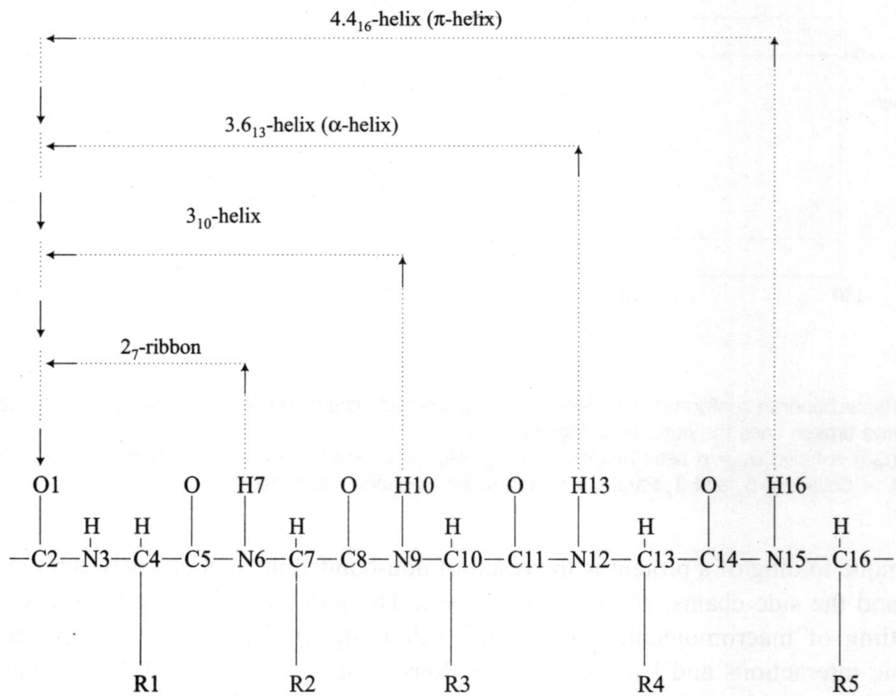

Fig. 2.14a Right-handed helices (2_7–, 3_{10}–, 3.6_{13}– and 4.4_{16}–) of a polypeptide backbone.

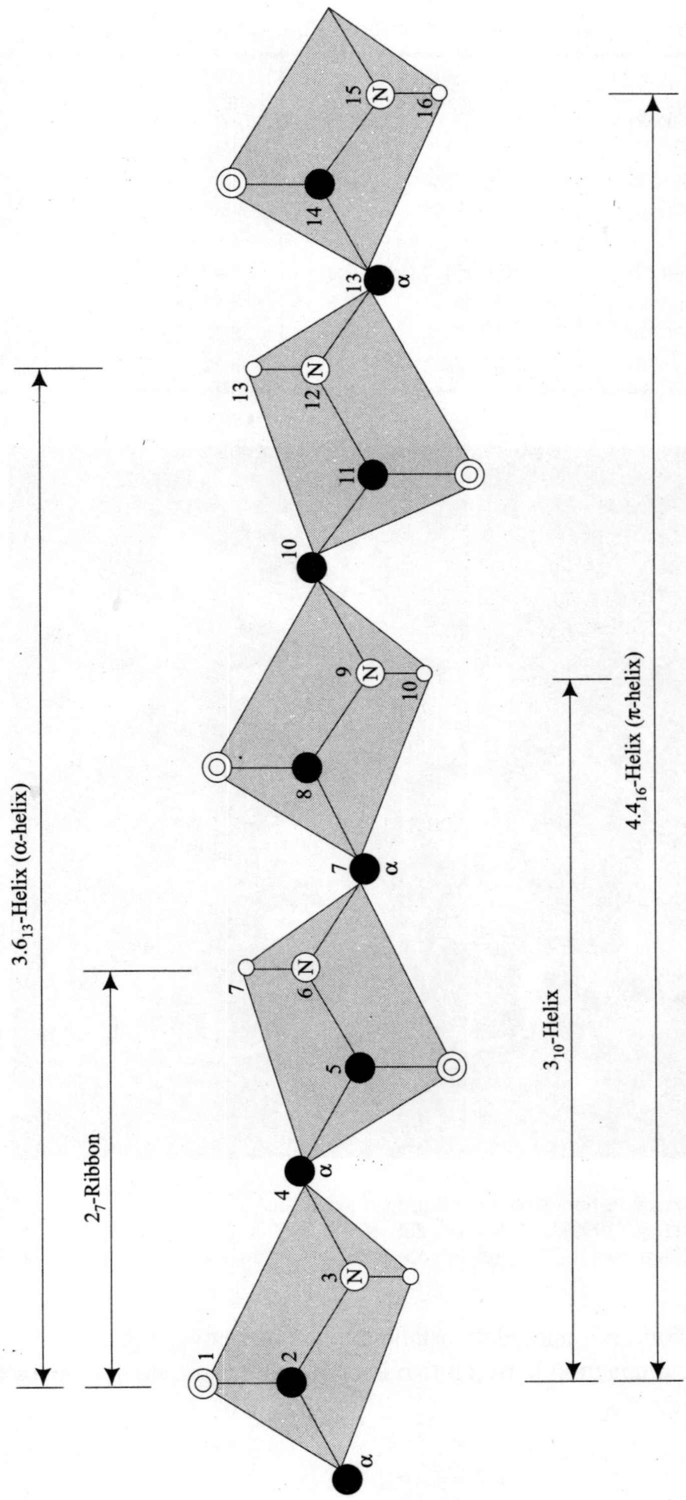

Fig. 2.14b Right-handed secondary structural segments of a polypeptide backbone.

Table 2.3 Structural parameters of ordered segments in polypeptides.

Type	$\varphi_{C_\alpha-N}$ *(deg)*	$\psi_{C_\alpha-C}$ *(deg)*	*n*	*d(Å)*	*P(Å)*	*Atoms in the loop*
2_7-ribbon	105(-75)	240(60)	2.	2.8	5.6	7
3_{10}-helix	130(-50)	155(-25)	3.0	2.0	6.0	10
3.6_{13}-helix (α_R-helix)	122(-58),	133(-47)	3.6	1.5	5.4	13
4.4_{16}-helix (π-helix)	120(-60)	110(-70)	4.4	0.8	3.5	16
β-sheet						
Parallel	60(-120)	295(115)	2.0	3.47	6.95	—
Antiparallel	40(-140)	315(135)	2.0	3.47	6.95	—
Polyproline L-helix	103(-77)	330(150)	–3.0	3.12	9.36	—
Collagen monomer	130(-50)	330(150)	–3.0	3.12	9:36	—

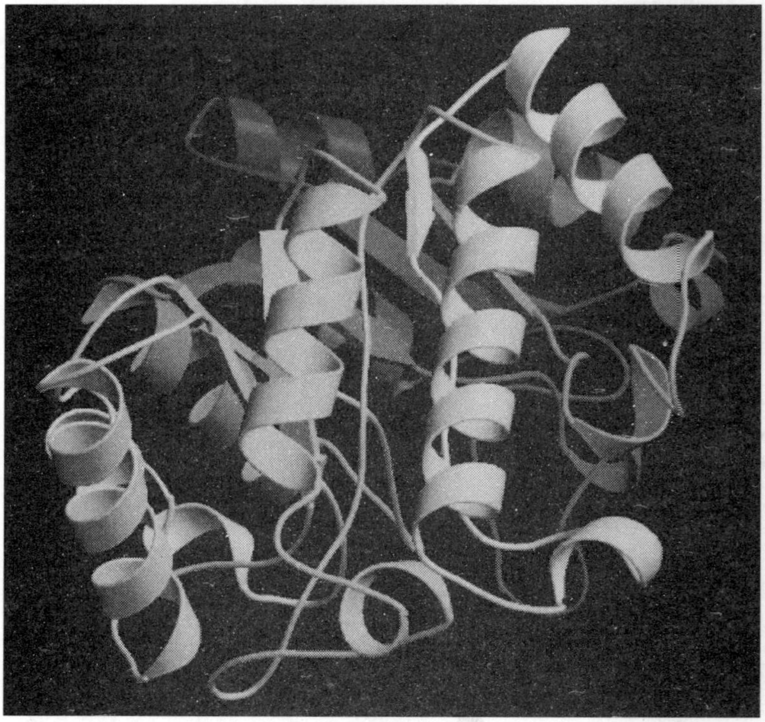

Fig. 2.15 Ribbon diagram of the tertiary structure of protein xylanase.
(Ref: Natesh, R. *et al.* (1999). *J Mol. Biol.*, 288, 999)
(Source: Protein Data Bank: 1TUX.pdb)

Ionic bond. The ionic bond is a pure electrostatic interaction between positive and negative ions or charges. The electrostatic interaction between two ions of charges q_1 and q_1, separated by distance, r, is given by **Coulomb's** equation

$$F = \frac{q_1 q_2}{\varepsilon r^2} \text{ and}$$

Table 2.4 Types of non-bonded interactions and their potentials.

Types of interaction	*Example*	*Potential*	*Energy ($kJmol^{-1}$)*
Ion/ion	NaCl	$1/r$	200
Ion/dipole	$Na(H_2O)_n$	$1/r^2$	20
Ion/induced-dipole		$1/r^4$	
Dipole/dipole	$SO_2.SO_2$	$1/r^3 - 1/r^6$	
Dipole/induced-dipole	$HCl.C_2H_6$	$1/r^6$	
Induced-dipole/induced-dipole (London forces)	He.He	$1/r^6$	<10
Hydrogen bond	$H_2O...H_2O$		~20

$$U = -\int F.dr = -\frac{q_1 q_2}{\varepsilon r} \quad ...(2.6a)$$

where F, U and ε are force, potential energy and dielectric constant, respectively. The binding energy, U_{ij}, which is the main contribution in ionic species is given by the **Madelung** equation

$$U_{ij} = \overset{\text{Coulomb}}{-\frac{q^2}{r_{ij}}} + \overset{\text{repulsive}}{\lambda \exp\left(-\frac{r_{ij}}{\rho}\right)} \quad (2.6b)$$

where λ and ρ are empirical parameters.

NaCl is an example of a pure ionic crystal (Fig. 2.16). Each Na^+ ion is surrounded by six Cl^- ions and each Cl^- ion, in turn, is surrounded by six Na^+ ions. There is no individual NaCl molecule. So Na^+Cl^- contains, like other ionic compounds, an infinite array or lattice, of positively and negatively charged ions. Balance of charge is the main criterion for stability in ionic compounds. There is no directionality to ionic interactions. Ion pairs play a minor role in the stability of macromolecular structure.

An example of ionic interaction in macromolecules is the binding of a substrate, e.g., glycyl-L-tyrosine to carboxypeptidase-A, an enzyme that cleaves carboxyl-terminal residues (Fig. 2.17). The negatively-charged terminal carboxylate group of the dipeptide substrate forms a salt bridge with the positively charged guanidinium group of an arginine residue of the enzyme. Similarly, in other cases, if the substrate is negatively charged, arginine or lysine residues of an enzyme can form salt bridges. On the other hand if the substrate is positively charged, it can form salt bridges with the carboxylate group of aspartate or glutamate residues of the enzyme.

Salt bridges play a structural role in allosteric cooperativity in multisubunit enzymes and maromolecules (e.g., aspartic carbamylase, and hemoglobin). As an example: the hemoglobin molecule consists of four polypeptide subunits, α_2, β_2, and exists in two structural states, deoxy- and oxy-hemoglobin. In the deoxy-state, the carboxyl terminal residues of all four polypeptide chains are anchored by salt bridges forming a more rigid structure, and therefore is called the tense-state (**T-state**). Transition from the deoxy-state to the oxy-state, upon binding of an oxygen ligand, leads to the breaking up of these salt bridges and lends freedom of rotation to the terminal carboxyl residues, and hence is called the relaxed-state (**R-state**). Concommitently there are structural changes in the α_1

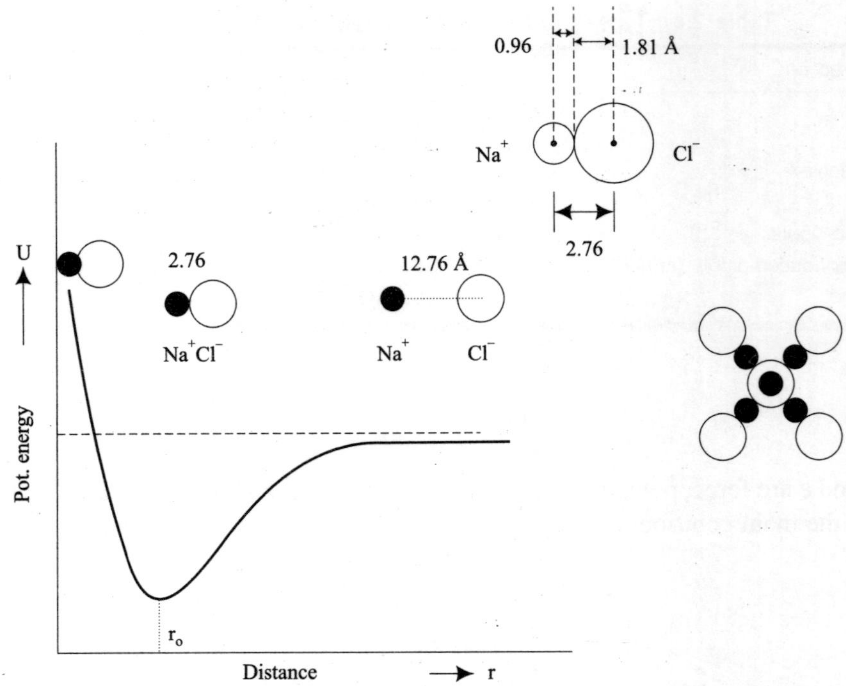

Fig. 2.16 Examples of ionic bond interaction (Na^+Cl^-) and packing

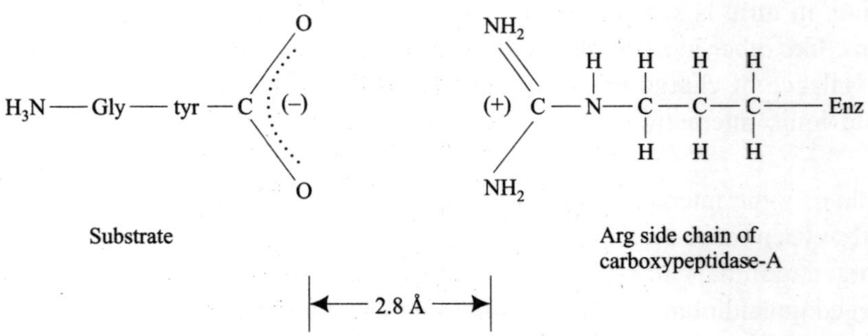

Fig. 2.17 Example of salt bridges in macromolecules

α_2 monomer contacts for the hemoglobin molecule and differences in the distances between Fe atoms in the subunits. Minor structural changes in the subunit interaction (quaternary interactions) lead to long-distance cooperative structural alterations, which in effect lead to allosteric functional features in allosteric proteins/enzymes.

Dipole/dipole interaction. Non-bonded molecular interactions—dipole/dipole, dipole/induced-dipole and London interactions are termed as van der Waals interactions. van der Waals interactions indicate how close the non-bonded atoms can approach each other (given by their van der Waals radii). van der Waals radii of some atoms relevant in biology are given in Table 2.5.

Table 2.5 van der Waals radii of some atoms relevant in biology.

Atom (group)	*van der Waals radius (Å)*
H	1.0
C, N and O	1.5
F	1.35
P and S	1.9
Cl	1.8
Br	2.0
I	2.2
$-CH_3$	2.0

Atoms or molecules with zero net charge ($q = 0$) can generate electrical fields if their centres of positive and negative charges do not coincide. A charge separation ads to a dipole moment ($\mu = ql$; $\mu \neq 0$). Molecules characterised by this type of charge separation are dipoles and are called polar molecules. That is, dipole-dipole interaction occurs only between polar molecules or molecules that possess permanent dipole moment ($\mu \neq 0$). Since the energy of a charged species is affected by an electrical field, the energy of an ion or dipole will be affected by the presence of other ions or dipoles. Electrostatic interaction between two polar molecules of dipole moments μ_1 and μ_2 at a distance, r. differs depending on the orientation of the molecules. For example, the energy of interaction, U, is

$$U = -\frac{2\mu_1\mu_2}{\varepsilon r^3} \quad \text{for prolate dipoles} \qquad \text{...(2.7)}$$

$$U = -\frac{\mu_1\mu_2}{\varepsilon r^3} \quad \text{for oblate dipoles}$$

The average energy of interaction of permanent dipoles randomly distributed is given by the **Keesom** equation

$$U = -\frac{2}{3}.\frac{\mu_1^2\mu_2^2}{\varepsilon r^6}.\frac{1}{KT} \qquad \text{...(2.8)}$$

Dipole/induced-dipole interaction. Ions or permanent dipoles can induce dipole moments in a neutral, but polarisable molecule (Fig. 2.18). The induced dipole moment, μ_{ind}, is

$$\mu_{ind} = \alpha E \qquad \text{...(2.9)}$$

and the energy of interaction, U, of a dipole and an induced dipole is given by the **Debye** equation

$$U = -\frac{2\alpha\mu^2}{\varepsilon r^6} \qquad (2.10)$$

where E is the electric field, α is polarisability of the molecule and ε is the dielectric constant of the medium (permitivity). If the dielectric constant of a substance is large, then its molecules are polar and are highly polarisable.

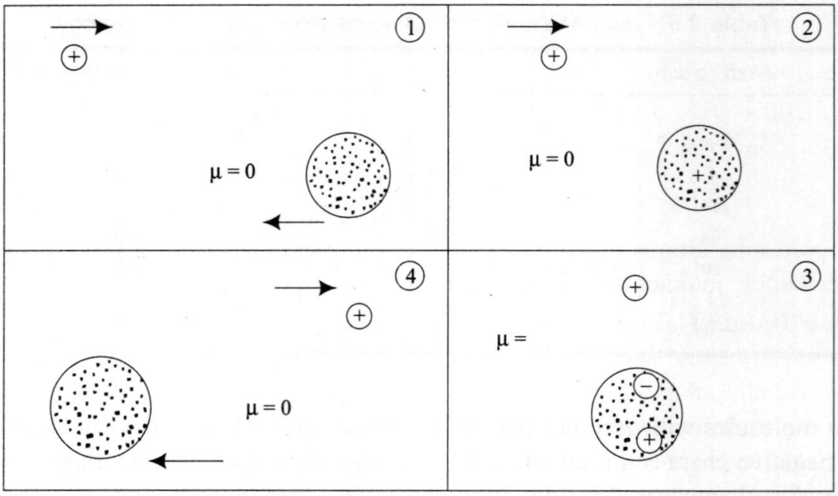

Fig. 2.18 Induction of dipole moment by an ion in a neutral molecule.

The presence of molecules, like HCl, with asymmetric charge distribution will affect neighbouring molecules. The molecular alignment is opposed by thermal motion, and the molecular polarisation, P_m, and relative permitivity ε_r (dielectric constant) are related according to the **Debye** and **Clausius-Mossotti** equations,

$$P_m = \frac{N_A}{3\varepsilon_0}\left(\alpha + \frac{\mu^2}{3K_BT}\right) \qquad \text{Debye Eqn.} \qquad ...(2.11a)$$

$$P_m = \left(\frac{\varepsilon_r - 1}{\varepsilon_r + 2}\right)\frac{M}{\rho} \qquad \text{Clausius-Mossotti Eqn.} \qquad ...(2.11b)$$

where M, ρ, N_A and ε_0 are respectively molecular mass, density of the sample, Avogadro's number and permitivity in vacuum, and K_B is the Boltzmann constant.

The P_m values can be evaluated from the measured values of ε_r of a molecular species at various temperatures. A plot of P_m against 1/T enables the determination of the permanent dipole moment, μ, from the slope, and the polarisability, α, from the intercept.

P_m can also be determined as a function of frequency. At high alternating frequencies the molecule has no time to align. Therefore, $P_\mu = 0$ and $P_m = P_\alpha$. A light beam can be employed as the rapidly alternating field. According to the **Maxwell** equation, the refractive index, n, is related to the dielectric constant.

$$\varepsilon = n^2 \qquad \text{Maxwell equation} \qquad ...(2.12a)$$

Therefore, the Clausius-Mossotti equation can be represented in terms of molecular refractive index, R_m.

$$R_m = \left(\frac{n^2 - 1}{n^2 + 2}\right).M/\rho \qquad \text{...(2.12b)}$$

$$\alpha = \frac{3\,\varepsilon_0}{N_A\,R_m} \qquad \text{...(2.12c)}$$

Dipole/dipole interactions are important for protein folding, hydration and solvent structure. Dipole/induced-dipole interaction is independent of the temperature.

Induced-dipole/Induced-dipole Interaction. The origin of cohesive forces that is, attractive interactions between neutral and nonpolar molecules, like helium, can be explained only by quantum physics. In neutral and nonpolar species, the electron charge distribution is spherically symmetrical, but there can be instantaneous changes (with time). This instantaneous assymmetry of charge distribution around an atom can create an induction in a neighbouring atom. For the induced-dipole/induced-dipole, the quantum mechanical treatment gives the energy, U

$$U = -\frac{3}{4}\frac{I\alpha^2}{r^6} \qquad \textbf{London}\text{ equation} \qquad \text{...(2.13)}$$

where I is the first ionisation potential of the molecular species.

These interactions, called dispersion forces or **London** interactions, differ from the other van der Waals interactions in that the London interactions are always attractive. But, at very short distances (nuclear distances), atoms are prevented from fusion due to repulsive forces that arise from the overlap of electron clouds and nuclei of different atoms. Such a potential (Eqn. 2.14) is called a **Lennard-Jones** potential (or 6 - 12 potential).

$$U = -\frac{A}{r^6} + \frac{B}{r^{12}} \qquad \text{...(2.14)}$$

where A and B are constants, and r = interatomic distance.

Hydrophobic Interactions. Hydrophobic interactions are nondirectional. Hydrophobic interactions in molecular associations and stability are least characterised but are of fundamental importance in understanding higher-order organisation in macromolecules. The formation of nonpolar interactions is one of the most important factors in macromolecular folding, stacking and structure assembly. That is, hydrophobic interactions, though weak, are important in stabilising macromolecular associations. Hydrophobicity depends on surface area and dipole content. A linear relationship exists between $\Delta G_{transfer}$ and the water-accessible surface area for completely nonpolar side chains. The driving force for the formation of hydrophobic environment is entropy (positive entropy change)

$$-RT \ln K = \Delta G_{tot} = \Delta H_{ch} = \Delta H_{sol} - T\,\Delta S_{ch} - T\,\Delta S_{sol} \qquad \text{...(2.15)}$$

where ch = chain and sol = solution.

For nonpolar region, ΔH_{ch} and ΔH_{sol} are small (ΔH_{ch} is positive and ΔH_{sol} is negative) and ΔS_{sol} is positive and large. That is, ΔS_{sol} from a nonpolar group gives the largest single contribution to protein stability. Whereas in polar regions, ΔS_{sol} is cancelled by ΔH_{ch} (that is $\Delta G \approx 0$). (Concepts of bioenergetic terms, enthalpy, entropy etc., are explained in Bioenergetics (Chapter 13)).

The Hydrogen Bond. The hydrogen bond is a special case of a permanent dipole interaction. It is largely polar in character and is formed whenever a polar bond containing a hydrogen atom (from donor groups such as O-H, N-H etc.) interacts with an electronegative acceptor atom(s) such as O, N, Cl or F. When a hydrogen bond is formed, the H-atom is shared by the other two electronegative atoms and the bond is linear and directional. The reason for hydrogen bond formation is that when a hydrogen atom (which has only a single electron) forms a covalent bond with an electronegative atom (such as O, N, F), its nucleus is partially unshielded. As a consequence, the proton interacts with another electronegative atom of a neighbouring molecule/group (Fig. 2.19a).

The hydrogen bond is a weak non-bonded interaction ($\approx$ 20 $kJ.mol^{-1}$), but is most important and plays a crucial role in determining the physico-chemical properties, structural stability and function of many compounds—such as water and macromolecules like proteins, nucleic acids etc.

(a) Water and its properties. Biological structures and process can only be understood in terms of the physical and chemical properties of water, as these molecules exist and carry out their functions in aqueous media. Water is a highly polar molecule, with its O-H bonds showing considerable ionic character due to the difference in the electronegativities of O and H atoms. In the water molecule the oxygen atom forms a tetrahedral configuration, with the O-H bonds and two lone-pair electron lobes (Fig. 2.19b).

The electrostatic attractions between the dipoles (H_2O) in bulk water tend to orient the water molecules, $(H_2O)_n$ in such a manner that each oxygen atom, of a water molecule, can have at least four tetrahedrally-arranged hydrogen bonds: two donor hydrogen bonds through O-H bonds of the molecule and two acceptor hydrogen bonds through the two lone-pair electron lobes thus forming a three-dimensional (spatial) network of clatharate-like (cage-like) giant water structure (Fig. 2.19c). That is, each H_2O molecule is surrounded by four neighbouring H_2O molecules tetrahedrally oriented in space. This cage-like spatial structure of water allows interstitial spaces in which other small ions and groups can enter. The directionality of hydrogen bonding and consequent formation of a clatharite-like structure allows water to have a more open structure, dynamically changing due to the formation and breakage of hydrogen bonds.

The physicochemical properties of ice and water are due to inter-molecular hydrogen bonding. The clatharate-like open lattice makes water expand on freezing and the density of ice is less than that of water. These properties of water have overwhelming consequences for the survival of organisms.

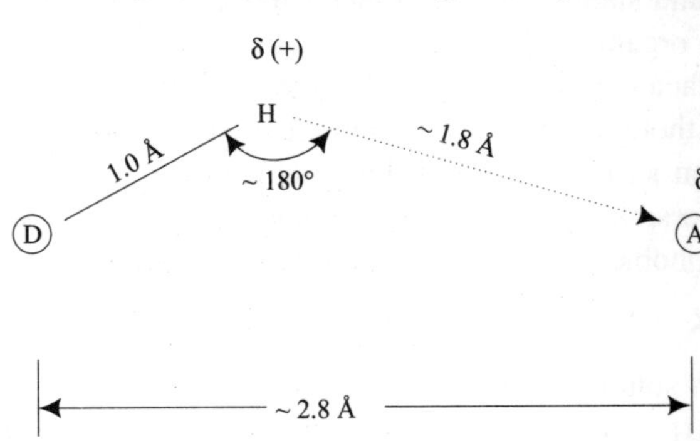

Fig. 2.19a The stereochemistry of the hydrogen bond. *D* and *A* are donor and acceptor atoms, respectively, of a proton

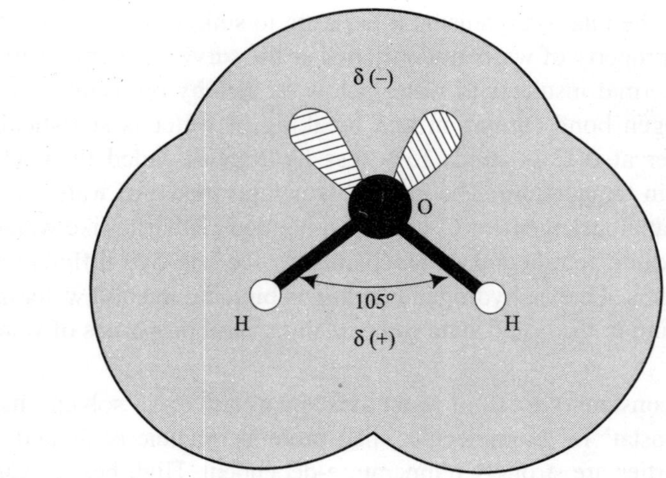

Fig. 2.19b Tetrahedral arrangement in the water molecule, with van der Waals surface (shaded). Lone-pair electron lobes are indicated by dashed lines.

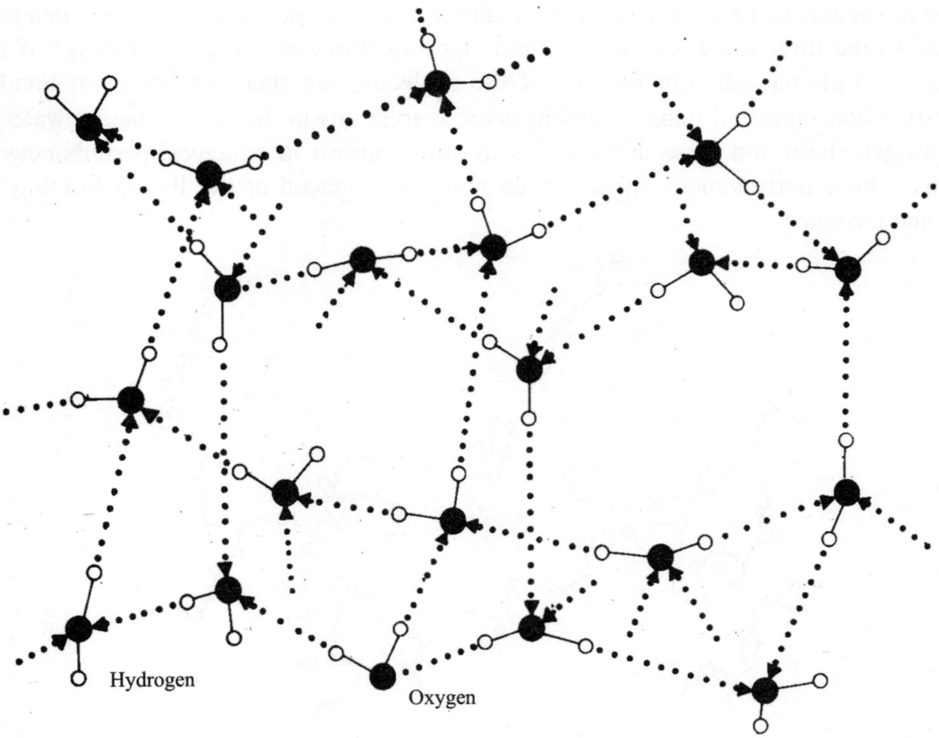

Fig. 2.19c Clatharate-like structure of ice/water. Solid lines are covalent bonds. Hydrogen bonds are indicated by dotted lines.

One could well imagine the fate of organisms if ice were to sink in oceans if its density were higher than that of water. This property of water not only makes the survival of organisms in water possible, but ice also acts as a thermal insulator to water below it, thereby protecting organisms from being frozen to death. Hydrogen bond formation and breaking in water is statistical and temperature-dependent. Liquid water at 0°C is about 15% less hydrogen-bonded than ice and the disorder increases with increase in temperature. The general structural model of water is that it consists of a randomly fluctuating spatial arrangement of hydrogen-bonded clatharite-like water molecules which, over short distances, exhibit tetrahedral stereochemistry. Ice has two different kinds of hydrogen bonds of different strengths. That is, hydrogen bonding is bimodal in solid water and the existence of bimodal hydrogen bonding in the liquid state can explain several properties of water, such as its large specific heat.

The high dielectric constant ($\varepsilon = 80$) of water makes it a 'universal' solvent. Its high heat capacity makes it a good 'thermostat' for biomolecules, like proteins, nucleic acids and membranes, where structure-function properties are strongly temperature-dependent. High heat of vaporisation of water helps to reduce the loss of water by evaporation and the cooling effect of evaporation helps to regulate the temperature of organisms.

Water plays an important role in biological systems and hydration of proteins plays a crucial role in the structure, assembly and function of biological molecules. Each protein is surrounded by a **hydration shell** of water molecules which directly interact with protein molecules. The majority of the water molecules that are invariant form multiple hydrogen bonds with protein molecules. These contribute to the three-dimensional structural stability (three-dimensional folding) of the protein molecule by bridging different regions of the molecule together via hydrogen bond networks (Fig. 2.20). Water-mediated transformations cause disruptions in the protein-bound water molecules (first hydrogen shell) and these disturbances in turn manifest in structural perturbances in protein molecules. These perturbances appear to be more pronounced at the ligand binding regions of proteins and enzymes.

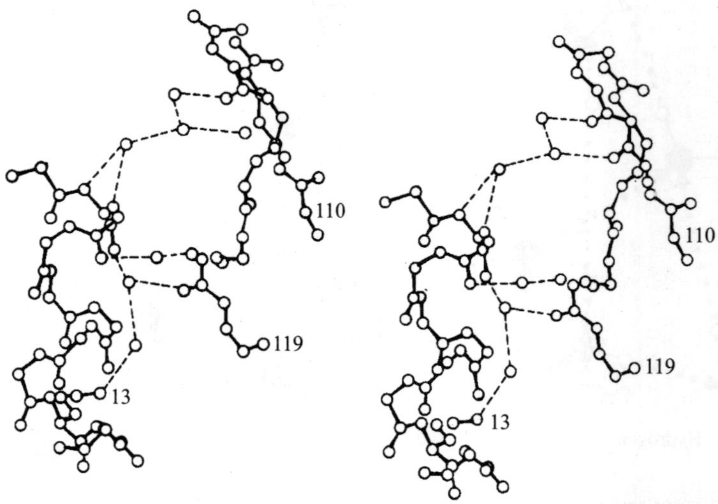

Fig. 2.20a Stereoview of invariant water molecules stabilising the N-terminal helix and C-terminal strand in ribonuclease **A** structure#.

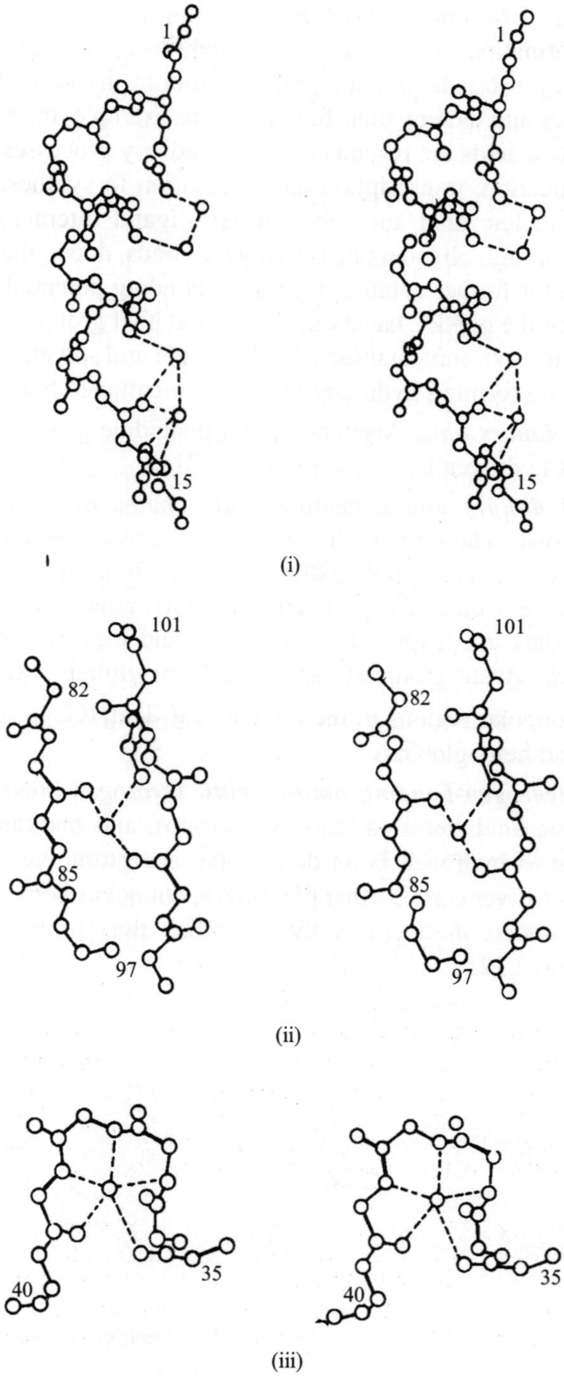

Fig. 2.20b Invariant water molecules stabilising (i) helix, (ii) sheet and (iii) loop regions in ribonuclease **A** structure.[#] (Ref: Radhakishan, K.V. *et al.* Acta Cryst D. (1995).
([#]Photographs: Courtesy of Prof. M. Vijayan, Indian Institute of Science, Bangalore).

(b) Structural stability and function. Hydrogen bonding is largely responsible for the structural stability and proper conformation of biological molecules, such as proteins and nucleic acids. Intra- and inter-molecular hydrogen bonds play a significant role in the secondary and tertiary structural stability of macromolecules and thereby their functions. For example, the hydrogen bonding networks between the bases in nucleic acids are responsible for hereditary processes in organisms— storage of the genetic code, its replication, transcription and translation to synthesise proteins. Specificity in protein-protein, protein-nucleic acid and other protein-ligand interactions (molecular-molecular recognition) are effected through oligodentate (hydrogen) bonds, due to the directionality of hydrogen bonds (refer to Chapter 3 for further details). Hydrogen bonding potentiality in proteins depends not only on the polar groups of the peptide backbone (C=O and N-H groups) and side chains, but also on their chemical forms due to environmental factors such as pH and solvent. Some of the characteristics of amino acid side chains in forming hydrogen bonds are mentioned below:

(i) ***Hydrogen bond donors only.*** Arginine with a guanidine group, lysine, the indole group of tryptophan act as hydrogen bond donors (Fig. 2.21a).

(ii) ***Hydrogen bond donors and acceptors.*** Side chains of asparagine, glutamine, serine, threonine and tyrosine can serve both as hydrogen bond donors and as acceptors (Fig. 2.21b). Also their hydrogen bonding potentiality depends on environment. For example, the side chains of tyrosine, as well as those of serine and threonine act as hydrogen donors as well as acceptors when they are 'exposed' to a solvent and they act only as donors of hydrogen bonds to the carboxylate group of aspartic acid or glutamic acid residues when they are 'buried' in the nonpolar regions of the protein (e.g. Tyr(HC2) –⟨O⟩– OH ··>O=C⟨ (FG4) in myoglobins and hemoglobins).

(iii) ***pH-dependant hydrogen-bonding amino acids.*** Hydrogen bonding capabilities of aspartic acid and glutamic acid (terminal carboxyl group), and histidine are pH-dependant. The groups can serve as hydrogen bond donors and acceptors over a certain pH-range and as acceptors or donors over certain other pH-ranges, but not as both simultaneously (Fig. 2.21c). Types of non-bonded interactions that stabilise the 'folding' in macromolecules are represented in Fig. 2.22.

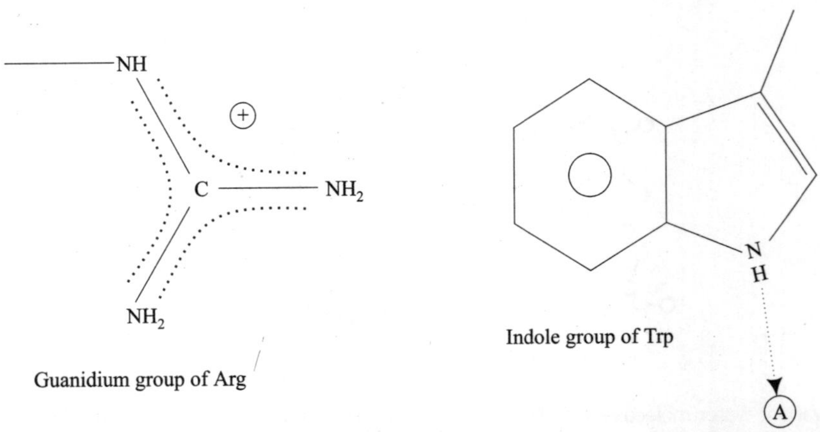

Fig. 2.21a Hydrogen bond donor side chains in amino acids.

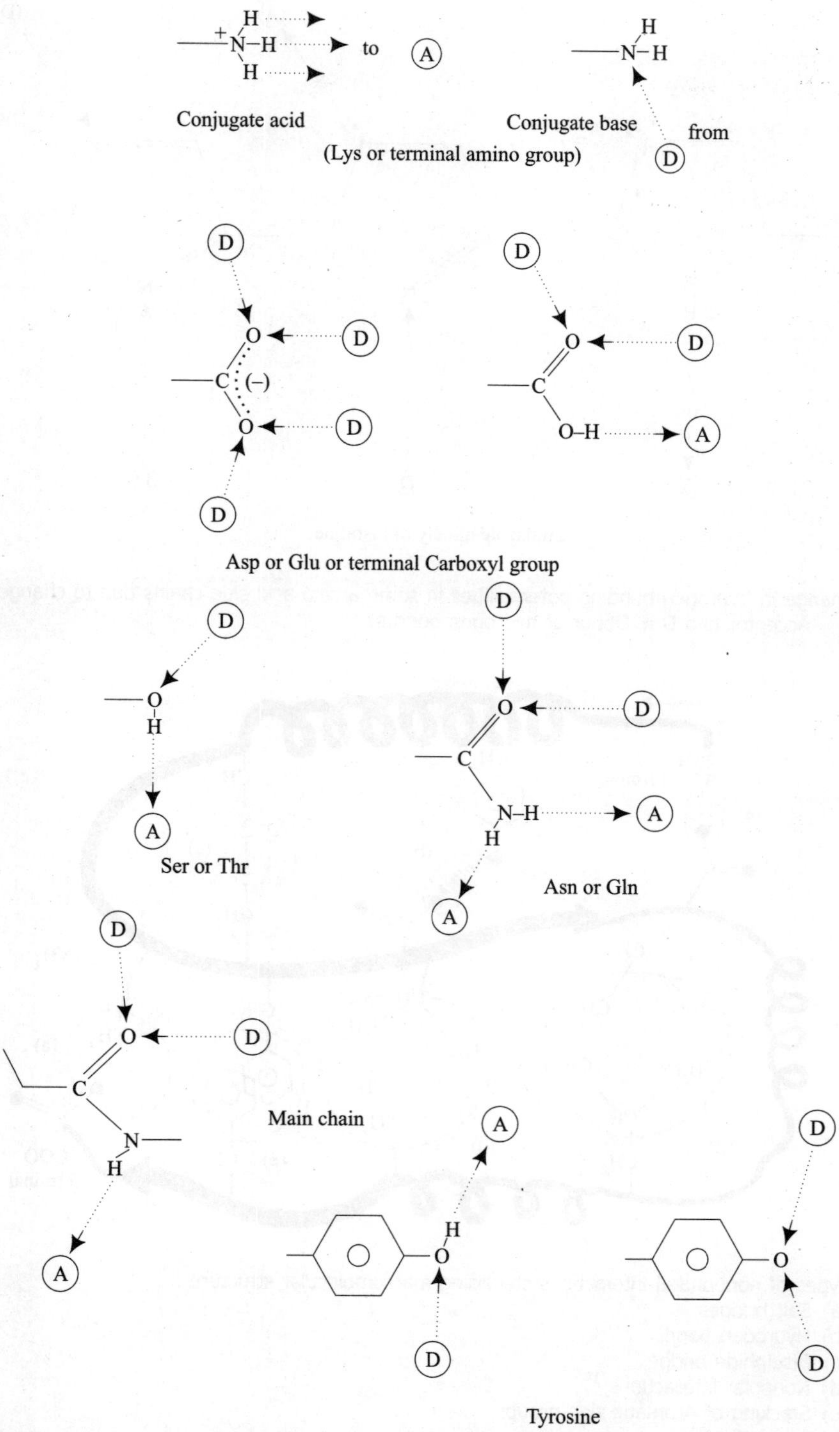

Fig. 2.21b Hydrogen bond donor and acceptor side chains in amino acids.

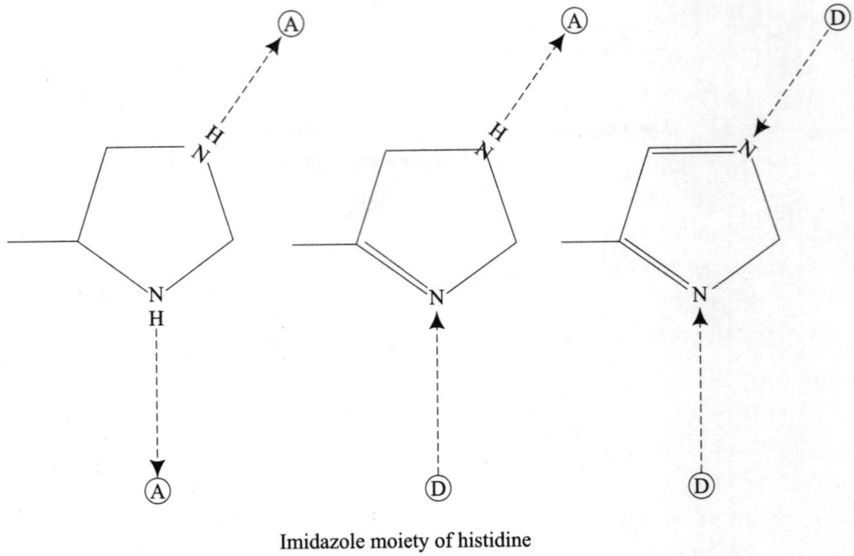

Fig. 2.21c Change in hydrogen bonding potentialities in some amino acid side chains due to change in pH. A = Acceptor and D = Donor of hydrogen bond(s)

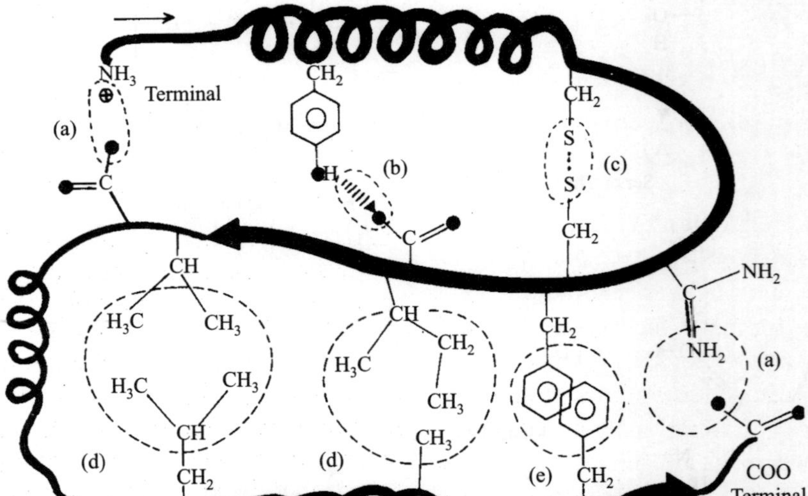

Fig. 2.22 Types of nonbonded interactions stabilising macromolecular structure
(a) Salt bridges,
(b) Hydrogen bond,
(c) Disulphide bridge,
(d) Nonpolar interactions,
(e) Stacking of Aromatic side groups,
[helix symbol] ≡ helical
[arrow symbol] ≡ β-sheet segments

2.2.4 The Quaternary Structure (4°-Structure) of Proteins

The structural organisation of proteins consisting of single polypeptides, such as myoglobins, lysozymes, cytochrome **c**, trypsin and chymotrypsin etc., is complete at the tertiary level. However, in the case of proteins and protein complexes that comprise two or more polypeptide chains (poly-subunit proteins), such as hemoglobins (Fig. 2.23a), cytochrome oxidase, ATPases and collagen and viruses etc., another level of organisation, namely quaternary structure, deals with the specific arrangements of the subunits with respect to one another in the protein complex and the interactions that stabilise these arrangements and associations. Non-polar interactions, which are weak, nondirectional and driven by entropy factors, play a major role in the structural organisation and stability of quaternary structures in macro-molecular complexes.

Minor and subtle alterations in the interactions between subunits (subunit associations) in a multimeric protein complex lead to large range structural changes and thereby to crucial functional changes. Such allosteric effects are common in multimeric protein complexes (hemoglobin, aspartic carbamylase etc.).

The differences in subunit arrangement could be due to extrinsic factors such as non-bonded interactions and solvent effects, or could be due to factors intrinsic to the protein complex (probably due to packing considerations) as is the case in the quaternary structural arrangement in legume lectins. The quaternary arrangement of peanut lectin is shown in Fig. 2.23b. Small alternations in tertiary structures apparently lead to different types of quaternary association, and consequently to functional differences.

2.3 PROTEIN CLASSES

From the structural perspective, proteins can be grouped under globular and fibrous molecules. Globular proteins, both single polypeptide chain proteins and enzymes (myoglobins, cytochrome **c**, carboxypeptidases, lysozymes, trypsins, papains, dehydrogenases etc.) and multi-subunit proteins and complexes (hemoglobins, ATPases etc.), have compact globular shapes due to chain reversals of polypeptide backbones (Fig. 2.24a) and packing of side chains and are, in principle, amenable to crystallisation in a single-crystalline form. Detailed three-dimensional structural information from X-ray diffraction for many such single-crystalline globular proteins and protein complexes is available or can be obtained. Fibrous proteins and protein-complexes (viruses), whose axial lengths are far greater than their widths, are not single-crystalline but are semicrystalline (periodic only along the fibre axis). Structural information obtainable from a fibrous molecule by X-ray diffraction and other methods is meagre, as a result of the inherent lack of three-dimensional order in such molecules.

2.3.1 Fibrous Proteins

Fibrous proteins can be divided into, i) the collagen group, ii) the keratin (**k-m-e-f**) group and iii) others.

2.3.1.1 The collagen group

Collagen is a structural protein and is abundant in vertebrates. It is found in skin, tendon, blood vessels, bone and teeth. The structural unit of the collagen fibril is tropocollagen which is rod-shaped of length ≈3000 Å, width ≈15 Å and molecular mass ≈285 kilodaltons. It comprises three polypeptide

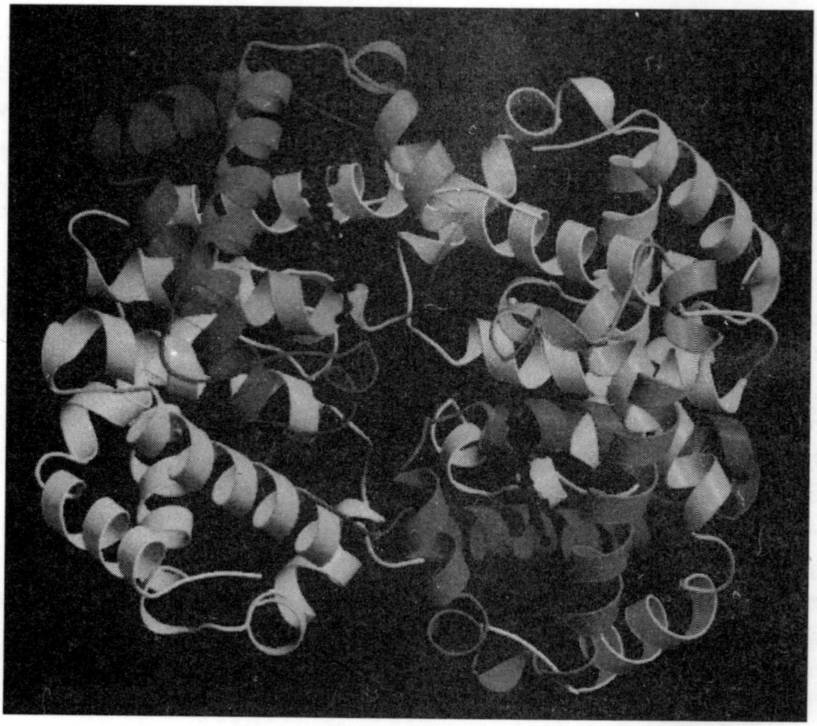

Fig. 2.23a Ribbon diagram of hemoglobin, a tetrameric ($\alpha_2\beta_2$-subunit) protein.
(Ref: Tame, J. & Vallone, B.)
(Source: Protein Data Bank: 1A3N.pdb)

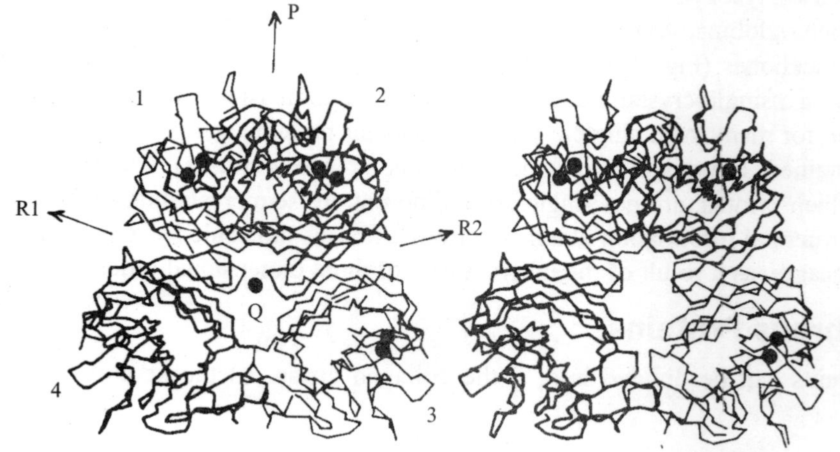

Fig. 2.23b Stereo view of the α-carbon backbone representation of the tetrameric subunits arrangement in peanut lectin.
(Ref. Banerjee, R. et al. Proc. Natl. Acad. Sci. U.S.A. 91; 227 (1994).
(Photograph—Courtesy Prof. M. Vijayan, Indian Institute of Science, Bangalore)

cains. The amino acid composition of each monomer is glycine ~33%, proline ~15-30% and hydroxyproline (Hypr) ~15 to 30%. The amino acid sequence of each monomer is **(Gly-X-Y)**$_{333}$ where **X** is generally proline and **Y** is generally 4-hydroxyproline. That is, every third residue is glycine and stereochemical restraints do not allow any other residue to be in its position. The conformation of the monomers is a left-handed polyproline II or polyglycine II type helix. Three left-handed helices are wound around each other forming a coiled-coil right-handed triple helix, a cable-like structure (Figs. 2.24b & c).

Trans-hydrogen bonds between neighbouring chain residues, proper docking of proline and hydroxyproline residues and other stabilising interactions such as hydroxylation and covalent cross-

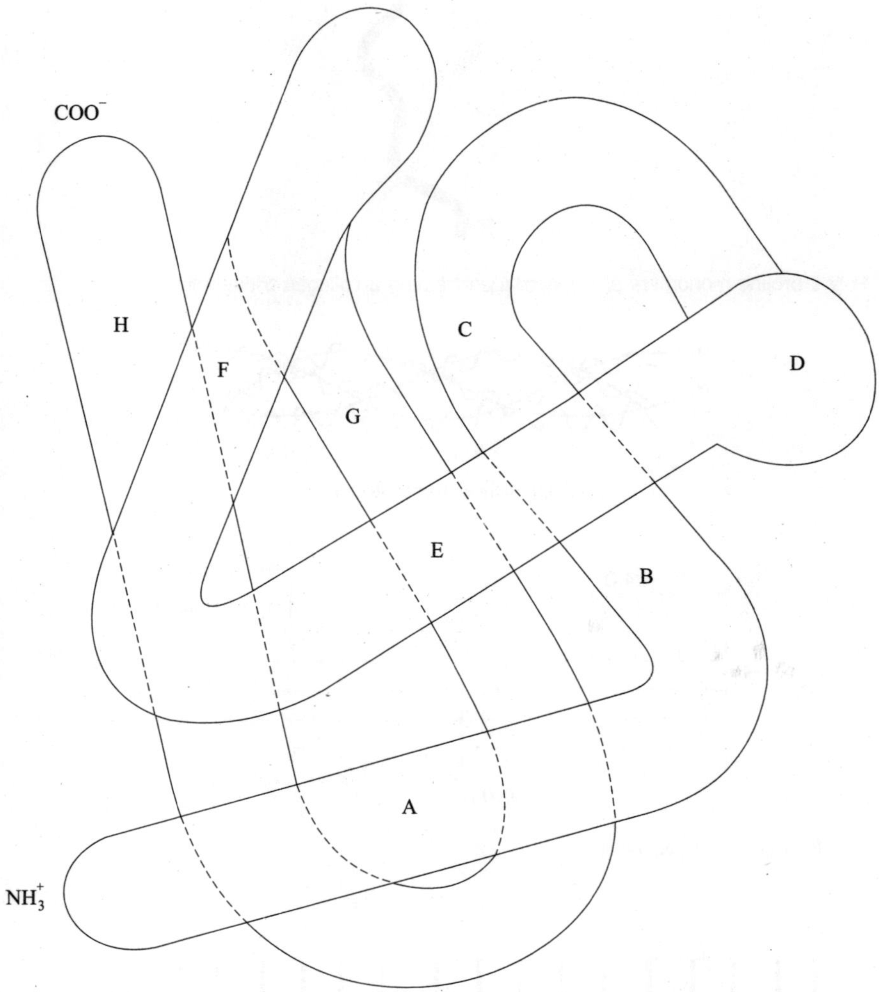

Fig. 2.24a Globular shape of myoglobin tertiary structure due to chain reversal. A, B, C, D, E, F, G & H are α-helical segments.
(After Dickerson, R.E. in "The Proteins", Vol. II. 2nd Edn. (Ed. Neurath. H). Acad Press/New York. (1964).

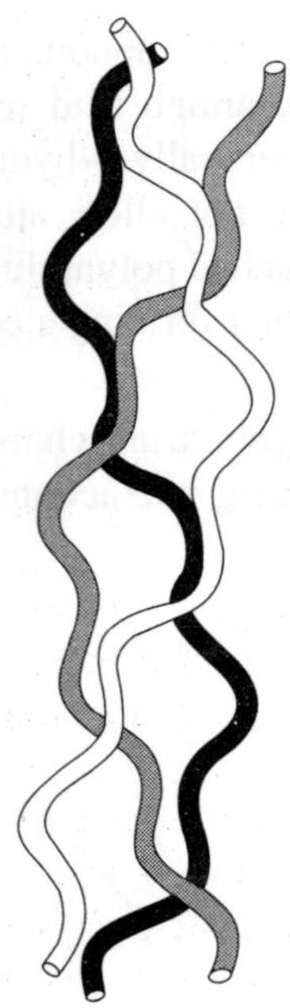

Fig. 2.24b Poly L-proline monomers of tropocollagen forming a collagen triple-helix.

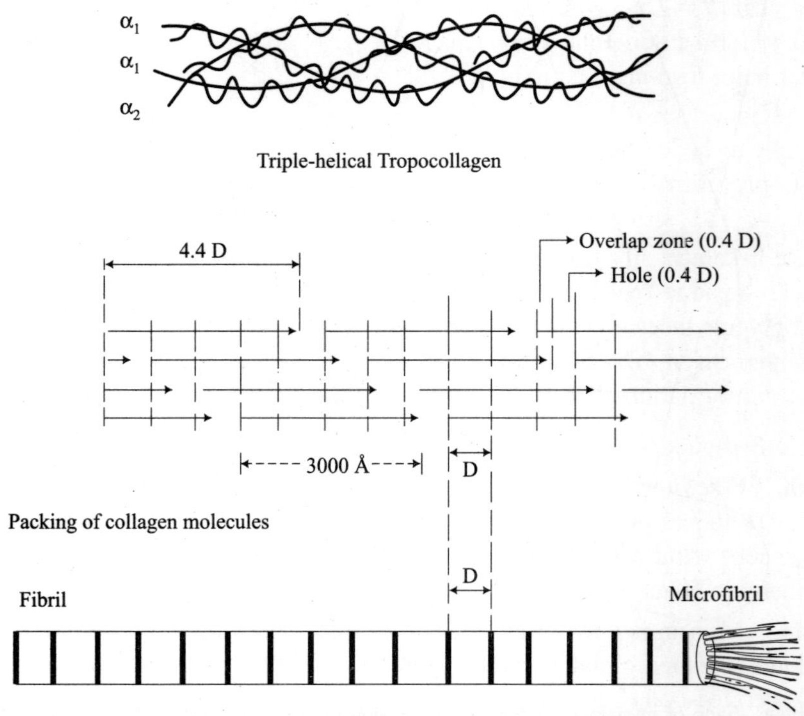

Fig. 2.24c Structural organisation in collagen fibres.

links by lysine side chains render structural strength to collagen fibres. Lack of proper packing, and defective hydroxylation or cross-links, lead to molecular diseases like scurvy, bone and skin deformities (Ehlers Danlos syndrome). The overall structural design of collagen fibres is a quarter-staggered arrangement of tropocollagen molecules with 680 Å periodicity (D-period) (Fig. 2.24c).

2.3.1.2 Keratin (K-m-e-f) group proteins

Keratin group (**K-m-e-f**) proteins include α-keratin of hair, wool, horns, myosin of muscle, epidermis of skin, and fibrinogen of blood. Silk fibroin is a β-keratin.

α-keratin: is predominantly found in skin, hair, nails, horns and feathers. The primary structure shows a general trend for a seven-residue repeat with a preponderence of nonpolar residues at the 1st and the 4th positions and of charged residues at the 5th and the 7th positions-

1	2	3	4	5	6	7	
$-(a$	$-b$	$-c$	$-d$	$-e$	$-f$	$-g-)_n$	(*refers to nonpolar residues).
*			*	±		±	

The most common amino acid residues in the heptad repeats of coiled-coil proteins are: **Ala, Ile, Leu** and **Val** at the 1st and 4th positions (nonpolar residues); Glu, Lys, Gin and Arg at the 5th and 7th positions (polar residues). Also, the residues at the 2nd, 3rd and 6th positions are generally polar residues (**Asp, Glu, Lys, Gln, Arg**) with **Ala** being the exception.

The conformation of α-keratin is α_R-helical and since the α_R-helix has 3.6 residues per turn, the seven-residue repeat ($7 = 2 \times 3.6$) leads to the alignment of the nonpolar 1st and 4th residues on one side of the helix to form a non-polar strip. This nonpolar face of the helix associates with the nonpolar face of the other helix forming a coiled-coil structure (Fig. 2.25a).

The polar residues act to stabilise the multistrand coiled-coil structure of α-keratin which is generally thought to be composed of nine protofibrils forming a (9 + 2) microfibril of ~80 Å diameter. Each protofibril has two or three α-helical strands entwined to form coiled-coils (Fig. 2.25b).

β-keratin. The β-keratin silk fibroin in *Bombyx mori* has an antiparallel β-pleated sheet structure with a pseudo six-residue repeat unit- (**Gly-Ser-Gly-Ala-Gly-Ala**)$_n$. Individual sheets are packed together so that glycine faces glycine and alanine or serine. To a first approximation *B. mori* fibroin is a coplanar arrangement of **Gly-Ala**, alternating with an occasional substitution of serine for alanine (Fig. 2.26). The conformation of the polypeptide in Tussah silk is also β-sheet.

Fibrous Muscle Proteins

(i) F-actin. F-actin is a major component of thin filaments. Seven monomers of **G**-actin, and a globular α-helical polypeptide chain constitute a unit of **F**-actin. Two polymers of **G**-actin (seven each) wind around each other forming a coiled-coil structure (7 + 7 = 14 **G**-actin monomers) in **F**-actin (Fig. 2.27). Actin has a high affinity for myosin.

(ii) Myosin. Myosin is a major component of thick filaments in muscle fibres. It is composed of a double-headed globular region joined to a fibrous structure which is a helical double-stranded coiled-coil (Fig. 2.28). The fibrous part of myosin, which is rod-like is 1350 Å long and ~20 Å wide. Myosin fibres meet tail to tail at the M-line in the H-zone of the muscle fibre.

(iii) Tropomyosin. Tropomyosin is a component of a thin filament in muscle. It is also a rod-shaped double-stranded molecule consisting of two similar α-helical polypeptide chains coiled around each other and is attached to the chains of the F-actin filament. It has a seven-residue repeat sequence, similar to keratin, with the 1st and 4th residues nonpolar and the 5th and 7th residues charged. The molecule is ~ 400 Å long and is divided into ~ 14 regions of 28 Å each. Within each of these regions there is a zone with a high concentration of charged side chain groups and a zone with a high concentration of nonpolar side chains. These zones provide binding sites for the other protein components of the muscle: actin and troponin. A troponin complex (**TnT**, **TnC** and **TnI**) bound to a molecule of tropomyosin regulates the activity of the seven G-actin monomers (Fig. 2.27). (Refer to Chapter 16 for further details on muscle proteins).

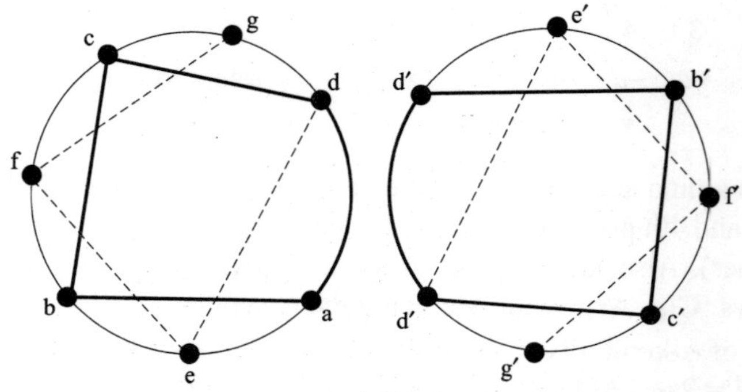

Fig. 2.25a Alignment of nonpolar edges (a-d′ and d-a′ residues) in an α-keratin coiled-coil.

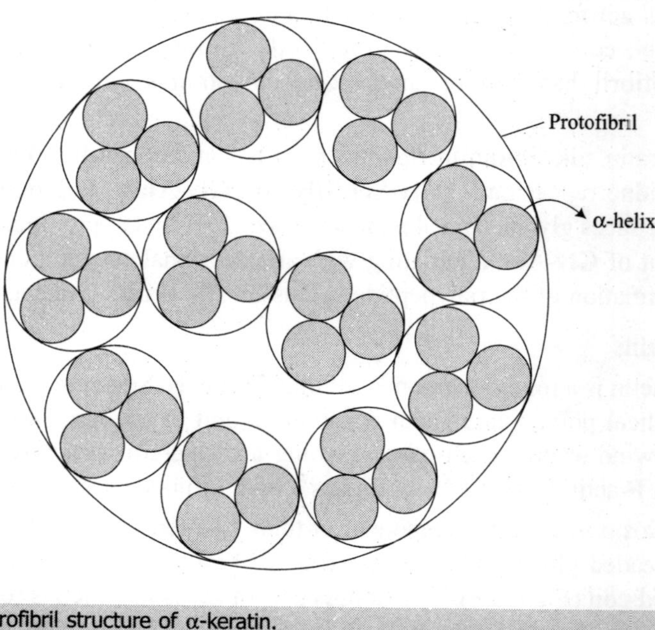

Fig. 2.25b (9 + 2) Microfibril structure of α-keratin.
Each protofibril has two or three α_R-helices wound around each other forming a coiled-coil

2.3.1.3 Other classes

Elastin. Elastin is a protein which resembles rubber in its elastic properties and occurs in intracellular connective tissues such as in lungs, walls of blood vessels, tendon and ligaments. It belongs neither to the collagen group nor to the **k-m-e-f** group of proteins. The amino acid composition reveals it to be an essentially nonpolar molecule consisting of ≈ 30% glycine and ≈ 30%

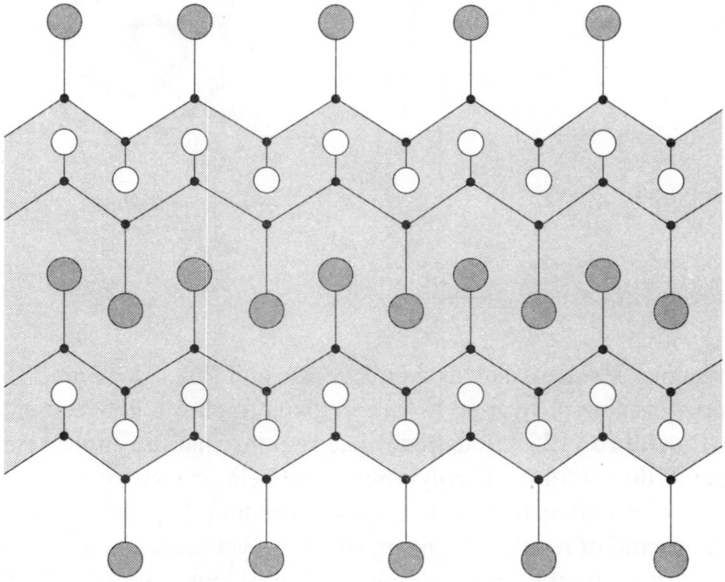

Fig. 2.26 Schematic diagram of β-pleated sheet in silk fibroin (shaded to highlight the pleated nature)
Side chains: ○—Hydrogen (of Gly)
●—Methyl group (of Ala)

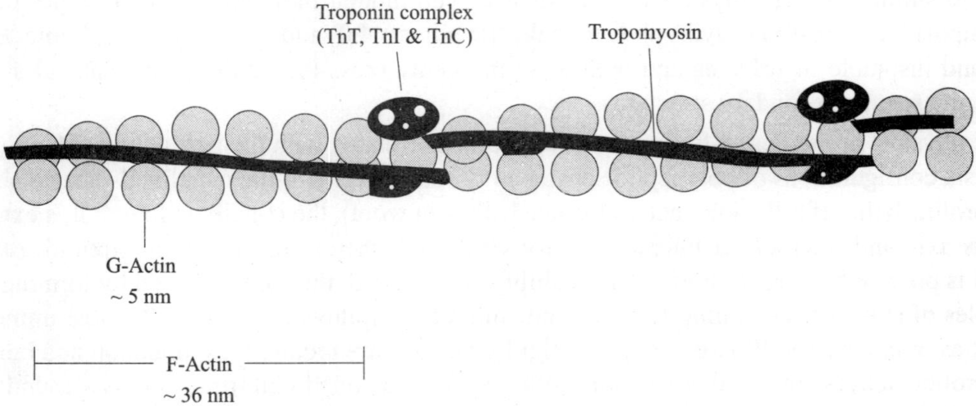

Fig. 2.27 Structural model of F-actin-tropomyosin-troponin complex

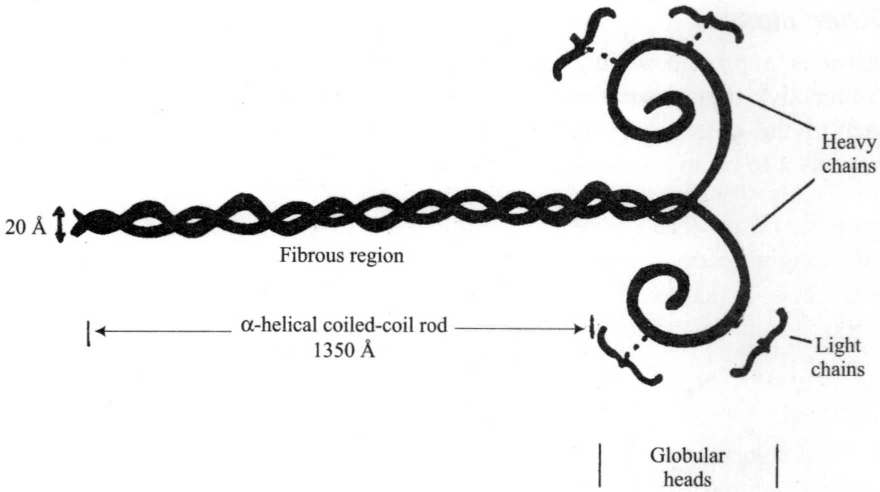

Fig. 2.28 Schematic structure of a myosin filament.

proline and other nonpolar residues such as **Ala, Leu, Ile** and **Phe**. There are few polar amino acids and there is no hydroxyproline or hydroxylysine. Regions of elastin between cross-links are rich in **Gly, Pro** and **Val**. Proelastin (125 kilodaltons) has regions that are highly rich in **Ala** and **Lys**. However, elastin strands do not form a L-polyproline type helix in spite of its amino acid composition which is similar to that of collagen. The actual conformation of elastin is still unknown. Elastin molecules seem to be devoid of regular secondary structure, yet possess a tertiary structure of a three-dimensional network formed by the covalent cross-linking of randomly linked polypeptide.

Resilin. Resilin is also an elastic protein found in the cuticle of insects. Its amino acid composition shows **Gly** ≈ 35%, **Pro** ≈ 8% and polar amino acids ≈ 34%. Hydroxyproline, cysteine, methionine and tyrosine are absent. It is readily digested by proteases, acids and bases. It swells in water, and nonpolar solvents do not penetrate. It forms a randomly coiled cross-linked structure. Resilin and elastin are similar in their physical characteristics, even though they have different types of amino acid compositions—resilin is hydrophilic (isoelectric point = 4.0) and elastin is hydrophobic. Both are elastic and insoluble in solvents and both form randomly cross-linked three-dimensional structures without any regular secondary structures.

The physical properties of fibrous proteins can be correlated with their structures. There are only three basic configurations of polypeptide chains in fibrous proteins—the α-helix, β-pleated sheet and L-polyproline helix. If a flexible material is needed (as in wool), the α-helix is found; it is extendable along its axis and interhelical interactions are weak and, therefore, the lateral strength of such a material is poor. Where extra strength and stability are required, they are achieved by forming coiled-coil cables of α-helices as in muscle fibres and coiled-coils followed by a (9 + 2) cable immersed in a matrix as in α-keratins. Where strong and rigid structures are required, as in tendon and bone, then L-polyproline helices entwined around each other forming a coiled-coil triple helix are found and the structure is further stabilised by inter-chain hydrogen bonds and covalent cross-links as in collagen. Where extra tensile strength is needed, β-pleated sheets are usually observed. Such structures provide lateral strength to the material due to extensive inter-chain hydrogen bonds, as in silk. Silk fibres, though strong compared to wool, cannot be stretched because the β-sheet is a fully-extended structure unlike the α-helix that is found in wool.

2.3.2 Globular Proteins

Globular proteins are compact in form compared to their polypeptide chain length due to folding of the polypeptide backbone. They exist in a wide variety of configurations/conformations and folds, with distinctive super secondary and tertiary, and multimeric protein complexes with quaternary structural hierarchies. Super secondary structural elements (motif, domains etc.), both structural and functional, forming unique tertiary structures (protein folds) are quite common in many classes of globular proteins.

2.3.2.1 Motifs

Aggregates of secondary structural elements are called motifs. The reason for the aggregation of secondary structural elements is that the formation of protein-protein interfaces removes contact area from the solvent ~15,000 $Å^2$) and 2/3 of this 'buried' area is hydrophobic. The packing interactions of secondary structural elements (ordered segments) are—helix-helix, helix-sheet and sheet-sheet interactions. Secondary structure elements are connected into simple motifs. α-helices are formed when the consecutive residues all have the φ, ψ-values ≈ –60 and –50°, respectively, α-helices can fold together forming coiled-coils (as in several fibrous proteins). Helix-loop-helix motifs, each with different structural topologies, are found in DNA-regulatory proteins (see Chapter 3) and in calcium-binding proteins (E-F hand), such as in parvalbumin, calmodulin, troponin-C etc.

(i) *Helix-helix packing.* The helix surface consists of parallel ridges separated by shallow grooves, Therefore, in helix-helix packing, the surface ridges of one helix pack into the grooves between the ridges of the other helix (ridges-into-grooves packing). α-helices with rod-like shapes pack with one face on the interior of the protein and the other face exposed to the solvent. Therefore, the packing geometries of α-helices around a hydrophobic core can be described by polyhedra—three-helix packing by octahedra, four-helix packing by dodecahedra, five-helix packing by hexadecahedra and six-nelix packing by icosahedra. The packing geometry of more then six helices can be described by 'layer' structure packing of α-helices (similar to β-sheets).

(ii) *Helix-sheet packing.* In helix-sheet packing, the helix packs on to a β-pleated sheet with the helix axis roughly parallel to the axis of the strands of the β-sheet, Therefore, the contact residues of the helix form a surface complementary to that of the sheet, with a slight twist due to the curl of the β-sheet structure, The β-α-β motif is generally found in protein structures with parallel β-strands (e.g., triosephosphate isomerase).

(iii) *Sheet-sheet packing.* As the β-sheet has a slight right-handed twist (τ), the packing of the β-sheets with twists, τ_1, τ_2, would be such that the orientation of the sheets, that is the angle between the axes of the strands of the sheets, Θ, is

$$\Theta = -2\,(|\tau_1 - \tau_2|) \qquad ...(2.16)$$

Greek-key motif is an example for adjacent antiparallel β-sheets of sheet-sheet packing that occurs in antiparallel β-sheet structures.

2.3.2.2 Domains

Domains are polypeptide chain regions which can be structurally sub-divided into two or more separate groups, and these are built from structural motifs. Most proteins with more than 150 amino acid residues generally form domain structures. Domains are usually joined together by a single strand of the chain and are, therefore, separated not only spatially but also along the chain. Since the neighbourhood correlation within the domain is much greater than those between the domains, the domains are expected to fold separately, independent of each other. Therefore, a domain, and not a polypeptide chain, should be regarded as the basic folding unit. Structural domains are in the range of ~ 20,000 daltons. Globular proteins consist of one or more structural domains. A protein can form a functional domain consisting of one or more structural domains. That is,

one polypeptide chain → one or more functional domains.

one functional domain → one or more structural domains.

For example, an 'active site' of a functional domain is generally located at the interface of two structural domains (e.g. immunoglobulin domain, NAD-binding domain etc.).

Protein structures can be grouped under several structural classes, based on their structural motifs (determined primarily by X-ray diffraction techniques). These are: a) α–; b) β–; c) α/β and d) other types of domain structures.

(a) α-Domain structures. In fibrous proteins, bundles of α-helices are wound around each other forming coiled-coil structures. As mentioned earlier, some of the muscle proteins and keratin type proteins are examples.

In globular proteins, short pieces of α-helices, interconnected by loops, are packed such that hydrophobic regions form the interior core and hydrophilic regions are exposed to the solvent. Some of the common structural motifs in α-domain proteins are — a simple '**up-down**' helix pair that is found in melittin, the 'up-down' 'four-helix' bundle that is found in hemerythrin, myohemerythrin, cytochrome $\mathbf{b}_{562}$, cytochrome *c′* and RNA-binding protein (ROP), and multiple α-helix bundles found in bacteriorhodopsin, and colicin etc. A more complex **helix bundle** motif is found in phospholipase, PL2 of human sinovial fluid.

Another type of domain, the '**globin-fold**', comprising α-helices wrapped around the active site, has been found in myoglobins, hemoglobins, and phytocyanins and similar proteins. In all these structures, the structural motifs and the domains are maintained by preserving the hydrophobic interior during evolution by allowing the interior side-chain mutations that preserve the geometry of the active site and the hydrophobic nature of the interior core by the overall movement of the helices (quaternary packing accommodations).

Trans-membrane proteins, such as bacteriorhodopsin, photoreaction centre proteins, Na^+/K^+.ATPases etc., have multiple α-helices spanning the membrane bilayer and packed and generally aligned at an angle to the membrane bilayer axis.

Some examples of structural motifs belonging to the α-domain structure are given in Fig. 2.29.

(b) β-Domain structures. Antiparallel β-structural domains have broadly three types of structural motifs: (i) Up-down β-sheets, (ii) Greek-key β-barrel and (iii) Jelly roll.

(i) *Up-down β-sheets.* Simple up-down β-sheet structures have one β-strand connected to the next by a short loop. Such structural motifs are found in snake venom neurotoxins, T4-

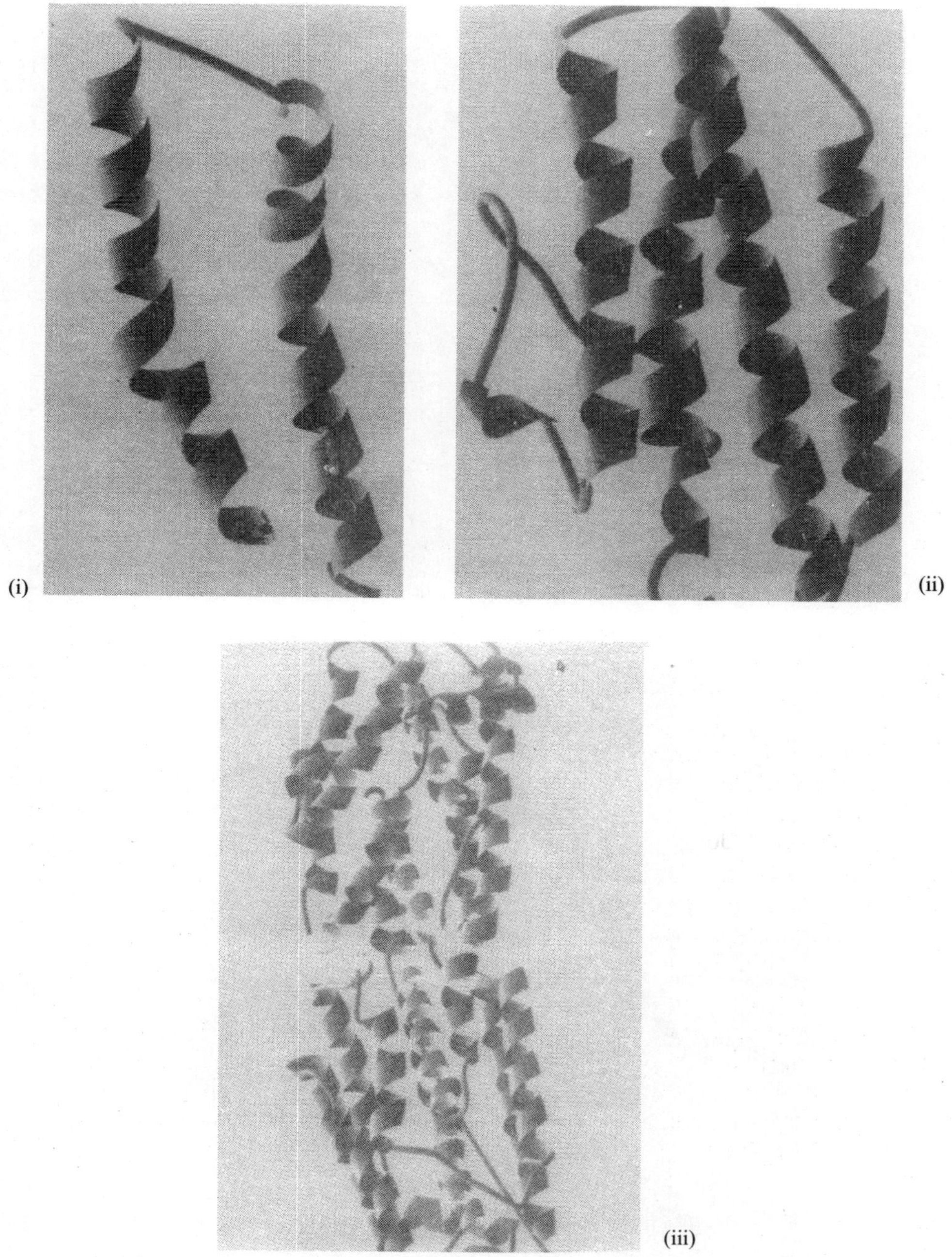

Fig. 2.29 (i) Simple up-down two-helix bundle motif in melittin (*).
(*Source:* Eisenberg D. et al. 2 MLT; 10/90).
(ii) Simple-up-down four-helix bundle motif in myohemerythrin (*).
(*Source:* Sheriff S. et al. 2 MHR; 4/87).
(iii) Multiple α-helix bundles in colicin (C-terminal domain) (*).
(*Source:* Parker, M.W. et al. 1 COL, 7/91

(Contd.)

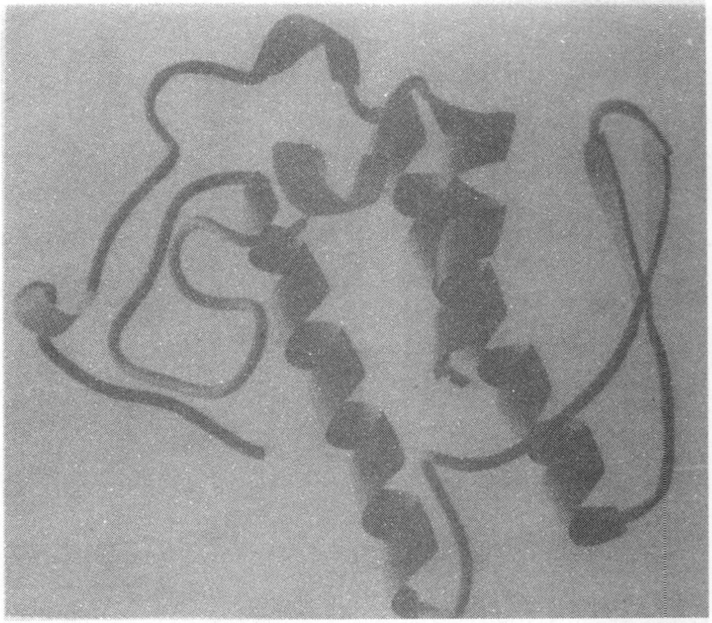

(iv)

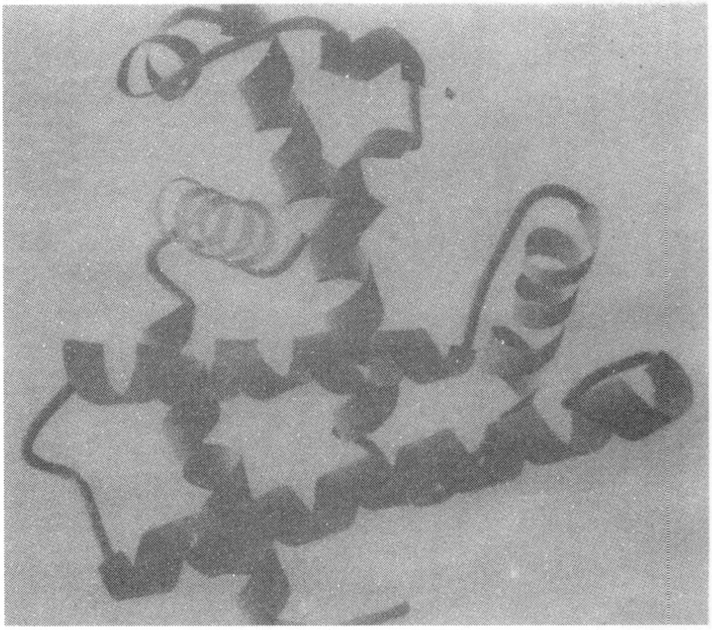

(v)

Fig. 2.29 (iv) Irregular α-helix bundle with helix, sheet and loop combination (*) in phospholipase A2, PLA2 (human sinovial fluid).
(*Source:* Scott, D.L. et al. 1 POD; 9/92).
(v) α-helix bundle forming "globin fold" with α-helices wraped around the core in different directions (*).
(*Source:* Watson, S.C. and Kendrew, J.C. 1 MBN; 4/73)

lysozyme, serine proteases, papain domain 1, plasma retinal-binding protein (RBP), biliverdin-binding protein, β-lactoglobulin and neuramidase from influenza virus etc. Some examples of structural motifs belonging to β-domain structures are presented in Fig. 2.30. Open-faced β-strand structure is found in bacteriochlorophyll A (Figs. 2.30 and 2.31).

(i)

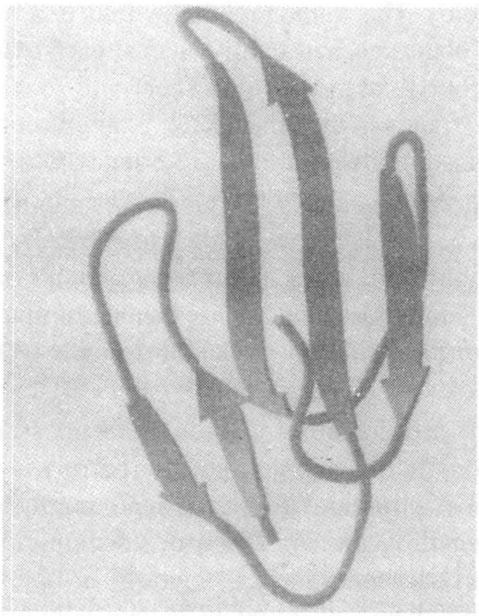

(ii)

Fig. 2.30 (i) Simple up-down β-sheet motif found in α-cobratoxin.
(*Source:* Saenger, W. and Walkinshaw, M.D. 1CTX; 4/82).
(ii) Up-down β-sheet motif found in erabutoxin A (Sea Snake).
(*Source:* Corfield, P.W.R. et al. 5 EBX; 12/89).

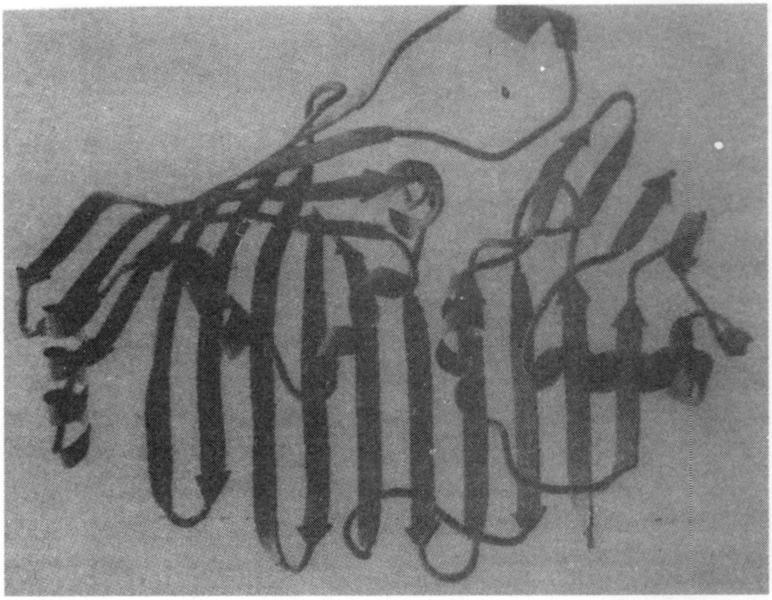

Fig. 2.31 Open-face sandwich β-sheet motif found in bacteriochlorophyll A Protein (*). (*Source:* Tronrud, D. et al. 3 BCL; 6/87).

(ii) The *Greek-key motif* has been found in diverse proteins such as plastocyanin, superoxide dismutase (SOD), trypsin domains 1 and 2, γ-crystallin and also in immunoglobulins (immunoglobulin fold). The immunoglobulin fold is a twisted β-barrel structure with two antiparallel β-sheet planes packed tightly against each other, joined together by a disulphide bridge. There are four β-strands in one sheet and three (in constant domain) or five (in variable domain) β-strands in the other sheet. Examples of Greek-key β-barrel structures and immunoglobulin folds are given in Figs. 2.32 and 2.33.

(iii) *Jelly-roll structures.* A variation of the Greek-key motif is called the jelly roll (*Swiss roll*) motif. The motif is barrel-shaped with the barrel usually divided into two sheets. The jelly roll structural motif has been found in viral coat proteins (southern bean mosaic virus, tomato bushy stunt virus—domains 2 and 3, sesbania mosaic virus etc.), conconvalin A, and hemeaglutinins from the influenza virus. An example of the jelly-roll structure is shown in Fig. 2.34.

(c) α/β-Domain structures α/β-domain structures are the most frequent and the most regular of the protein structures. They consist of central β-sheets (barrel) with α-helices wound around. The structure is built up by βαβ structural motifs (found in nucleotide-binding proteins) and the *Rossmann fold* (found in dehydrogenases). The structures are either parallel β-sheet barrel with α-helices wound around (as in triosephosphate isomerase) or open-twisted parallel β-sheets with α-helices on either side of the sheet (as found in lactate dehydrogenase domain 1). Some of the structures that belong to the α/β family are: aldolase dehydrogenases, kinases, taka amylase, aldolase, enolase, carboxypeptidase, xylose isomerase, triosephosphate isomerase, flavodoxin etc. Some examples of α/β domain structures are presented in Fig. 2.35.

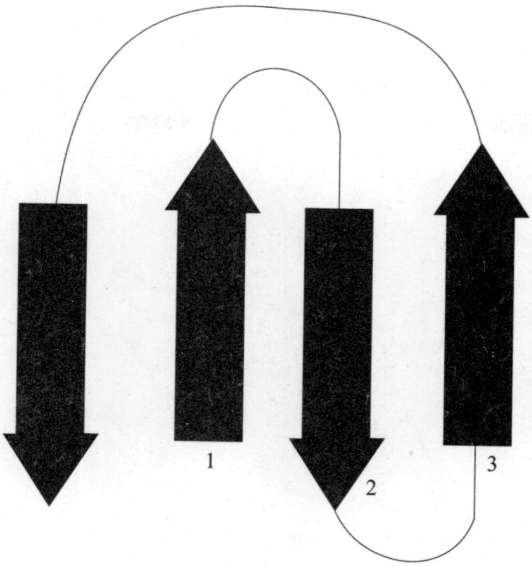

(i)

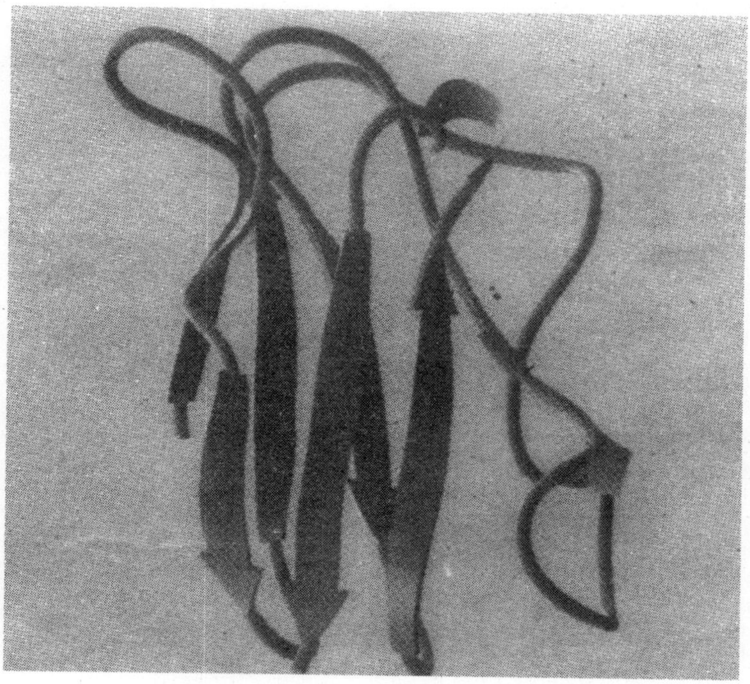

(ii)

Fig. 2.32 (i) Topology of Greek-key motif.
(ii) Greek-key β-barrel motif found in Plastocyanin (*).
(*Source:* Fields, B.A. et al. 1 PNC; 9/93).

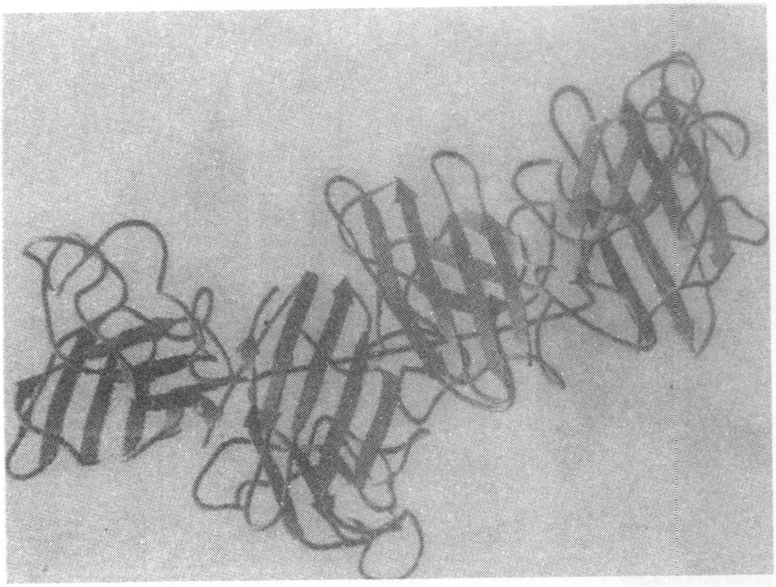

(iii)

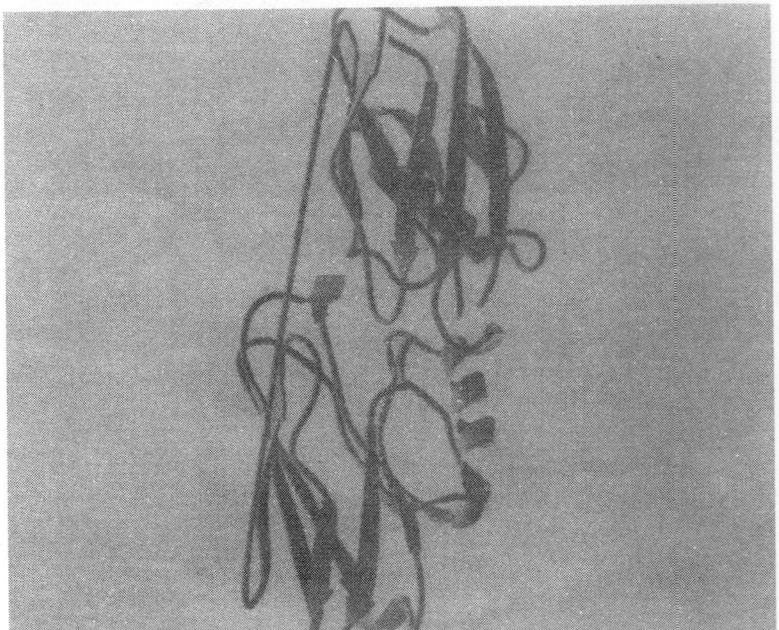

(iv)

Fig. 2.32 (iii) Greek-key β-barrel structural motifs found in superoxide dismutase (Cu.Zn) (*).
(*Source:* Kitagawa Y. and Katsube, Y. 1 SRD; 4/93).
(iv) Greek-key β-barrel structural motifs found in azurin (*).
(*Source:* Baker, E.N. et al. 1 AIZ; 11/93).

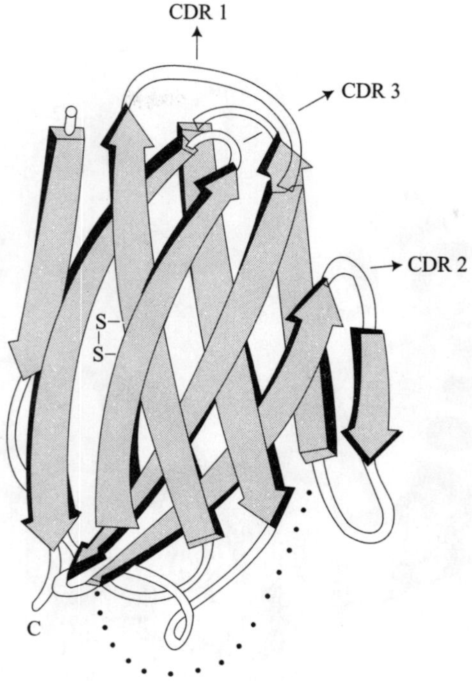

Fig. 2.33 Immunoglobulin fold (a version of Greek-key Motif).
Two layers of antiparallel β-sheet are folded and stabilised by an S–S bridge to form a sandwich structure. There are four β-strands in one sheet and five strands in other sheet in V_L domains and three strands in C domains (two strands at the right edge, CDR2 strands are missing. The connecting 100 p is indicated by dotted line). (After: Richardson. *J. Adv. Proto Chem.* 34; 167 (1981).

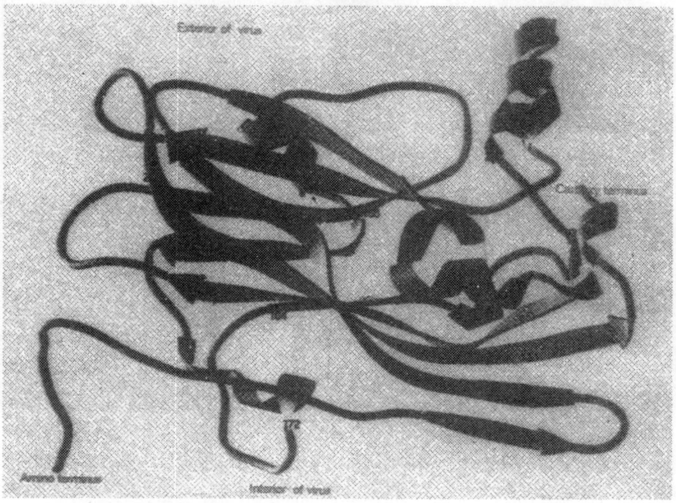

Fig. 2.34 "Jelly-roll" Greek-key β-barrel structural motif found in sesabania mosaic virus (Photograph: Courtesy of M.R.N. Murthy, Indian Institute of Science, Bangalore).

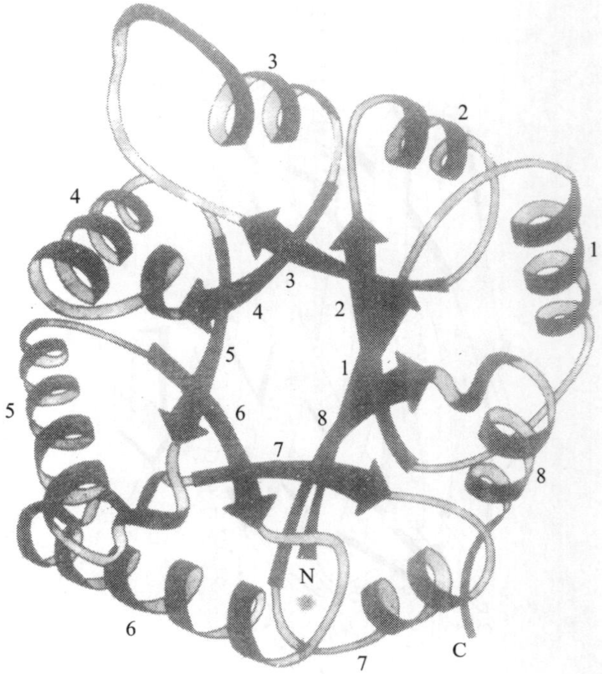

(i)

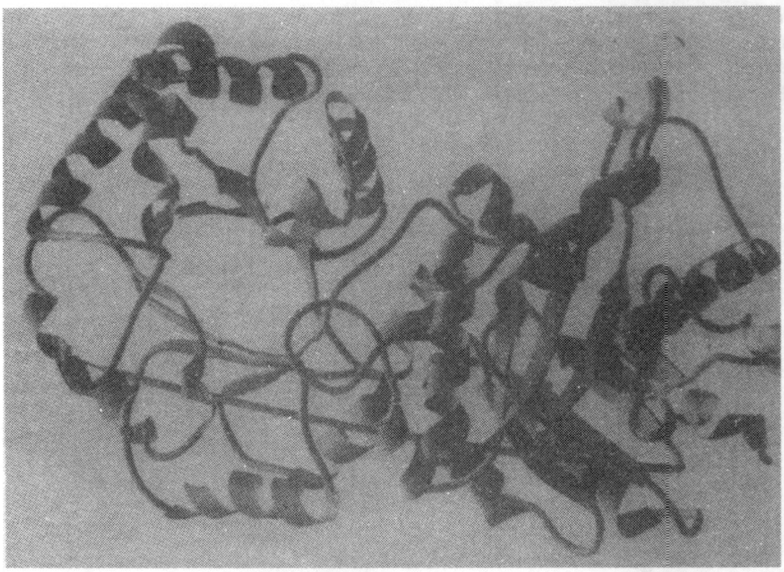

(ii)

Fig. 2.35 (i) Parallel α/β domains. 8-stranded β-barrel surrounded by α-helical cylinders with +1 X topology in Triosephosphate isomerase.
(After Richardson, *J.S. Adv. Prot. Chem.* 34; 167. (1981)).
(ii) Parallel α/β structural domains found in triosephosphate isomerase (yeast) (*).
(*Source:* Alber. T. et al. 1 YPI; 5/90).

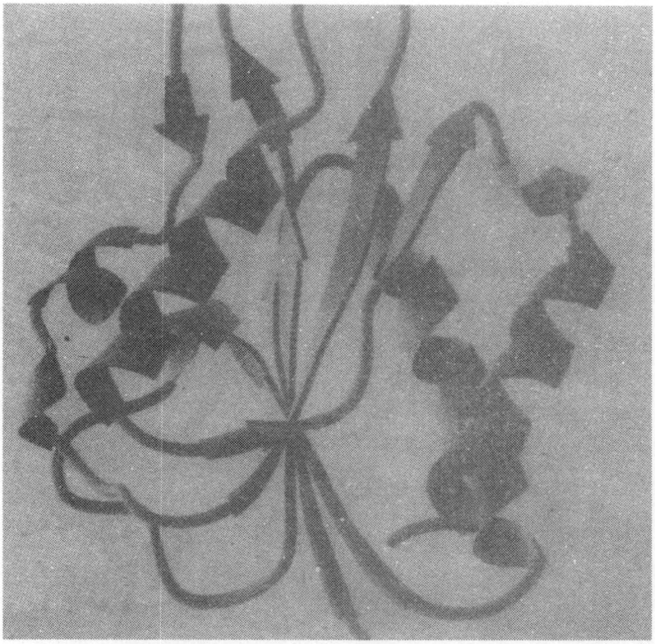

(iii)

(iv)

Fig. 2.35 (iii) Open-twisted α/β domain found in flavodoxin (Vulgaris) (*).
(*Source:* Watt, W. and Watenpaugh, K.D. 2 FX2; 10/91).
(iv) Open-twisted α/β structural domains found in D-galactose-binding protein (*).
(*Source:* Vyas, N.K. et al. 2 GBP; 2/89)

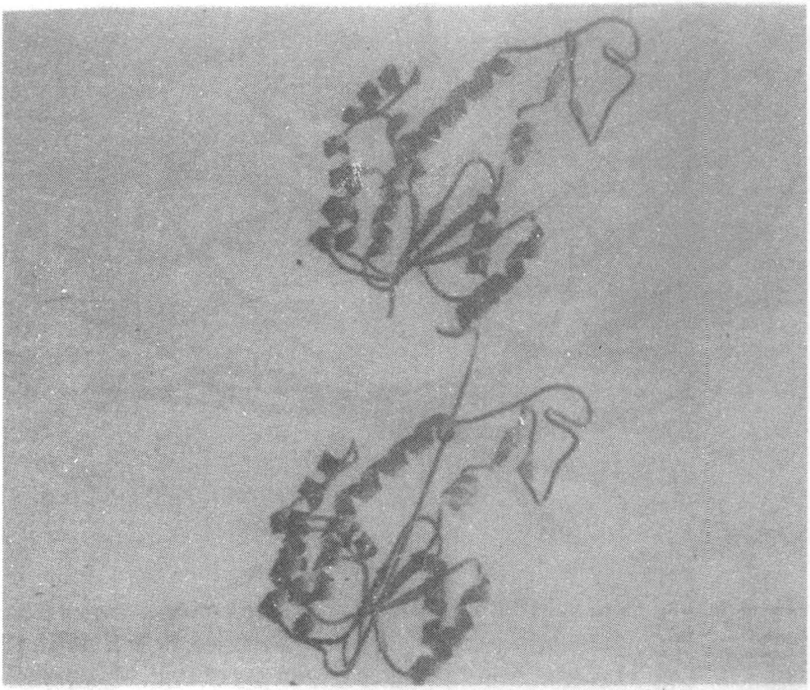

(v)

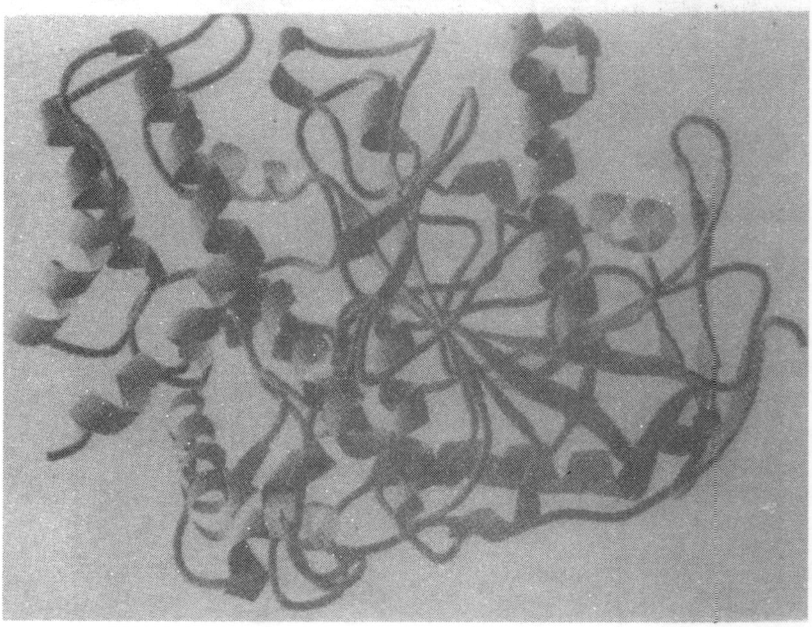

(vi)

Fig. 2.35 (v) Open-twisted α/β structural domains found in adenylate kinase (isozyme 3) (*).
(*Source:* Diderichs, K. and Schulz, G.E. 1 AK3; 1/90)
(vi) Open-twisted structural domains found in acetylcholine esterase (T. california) (*).
(*Source:* Sussman, J.L. et al. 1 ACE; 10/91).

2.3.2.3 *Other types of structural motifs*

There are structural motifs with a combination of secondary structural elements which are not easily classifiable under the more common structural categories. One example is the structural motif found in legume lectins, which is a combination of a flat six-stranded β-sheet and a curved seven-stranded β-sheet interconnected by loops of various lengths (Fig. 2.36). Another example is the motif that may be common to scorpion venom toxins, with helix and β-strand regions stabilised by two disulphide bonds (Fig. 2.37).

DNA-recognition proteins have **helix-turn-helix (αtα)**, **zinc-finger** and **leucine-zipper motifs** (refer to Chapter 3 for further details). In DNA-recognition proteins, the helix-turn-helix is a structural motif of approximately 20 amino acids which functions as the specific DNA-binding region. Stereochemical constraints dictate the correct topology in the DNA-recognition region: the 4th and 5th residues must be hydrophobic and buried, 8th and 10th residues must also be hydrophobic (but not necessarily buried), the 5th residue should not possess a branched-side chain and the 9th residue must be glycine. A helix-turn-helix motif (Fig. 2.38a) with different topology (**EF-hand**) has been found in calcium-binding proteins, such as parvalbumin, calmodulin, troponin **c** etc. Metals such as Ca^{2+} can also bind at the domain or subunit interfaces stabilising the motif domain/subunit interactions (Fig. 2.38b).

Fig. 2.36 Structural motif with flat six-stranded and curved seven-stranded β-sheets interconnected by loops found in peanut lectin. (Ref. Banerjee, R. et al. Proc. Natl. Acad. Sci. USA, 91: 227 (1994) (Photograph: Courtesy of Prof. M. Vijayan, Indian Institute of Science, Bangalore).

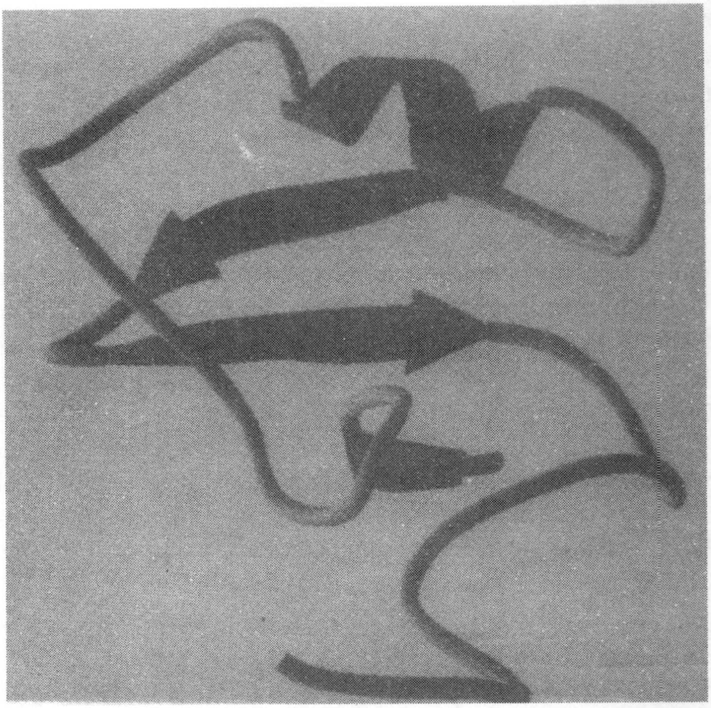

Fig. 2.37 Structural motif with helix-sheet stabilised by two disulphide-bridges (common to scorpion venom toxins) found in scorpion toxin V1 (*).
(*Source:* Lew, W. and Krishna, N.R. 1 VNB; 10/93).

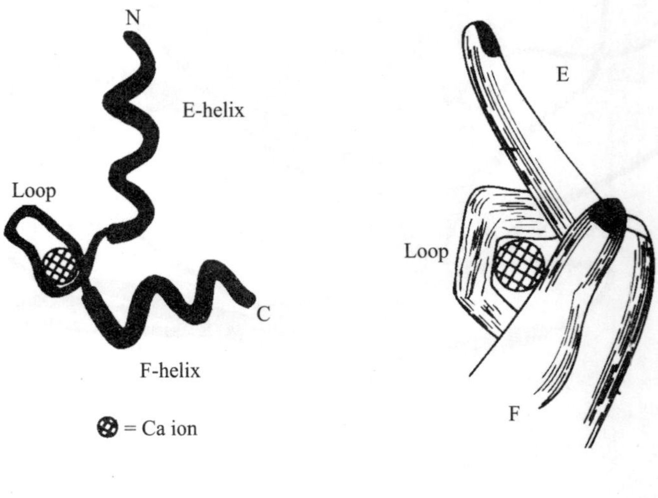

Fig. 2.38a Schematic representation of EF-hand motif in Ca^{2+}-binding protein. Helix E represents right hand forefinger, helix F the thumb and the folded middle finger and the loop holding the Ca^{2+}-ion. (After Krestsinger, R.H. CRC Crit. Rev. Biochem. 8; 119 (1980)).

Fig. 2.38b Metal binding at the interfaces of the subunits/domains, stabilising the subunit interactions in sesabania mosaic virus.
(Photograph: Courtesy of Prof. M.R.N. Murthy, Indian Institute of Science, Bangalore).

Suggested Reading

Books

Barrett, G.C. (Ed.). *"Chemistry and Biochemistry of Amino Acids"*. Chapman-Hall/London. (1985).

Baum, S.J. & Scaife, C.W.J. "*Chemistry— A Life Science Approach*". McMillan/New York. (1980).

Branden, C.J. & Tooze, I. *"Introduction to Protein Structure (2nd Edn)"*. Garland Press/New York. (1999).

Cantor, C.R. Schimmel, P.R. *"Biophysical Chemistry"*. Freeman/New York. (1980).

Chang, R. *"Physical Chemistry with Applications to Biological Systems"*. McMillan/New York. (1977).

Creighton, T.E. "*Proteins— Structures and Molecular Properties" (2nd Edn)*. Freeman/New York. (1982).

Haschmeyer, R.H. & Haschmeyer, A.E.V. "*Proteins*". Wiley/New York. (1973).

Lesk, A.M. *"Protein Architecture: A Practical Approach"*. IRL Press/London. (1991).

Lippard, S.J. & Berg, J.M. "*Principles of Bioorganic Chemistry*". University Science Books/Mill Valley, CA. (1994).

Mahler, H.R. & Cordes, E.H. "*Biological Chemistry*". Harper & Row/New York. (1971).

Neurath, H. (Ed). "*The Proteins" (Vols. II & IV)*. Academic Press/New York. (1964 & 1966).

Perutz, M.F. "*Protein Structure and Function*". Freeman Press/New York. (1991).

Ramachandran, G. (Ed). *"Aspects of Protein Structure"*. Academic Press/New York. (1963).

Schulz, G.E. & Schirmer, R.H. "*Principles of Protein Structure*". Springerverlag/Berlin. (1979).

Stryer, L. *"Biochemistry" (4th Edn)*. Freeman/New York. (1995).

Tanford, C. *"The Hydrophobic Effect"*. Wiley/New York. (1990).

Voet, D. & Voet, J.D. "*Biochemistry*". John Wiley/New York. (1990).

Articles

Bork, P. & Koonin, E.V. *"Protein Sequence Motifs"*. Curr. Opin. Struct. Biol., 6(3); 366. (1996).

Chothia, C. *"Principles that Determine the Structure of Proteins"*. Annu. Rev. Biochem. 53; 537. (1984).

Cohen, C. & Perry, D.A.D. *"A-Helical Coiled Coils– A Widespread Motif in Proteins"*. Trends. Biochem. Sci., 11; 245. (1993).

Doolittle, R.F. "*Proteins*". Sci. Amer., 253(4); 88. (1985).

Klotz, I.M. et al. *"Quaternary Structure of Proteins"*. Annu. Rev. Biochem., 39; 25. (1970).

Kyte, J. & Doolittle, R.F. "*A Simple method for Displaying the Hydrophobic Character of a Protein*". J. Mol. Biol. 157; 153. (1982).

Levitt, M. & Chothia, C. *"Structural Patterns in Globular Proteins"*. Nature. 261; 552. (1976).

Richardson, J.S. "*The Anatomy and Taxonomy of Proteins Structure*". Adv. Prot. Chem. 34; 167. (1981).

Richardson, J.S. "*Describing the Patterns of Protein Tertiary Structure*". Methods. Enzymol. 115; 341. (1985).

Rose, G.D. et al. *"Hydrophobicity of Amino Acid Residues in Globular Proteins"*. Science, 229; 834. (1985).

Struhl, K. *"Helix-turn-helix, Zinc-finger, and Leucine-zipper Motifs for Eucaryotic Transcriptional Regulatory Proteins"*. Trends. Biochem. Sci., 14; 137. (1989).

Weinberg, R.A. et al. *"The Molecules of Life"*. Sci. Amer., 253(4); 48. (1985).

3 Nucleic Acids

Synopsis: Nucleic acids—DNAs and RNAs (except tRNAs) are long, thread-like macromolecules that play central roles in all the hereditary processes—information storage, replication, transcription and translation.

The basic repeating unit of a nucleic acid, in general, is a nucleotide. The nucleic acid bases are purines (R)—adenine (A) and guanine (G); and pyrimidines (Y)—cytosine (C), thymine (T) and uracil (U). The bases are planar and occur only in keto/amino tautomeric forms. The sugars are furanosides—ribose in RNAs and deoxyribose in DNAs. The five-membered furanose ring is puckered and occurs predominantly in either C2′-*endo* or C3′-*endo* conformation. The nucleic acid base is covalently linked to the sugar moiety at the C1′-position by the glycosyl bond (C1′ – N) thus forming a nucleoside. The conformation of the base relative to the sugar moiety is represented by the torsion angle, χ, about the glycosyl bond. Sugar is phosphorylated at the C5′-position. The sugar and a nucleoside constitute a nucleotide. In polynucleotides, the nucleotide units are linked via 3′, 5′-phosphodiester bonds. The conformation of the ribose-phosphate backbone of a nucleotide unit is represented by six torsion angles. In nucleic acids, all torsion angles are correlated; that is, structural changes follow a concerted motion.

The **Watson-Crick** model for DNA structure—a right-handed double-helical structure, formed by two antiparallel DNA helical strands linked via sequence-specific base-pairing (Watson-Crick type) between the bases of one strand and the bases of the other strand and with the sugar-phosphate backbone at the periphery of the double helix—explains all the essential features of genetic processes at the molecular level, namely, information storage, replication, transcription and translation.

Double-helical nucleic acids are structurally classified under A, B and Z-families. The A-family has structural conservatism and all RNAs and RNA-DNA hybrids occur in the A-family. The right-handed DNA duplex has more structural flexibility. It can occur in the A- as well as in the B-family (A-DNA, B-DNA, C-DNA and D-DNA). Whereas, A- and B-families of nucleic acids are right-handed double helices, with the repeating unit being a nucleotide, DNA (poly (dG-dC)) can, under proper solvent and salt concentrations, occur in a left-handed zig-zag form, Z-DNA, where the repeating unit is a dinucleotide.

Nucleic acids do not exist in isolation in cells. Protein-nucleic acid interactions are an important part of molecular associations and are the basis for regulation and transformation of genetic information and cellular functions and biochemical reactions. Many of the transcriptional regulatory proteins have distinct structural motifs (helix-turn-helix, zinc-finger and lucine-zipper) that facilitate nonspecific and specific protein-nucleic interactions at molecular level.

3.1 NUCLEIC ACIDS

Generally, the basic unit of a nucleic acid is a nucleotide. In other words, nucleic acids are linear polymers of nucleotide units. A nucleotide comprises (i) a base (ii) a sugar and (iii) a phosphate group as shown in Fig. 3.1. A polynucleotide has a ribose-phosphate backbone, linked via 3′, 5′-phosphodiester linkages and the base sequences are random.

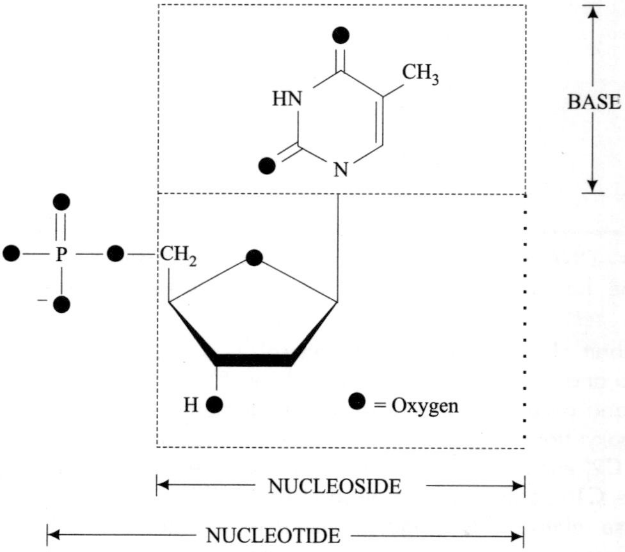

Fig. 3.1 Composition of a nucleotide.

3.1.1 Bases

There are two classes of nucleic acid bases. They are: (a) purines (R)—guanine (G) and adenine (A); and (b) pyrimidines—cytosine (C), thymine (T) in DNAs and uracil (U) in RNAs. The bases are planar and, in all nucleic acids that are genetically relevant, only the tautomeric forms as shown in Figs. 3.2 & 3.3 are observed. The exocyclic substituents—NH_2 and C=O groups are in amino and keto forms, never in imino and enol forms. This condition of proper tautomeric forms of the bases has fundamental genetic and functional implications. Only the bases with 'proper' tautomeric forms can form the 'correct' hydrogen bonding patterns (***Watson-Crick* base-pairing**) which are essential for the storage and transmission of genetic information.

A nucleic acid base is covalently attached to a sugar moiety by the β-glycosyl linkage (C1'–N bond) in all naturally occurring nucleic acids. A nucleic acid base with a sugar moiety is called a nucleoside (Fig. 3.4).

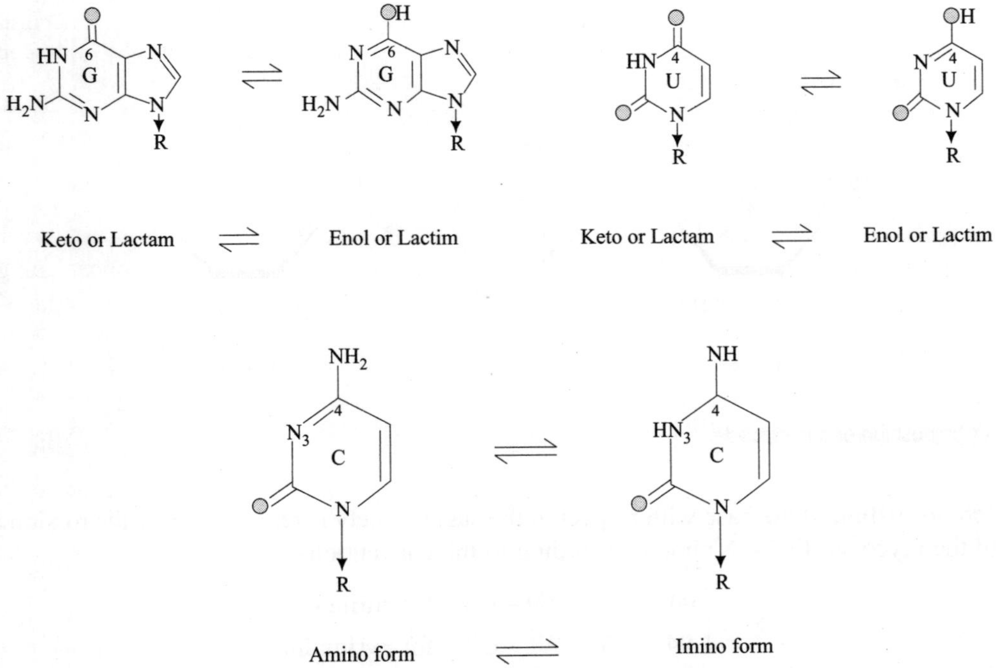

Fig. 3.2 Tautomeric forms in purine/pyrimidine bases. (Keto or lactam/enol or lactim; and amino/imino tautomers).

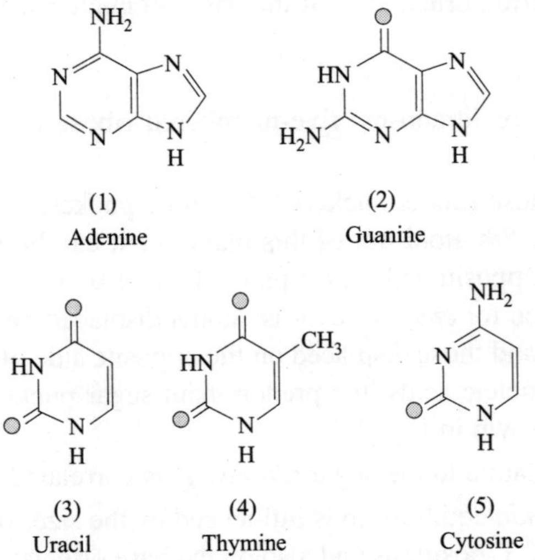

Fig. 3.3 Tautomeric forms (keto or lactam; and amino) of purine/pyrimidine bases that occur in nucleic acids. (1) A = 6-aminopurine; (2) G = 2-amino-6-oxypurine; (3) U = 2, 4-dioxypyrimidine; (4) T = 5-methyl uridine; (5) C = 4-amino-2-oxypyrimidine.

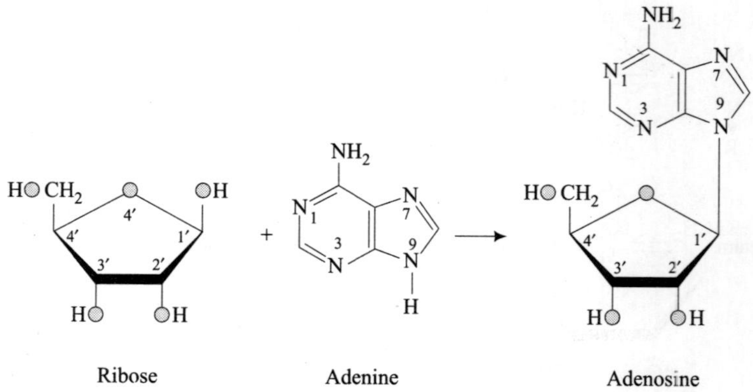

Fig. 3.4 Composition of a nucleoside.

The conformation of the base with respect to the sugar moiety is represented by the torsion angle, χ, around the glycosyl (C1' – N) bond, according to the convention-

$$\chi = \begin{cases} 04' - C1' - N9 - C4 & \text{for purines} \\ 04' - C1' - N1 - C2 & \text{for pyrimidines} \end{cases}$$

Two main conformations of the bases are **syn (*cis*)** ($\chi \approx 0$) and **anti (*trans*)** ($\chi \approx 180°$) as shown in Fig. 3.5.

There is relatively no steric hindrance between the base and the sugar ring in χ *anti* conformation and generally this is the preferred orientation of the bases in nucleic acids.

3.1.2 Sugars

The sugars in nucleic acids are furanoses: five-membered ribose in RNAs and 2-deoxyribose in DNAs (Fig. 3.6).

The five-membered furanose ring is puckered. It can be puckered in an envelope (E) form with four atoms in a plane and the 5th atom out of this plane; or it can be in a twist (T) form with two adjacent atoms displaced on opposite sides of a plane defined by the other three atoms of the five-membered ring. The convention for *endo-* or *exo-* is: atoms displaced on the same side of the plane as the C5′ atom are called *endo*, and those displaced on the opposite side of the plane with respect to the C5′ atom are called *exo.* In nucleic acids, the predominant sugar puckers are—C2′-*endo* (in DNAs) and C3′-*endo* (in RNAs) as shown in Fig. 3.7.

Orientation of the base relative to the sugar moiety, χ, is correlated with the sugar pucker.

The C2′-*endo* $\rightleftharpoons$ C3′-*endo* equilibrium is influenced by the size, shape and electronegativity of the substituent at the C2′-and C3′-positions and also on the base orientation (*syn* correlating with C2′-*endo*). The preference for C3′-*endo* increases linearly with the electronegativity of the subtituents at the C2′-position. Accordingly, deoxynucleotides adopt C2′-*endo* pucker because the substitution of H at the C2′-position is less electronegative than the O2′H substitution in RNAs. Sugar pucker is also influenced by base modification—bulky substituents at C8 (in purines) and C6 (in pyrimidines) shift

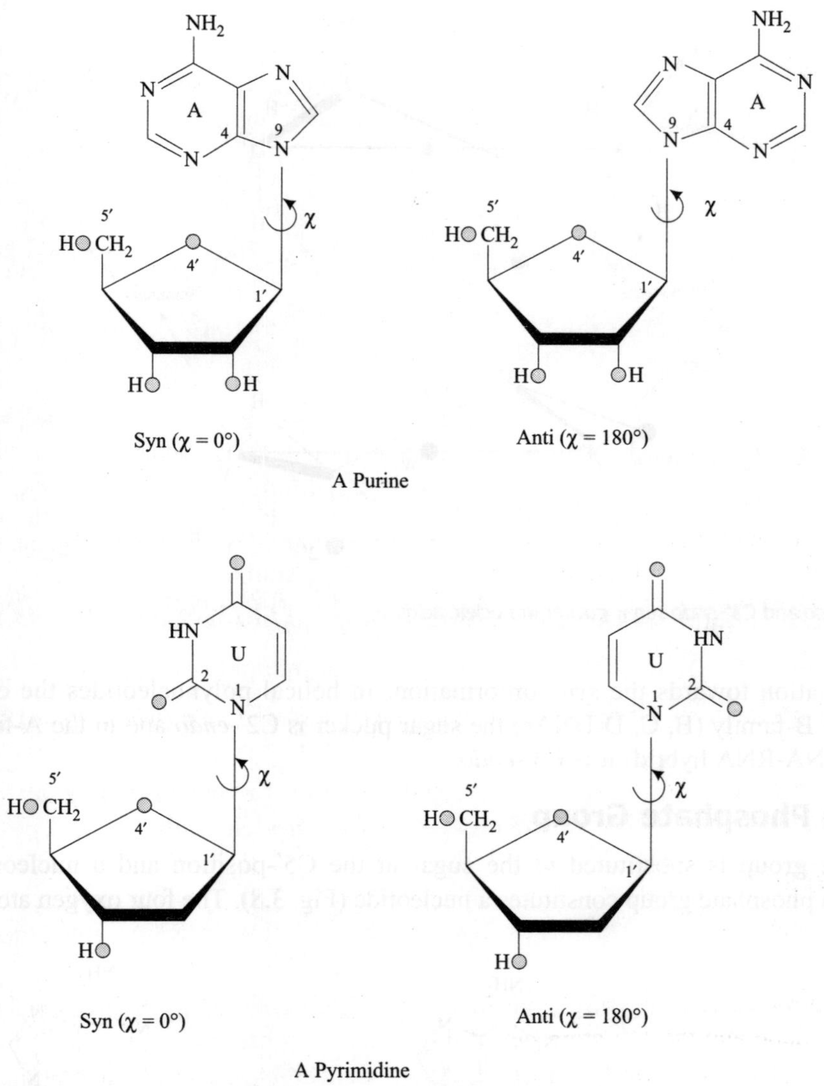

Fig. 3.5 *Syn* (χ = 0) and *anti* (χ = 180°) conformation of the base relative to the sugar moiety in nucleic acids. **Note:** In some of the diagrams H-atoms are not included for clarity and convenience.

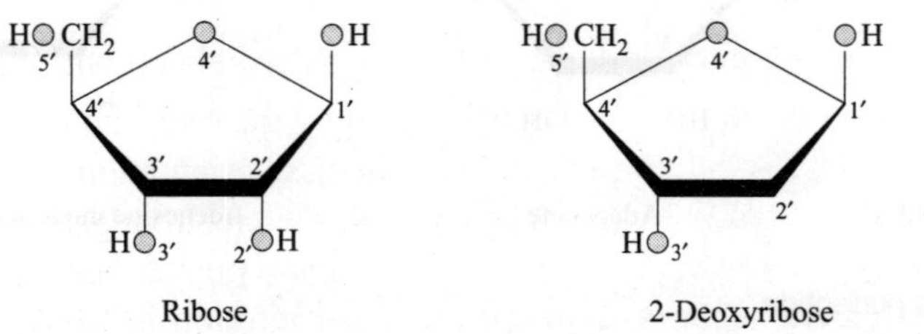

Fig. 3.6 Representation of ribose and 2-deoxyribose.

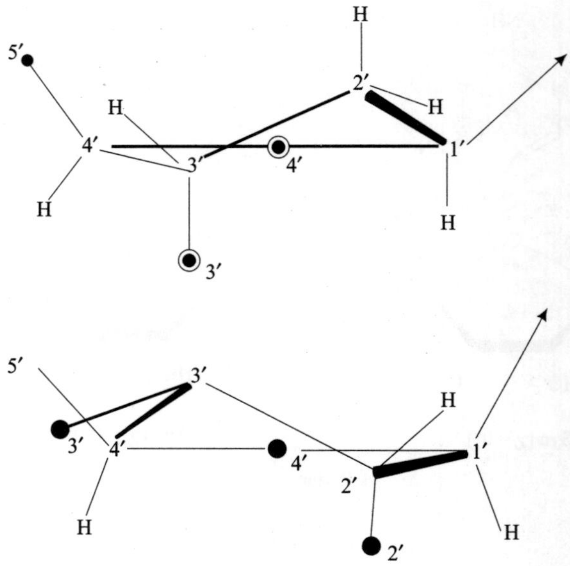

Fig. 3.7 C2′-*endo* and C3′-*endo* sugar pucker in nucleic acids.

the base orientation towards the *syn* conformation. In helical polynucleotides the equilibrium is 'frozen'. In the B-family (B, C, D-DNAs) the sugar pucker is C2′-*endo* and in the A-family (RNAs, A-DNA and DNA-RNA hybrid) it is C3′-*endo.*

3.1.3 The Phosphate Group

The phosphate group is substituted to the sugar at the C5′-position and a nucleoside with the attachment of a phosphate group constitutes a nucleotide (Fig. 3.8). The four oxygen atoms are linked

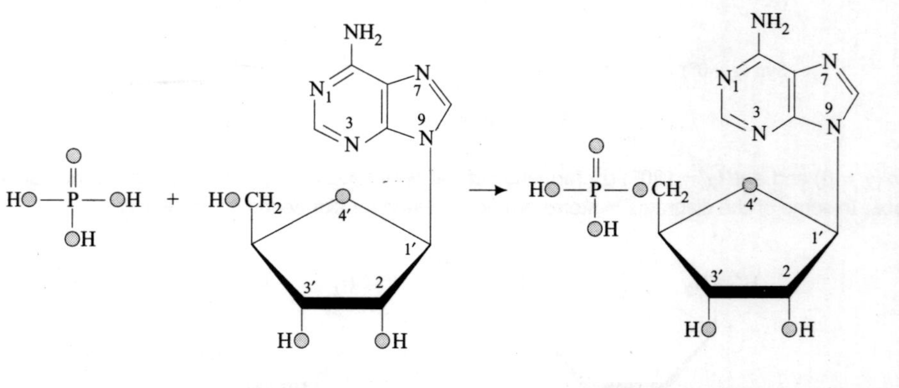

Fig. 3.8 Constituents of a nucleotide.

to the phosphorus atom by sp^3 bonds (σ-bonds) and are tetrahedral. The orientation of the phosphate group relative to the sugar moiety is given by the torsion angle about the C5′ – C4′ bond (γ or ψ). The exocyclic torsion angle can have three staggered conformations—g^+ (60°), t (180°) and g^- (–60° or 300°). In polynucleotides with χ *anti* orientation g^+ conformation is favoured for the phosphate group, partly due to the $C_{\binom{6}{8}}$ – H... O5′ hydrogen bond.

3.1.4 The Ribose-Phosphate Backbone

The conformation of the ribose-phosphate backbone of a nucleotide unit is represented by six torsion angles as shown in Fig. 3.9 and Table 3.1. The rotations about C-O bonds are restricted largely due to the steric hindrance between the phosphate group and the sugar moiety. In contrast, the rotations about the phosphoester bonds (P-O) are more flexible. The right-handed helical conformation of ribose-phosphate is feasible only if the conformations of the phosphoester bonds (torsion angles α, ξ/ω, ω′) are in the range g^-(270°). In summary, in the right-handed double-helical polynucleotides, all torsion angles are correlated so that structural changes follow a concerted motion and the rotations are restricted to the range—

α & ξ (ω & ω′) ($O5' \overset{\alpha}{—} P \overset{\zeta}{—} O3'$) 300° ($g^-$)
γ (about C5′—C4′) 60° (g^+)

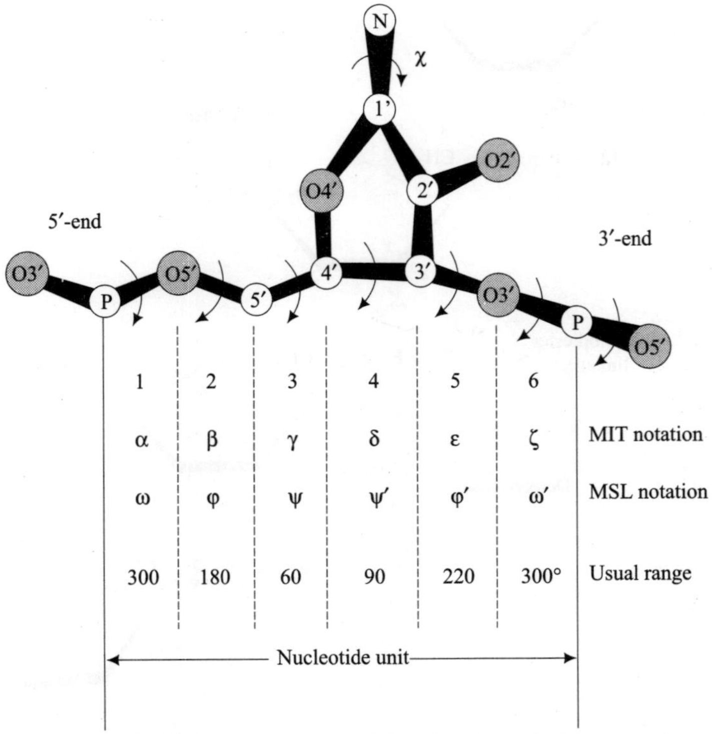

Fig. 3.9 Conformation of the ribose-phosphate backbone of a nucleotide.

Table 3.1 Torsion angles of the ribose-phosphate backbone of a nucleotide unit.

Torsion angle	*Notation*		*Usual value in°*
	MIT	*MSL*	
O′ – P – O5′ – C5′	α	ω	300 (g^-)
P – O5′ – C5′ – O4′	β	φ	180 (t)
O5′ – C5′ – C4′ – C3′	γ	ψ	60 (g^+)
C5′ – C4′ – C3′ – O3′	δ	ψ'	90
C4′ – C3′ – O3′ – P	ε	ϕ'	220
C3′ – O3′ – P – O5′	ξ	ω'	300 (g^-)

MIT: MIT notation; MSL: Sundaralingam notation; g = *gauche* and t = *trans*.

β & ε (ϕ′ & φ′) (about C5′ – O5′ & C3′ – O3′) 180° (t)
Glycosyl torsion angle, χ 180° (*anti*)
Sugar pucker: either C2′-*endo* or C3′-*endo*.

A section of a polypeptide backbone with phosphodiester linkages is given in Fig. 3.10.

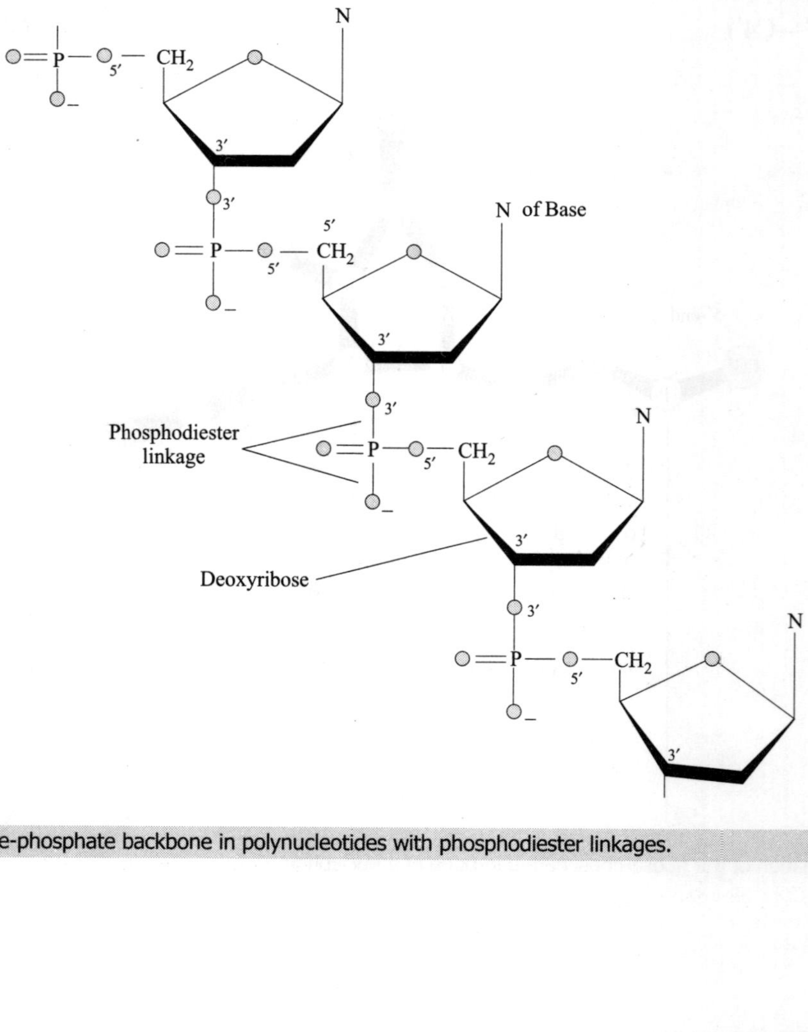

Fig. 3.10 Ribose-phosphate backbone in polynucleotides with phosphodiester linkages.

3.2 PRINCIPLES OF BASE-PAIRING/BASE-STACKING

Watson-Crick type base pairing is essential for the storage and transmission of genetic information from the right-handed DNA double helix. In addition, non-Watson-Crick type base-pairing (hydrogen bonding) schemes— reverse Watson-Crick, Hoogsteen, and reverse Hoogsteen exist in nucleic acids.

3.2.1 The Watson-Crick Hypothesis of DNA Structure

(i) DNA (B-form) is a right-handed double helix, of 20 Å diameter, formed by two antiparallel right-handed helical strands wound around each other.

(ii) The planar purine and pyrimidine bases are stacked inside the double helix, 3.4 Å apart, perpendicular to the central helix axis. The sugar phosphate backbone is at the periphery of the double helix, the structure analogous to a staircase, with steps pointing inwards and the winding frame at the periphery.

(iii) Each base of a particular strand is paired to a base of the second strand by sequence-specific hydrogen bonding (Watson-Crick type), forming a planar base-pair normal to the helix axis. Adenine (A) always pairs with thymine (T) in DNAs and with uracil (U) in RNAs, and

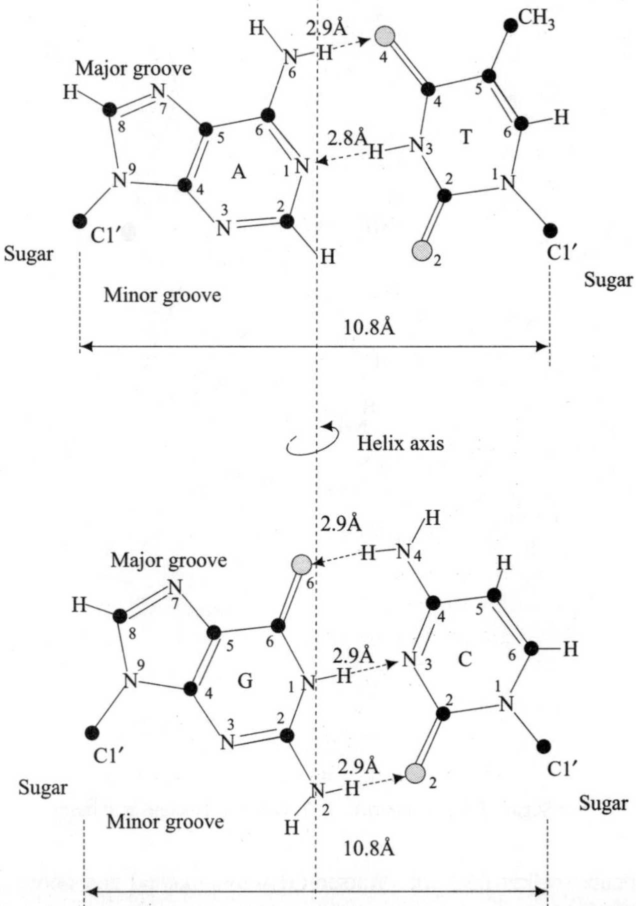

Fig. 3.11 Watson-Crick base-pairing between A = T (U) and G ≡ C bases in double-helical nucleic acids.

guanine (G) always pairs with cytosine (C) (Fig. 3.11).

Watson-Crick base-pairing and base-stacking are possible only if the tautomeric forms of bases are in amino/keto form and the bases are planar. The base-pairing specificity is a consequence of hydrogen bonding between bases and structural restrictions imposed on the double helix within which base-pairing occurs. The $C'_1...C'_1$ distance of 10.8 Å (Fig. 3.11) can accommodate only the **purine...pyrimidine** type base-pairing.

The specificity of base-pairing—A ::: T (or U) and G ⁝⁝⁝ C explains **Chargaff's rules:**

(a) A + C = G + T — 6-amino residues = 6-keto residues

(b) A + G = C + T — Sum of purines = Sum of pyrimidines

(c) A/T = G/C = 1.0

The specific base sequence in one strand uniquely determines the base sequence in the second strand. This complementarity of the two strands of the DNA-double helix lends a logical explanation to the mechanism of DNA-replication (precise reproduction of stored information). (Figs. 3.12 & 3.13)

Further, base-stacking interactions, which are hydrophobic and enthalpically favoured, are also important in stabilising nucleic acid structures.

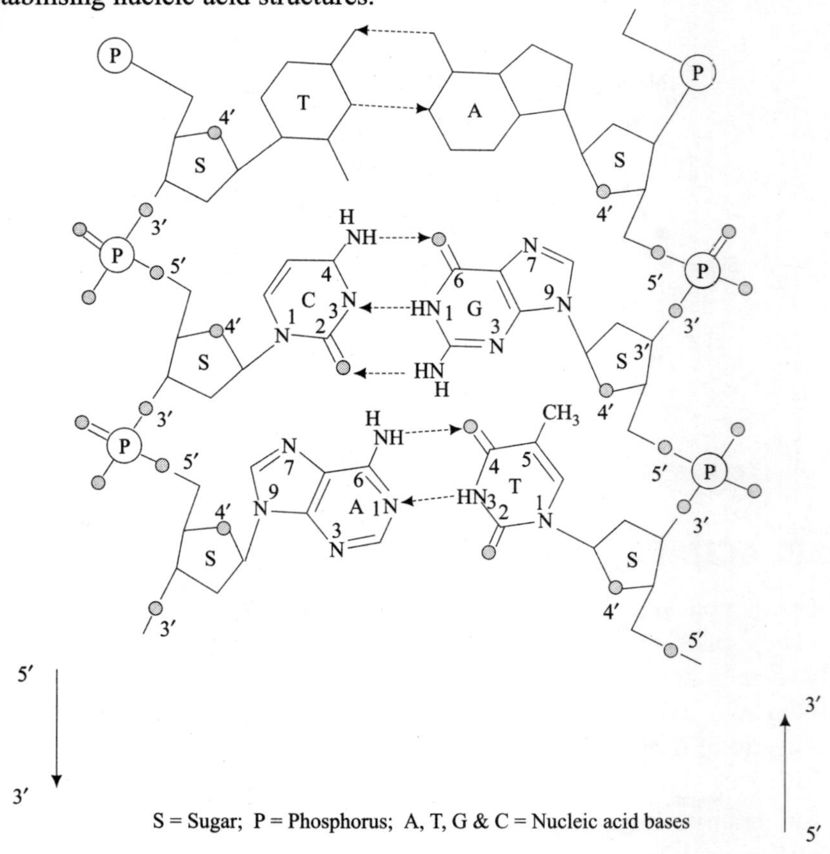

Fig. 3.12 Right-handed double-helical DNA with Watson-Crick base-pairing and ribose-phosphate backbones (in antiparallel directions).

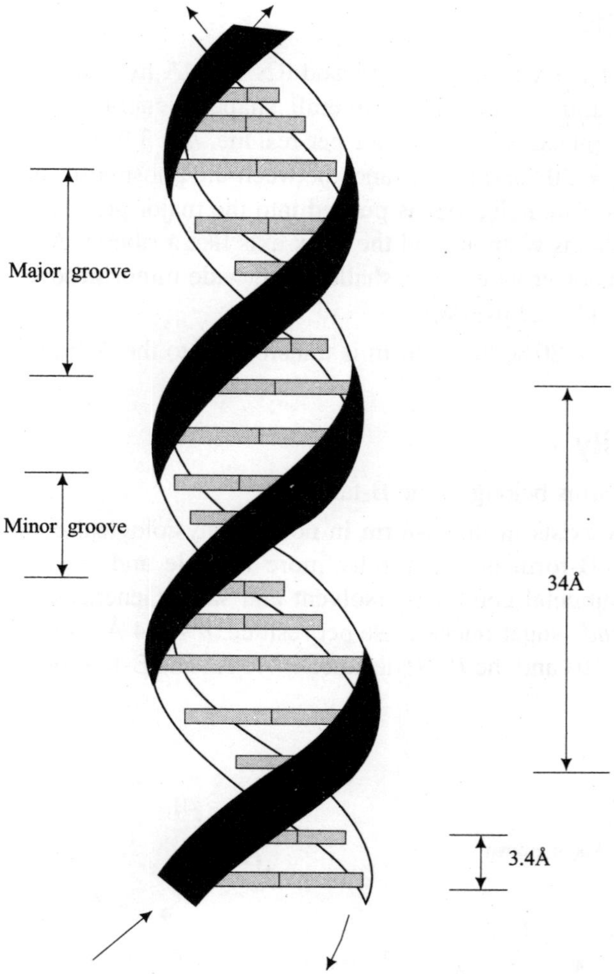

Fig. 3.13 Right-handed DNA double helix (B-form) indicating major (wide) and minor (narrow) grooves.

3.3 NUCLEIC ACID FAMILIES (A, B AND Z)

In the Watson-Crick model of the double-helical nucleic acids, some of the assumptions—the coplanarity of the base-pairs, and the base-stacking perpendicular to the helix axis are not quite true. In reality, the base-pairs are not coplanar, but have a propeller twist, Θ_p; and the bases are not perpendicular to the helix axis, but are tilted, Θ_T, and rolled Θ_R, In addition, the helix axis does not pass through the centre of the base-pairs in all the forms and the base-pairs are dislocated from the helix axis ($\Delta \neq 0$).

Depending on relative humidity, uniqueness of the base sequences and solvent and salt concentrations, nucleic acids exist in several structural forms and they can be classified under A-, B- and Z-families, based on structural features, like—base orientation (χ), sugar pucker, ν, base-pair tilt, Θ_T, and base-pair dislocation, Δ, etc. Due to subtle variations in their structural parameters helical arrangements in different families are macroscopically different and distinct.

3.3.1 The A-family

To this family belong RNAs (A-form), A-DNA and RNA-DNA hybrid (A′-form). The A-family has structural conservatism and is uniform in overall shape. General structural features are: base orientation, χ *anti,* sugar pucker C3′-*endo,* rise per residue, $h \approx 3.0$ Å, number of residues per turn, $N \approx 11$, base-pair tilt, $\Theta_T \approx 20°$ and the distance between the phosphorus atoms of the neighbouring residues, $P......P \approx 6.0$ Å. The helix axis is pushed into the major groove, $\Delta = 4.5$ Å (see Fig. 3.14) and the polynucleotide chains wrap around the helix axis like a ribbon. As a consequence, there is a very deep and narrow major groove, but a shallow and wide minor groove. The A-family has both inter-strand and intra-strand base overlap.

For salt concentration > 20%, the A-form is transformed to the A′-form. The RNA-DNA hybrid exists in A′-form.

3.3.2 The B-family

B-, C-, D- and T-DNA forms belong to the B-family.

B-DNA: Normal DNA exists in the B-form in normal physiological conditions and high relative humidity (> 90%). The B-form is structurally more flexible and is sensitive to sequence, base composition and environmental conditions (solvent and salt). General structural features are: base orientation, χ *anti, C2′-endo* sugar pucker, rise per residue, $h = 3.4$ Å, number of residues per turn, $N \approx 10$, base pair tilt $\Theta_T \approx -10°$ and the $P...P$ distance = 7.0 Å. The B-family has only intra-strand base-stacking.

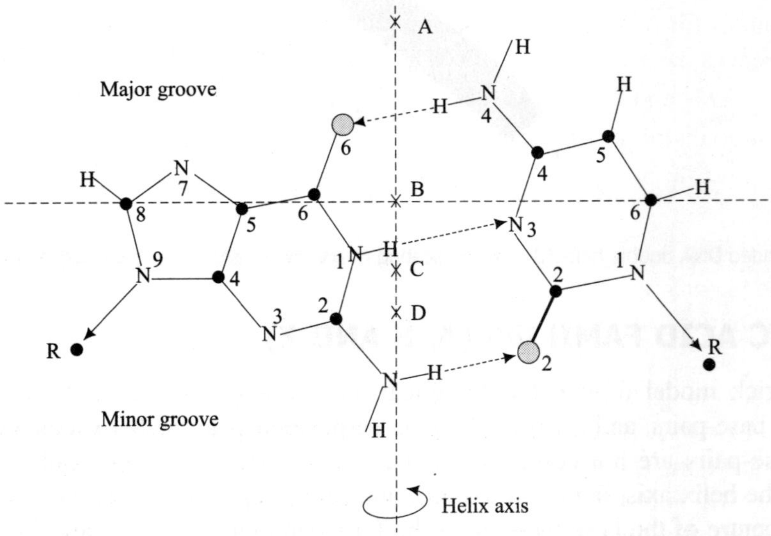

Fig. 3.14 Helix dislocation, Δ, in nucleic acids, A, B, C & D.
Δ = +4.5 Å for A-form;
= 0.0 Å for B-form;
= −1.0 Å for C-form;
= − 1.8 Å for D-form.

In B-DNA the helix axis passes through the centre of base-pairs (see Fig. 3.14), $\Delta = 0$. Therefore, the major and minor grooves are of equal depth, but of unequal widths.

When the relative humidity is reduced to < 75%, B-DNA undergoes a reversible transformation to the A-form. The transformation is dependent on base composition.

(dG + dC) < 30% *B* to A-form transition is greatly inhibited;

(dG + dC) > 30% *B* to A-form transition is favoured.

Similarly, (A – T) rich regions of DNA (> 60%) do not favour B- to A-transition.

In actuality, the DNA double helix is not a uniform, rod-like molecule, but has systematic sequence-dependent structural modulation, which may be needed for nucleic acid-ligand recognition and interaction processes. For example, in the A-family the rise per residue, **h**, varies from 2.6 to 3.3 Å. If **h** is small (2.6 Å), the major groove is narrow and accessible only to metal ions and water. If **h** is maximum (3.3 Å), then the major groove, though still deep, is accessible to accommodate larger ligands. In RNA, an antiparallel β-strand peptide can be fitted in the shallow minor groove with the RNA-ligand complex stabilised by the O2′H (RNA)...O = C (peptide) hydrogen bonds. Similarly, histone, H2A can be fitted into the narrow minor groove of B-DNA and stabilised by the peptide *NH* and the polynucleotide O3′ hydrogen bonds (NH...O3′).

Not only steric hindrance but also electronic complementarity play a major pole in nucleic acid-ligand interactions. In the minor groove negative electrostatic potential prevails in the case of an *A-T* sequence (both N3 of *A* and O2 of *T* contribute to the negative potential), but in the case of a G-C sequence an electrostatically neutral environment prevails (negative charges of N3 of G and O2 of C are counter-balanced by the positive charge of the N2 group of *G*). In the major groove also the electrostatic potentials for A-T and G-C sequences are different. Looking into the major groove, both *A-T* and *T-A* sequences have symmetrical electrostatic potential (– + –). That is, electrostatically, both A-T ($\overset{-}{N}7.\overset{+}{N}6$–H......$\overset{-}{O}4$) and T-A ($\overset{-}{O}4$......H–$\overset{+}{N}6.\overset{-}{N}7$) are similar. In the case of G – C and C – G sequences, the charge distributions are different.

For G-C it is (– – +) ($\overset{-}{N}7$.O6......H–$\overset{+}{N}4$) and

for C-G it is (+ – –) ($\overset{+}{N}4$–H......$\overset{-}{O}6.\overset{-}{N}7$)

That is, in the major groove the *G – C* and *C – G* sequences can be clearly distinguished by their asymmetric distribution of electrostatic potentials, while *A-T* and *T-A* are electrostatically indistinguishable.

C-DNA: C-DNA belongs to the B-family and is an intermediate between A- and B-forms. C-DNA is produced when Li^+-DNA (B-form) salt is at low humidity (45 to 65%). Low *d* (*G* + *C*) content also favours $B \rightarrow C$ transition. The structural features are: base χ *anti*, C2′-*endo* sugar pucker, $h = 3.3$ Å, $N = 9.33$ and $\Delta = -1.0$ Å (helix axis passes through the minor groove) (see Fig. 3.14). The grooves are similar to B-DNA, but the major groove is wider in C-DNA.

D-DNA: Natural DNAs do not adopt the D-form, phage T2-DNA being the exception (because of 5-hydroxymethyl cytosine base). Features are: base χ *anti*, sugar is C2′-*endo*, $h = 3.0$ Å, ~N = 8 and $\Delta = -1.8$ Å (Fig. 3.14). As the helix axis is in the minor groove, the major groove is shallow, wide and open and the minor groove is very deep and narrow.

In all the right-handed, double-helical nucleic acids, both in A- and B-families, the repeating unit is a mononucleotide.

3.3.3 The Z-family

At low salt concentrations, poly ($dG - dC$) exists in the right-handed B-DNA form. On increase in salt concentration (0.7 M $MgCl_2$ or 2.5 M NaCI) and/or addition of alcohol, the B-form structure transforms to the left-handed **zig-zag** form, Z-DNA. The geometries of the individual nucleotides, namely, *dG* and *dC* in the left-handed Z-DNAs differ greatly from each other and, also, the phosphate groups are in chemically nonequivalent environments (Figs. 3.15 & 3.16). Features of the Z-family are:

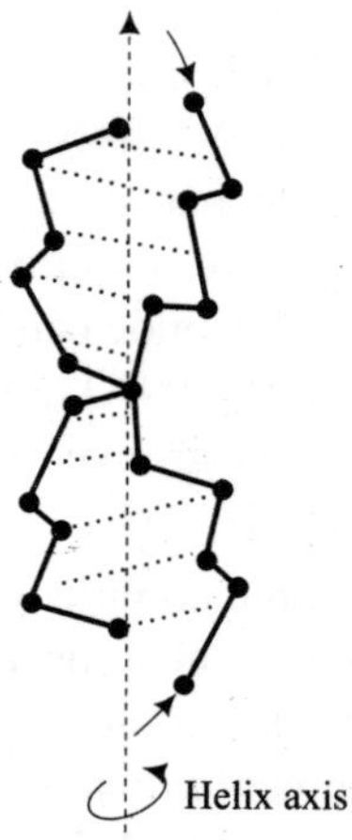

Fig. 3.15 Zig-zag course of left-handed Z-DNA. The line connecting phosphate groups is vertical when passing guanines and horizontal when passing cytosines.

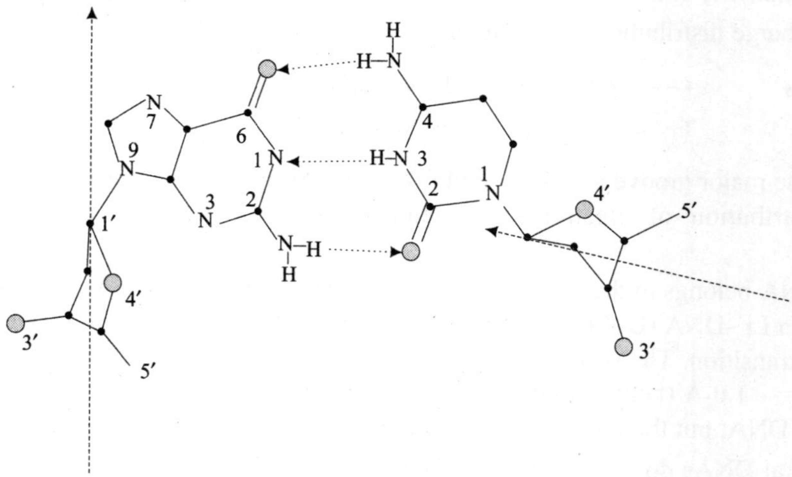

Fig. 3.16 Environs of *dG-dC* base-pairing in Z-DNA (**Watson-Crick** type indicated by dots). Arrows indicate different orientations of sugar moieties of *dG* and *dC*.
— — → directions of sugar moiety alternating in zig-zag manner.
(*Source:* Zimmermann, S.B. & Pheiffer, B.H. *J. Mol. Biol.* 135; 1023 (1979))

(i) The repeat unit is a dinucleotide (*dG* + *dC*), and not a mononucleotide.

(ii) The two strands of the double helix are antiparallel with **Watson-Crick** type base-pairing.

(iii) The double helix is left-handed and has **zig-zag** form.

(iv) The conformations of the *dC, dG* and phosphate moieties are different. While *dC* adopts a standard conformation—base orientation χ *anti,* C2′ *endo* sugar pucker and the phosphate group orientation $\gamma(\psi)$ (about C4′ – C5′ bond) is g^+; the conformation is different in the case of *dG*—base orientation is *syn,* C3′-sugar pucker and the phosphate group conformation is **t** (*trans*).

(v) Plot of $\chi - \delta$ ($\chi - \psi'$) conformations does not show a linear dependence as in the case of the B-family.

(vi) The helix axis is outside the basepairs and is in the minor groove. There is no major groove; it is a convex surface with *N*7 and *C*8 atoms of *dG* exposed to the solvent (as a consequence of χ *syn*).

(vii) No proper base stacking—cytosine bases stack on each other and guanines stack on the endo-cyclic sugar oxygen atom, O4′ of the preceding *dC*s.

Transformation of the B-form to Z-form of DNA does not require strand separation. $B \rightarrow Z$ transformation can travel along the helix as a bubble. Z-DNA can be stabilised by chemical modification. Methylation at *N*7 (in *G*) and at *C*5 (in *C*) and bromination at *C*8 (in *G*) convert the B-form to the Z-form near physiological conditions. In general, the *N*2-amino group of *G* and alternating *dG dC* sequence appear to be prerequisites for Z-family.

Z-DNA may be biochemically important—Z-DNA reduces super-coiling in plasmids, and thereby influences overall topology. $B \rightarrow Z$ transition modulates overall DNA structure and could influence regulation of DNA expression. Z-DNA occurs in chromosomes, but is not able to form a chromatin assembly.

The general structural features of double-helical nucleic acid families (A, B & Z) are tabulated in Table 3.2, and conditions for transitions of these families to one another are given in Fig. 3.17.

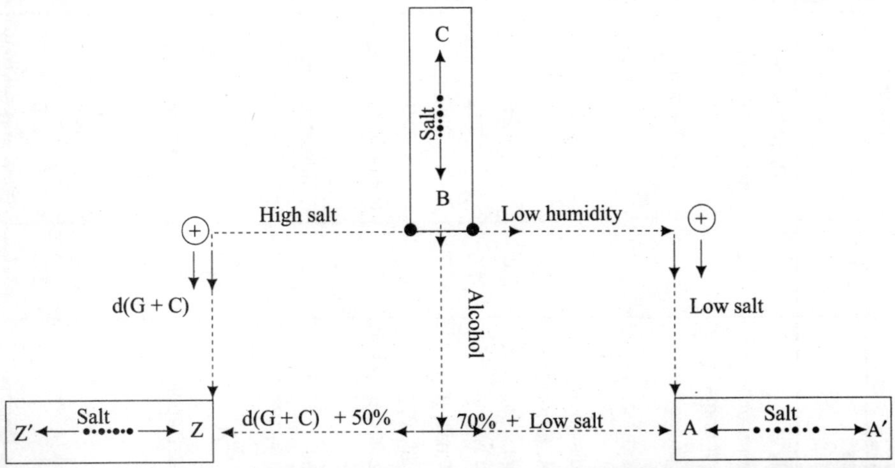

Fig. 3.17 Influence of humidity and salt concentrations on the transformations of nucleic acid families (A, B and Z). Transition within the members of the same family (A......A′), (Z......Z′) and (B......C) etc., are influenced by salt concentrations. Transitions between families — B → A, B → Z etc., are influenced by humidity and salt concentrations.

Table 3.2 Structural features of double-helical polynucleotides.

Family	*A-Family*			*B-Family*			*Z-Family*
Species	*A-RNA*	*A-DNA*	*A′-RNA (DNA-RNA*	*B-DNA*	*C-DNA*	*D-DNA*	*Z & Z′-DNA*
Chirality		Right-handed			Right-handed		Left-handed
Repeating unit		Mononucleotide			Mononucleotide		Dinucleotide
Glycosyl angle, χ		*Anti*			*Anti*		*Anti* at C & *Syn* at G
Sugar pucker, ν		C3′-*endo*			C3′-*endo*		C2′-*endo* at C C3′-*endo* at G C1′-*exo* at G (in Z′-DNA)
P…P distance		6.0 Å			7.0 Å		—
No. res. turn, N	11_1	11_1	12_1	10_1	9.33 (28_3)	81	6 × 2 = 12
Rise per. res., h	2.8	2.6	3.0 Å	3.38	3.3	3.4 Å	3.7 × 2 = 7.4 Å
Pitch P (Å)	31.0	28.5	36.0 Å	33.8	31.0	25.0 Å	45 Å
Rotn. Res°, (θ_R)	32.7	32.7	30.0°	36.0	38.6	45.0°	(–60°) per dinucleotide
Base tilt°, θ_τ	20	20	10°	–6	–8	–16°	–7°
Helix axis dislocation		In the Major groove		In the centre	In the Minor groove		In the Minor groove
Δ(Å)	4.5	4.5	4.9 Å	–0.1	–1.0	–1.8 Å	
Grooves:		Major groove			Major groove		No. Major groove. Convex surface
Width } Å	3.5	3.5	3.5	11.7	10.5	9.0	
Depth } Å	13.5	13.5	13.5	8.5	7.5	5.8	
Width & Depth		Narrow and Deep		Wide & Equal	Wider & Equal	Wider & Shallow	
		Minor groove			Minor groove		Only Minor groove
Width } Å	11.0	11.0	11.0	5.7	4.8	1.3	
Depth } Å	2.8	2.8	2.8	7.5	8.0	6.7	
Width & Depth		Wide and Shallow		Narrow & Equal	Narrower & Equal	Narrower & Deep	

3.4 PROTEIN–LIGAND INTERACTIONS

The function of a cell or an organism depends on the various interactions amongst its components. Of central importance to the structural as well as functional features of a cell or organism are proteins and enzymes and their interactions or associations with other ligands; from small substrates to other macromolecular constituents, like nucleic acids, polysaccharides and membranes. Molecular interactions/associations are governed by (i) the size and shape of the ligand that imposes the steric hindrance and (ii) electronic potential at the site of interaction. The predominant forces of importance coming into play in macromolecular associations are non-covalent interactions; as such complexes should be liable enough to undergo the conformational changes necessary for their functional roles with slight changes in temperature, pH, ion concentrations and other changes in the solvent conditions. More quantitative structural data are available in the case of protein-protein and protein-nucleic acid interactions than in the case of protein-membrane and protein-polysaccharide associations. Interactions between functional groups in protein-nucleic acid complexes can serve as the general type of interactions in all macromolecular associations.

3.4.1 Protein–Nucleic Acid Interactions

Various types of protein-nucleic acid interactions, both nonspecific and specific, have been established at the molecular level, from three-dimensional structural data, obtained from X-ray diffraction methods on 'model' compounds and complexes and nucleic acid-protein complexes. These non-bonded interactions play a central role in all biological processes and the level of our understanding of these interactions will determine our understanding of life processes—reproduction, development, gerontology and evolutionary characteristics of organisms. Some of these interactions may be nonspecific, like in histone-DNA interactions in chromatin, while others are specific, like in restriction endonucleases and their conjugate DNA-complexes. Much of the structural information has been derived from model compounds and DNA-regulatory protein complexes—such as specific and nonspecific interactions, the existence of structural motifs (**helix-turn-helix**, **zinc-finger** and **leucine-zipper**) and their structural and functional roles.

Interactions between functional groups in protein-nucleic acid complexes can be classified under (i) electrostatic, (ii) hydrogen bonding and (iii) hydrophobic and stacking interactions.

Electrostatic Interactions. Most protein-nucleic acid complexes involve electrostatic bonds between positively-charged **lysyl** or **arginyl** side chains and negatively-charged phosphate moieties of nucleic acids, forming ion pairs. Histidine in its protonated form could also engage in electrostatic bonding with the phosphate groups. In these interactions between groups, pKs might shift (e.g. free Lys has $pK_a = 10.5$, but its pK_a is 8.0 in RNA polymerase. Similarly, the pK_a of free ASP is 4.5, but is 6.8 for ASP102 in α-lytic protease).

Electrostatic interactions are also mediated by cations (metal ions). Metal ion coordination sites on nucleotides are shown in Fig. 3.18 and preferred binding sites of metal ions with nucleic acid constituents are given in Table 3.3. Polynucleotides do not interact with peptide residues, like aspartic and glutamic acids at neutral pH, because they are also negatively charged. Divalent cations, like Ca^{2+}, Zn^{2+}, can induce the formation of ternary complexes. A metal coordination complex of a metal cation between a protein and a nucleic acid constituent (the metal ion acting as a bridge) could be a

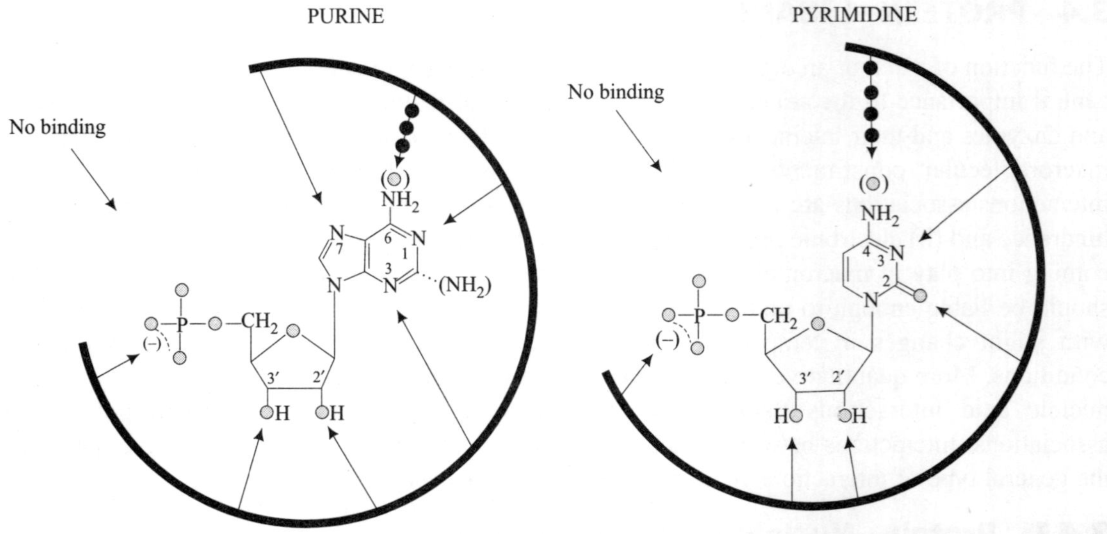

Fig. 3.18 Metal ion coordination sites in nucleic acid constituents (•••→ coordination sites for guanine and uracil/thymidine).

Table 3.3 Preferred binding sites of metal ions with nucleic acid constituents

Metal ions	*Nucleic acid-binding site*
Alkaline: Li$^+$, Na$^+$, K$^+$, Cs$^+$, Rb$^+$ *Alkaline earth:* Mg^{2+}, Ca^{2+}, Sr^{2+}, Ba^{2+}	Nucleic acid base, ring nitrogen and keto oxygens + sugar hydroxyl + phosphate oxygen groups.
Transition Metals: Mn^{2+}, Ni^{2+}, Co^{2+}, Co^{3+}, Zn^{2+}, Cd^{2+} Pt^{2+}, Pd^{2+}, Cu^{2+}, Ag$^+$, Au^{3+}, Hg^{2+} Ru^{3+}, Os^{6+}	Nucleic acid base, ring nitrogen and keto oxygens + phosphate oxygen groups.

general feature of many nucleic acid-metal-protein complexes, as has been shown in the case of Staphylococcal nuclease thymidine – 3′, 5′ diphosphate complex (Fig. 3.19).

Site-bound metal ions may also take part in molecular cleavage, depending on the local environment, and pH, acting as local metallo-enzymes. While the pK_a values of most metal-bound water complexes are in a higher pH range (pH > 8.5), the pK_a of Pb (II)-bound water is pH ~ 7.4); and water-mediated site-specific cleavage of Pb (II) (Fig. 3.20) in the physiological pH range can be cited as an example of the action of metal ions on nucleic acids, as a function of local environment and pH.

Hydrogen Bonding. Free energy of stabilization of DNA-protein complexes comes primarily from relatively nonspecific interactions, while specific interactions are largely via hydrogen-bonded networks, between protein side chains and the parts of the base moieties exposed in the grooves of the nucleic acids. Both ammo acids and nucleic acid constituents can act as acceptors as well as donors of hydrogen bonds. The propensities of amino acid residues to form hydrogen bonds can be pH-

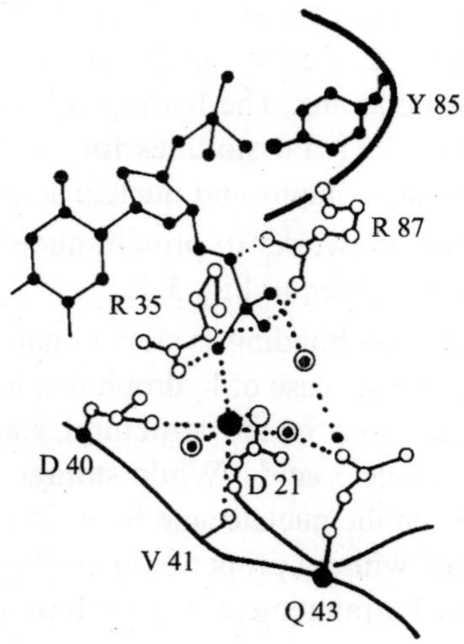

Fig. 3.19 Example of a metal ion acting as a bridge between protein an nucleic acid constituents in Staphylococcal nuclease–Ca^{2+}-pdTp complex.
Light: Deoxythymidine-3′, 5′-diphosphate. Bold: Protein backbone (and side chains).
◉: Water molecules ◖: Ca^{2+} ion.
Coordination and hydrogen bonds are given as dotted spots.
(After: Cotton, F.A. et al. PNAS. USA. 76; 2551 (1979).

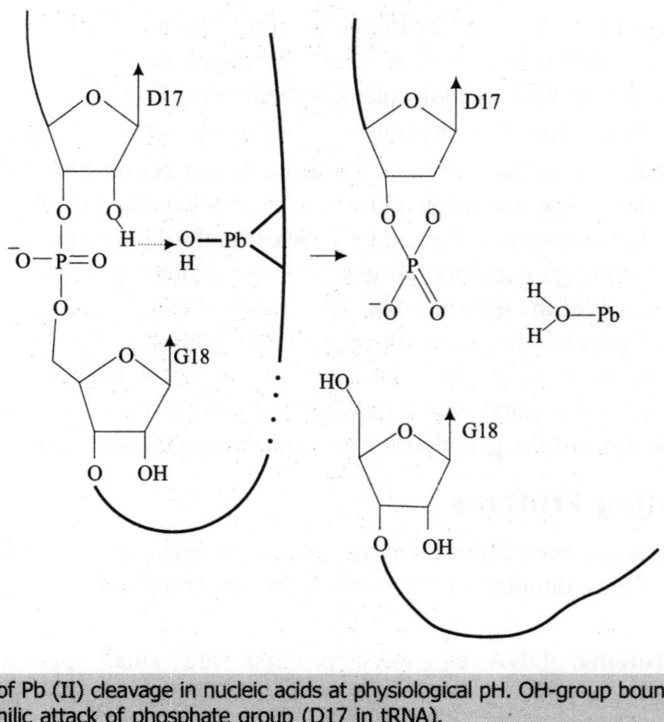

Fig. 3.20 Mechanism of Pb (II) cleavage in nucleic acids at physiological pH. OH-group bound to Pb (II) is involved in the nucleophilic attack of phosphate group (D17 in tRNA).
(*Source:* Brown, R.S. et al., *Nature*; 303; 5430 (1983)).

dependent or pH-independent. Nucleic acid bases can form hydrogen bonds and those with oligodentate hydrogen bonds network with the amino acid side chains of the polypeptides to impart specificity to protein-nucleic acid associations. The hydrogen-bonding potentialities of nucleic acid bases, A, T, G and C are given in Fig. 3.21. Possibilities for complex formation involving 'double-hydrogen bonds' between amino acid side chains and nucleic acid bases are given in Table 3.4.

Several possible hydrogen-bonded networks in protein-nucleic acid complexes (some of them established from model compounds) are given in Fig. 3.22.

Stacking Interactions. While hydrogen bonding is directional, hydrophobic interactions are not. But stacking of planar molecules is a special case of hydrophobic interactions imparting directionality. These interactions play an important role in the structural stability and function of biological molecules, such as base-stacking in nucleic acids. While storage of genetic code information is by Watson-Crick type base-pairing between the nucleic acid bases in DNA, such a planar hydrogen-bond network would not have been possible without proper base-stacking. The linear, rod-like structure of nucleic acids is also due to the stacked arrangement of their base-pairs.

One of the ways DNA interacts with molecules with aromatic groups is by intercalation. The conformational flexibility of the DNA molecule allows planar moieties of the ligand to intercalate between DNA base pairs. The DNA molecule is able to accommodate non-Watson-Crick basepairs with local unwinding of the double-helix and thus facilitating interaction of DNA with other molecules, such as drugs, dyes and antibiotics with planar moieties.

Polypeptides with aromatic side chains, such as **Phe**, **Trp**, and **Tyr**, can interact with nucleic acids by intercalation, and be stabilised by hydrogen bonding between the peptide and the nucleic acid. Such a combination of stacking and hydrogen bonding favours specific recognition by an amino acid of particular nucleic acid bases, since π–π and hydrogen bonding interactions can select the most suitable moieties among various molecular combinations. Therefore, aromatic amino acids in proteins may have selective recognition propensities for nucleic acids.

Base-stacking interactions appear to have an influence on **codon-anticodon** interactions. The genetic code (triplet code) is degenerated as more than one codon can code for the same amino acid. Although the '**wobble**' hypothesis put forward by **Crick** (the third base of the codon triplet can have non-Watson-Crick base-pairing) explains the genetic code degeneracy, it cannot fully explain the variability in the degeneracy of the genetic code. The degree of base-stacking interactions of the base pairs of the triplet would certainly have an influence on the variability of wobble. There appears to be a hierarchy in the variability of wobble (C–G > G–C > U–A > A–U). The greater the base-stacking the less the variability of wobble and *vice versa*, which means that the variability of wobble in the codon-anticodon interactions of the genetic code is sequence-dependent.

3.4.2 DNA-Binding Proteins

DNA-regulatory proteins and transcription factors, restriction endonucleases, DNA-polymerases and nucleosome are some of the examples which highlight the structural and functional features of DNA-protein interactions.

DNA-Regulatory Proteins. DNA-regulatory proteins are small proteins associated with transcriptional control. These proteins, by binding to specific DNA sequences help in switching ***off*** or

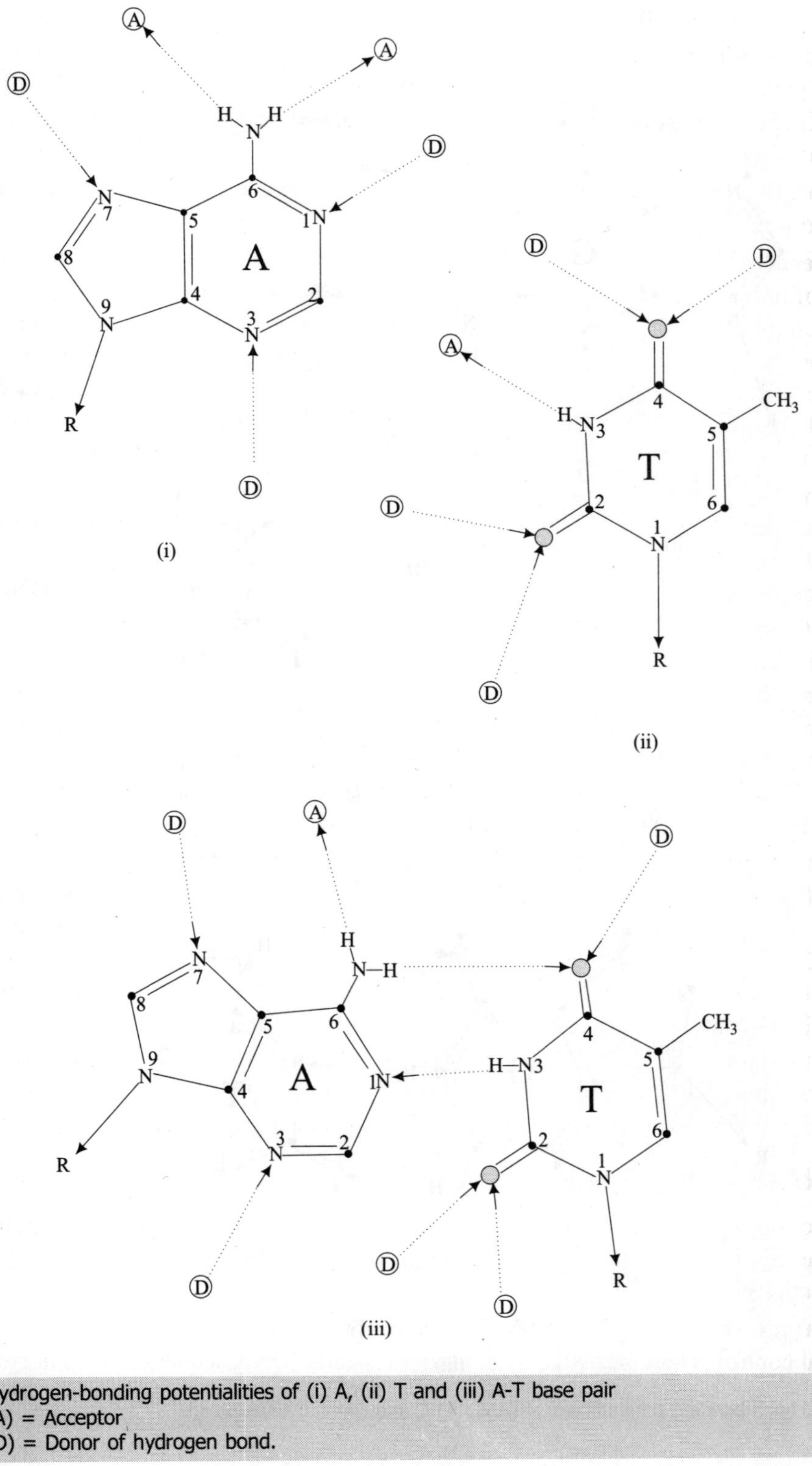

Fig. 3.21a Hydrogen-bonding potentialities of (i) A, (ii) T and (iii) A-T base pair
(A) = Acceptor
(D) = Donor of hydrogen bond.

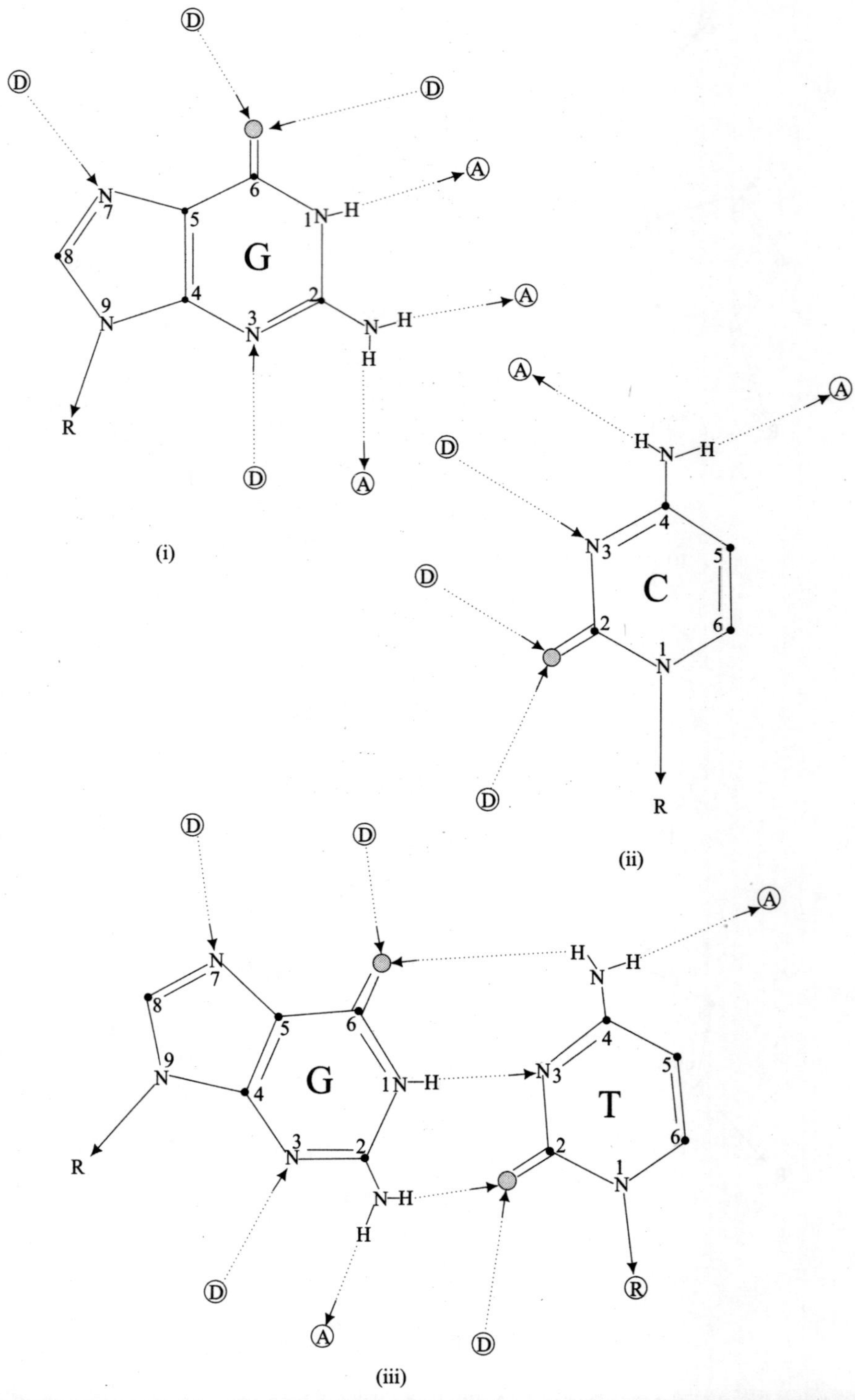

Fig. 3.21b Hydrogen bonding potentialities of (i) G, (ii) C and (iii) G-C base pair.

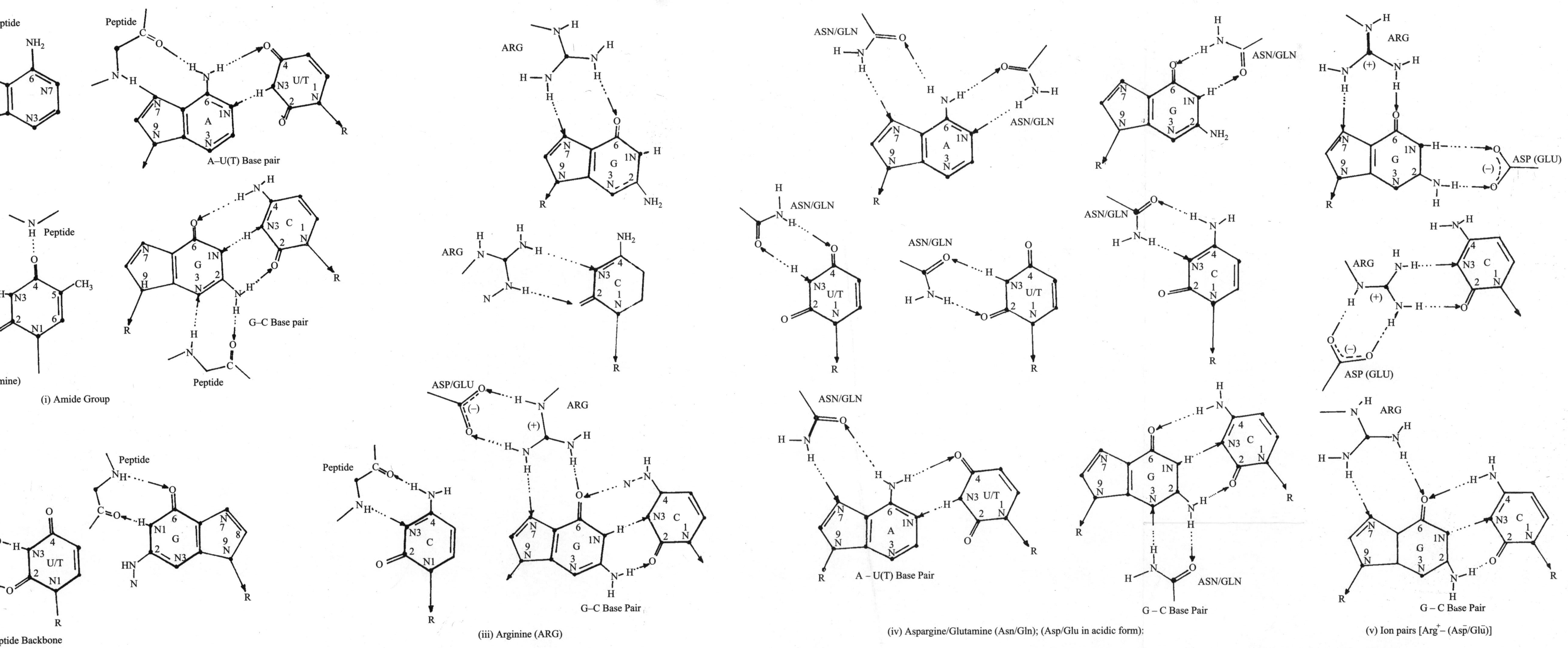

interactions (with nucleic acid bases).

one, (iii) Arginine (Arg), (iv) Asparagine/glutamine (Asn/Gln) (Asp/Glu in acidic form)], (v) Ion pairs ($Arg^{(+)} - Asp^{(-)}/Glu^{(-)}$).

Table 3.4 Possibilities of complex formation involving "double-hydrogen bonds" between amino acid side chains and nucleic acid bases.

Amino acid	*A*	*G*	*T(U)*	*C*
Neutral				
Asn, Gln	++	++	+	+
Ser, Thr, Tyr	++	++	+	+
Cys	++	++	+	+
Ionised				
Asp, Glu	–	+	–	–
Arg	–	++	–	+

– Not possible.
\+ Complex formation is possible in single-stranded nucleic acids.
++ Complex formation is possible in double-stranded nucleic acids also.

on genes as required. Many of these proteins have identifiable structural motifs (helix-turn-helix, zinc-finger and leucine zipper) which are involved in their functions and, therefore, those amino acid residues that are needed to maintain specific structural motifs are conserved among different proteins.

(i) **Helix-Turn-Helix Motifs.** Most of the prokaryotic transcriptional regulatory proteins (e.g. **λ repressor**, **Cro**, **CAP**, **lac repressor**, **trp repressor**) in general have the **helix-turn-helix** (HTH) structural motif (Fig. 3.23), which is associated with DNA-protein function. Generally, the sequence-specific DNA (operator) regions are are palindromic (or nearly so) and have two-fold rotational symmetry. The functional protein is a dimer.

As a general example of switch *on* and *off* processes in prokaryotic systems: bacteriophage λ contains a long stretch of DNA molecule containing structural genes coding both for repressor protein (upstream) and for **Cro** (control repressor) protein (downstream), separated by an operator region, OR. The operator region contains three protein-binding sites, OR1, OR2 and OR3, each approximately 17 base pairs long with ≈ 2-fold symmetry. Repressor protein binds to sites OR2 and OR1 (thus blocking the *Cro* promoter site, OR1) which results in the switching *off* of the **Cro** gene and switching *on* of the synthesis of repressor protein (lysogenic state); binding of *Cro* protein to sites OR2 and OR3 (blocking repressor-promoter site) leads to the switching *off* of repressor gene and the synthesis of *Cro* protein (lyctic state) (Fig. 3.24). That is, the binding of repressor or *Cro* at the specific operator sites of DNA provides a molecular switch to control lysogenic and lyctic pathways in many bacteria.

The molecular interactions that encompass DNA-protein binding involve (i) sequence-specific interactions to recognise operator regions, (ii) non-specific interactions for proper alignment of DNA-protein complex to maximise the association and (iii) conformational changes in the DNA, modulated by the local DNA base sequence in the middle of the operator regions, enabling differential binding affinities.

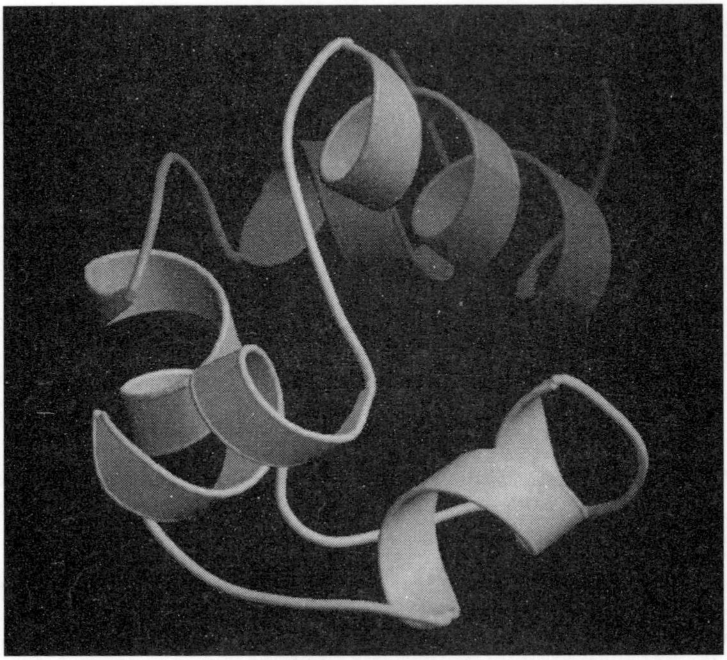

Fig. 3.23 Structure of Phage 434 Cro Protein with "Helix-Loop-Helix" Structural Motif. (Ref: Mondragon, A., Wolberger, C. & Harrison S.C. (1989) *J .Mol. Biol.*, 205; 179) (*Source:* Protein Data Bank: 2CRO.pdb).

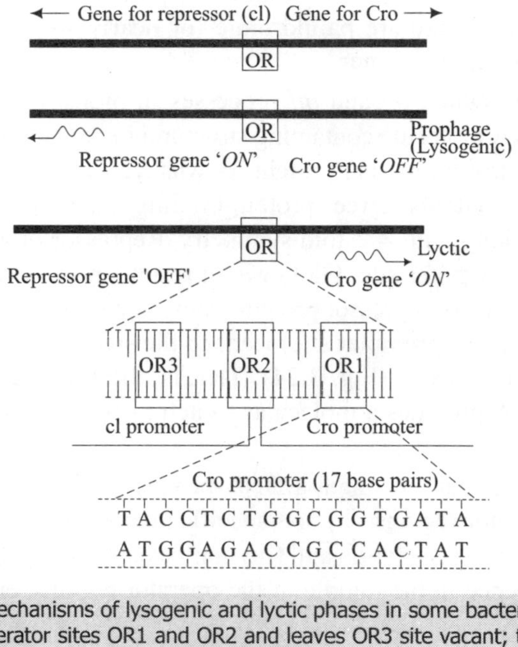

Fig. 3.24 Switch *on* and *off* mechanisms of lysogenic and lyctic phases in some bacteriophages. The repressor protein (dimer) occupies operator sites OR1 and OR2 and leaves OR3 site vacant; the Cro protein (dimer) occupies operator sites OR2 and OR3 and leaves OR1 site vacant.

Operator regions of DNA are recognised by these proteins through the helix-turn-helix structural motif. Each monomer has an HTH motif of ≈ 20 residues long (a seven-residue helix, a short hair-pin loop and a nine-residue helix). The second helix of the motif, called the *recognition helix*, fits into the major groove of B-DNA. Similarly, the 2-fold related recognition helix of the dimer, which is ~ 34 Å apart, can fit into the next major groove of the DNA double helix (Fig. 3.25a). That is, a pair of 2-fold related α-helices (recognition helices) of the dimeric protein, lying in successive major grooves of a B-DNA is a major determinant factor in sequence-specific DNA-protein recognition. The specific amino acid sequence of the recognition helix provides sequence-specific DNA-protein interactions. Therefore, invariant residues maximising sequence-specific interactions are important in such associations, for proper recognition of DNA operator regions. Non-specific interactions between the sugar-phosphate backbone of the DNA and the protein (mainchain NH-groups) are important in the proper alignment of protein domains and maximising DNA protein associations. These non-specific interactions extend to the entire DNA-protein binding region. Formation of a DNA-protein complex leads to conformational changes in the DNA molecules, leading to differential binding of the protein to different operator sites. These conformational changes are modulated (fine-tuned) by local DNA sequences at the operator sites (A-T base pairs are more flexible for change than G-C base pairs).

Some prokaryotic DNA-binding proteins are activated by the binding of an allosteric effector molecule (e.g. **Trp**, **Lac**). The binding of the allosteric effector molecular alters the orientation of the recognition helices to align them properly to fit into the major grooves of the B-DNA, 34 Å apart (Fig. 3.25b). Such ligand-induced conformational changes provide the molecular mechanism of functional switches in nucleic acid-protein interactions.

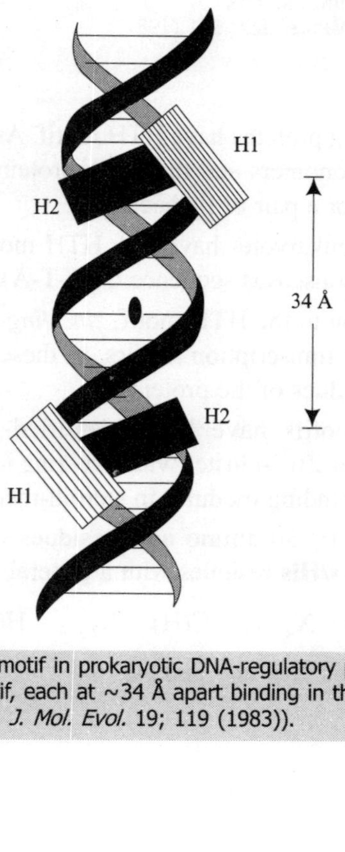

Fig. 3.25a Representation of the HTH motif in prokaryotic DNA-regulatory protein dimer-DNA associations. H2 is the recognition helix of the motif, each at ~34 Å apart binding in the major groove. ⬮ = Two-fold axis. (After: Ohlendof, D.H. et al. *J. Mol. Evol.* 19; 119 (1983)).

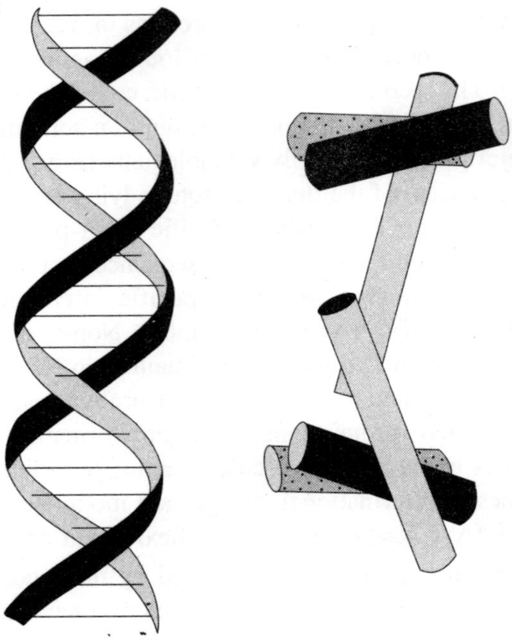

Fig. 3.25b Control mechanism of allosteric effector in DNA-binding proteins. Black cylinders: Trp-repressor in inactive state, without allosteric effector, **Trp.** Recognition helices (black) of the dimer are tilted. ......cylinders: Recognition helices of the Trp-repressor dimer oriented properly (......) at 34 Å apart upon the binding of the allosteric effector, **Trp.**
(*Source:* Zhang, R.O. et al. *Nature.* 327; 591 (1987)).

Not all prokaryotic DNA-binding proteins have HTH motif. As an example, *E. coli* Met-repressor protein is a dimer of intertwined monomers and the DNA-protein interactions are presumably via a pair of symmetry-related α-helices or a pair of β-sheets.

The homeodomain proteins of eukaryotes have also HTH motifs, but they bind as monomers to non-palindromic DNA sequences (consensus sequence: A-T-T-A).

(ii) **Zinc-finger Motifs.** In addition to the HTH motif, *zinc-finger* motifs and *leucine-zipper* motifs are also found in several eukaryotic transcription factors. In these the Zn^{2+} ion forms a coordination moiety (moieties) with **Cys/His** residues of the protein.

Three types of zinc-finger motifs have been identified—a single Zn^{2+}-ion tetrahedrally coordinated (C_2H_2-motif) a binuclear Zn^{2+}-cluster with two zinc ions tetrahedrally coordinated to six **Cys** residues (C_6-motif) and DNA-binding modules in steroid-receptor(s) (C_4-motif).

In a regular zinc-finger motif (≈ 30 amino acid residues (Figs. 3.26a & b) the Zn^{2+} ion is tetrahedrally coordinated to four **Cys/His** residues with a general sequence of the form

$$- C(H) - X_{2-5} - C(H) - X_{4-13} - C(H) - X_{2-4} - H(C) - X_{2-6}$$

where C(H) are cysteine/histidine and X is any residue.

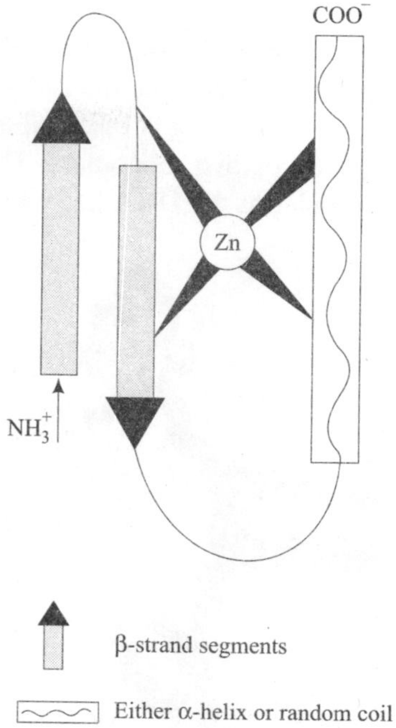

Fig. 3.26a Schematic diagram of Zn-finger motif in DNA-regulatory proteins, with tetrahedral coordination of Zn^{2+} with Cys/His residues.

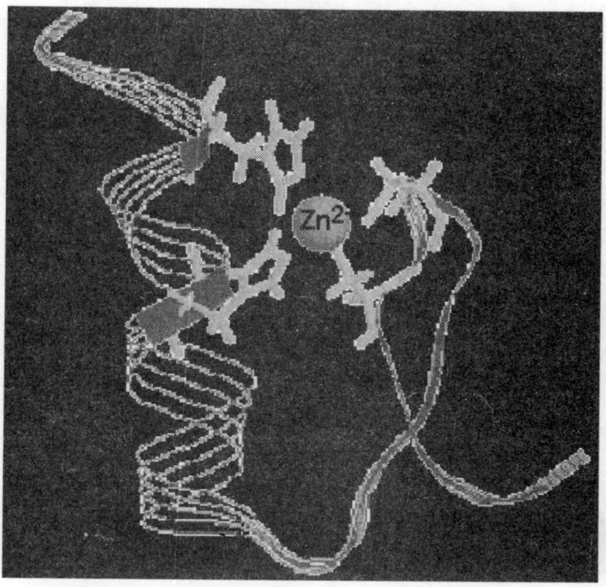

Fig. 3.26b Example of C_2H_2 zinc-finger motif found in SP1.
(*Source:* Protein Data Bank: 1SP1.pdb)

Binuclear Zn-cluster motif (C_6-motif) (≈ 30 amino acid residues with two Zn^{2+} ions sharing six cysteine residues has been found in GAL4 protein (Figs. 3.26c & d). The protein binds as a dimer to a seventeen base pairs palindromic sequence of DNA and the Zn^{2+}-containing domain recognizes the CCG triplet at each end of the site. The DNA binding is essential to nucleate the dimerisation of the protein, resulting in cooperative association of two monomers. The α-helical coiled-coil protrudes above the DNA minor grooves, perpendicular to the DNA helix axis.

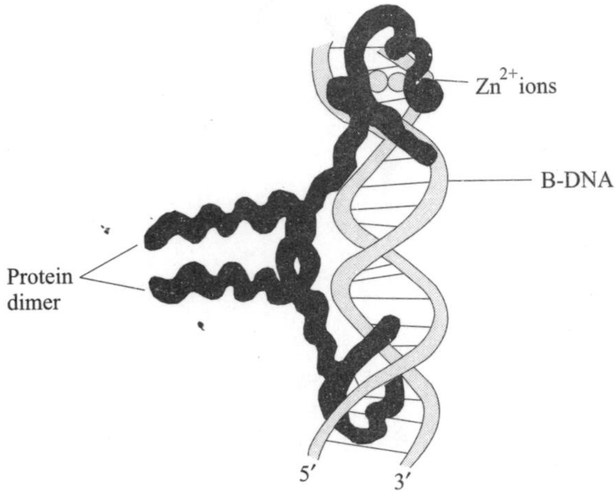

Fig. 3.26c Bi-nuclear **zinc-finger** motif in GAL4-DNA complex. Binding of DNA nucleates the dimerisation of protein with distinct DNA-recognition linker and α-helical coiled-coil regions. The coiled-coil is perpendicular to the DNA helix axis.
(From. Marmorstein, R. et al. *Nature* 356; 408 (1992)).

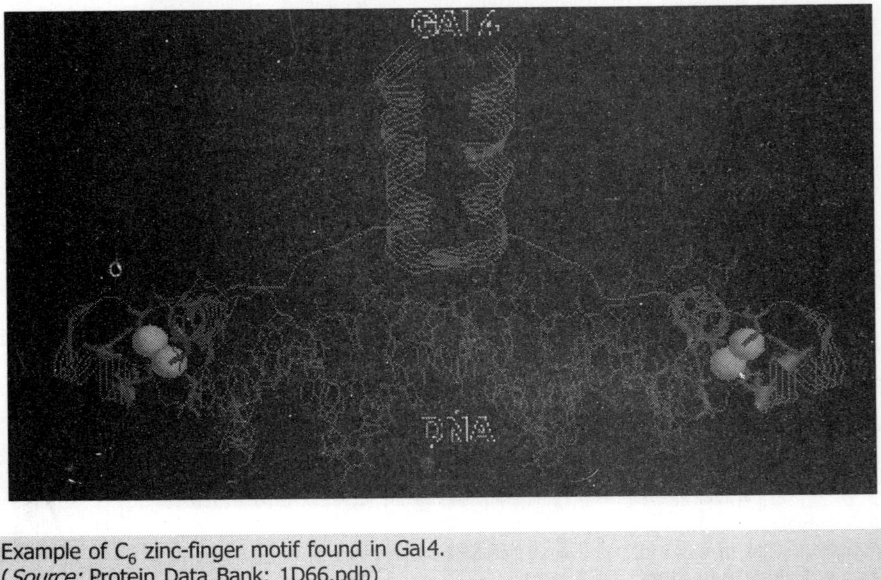

Fig. 3.26d Example of C_6 zinc-finger motif found in Gal4.
(*Source:* Protein Data Bank: 1D66.pdb)

In nuclear-receptor zinc modules (≈70 amino acid residues), two Zn^{2+}-ions are tetrahedrally bound, each by four ligands separately to form a compact domain twice the size of mononuclear or binuclear Zn^{2+}-domains.

(iii) **Leucine-Zipper Motif.** The **leucine-zipper** motif (≈30 amino acids) consists of leucine (or isoleucine) at seven residue intervals (heptad spacing) for at least eight turns of α-helix. The basic motif is

$$(-L - X_6 — L - X_6 — L - X_6 — L - X_6 - L -).$$

Leucine residues form a ridge on one side of the α-helix and mediate the dimerisation—two parallel α-helices forming a coiled-coil. The dimerisation interface is similar to that in the binuclear Zn-finger domain, of the GAL4-DNA complex (Fig. 3.27a).

The **'scissors-grip'** model, with Y-arms of the coiled-coil binding to the major grooves of the DNA and the 'arms' of the protein protruding above the DNA (perpendicular to the helix axis) as 'scissors-grip' has been proposed for the DNA recognition of this motif (Fig. 3.27b).

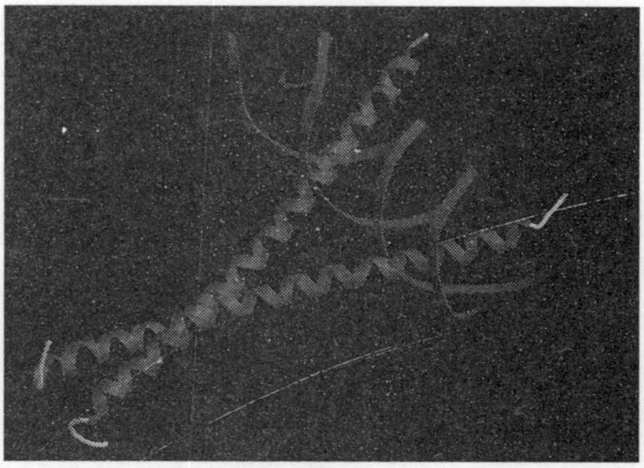

Fig. 3.27a Example of "Leucine-zipper" structural motif found in DNA-regulatory proteins. (Ref: Vinson, C.R. et al. (1989), *Science*, 246; 211).

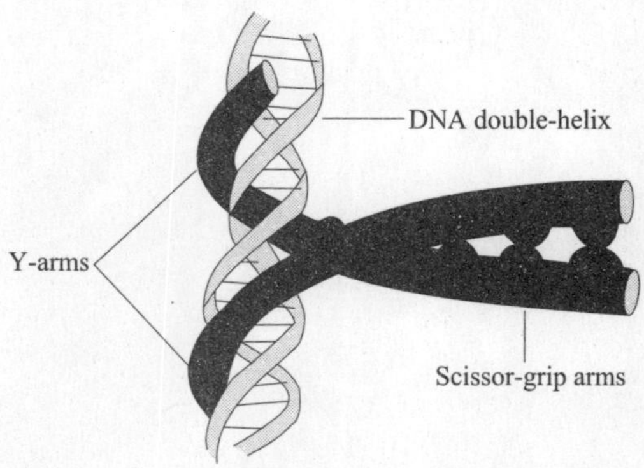

Fig. 3.27b Schematic representation of the **'scissor-grip'** model of **leucine-zipper** motif proteins binding to DNA double helix.
(Y-arms of the proteins dimer bind to the major grooves of the DNA and scissor-grip arms with leucine residues extend out from the DNA duplex.
(After: Vinson, C.R. et al. *Science*, 246; 211. (1989).

Restriction Endonucleases. Restriction endonucleases are also DNA-binding proteins which recognise specific DNA sequences and cleave them. The sequence specificity and, hence, structural specificity are due to the relative positioning and identity of the nucleic acid bases as well as to the conformation of the specific side chains of the amino acid residues of the endonuclease at the DNA-protein interface. Three-dimensional X-ray structural information on the enzyme, **EcoRI endonucleases** (Fig. 3.28a) bound to its cognate hexanucleotide DNA piece, GAATTC, has enabled the mechanism of the sequence-specific interactions, at the atomic level, of the protein and its cognate hexanucleotide to be established.

EcoRI is a symmetrical dimer and the cognate DNA sequence is palindromic. The α-helix of the protein fits into the major groove of the B-DNA at an angle. The formation of DNA-enzyme complex introduces a kink in the DNA which helps EcoRI to nick the DNA at a particular nucleotide.

In the endonuclease-DNA interactions, the binding of the protein seems to induce structural distortion of its DNA while forming the sequence-specific protein-nucleic acid complex. The sequence specificity is achieved by the formation of bidentate hydrogen bond networks between specific amino acid residues of the protein and the bases of the cognate DNA piece (Fig. 3.28b). Maximum cleavage, that is, proper function of the protein, occurs only when the bidentate hydrogen bond pairs are maximum. Any alterations in the base sequence of the cognate oligonucleotide, either

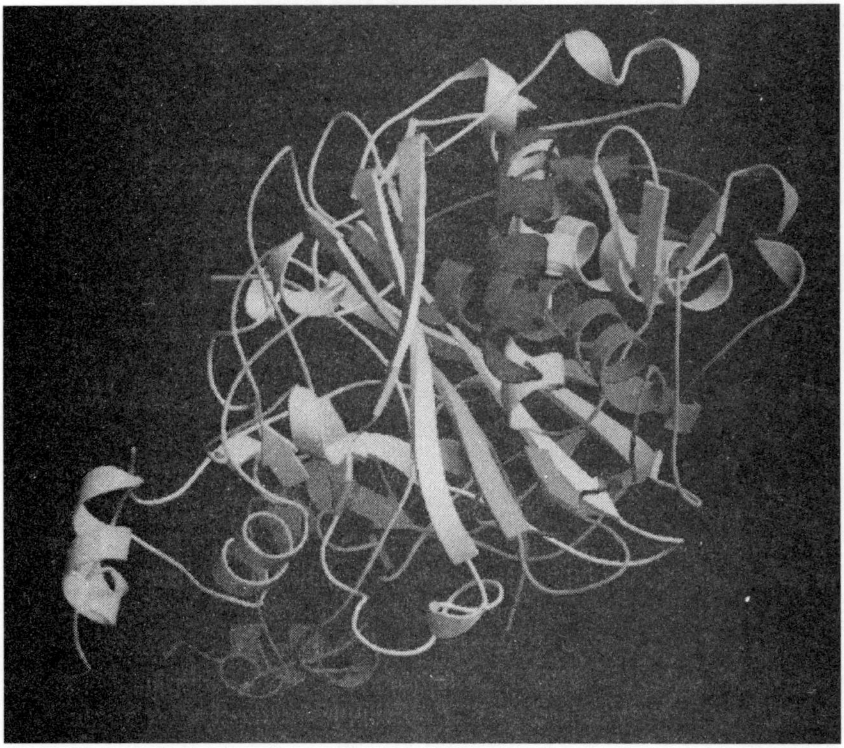

Fig. 3.28a Structure of EcoRI endonuclease-DNA complex.
(Ref: Kostrewa, D. & Winkler, F.K. (1995) Biochemistry 34; 683)
(*Source:* Protein Data Bank: 1RVA.pdb)

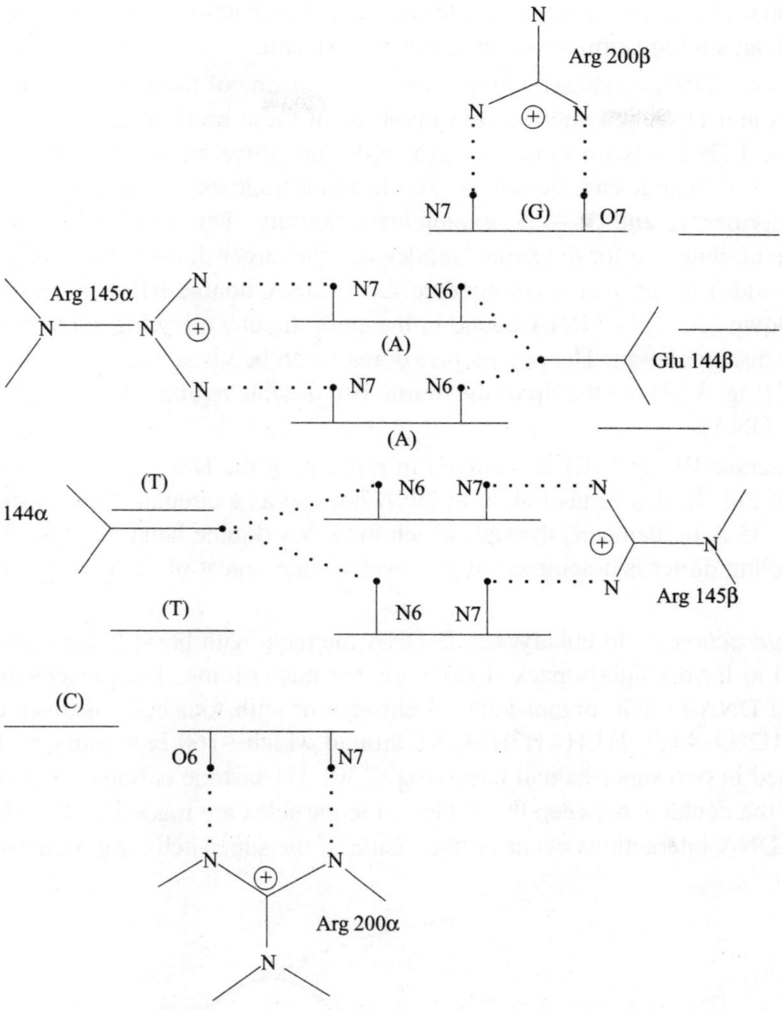

Fig. 3.28b Sequence-specific hydrogen bond networks in EcoRI endonuclease-cognate DNA (GAATTC) complex. ... = hydrogen bonds; α and β refer to identical subunits of the enzyme. (Ref.: Maclarin, J.A. et al. *Science*, 234; 1526 (1986).

by replacement or alterations in their positions, that reduce the number of hydrogen bond pairs involved in the sequence-specific protein-nucleic acid interactions would drastically reduce the function of the protein. Even minor modifications like methylation of the base(s) of the cognate oligonucleotide would perturb the delicate balance of the electrostatic charge distribution in that region, causing the perturbation of the entire recognition interface. This fine-tuning of molecular interactions (combination of structural and electrostatic features of the moieties) makes the molecular recognition highly specific. For example, whereas particular DNA-methylase can recognise its methylated cognate oligonucleotide it does not cleave the nucleotide. Besides, these sequence-specific interactions are cooperative; that is, formation of some correct protein-nucleic acid associations facilitate additional correct interactions, whereas incorrect interactions with noncognate residues have inhibitory effect, an anology similar to the *zipper* mechanism.

DNA Polymerases. DNA synthesis with proper editing is one of the most fundamental reactions in living organisms and DNA polymerases are involved in these multifunctional enzymatic reactions. DNA polymerase-I (Pol I) (with three modular units) has three enzymatic functions—polymerase, $3' \to 5'$ and $5' \to 3'$ exonuclease activities. The Klenow fragment, containing a larger fragment of Pol I retains polymerase and $3' \to 5'$ exonuclease activity. The smaller domain of the Klenow fragment has the binding site for deoxynucleotides and the larger domain has a very deep cleft (30 Å deep and 20 Å wide) which can accommodate the B-DNA double helix. A flexible region of the domain closes down 'on top' of DNA bound in the cleft, thereby allowing nucleotide polymerisation without enzyme disassociation. The polymerase domain can be visualised as a right hand at the point of holding a rod (Fig. 3.29). At the tip of the thumb is a flexible region which can close in to hold the rod in a grip (B-DNA).

DNA polymerase-III, (Pol III) is involved in replicating the DNA genome during cell division. The β-subunit of Pol III, that binds tightly to DNA does so as a circular dimer, forming a molecular '*curtain ring*' (~ 35 Å in diameter) through which the DNA double helix can slide like a curtain rod, The DNA-encircling dimer is assembled by the twelve-time repeat of the unit (β – α – β – β – β) to form a circle.

DNA-Histone Interactions. In eukaryotes, B-DNA interacts with basic histone proteins (H1, H2A, H2B, H3 & H4) to form a tightly-packed structure, the nucleosome. The nucleosome is the primary repeating unit of DNA-protein organisation in chromatin, with four core histones complexed as an octomer, (H2A.H2B.H4.H3: H3.H4.H2B.H2A), around which ~160 base pairs of double helical B-DNA are wrapped in two super-helical turns (Fig. 3.30). H1 histone is bound outside this particle as a plug. Most of the contacts between the nucleosome particles are made by DNA-DNA interactions and the histone-DNA interactions occur on the inside of the super-helix, through non-specific sugar-

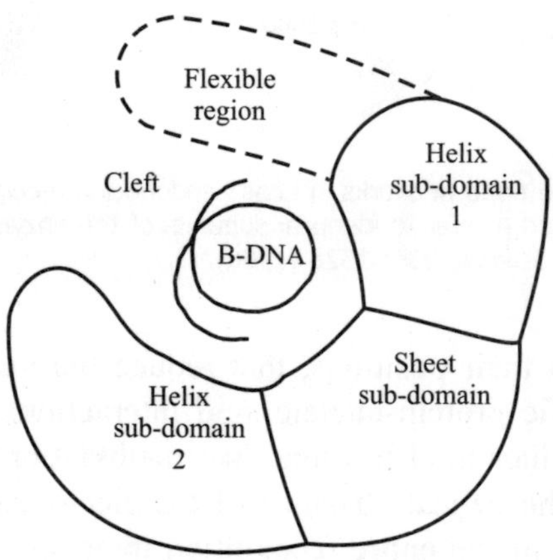

Fig. 3.29a Schematic representation sub-domain organization of the polymerase domain of the **Klenow** fragment. The helix sub-domain 1 with flexible region is analogous to right hand thumb, β-sheet sub-domain to palm and helix sub-domain 2 to folded fingers forming a cleft to hold the object (B-DNA). (After: Ollis, D.L. et al. *Nature* 313; 762 (1985)).

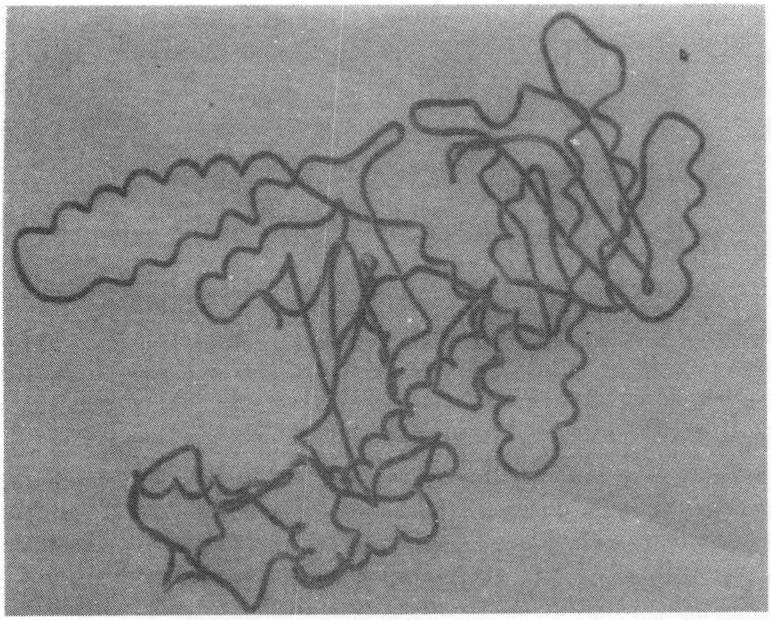

Fig. 3.29b Ribbon diagram of DNA polymerase I (Klenow fragment/dCMP)*. Analogous to right hand thumb, palm and folded fingers. The ribbon diagram was generated at the Bioinformatics Centre, Indian Institute of Science, Bangalore, from the PDB coordinates of the indicated source).
(*Source:* Beese, L. et al. 1 DPI: 8/87).

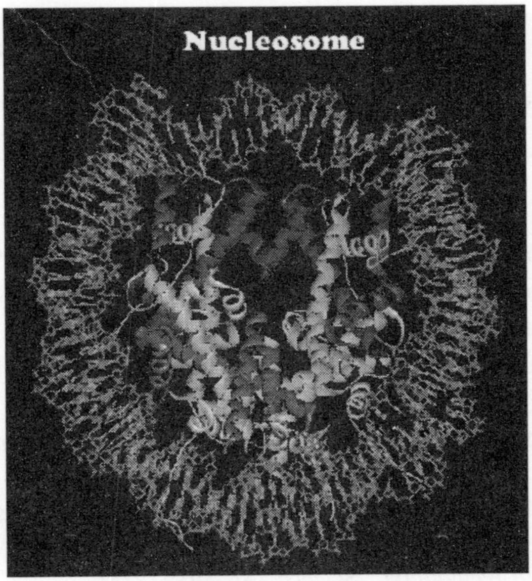

Fig. 3.30 Nucleosome structure.
(Ref: US Dept. of Energy: Genomes to Life Project)
(http://doegenomes to life.org)

phosphate backbone interactions. There are substantial contacts between the $(H3.H4)_2$ tetramer and the central turn of the DNA super-helix. The DNA super-helix has kinks as it wraps around the histone octomer. These deformations lead to alterations in groove widths and sequence-specific propensities—sequences AAA, or TTT tend to occur in the nucleosome at intervals of ~10 base pairs if the minor groove is facing inward (towards the histone core) and GGC or GCC if the minor groove is facing outward.

Suggested Reading

Books

Baum, S.J. & Scaife, C.W.J. "*Chemistry - A Life Science Approach*". MacMillan/New York (1980).

Bloomfeld, V.A., Crothers, D.M. & Tinoco, I. Jr. "*Physical Chemistry of Nucleic Acids*" Harper & Row/New York (1974).

Calladine, C.R. & Drew, H.R. "*Understanding DNA*". Academic Press/New York (1992).

Freifelder, D. "*Molecular Biology*", 2nd Edn. Jones & Bartlet Pub/Boston (1987).

Saenger, W. "*Principles of Nucleic Acid Structure*". Springer Verlag/Berlin (1984).

Saenger, W. & Heinemann, U. (Eds). *"Protein-Nucleic Acids Interactions"*. MacMillan/Houndmills. (1989).

Travers, A.A. "*DNA-Protein Interactions*", Chapman-Hall/London (1990).

Volkenstein, M.V. "*Molecule and Life*" An introduction to Molecular Biology". Plenum Press/New York (1970).

Watson, J.D. et al. "*Molecular Biology of the Gene*", 4th Edn. Benjamin-Cummings/Menlo Park (1987).

Articles

Baltimore, D. & Berg, A.A. "*DNA-A-binding Proteins*". Nature 373; 287 (1995).

Beese, L.S., et al., "*Structure of DNA Polymerase I Klenow Fragment Bound to Duplex DNA*". Science 260; 352 (1993).

Brennan, R.G. & Matthews, B.W. "*Structural Basis of DNA-protein Recognition*". Trends Biochem. Sci. 14; 286. (1989).

Brennan, R.G. & Matthews, B.W. "*Structural Basis of DNA-Protein Recognition*". TIBS. 14; 286 (1989).

Churchill, M.E., A. & Travers, A.A. "*Protein Motifs that Recognise Structural Features of DNA*". TIBS. 16; 92 (1991).

de Duve, C. "*The Second Genetic Code*". Nature, 333; 117 (1988).

Dickerson, R.E. "*The DNA Helix and how it is Read*". Sci. Amer., 249 (6); 86 (1983).

Dickerson, R.E. et al. "*The Anatomy of A-, B- and Z-DNA*". Science 216; 475 (1985).

Felsenfeld, G. "*DNA*" Sci. Amer. 253 (4); 58 (1985).

Harrison, S.C. "*A Structural Taxonomy of DNA-binding Domains*" Nature 353; 715 (1991).

Hayes, J.J. et al. "*The structure of DNA in a Nucleosome*". Proc. Natl. Acad. Sci. USA. 87; 7405 (1990).

Hunter, C.A. "*Sequence-dependent DNA Structure; the Role of Base Stacking Interaction*". J. Mol. Biol. 230; 1025 (1992).

Johnson, P.F. & McKnight, S.L. "*Eukaryotic Transcriptional Regulatory Proteins*". Annu. Rev. Biochem. 58; 799 (1989).

Jordon, R.S. & Pabo, C.O. "*DNA-recognition*". Science 241; 893 (1988).

Klug, A. & Rhodes, D. "*Zinc-fingers: a Novel Protein Motif for Nucleic Acid Recognition*" TIBS 12; 464 (1987).

Konforti, B. "*Rules for Protein-DNA Interactions*". Nature Struct. Biol. 6; 505 (1999).

McClarin, J.A. et al. "*Structure of the DNA-EcoRI Endonuclease Recognition Complex at* 3Å *Resolution*". Science 234; 1526 (1986).

Pabo, C.O. &. Sauer, R.T. "*Transcription Factors: Structural Families and Principles of DNA Recognition*". Annu. Rev. Biochem. 61; 1053 (1992).

Pavletich, N.P. & Pabo, C.O. "*Zinc Finger-DNA Recognition*". Science 252; 809 (1993).

Ptashne, M. "*How Eukaryotic Transcriptional Activators Work*". Nature, 335; 683. (1988).

Ptashne, M., Johnson, A.D, & Pabo. C.O. "*A Genetic Switch in a Bacterial Virus*", Sci. Amer. 247, 128; (1982).

Rhodes, D. & Klug, A. "*Zinc Fingers*". Sci. Amer. 268(2); 32 (1993).

Steitz, T.A. "*Structural Studies of Protein-nucleic Acid Interaction: The Sources of Sequence-specific Binding*". Quart. Rev. Biophys. 23; 205 (1990).

Struhl, K, "*Helix-turn-helix, Zinc-finger and Leucine-zipper Motifs for Eukaryotic Transcriptional Regulatory Proteins*". TIBS 14; 137 (1989).

Tijan, R. "*Molecular Machines that Control Genes*". Sci. Amer. 273(3); 7 (1995).

von Hippel, P.H. "*Protein-DNA Recognition*". Science 263; 769 (1994).

Wang, A.H.J. et al. "*Left-handed Double-helical DNA: Variations in the Backbone Conformation*". Science 211; 171 (1981).

Watson, J.D. & Crick, F.H.C. "*Genetical Implications of the Structure of Deoxy-ribonucleic Acid*", Nature 171; 964 (1953).

Watson, J.D. & Crick, F.H.C. "*The Molecular Structure of Nucleic Acids*". Nature 171; 737 (] 953).

Westhof, E. "*Water: An Integral Part of Nucleic Acid Structure*". Annu. Rev. Biophys. Biophys. Chem. 17; 125 (1988).

4 Carbohydrates

Synopsis: Saccharides and carbohydrates are essential constituents of all living organisms. They are sources of structural support and energy. Saccharides exist in nature mainly as mono-, di- and polysaccharides. Glucose, fructose, ribose and arabinose are examples of monosaccharides. Major disaccharides are sucrose, lactose and cellobiose. Polysaccharides are linear as well as branched-polymers of monosaccharide units, linked by glycosidic bonds (C-O-C). Starch is a polysaccharide of a mixture of a linear α (1 → 4)-linked polymer, α-amylose, and an α (1 → 4)-linked and α (1 → 6)-branched polymer, amylopectin. Glycogen, the animal food-storage polysaccharide is similar in structure to starch, but is more highly branched. The structural polysaccharide of plant cell walls, cellulose, is a linear polymer of β (1 → 4)-linked D-glucose residues. An intra-hydrogen bond network imparts mechanical strength, which is further strengthened by lignin. Glycoproteins have ubiquitous functional roles. Proteoglycans are found mostly in extracellular sites, such as in cartilage and tendon; the properties of these can be explained on the basis of the content of glyco-proteins. Bacterial cell walls consist of chains of alternating β (1 → 4)-linked N-acetylglucoseamine (NAG) and N-acetylmuramic acid (NAM), which are cross-linked by short peptides to form a bag-shaped peptidoglycan molecule. Lysozyme cleaves the glycosidic linkages between NAG and NAM of peptidoglycan. Oligosaccharides attach to protein in limited ways. In N-glycosidic attachments, an NAG is invariably bound to the amide N-atom of **Asn** of the sequence **Asn-X-Ser(Thr)**. O-Glycosidic attachments are made to **Ser(Thr)** in most proteins and to hydroxylysine in collagen. Glycoproteins are important mediators of cell-cell recognition.

4.1 MONOSACCHARIDES

Monosaccharides are the basic units of carbohydrates. The chemistry of monosaccharides is largely that of their hydroxy and carbonyl group (–OH and C=O). The classification of monosaccharides is according to the chemical nature of their carbonyl group. Carbohydrates that contain an aldehyde group are called aldoses and those that contain a ketone group are called ketoses (Fig. 4.1).

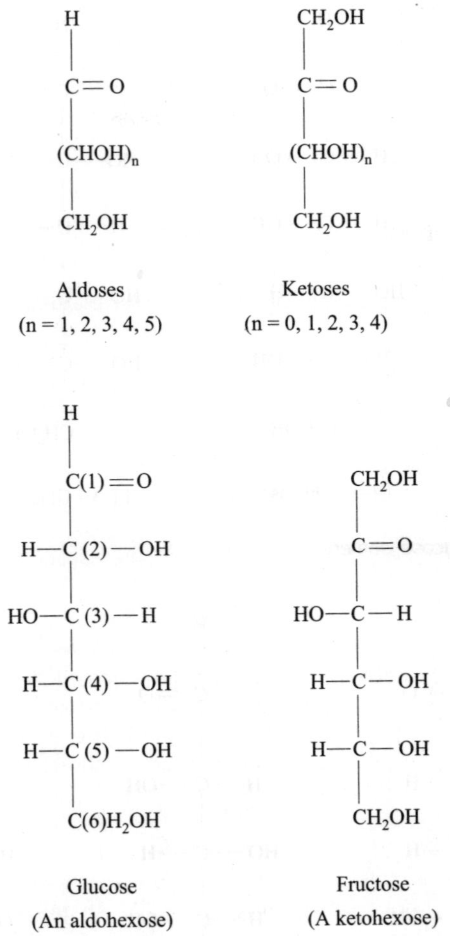

Fig. 4.1 Chemical formulae of aldoses and ketoses. Glucose and fructose are specific examples of an aldose and a ketose.

Saccharides can exist as D- or L-isomers. D-saccharide has the same configuration about the penultimate carbon atom (farthest from the keto or aldehyde group) as does D-glyceraldehyde (Fig. 4.2.)

Saccharides (a pair of diastereoisomers) that differ only in the configuration about a single carbon atom are known as epimers. For example, D (+) mannose and D (+) galactose differ from D (+) glucose only in the configuration about one asymmetric carbon atom, at C2 and C4, respectively. Therefore, D (+) mannose and D (+) galactose are **epimers** of D (+) glucose (Fig. 4.3).

Monosaccharides occur in cyclic forms: a saccharide with a 6-membered ring is called pyranose and a saccharide with a 5-membered ring is called furanose. (Fig. 4.4). The cyclic forms of sugars can be represented according to the **Haworth** convention:

(1) Darkened edges point towards the viewer.
(2) Atoms on the right in **Fischer** projection appear below the plane of the ring and any group written to the left appears above the plane.

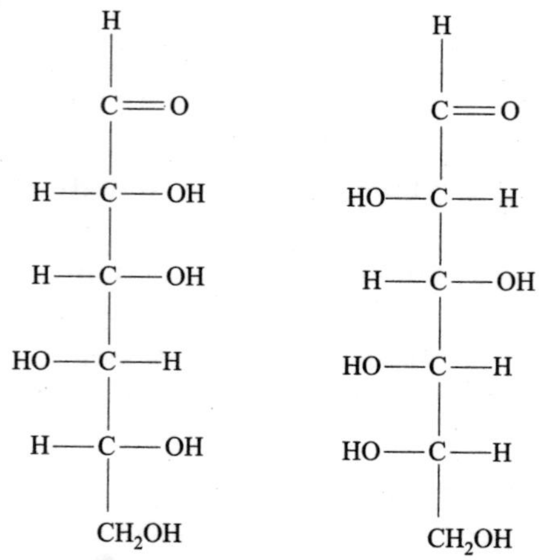

D(–) – Glucose L(–) – Glucose

Fig. 4.2 Formula of D- and L-glucose isomers.

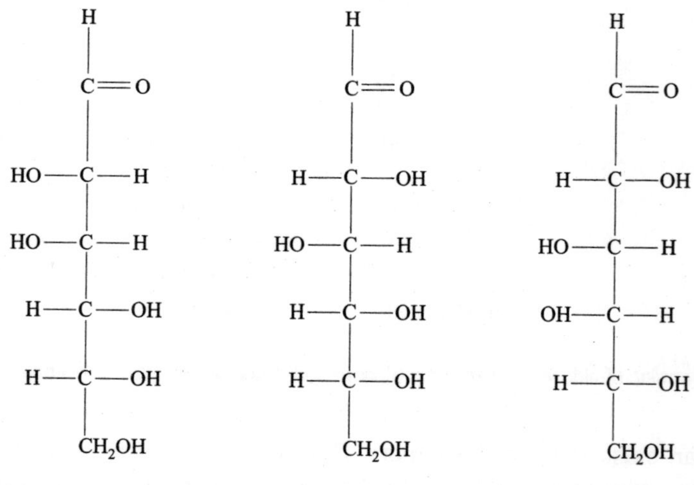

D(+) Mannose D(+) Glucose D(+) Galactose

Fig. 4.3 Epimers of D(+) glucose (D-mannose and D-galactose).

(3) In an α-isomer the substituent at the anomeric carbon is below the plane of the ring and is above the plane of the ring in the β-configuration.

The cyclic form of monosaccharides renders the former carbonyl carbon asymmetric. The resulting pair of diastomers are called anomers. In the α-anomer, the OH-substituent to the anomeric carbon is on the opposite side of the sugar ring from the CH_2OH-group at the chiral centre that designates D- or L-configuration (e.g. C5- in hexoses). The pyranose ring may assume boat or chair conformation.

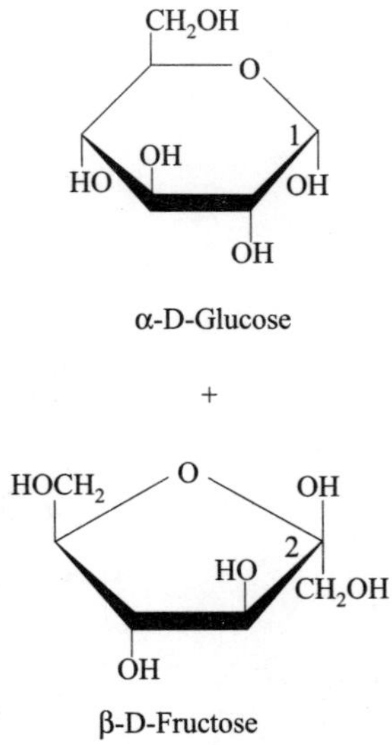

Fig. 4.4 Pyranose and furanose ring structures of monosaccharides.

4.1.1 Hexoses

Glucose is the most abundant sugar in nature. Fructose is the only naturally occurring ketohexose. Galactose is formed by the hydrolysis of lactose, a disaccharide comprising a glucose and a galactose unit; In amino sugars (hexoses), one or more -OH groups have been replaced by an amino group (on the C2-atom of aldohexoses). Two representatives of this class, D-glucosamine and D-galactosamine, are found in nature largely as N-acetylated compounds (Fig. 4.5). The aminosugar derivatives of N-acetylglucosamine (NAG) and N-acetylmuramic acid (NAM) are prominent components of bacterial cell walls.

4.1.2 Pentoses

The most important pentoses in biology are ribose and deoxyribose, found in nucleic acids. In deoxyribose, the 2′-OH group is replaced by H (Fig. 4.6).

4.2 DISACCHARIDES

Monosaccharide units form di-, oligo- and polysaccharides, *via* glycosidic linkages (anomeric carbon to the acetyl oxygen atom). The glycosidic bond is the carbohydrate analogue of the peptide bond in proteins. Many important carbohydrates are disaccharides—sucrose, maltose, cellobiose and lactose.

4.2.1 Sucrose

Sucrose (common sugar) is a disaccharide of α-D-glucose and β-D-fructose. The linkage is an α (1 → 2) glycosidic bond (Fig. 4.7).

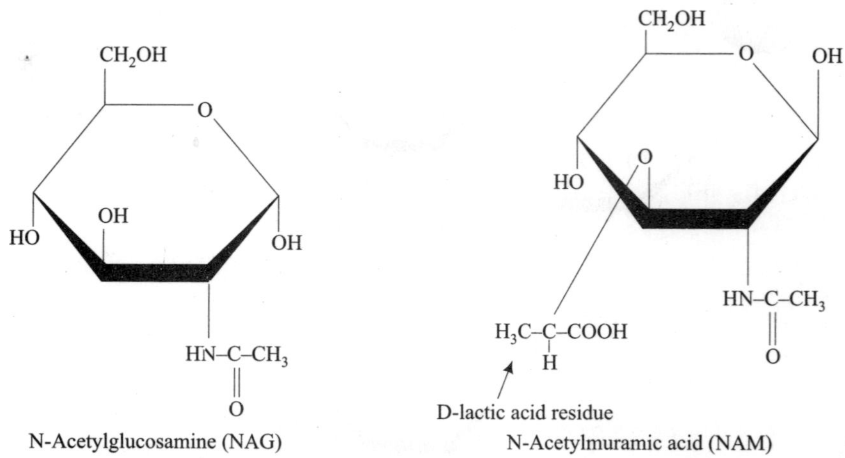

Fig. 4.5 Amino sugars NAG and NAM.

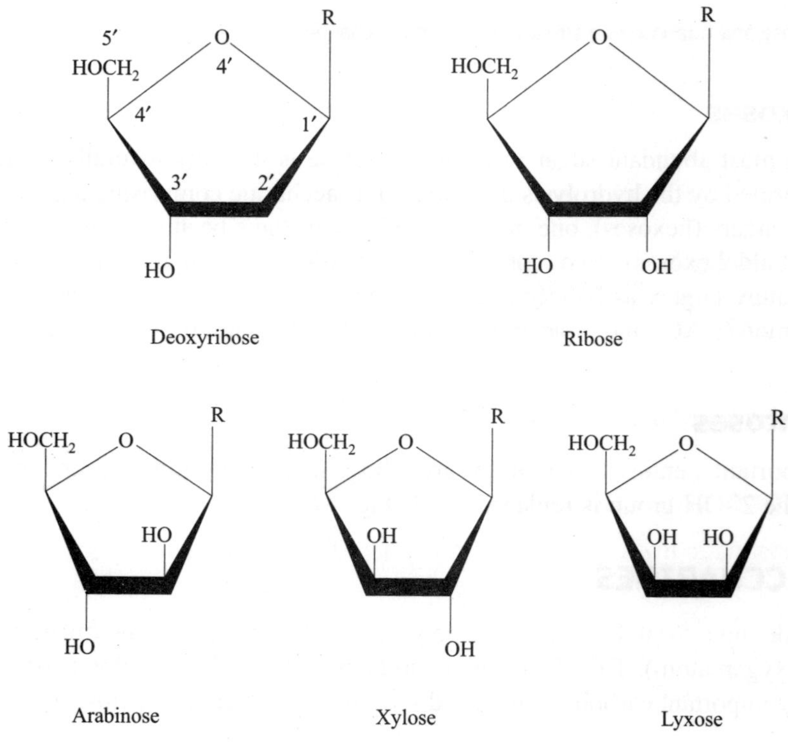

Fig. 4.6 Some important pentose sugars.

4.2.2 Lactose

Lactose is a disaccharide comprising a galactose and a glucose molecule linked by a β (1 → 4) glycosidic linkage. It occurs naturally only in milk (Fig. 4.8).

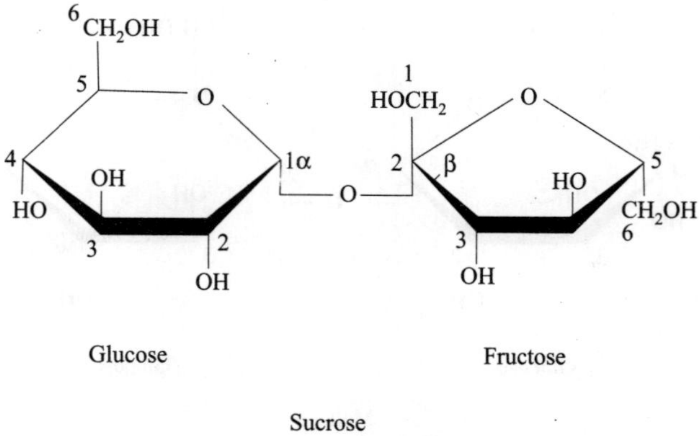

Fig. 4.7 Glucose and fructose units of sucrose.

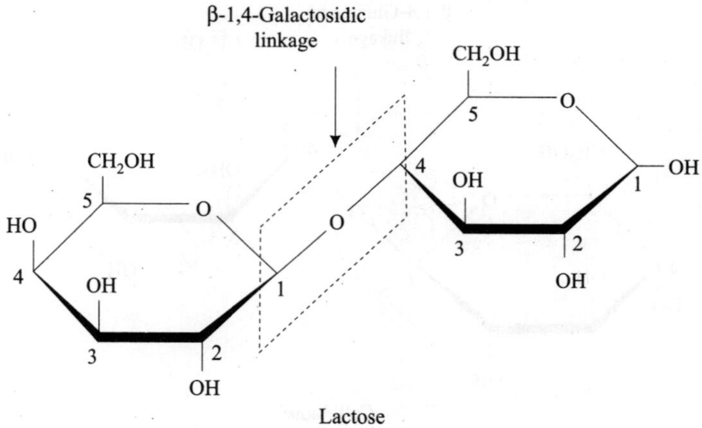

Fig. 4.8 Galactose and glucose units of lactose.

4.2.3 Maltose and Cellobiose

Maltose is an enzymatic hydrolysis product of starch. It is a disaccharide of two glucose molecules, linked by an α (1 → 4) glycosidic linkage (Fig. 4.9a). Cellobiose, the repetitive disaccharide of cellulose, is also a disaccharide of two glucose molecules, but the linkage is a β (1 → 4) glycosidic linkage. That is, cellobiose is a stereoisomer of maltose (Fig. 4.9b). Human beings do not have the enzyme apparatus for the hydrolysis of cellobiose to glucose units.

4.3 POLYSACCHARIDES

Polysaccharides serve as storage of food and energy (e.g. starch and glycogen), and as structural components (e.g. cellulose).

4.3.1 Starch and Glycogen

Starch is the most important source of carbohydrates in the human diet. It serves as a food reserve in plants and is a major nutrient for many organisms. Starch is a mixture of two polymers—α-amylose

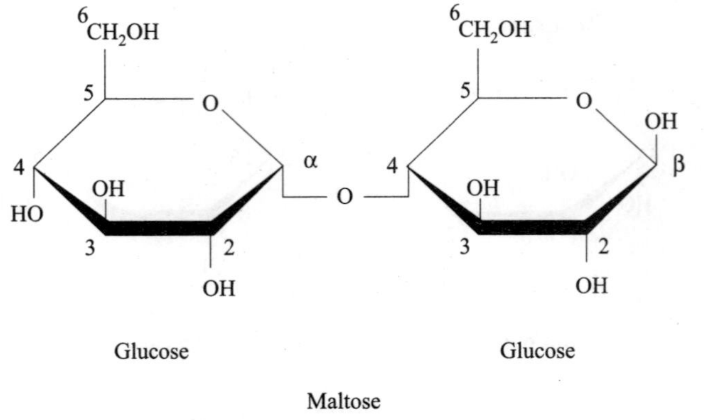

Fig. 4.9a Two glucose units of maltose (α (1 → 4)-linked).

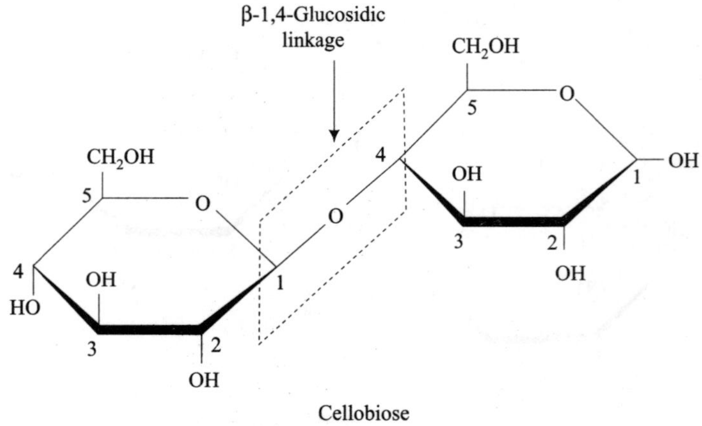

Fig. 4.9b Two glucose units of cellobiose (β (1 → 4)-linked).

and amylopectin. α-amylose is a linear polysaccharide consisting of several thousand D-glucose residues joined by α (1 → 4) glycosidic linkages (Fig. 4.10). The α-glycosidic linkages cause an irregular helically-coiled conformation. The other polymer, amylopectin, is a branched-chain polysaccharide composed primarily of α (1 → 4)-linked glucose units, but with occasional α (1 → 6)-linked branched points at about every thirty glucose residues.

Glycogen is called animal starch. It is the storage polysaccharide of animals and is present in skeletal muscle and liver. The primary structure of glycogen is similar to that of amylopectin, but glycogen is more highly branched (branching points ≈ at every 10 glucose residues). In the cell, the glycogen is degraded metabolically by glycogen-phosphorylase to yield glucose-1-phosphate.

4.3.2 Cellulose

Cellulose is the fibrous carbohydrate (polysaccharide) that serves as the primary structural component of plant cell walls. It is a linear polymer with cellobiose repeating units (two D-glucose units linked via β (1 → 4) linkages) (Fig. 4.11a). An extensive intra-hydrogen bond network renders exceptional

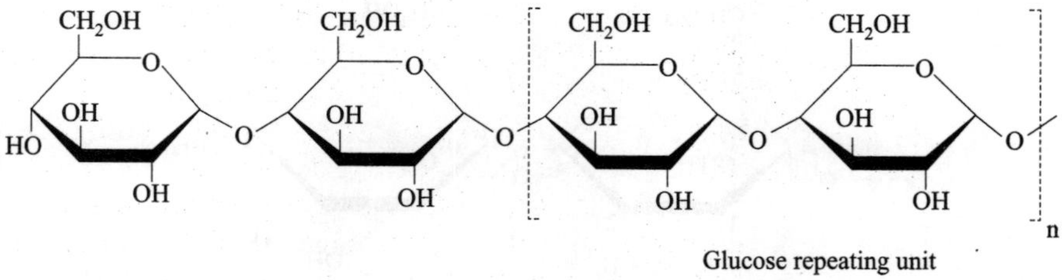

Amylose
(a)

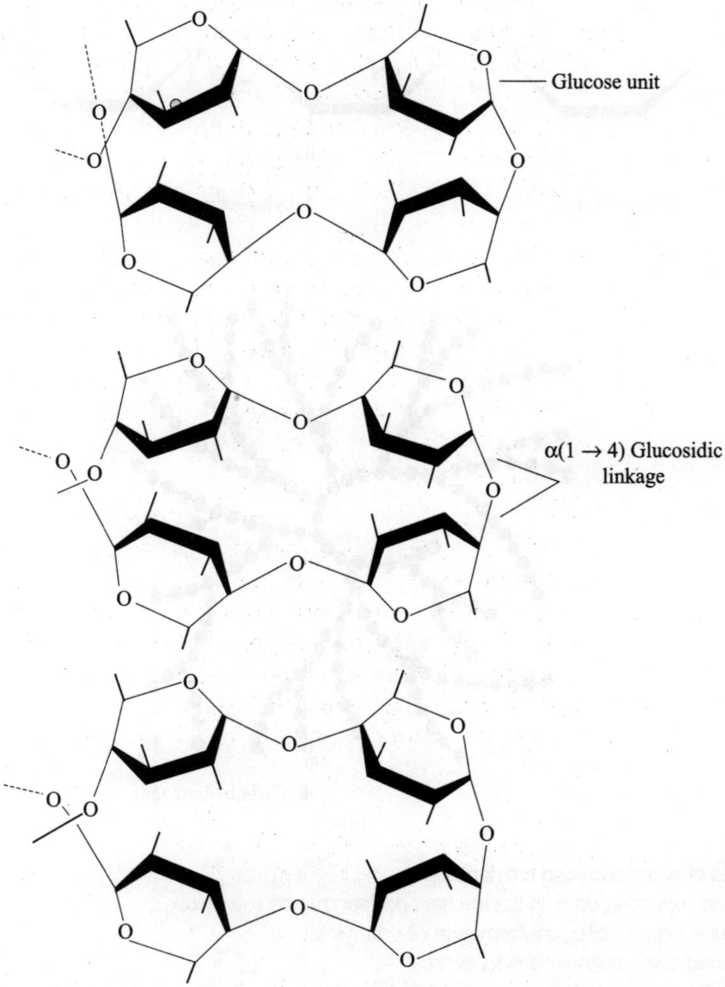

(b)

(Contd.)

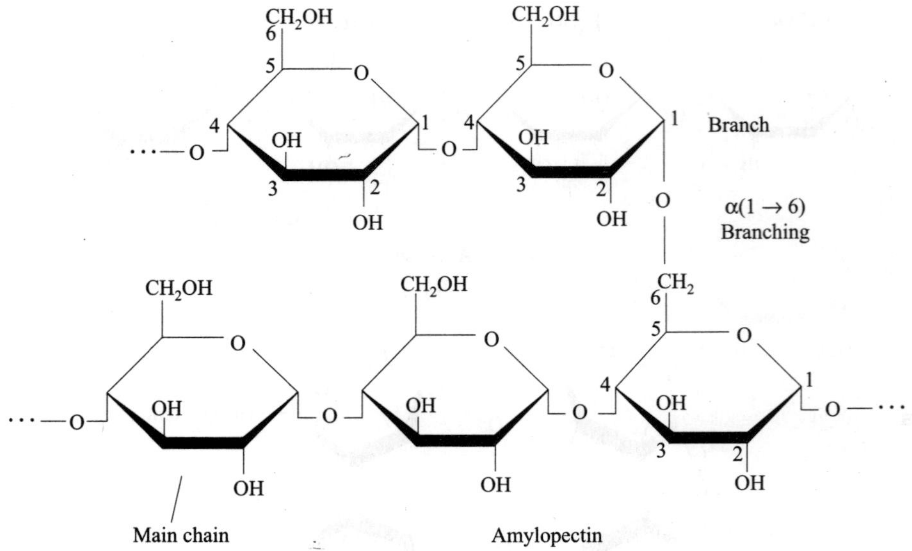

(c)

(d)

Fig. 4.10 Structures of α-amylose and amylopectin.
(a) Glucose repeating units in linear-chain polysaccharide α-amylose
(b) Irregular helical-coiled conformation of α-amylose
(c) Branched-chain polymer amylopectin.
(d) Bush-like structure in amylopectin and glycogen

strength to cellulose fibres and makes them water insoluble (Fig. 4.11b). In plants, the cellulose fibres are embedded in a lignin matrix, thereby adding additional strength to plant cell walls to withstand large stresses. Vertebrates lack the enzyme capable of hydrolysing the β (1 → 4) glycosidic linkages of cellulose.

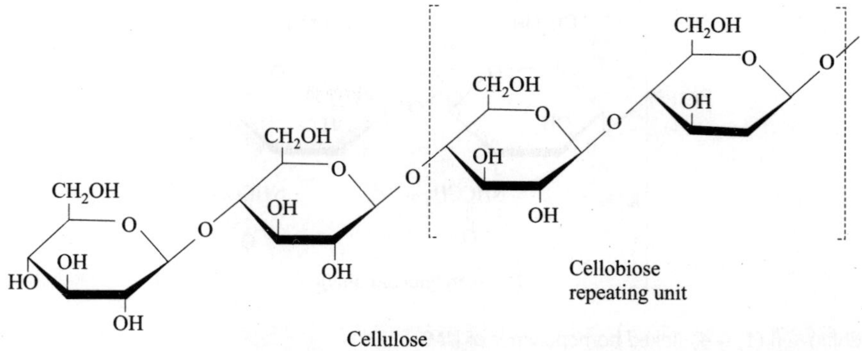

Fig. 4.11a Structure of cellulose.

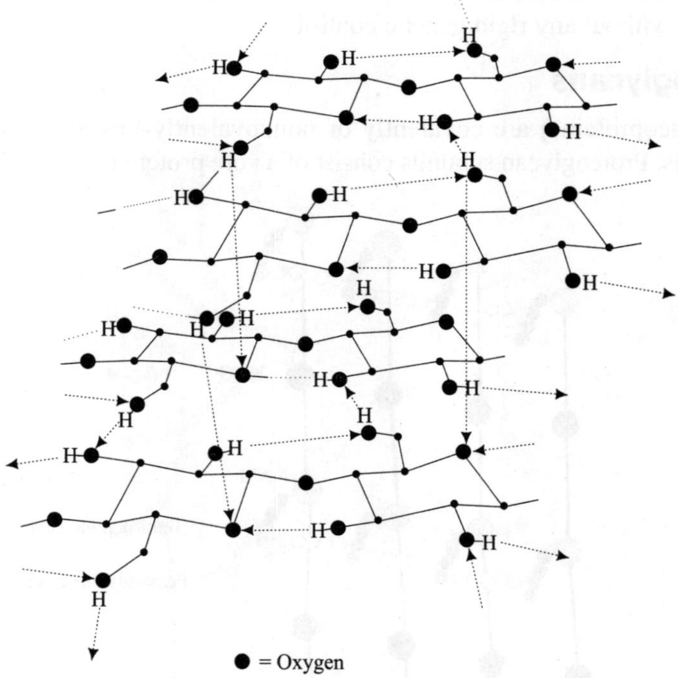

Fig. 4.11b Intra-chain hydrogen bond network in cellulose.

Chitin is a structural component of crustaceans, insects and spiders. It is a homopolymer of $\beta(1 \rightarrow 4)$-linked N-acetyl-D-glucosamine residues. The structure is similar to cellulose (Fig. 4.12).

4.4 GLYCOPROTEINS

Glycoproteins are proteins with covalent or noncovalent attachment of carbohydrates with variable carbohydrate content. Glycoproteins are abundant and span the gamut of protein functions—enzymes,

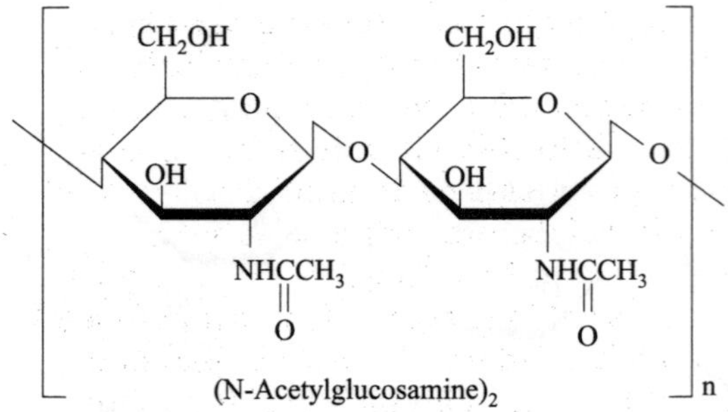

Fig. 4.12 Chitin—β (1 → 4)-linked homopolymer of NAG.

transport proteins, hormones and structural proteins. The polypeptide chains of glycoproteins are template-controlled, but their carbohydrate chains are enzymatically-controlled and linked to the polypeptide chains without any rigid genetic control.

4.4.1 Proteoglycans

Proteoglycans (mucoproteins) are covalently or noncovalently-linked aggregates of proteins and glycosaminoglycans. Proteoglycan subunits consist of a core protein to which glycosaminoglycans are

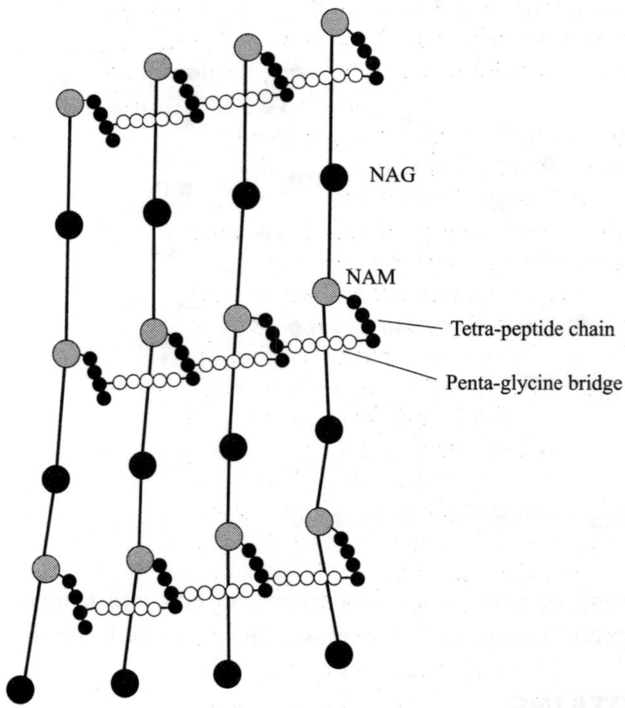

Fig. 4.13a Basic structural unit of a peptidoglycan.
NAM = N-acetylmuramic acid.
NAG = N-acetylglucosamine.

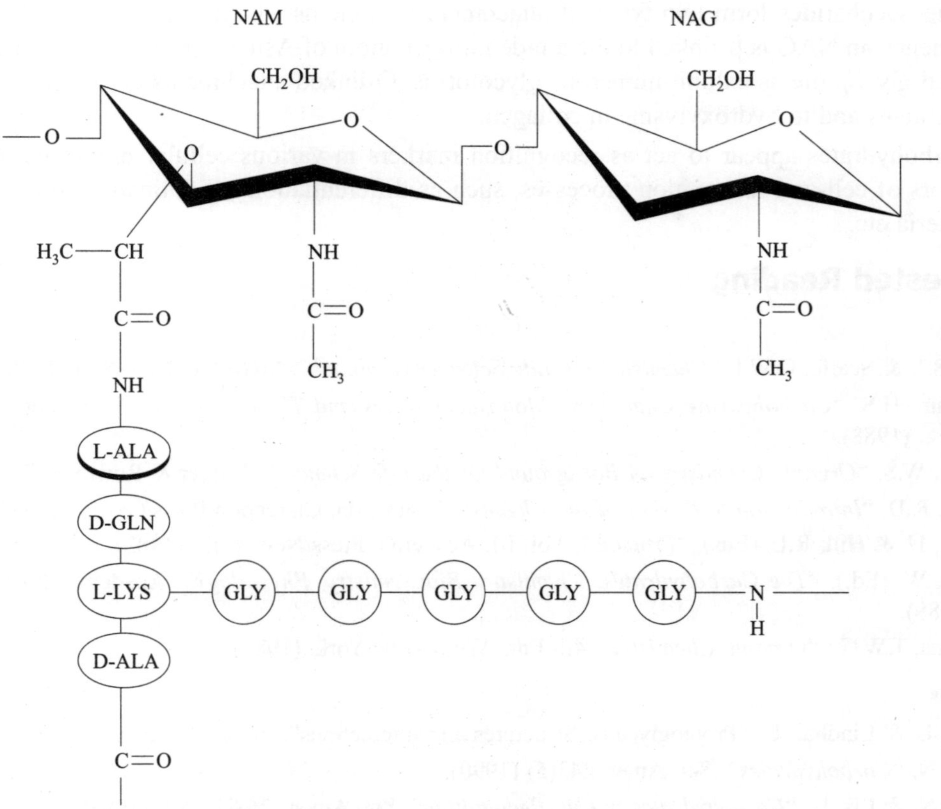

Fig. 4.13b Schematic representation of peptidoglycan.
The repeating unit is an NAG-NAM disaccharide. The tetrapeptide chain is linked covalently to the NAM sugar moiety *via* a lactyl side chain and to the adjacent tetrapeptide *via* a pentaglycine bridge.

attached. The extracellular spaces, such as cartilage, tendon and blood vessel walls, consist of collagen and elastin fibres embedded in a gel-like matrix composed largely of proteoglycans. In cartilage, for example, the core protein components and chondroitin sulphate interact with collagen. The tensile strength of cartilage is due to its collagen content and its resilient characteristics are due to proteoglycan content. The extended brush-like structure of proteoglycans, together with polyanionic properties, causes the complex to be highly hydrated. Under pressure, water is squeezed out from the charged regions and when pressure is released the molecules are again hydrated.

4.4.2 Peptidoglycans

The cell walls of bacteria consist of covalently-linked polysaccharide and polypeptide chains. Its polysaccharide component consists of linear chains of alternating β $(1 \rightarrow 4)$-linked N-acetylglucosamine (NAG) and N-acetylmuramic acid (NAM) (Fig. 4.13). The lactic acid moiety of NAM forms an amide bond with a D-amino acid-containing tetrapeptide to form a peptidoglycan repeating unit. Lysozyme catalyses the hydrolysis of the β $(1 \rightarrow 4)$ glycosidic linkages between NAM and NAG.

Oligosaccharides form two types of attachment to proteins—N-linked or O-linked. In N-linked attachments, an NAG is β-linked to the amide nitrogen atom of Asn in the sequence Asn-X-Ser(Thr). N-linked glycoproteins exhibit numerous glycoforms. O-linked attachments are made to Ser(Thr) in most proteins and to hydroxylysine in collagen.

Carbohydrates appear to act as recognition markers in various cellular processes. They act as mediators of cell-cell recognition processes, such as differentiation, agglutination, infection of cells by bacteria etc.

Suggested Reading

Books

Baum, S.J. & Scaife, C.W.J. "*Chemistry—A Life Sciences Approach*". McMillan/New York. (1980).

Elkhadam, H.S. "*Carbohydrate Chemistry: Monosaccharides and Their Oligomers*". Academic Press/New York. (1988).

Gardner, W.S. "*Organic Chemistry: A Background for the Life Sciences*". Harper & Row/New York. (1975).

Guthrie, R.D. "*Introduction to Carbohydrate Chemistry*", 4th Edn. Clarendon Press/Oxford. (1974).

Neurath, H. & Hill, R.L. (Eds). "*Proteins*", Vol. III. Academic Press/New York. (1982).

Pigman, W. (Ed.). "*The Carbohydrates: Chemistry, Biochemistry, Physiology*". Academic Press/New York. (1988).

Solomons, T.W.G.: "*Organic Chemistry*", 4th Edn. Wiley/New York. (1988).

Articles

Kjellen, L. & Lindhal, U. "Proteoglycans: Structures and Interactions". *Annu. Rev. Biochem., 60; 443. (1991).*

Sharon, N. "*Carbohydrates*". Sci. Amer. 243(5) (1990).

Sharon, N. & Lis, H. "*Carbohydrates in Cell Recognition*". Sci. Amer., 268(1); 82. (1993).

Vyas, N.K. "*Atomic Features of Protein-carbohydrate Interactions*". Curr. Opin. Str. Biol., 1; 732. (1991).

5 Lipids and Membranes

Synopsis: Lipids are heterogeneous organic compounds that are important constituents of animal and plant tissues. Lipids contain fatty acids, which are of long-chain (approx. 16 to 18 C-atoms), that may be saturated or have one or two double bonds (usually in *cis* configuration). Membrane lipids contain phospho- and glycolipids and cholesterol. They are amphipathic molecules and form lipid bilayers in aqueous media and act as highly impermeable barriers to ions and polar molecules. Biological membranes have membrane proteins embedded in the lipid bilayer, either peripherally or integrally. Lipid and membrane proteins are organised asymmetrically in biological membranes which facilitates the vectorial processes of membrane-mediated functions. While the lipid bilayer provides the structural framework, acting as a barrier to most water-soluble molecules, membrane proteins mediate specific functions, like mass transfer, charge transport, and signal transduction. Lipids and proteins can diffuse freely along the plane of the membrane (lateral diffusion). In the plasma membrane of eucaryotic cells, most of the proteins exposed on the cell surface have oligosaccharide chains covalently attached to them. The function of these carbohydrates is presumed to be in cell-cell and cell-matrix recognition processes.

5.1 LIPIDS—CLASSIFICATION AND STRUCTURE

Lipids are heterogeneous organic compounds that are important constituents of animal and plant cells. Lipids are the most important energy storage compounds in animal cells, in contrast to plant cells which store their energy in the form of starch.

Lipids are not polymers—they do not have repeating monomeric units as is the case with proteins and nucleic acids. Fats, oils and nonprotein membrane components are lipids. They can be classified under: i) simple lipids, ii) compound lipids and iii) steroids.

5.1.1 Simple Lipids

Simple lipids are a) fats and oils, which yield fatty acids and glycerol and b) waxes, which yield fatty acids and long-chain alcohols, upon hydrolysis. Fats and oils are triglycerols—esters with three fatty joined to a glycerol (a trihydroxy alcohol) (Fig. 5.1).

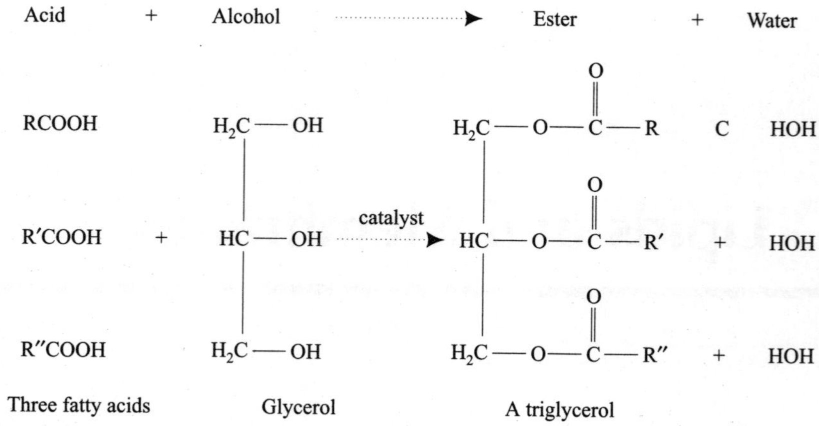

Fig. 5.1 Chemical formulae of glycerol and a fatty acid.

Fatty acids are long-chain (hydrocarbon chain) monocarboxylic acids (Fig. 5.2).

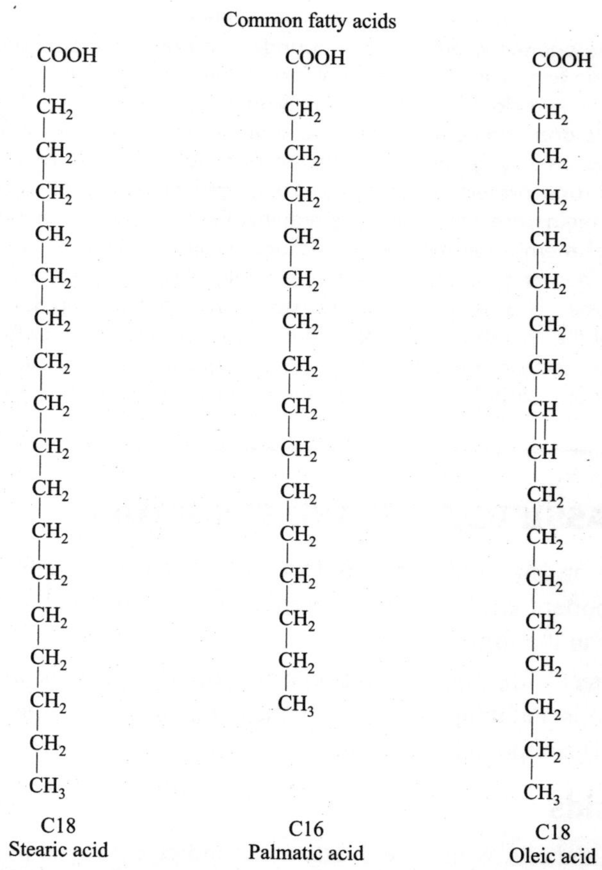

Fig. 5.2 Chemical formula of a general fatty acid.

Fatty acids are either saturated or unsaturated and the physical properties of fatty acids vary with their degree of unsaturation. Some common fatty acids are mentioned in Table 5.1.

Saturated fatty acids are highly flexible molecules which lead to a wide variety of conformational states of the hydrocarbon chains, due to free rotation about their single C-C bonds. The double-bonds of unsaturated fatty acids are always in *cis*-configuration (Fig. 5.3).

Fats are solids at room temperature due to greater content of saturated fatty acids (of general formula: $C_nH_{2n+1}COOH$; n = 11, 13, 15, 17 ...), whereas oils are liquids at room temperature due to

Table 5.1 Common fatty acids

Name	*Systematic name*	*Number of double bonds*	*Condensed structural formula*	*Abbreviated formula*
Saturated fatty acid				
Lauric acid	n-Dodecanoic	–	CH_3 $(CH_2)_{10}COOH$	$C_{11}H_{23}COOH$
Myristic acid	n-Tetradecanoic	–	$CH_3(CH_2)_{12}COOH$	$C_{13}H_{27}COOH$
Palmatic acid	n-Hexadecanoic	–	$CH_3(CH_2)_{14}COOH$	$C_{15}H_{31}COOH$
Stearic acid	n-Octadecanoic	–	$CH_3(CH_2)_{16}COOH$	$C_{17}H_{35}COOH$
Unsatured fatty acid				
Oleic acid		one	CH_3 $(CH_2)_7$ CH = CH $(CH_2)_7$ COOH	$C_{17}H_{33}COOH$ or $C_nH_{2n-1}COOH$; n = 17
Linoleic acid		two	$CH_3(CH_2)_4CH$ = $CHCH_2CH$ = $CH(CH_2)_7COOH$	$C_{17}H_{31}COOH$ or $C_nH_{2n-3}COOH$; n = 17
Linolenic acid		three	CH_3CH_2CH = $CHCH_2CH$ = $CHCH_2CH$ = $CH(CH_2)_7COOH$	$C_{17}H_{29}COOH$ or $C_nH_{2n-5}COOH$; n = 17
Archidonic acid		four	$CH_3(CH_2)_4$ $(CH = CHCH_2)_4$ $(CH_2)_2$ COOH	$C_{17}H_{27}COOH$ or $C_nH_{2n-7}COOH$; n = 17

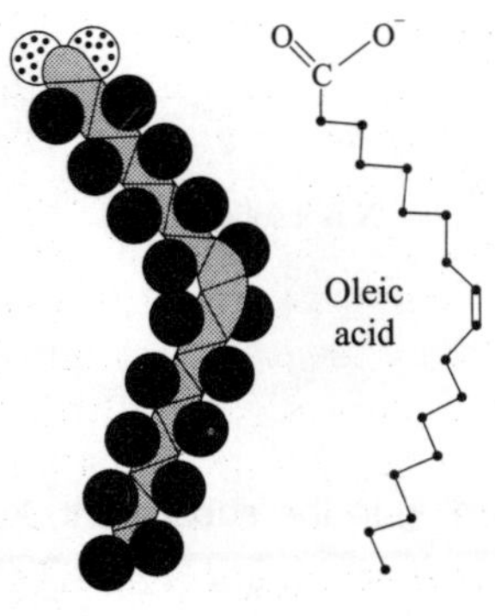

(a) Unsaturated

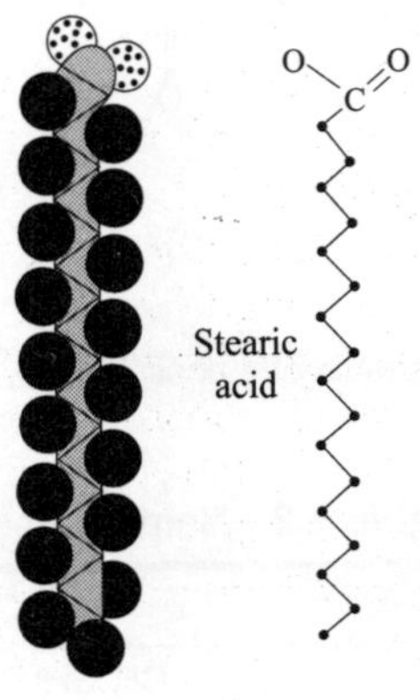

(b) Saturated

Fig. 5.3 Unsaturated and saturated fatty acids—configuration.
⊛ = Oxygen; ● = Hydrogen.
Double bonds introduce kinks and rigidity in those regions of the chain.

a greater percentage of unsaturated fatty acids (of general formula: $C_nH_{2n-m}COOH$, where n = 17; and m = 1, 3, 5 ... etc).

5.1.2 Compound Lipids

Compound lipids are: (i) phospholipids, which yield fatty acids, glycerol, phosphoric acid and a nitrogen-containing alcohol; (ii) glycolipids, which yield fatty acids, glycerol or sphingosine and a carbohydrate and (iii) sphingolipids, which yield fatty acids, sphingosine, phosphoric acid and an alcohol, upon hydrolysis.

Phospholipids. Phopholipids are derivatives of glycerolphosphate. They are esterified at the C1 and C2 positions to fatty acids (common fatty acids are 16 and 18 C-atom chains) and the C3 position to the fatty acid. The phosphoryl group is esterified to a substituent, X (Fig. 5.4).

Phospholipids are the major lipid components of biological membranes. They are found in liver, brain, spinal tissues and the outer membrane of most cells. Phospholipids (phosphoglycerides) are amphiphilic molecules with hydrophobic aliphatic 'tails' and polar phospholipid (–X) 'heads'. There are several phospholipids, the most common being phosphatidylcholine. The name of a phos- pholipid is designated according to the substituent, X (Table 5.2).

(i) *Lecithin.* Most common of the phospholipids is lecithin (phosphatidylcholine), whe!e choline is attached to a phosphoric acid moiety via an ester linkage (Fig. 5.5).

(ii) *Cephalins.* In cephalins, choline is replaced either by an ethanolamine or by serine (Fig. 5.6). Phosphatidylserine carries a net negative charge whereas phosphatidylethanolamine is zwitter-ionic at physiological pH.

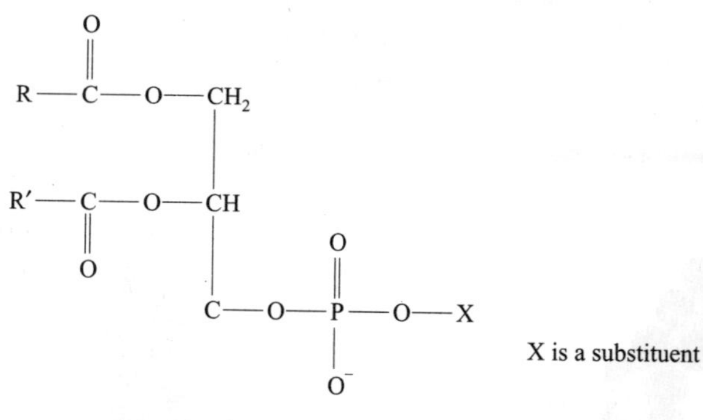

Fig. 5.4 Chemical formula of a phospholipid. X is substituent.

Table 5.2 Name of the phospholipid, according to the substituent, X.

Name	*–X*	*Name of the phospholipid*
Glycerol	$-CH_2CH(OH)CH_2OH$	Phosphatidylglyerol
Choline	$-CH_2CH_2N^+(CH_3)_3$	Phosphatidylcholine (lecithin)
Ethanolamine	$-CH_2CH_2NH_3^+$	Phosphatidylethanolamine
Serine	$-CH_2CH_3(NH_3)^+ COO^-$	Phosphatidylserine

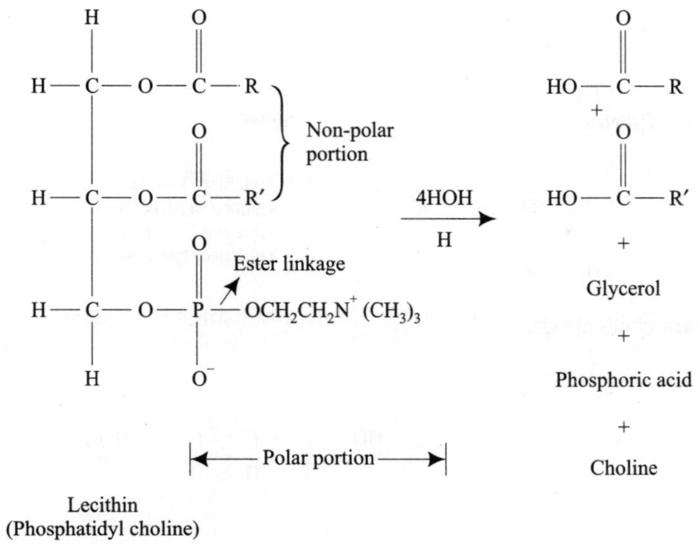

Fig. 5.5 Chemical formula of phosphatidylcholine (lecithin).

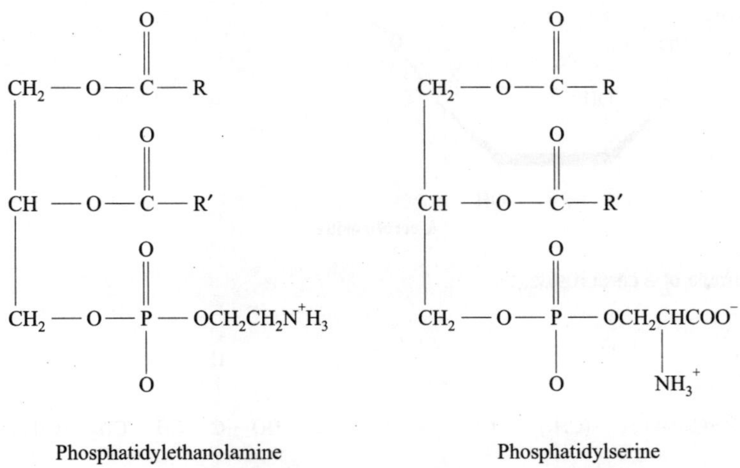

Fig. 5.6 Chemical formulae of cephalins.

Glycolipids. Glycolipids are carbohydrate-containing lipids (Fig. 5.7). The simplest glycolipid is a cerebroside (Fig. 5.8) which can also be classified as a sphingolipid. Complex glycolipids like gangliosides, which are ceramide oligosaccharides, contain a branched chain carbohydrate.

Sphingolipids. Sphingolipids are derivatives of the C18-amino alcohols, sphingosine or dihydrosphingosine (Fig. 5.9). The most abundant sphingolipids are sphingomyelins, which are ceramides with phosphocholine or phosphoethanolamine head groups (Fig. 5.10).

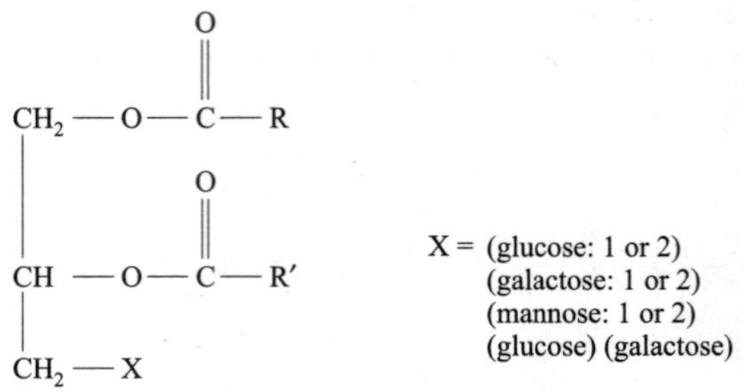

Fig. 5.7 Chemical formula of glycolipids.

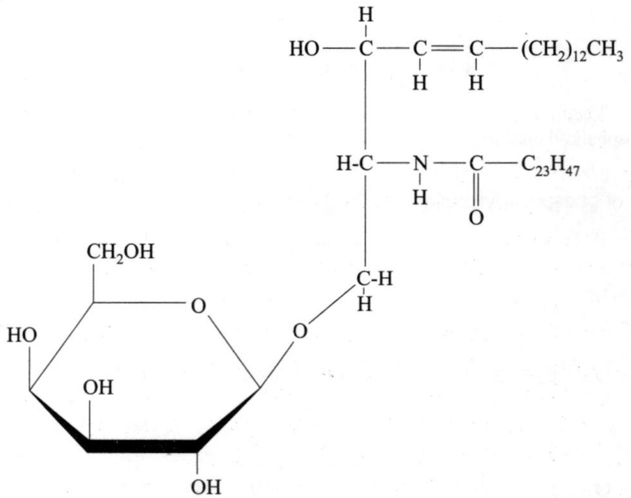

A cerebroside

Fig. 5.8 Chemical formula of a cerebroside.

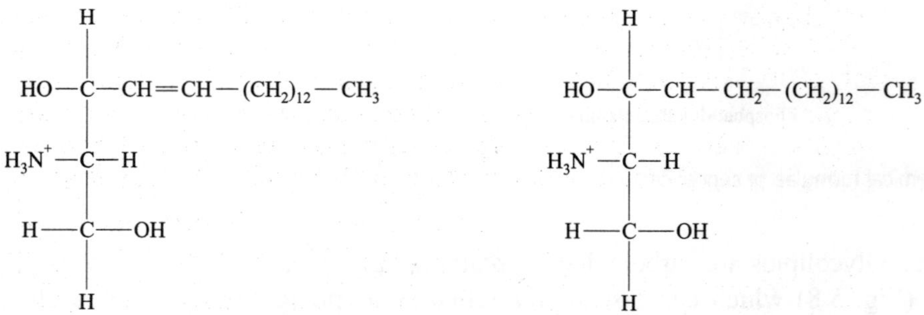

Sphingosine   Dihydrosphingosine

Fig. 5.9 Formula of sphingosine and dihydrosphingosine.

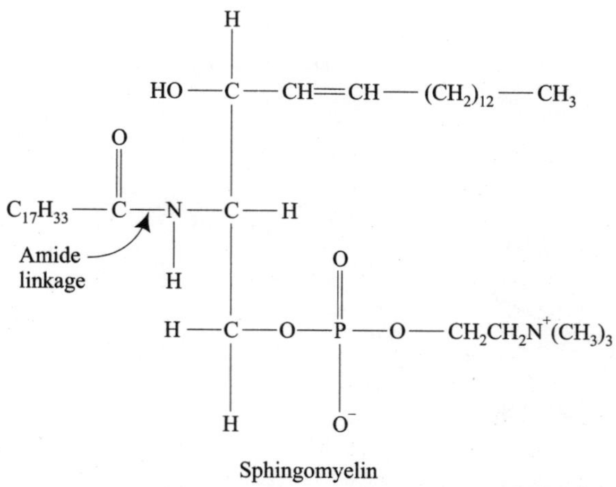

Fig. 5.10 Chemical formula of sphingomyelin.

5.1.3 Steroids

All steroids have a perhydrocyclopentanophenanthrane skeleton, which consists of a completely saturated phenanthrane moiety fused to a cyclopentane ring (Fig. 5.11). Steroids do not undergo saponification. Cholesterol, a sterol, is a major lipid in animal plasma membranes (Fig. 5.12). Meat, eggs, butter etc., are rich in cholesterol. It is a metabolic precursor in the biosynthesis of sterols, hormones and vitamin D. Cholesterol does not occur in plants.

5.2 MEMBRANES

In aqueous solutions, amphiphilic molecules, containing polar and nonpolar moieties, such as soaps and detergents, form micelles (Fig. 5.13). Single-tailed amphiphiles (soap anions) form spheroidal or ellipsoidal micelles.

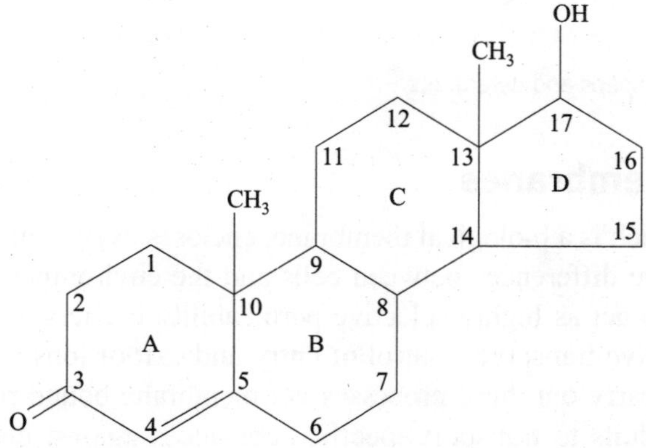

Fig. 5.11 Perhydrocyclopentanophenanthrane skeleton.

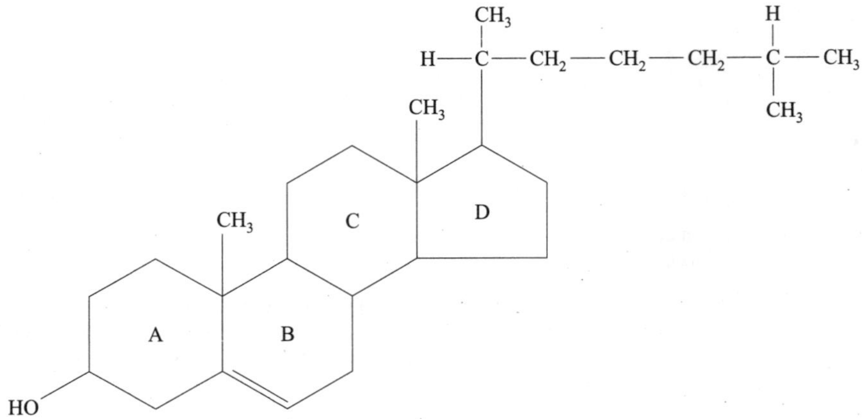

Fig. 5.12 Chemical formula of cholesterol.

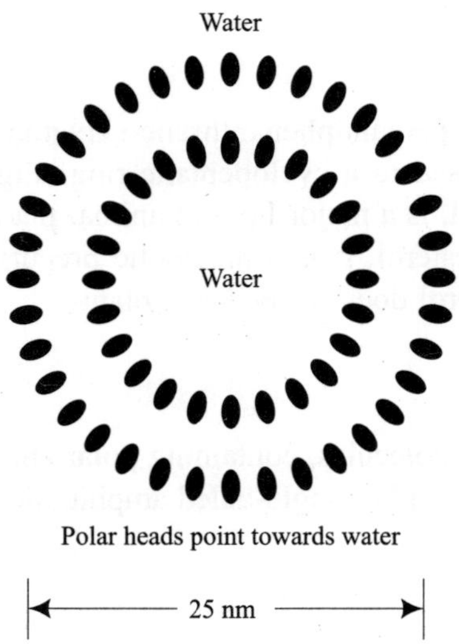

Fig. 5.13 Micelle formation in soaps and detergents.

5.2.1 Biological Membranes

The plasma membrane, which is a biological membrane, encloses every cell, compartmentalising cells and thereby maintaining the differences between cells and the environment. The primary role of all biological membranes is to act as highly selective permeability barriers to water-soluble molecules. They act as devices for active transport, control of entry and exit of ions and molecules, energy and signal transduction. They carry out these processes *via* membrane-bound proteins — '**pumps**' and '**gates**', which act as conduits to transport specific substances against electro-chemical gradients. Membrane lipids, due to distinct polar and nonpolar moieties, tend to form intrinsically bilayered structures. This is because of the thermodynamic stability of amphiphilic lipid which results from exclusion of hydrocarbon regions of fatty acids by water molecules.

There are three classes of lipid molecules in biological membranes — phospholipids, glycolipids and cholesterol, with discernible polar and nonpolar moieties (Table 5.3).

All biological membranes are protein-lipid complexes. The ratio of protein:lipid varies depending on the function of the membrane—from 1:4 (by weight) in myelin, 1:1 in plasma membranes to 3:1 in mitochondrial inner membranes; almost all membranes contain a substantial portion of phospholipids (mostly phosphoglycerides). Nearly all fatty acyl chains found in biological membranes have an even number of carbon atoms (16, 18 or 20). Unsaturated fatty-acyl chains have, in general, one double bond, though two or three double bonds are also found. All such double bonds are in the *cis* configuration. An important difference among phospholipids is the charge carried by the polar head groups at neutral pH. Phosphatidylcholine and phosphatidylethanolamine have no net charge, while cardiolipin, phosphatidylglycerol and phosphatidylserine have a net negative charge and some phospholipids possess a net positive charge at neutral pH. Subtle variations in bilayer structure, membrane fluidity and other characteristics are influenced by the presence and amount of such polar headgroups.

5.2.2 Membrane Fluidity

A lipid bilayer can be considered as a two-dimensional fluid. The degree of fluidity of a lipid bilayer depends on several factors—its composition (chain length and saturation of the hydrocarbon chains etc.), the nature of the distribution of lipid molecules in the bilayer, temperature and solvent etc.

(i) Short hydrocarbon-chain fatty acids, as well as unsaturated hydrocarbon-chain fatty acids are difficult to pack together closely and thereby increase the fluidity and also decrease the permeability of lipid bilayers to small water-soluble molecules. Optimal fluidity of the membrane is biologically very important.

(ii) The assembly of lipids in the bilayer is asymmetric and this asymmetry influences fluidity. For example, lipids that have choline, $(CH_3)_3N^+CH_2CH_2OH$, in their head groups (phosphotidylcholine and sphingomyelin) are in the outer half of the lipid bilayer, whil~ the phospholipid molecules with a terminal amino group (cephalins—phasphatidylethanolamine and phosphatidylserine) are in the inner half of the lipid bilayer. In addition, the fatty acid tails of phosphatidylcholine and sphingomyelin are more saturated than those of cephalin. The asymmetry in head groups as well as the asymmetry in the distribution of hydrocarbon tails makes the inner monolayer more fluid than the outer monolayer of the membrane.

Another noteworthy point is that whenever phosphatidylserine is present (in the inner monolayer), there exists a difference in charge between the two halves of the bilayer, due to negative charge of the phosphatidylserine moiety.

Table 5.3 Polar and nonpolar moieties in membrane lipids.

Lipid	*Polar moiety*	*Nonpolar moiety*
Glycerophospholipid	Phosphorylated alcohol	Fatty acid hydrocarbon chains
Glycolipid	Carbohydrates	One fatty acid and one sphingosine hydrocarbon chains
Sphingomyelin	Phosphorylcholine	One fatty acid and one sphingosine hydrocarbon chains
Cholesterol	OH-group at the C3-position	Cyclopentyl ring

(iii) The lipid bilayer undergoes a phase-transition below a characteristic transition temperature and becomes a gel-like solid. The transition temperature of a bilayer depends on the composition of the lipids. It increases with the chain length and the degree of saturation of the fatty acid hydrocarbon chains. Some of the bilayers may undergo lipid to liquid-crystalline to gel-like solid transitions with gradual decrease of temperature.

5.2.3 Membrane Structure

Biological membrane structures are assemblies of lipid and protein molecules, held together mainly by non-bonded interactions. Many models have been proposed to explain the structure of membranes, compatible with the experimental data. The lipid bilayer structure model of Danielli-Davson-Robertson takes into account the polar and nonpolar characteristics of the lipid constituents, but fails to offer a correct picture of the distribution of the membrane proteins in the lipid bilayer. In this model the membrane proteins are visualised as globules embedded either partially or completely at both outer surfaces of the lipid bilayer. The '**fluid mosaic**' model, proposed by **Singer** and **Nicolson**, offers a better picture of the structure of biological membranes and membrane-proteins (Fig. 5.14), consistent with the experimental data obtained from spectroscopic, X-ray diffraction and freeze-fraction electron microscopic methods.

Salient features of the 'fluid mosaic' model of biological membranes are:

(i) Biological membranes generally form bilayers in aqueous media and form two-dimensional fluids.

(ii) The membrane proteins associated with the bilayer are either peripheral (embedded on the surface of the membrane) or integral (traverse the lipid bilayer).

(iii) Individual lipid molecules as well as membrane proteins can move freely along the plane of the membrane (lateral diffusion). Transverse diffusion (flip-flop) is energetically unfavourable.

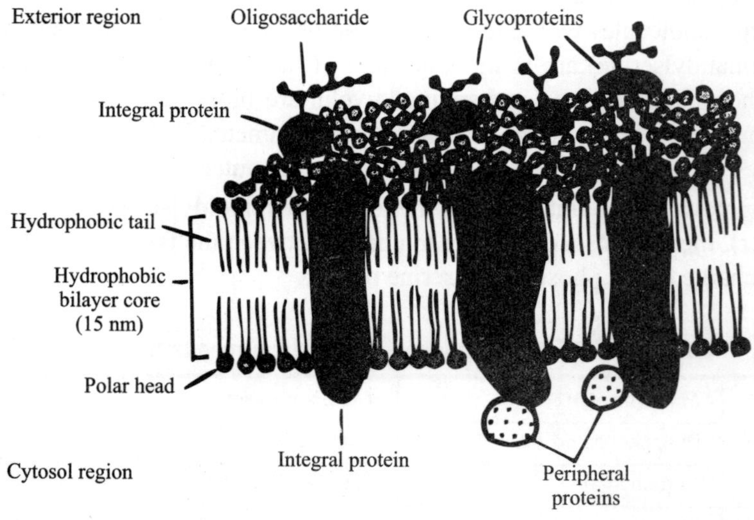

Fig. 5.14 Schematic diagram of the 'fluid mosaic' model of membrane structure, according to Singer and Nicolson. (Science, 175; 720 (1972)).

A variety of physical techniques have been employed to study structural and functional aspects of biological membranes. Whereas X-ray diffraction and electron microscopic studies provide a static picture of membrane structures, the structural aspects of membrane dynamics, temperature and solvent dependence etc., have been studied by fluorescence, NMR, IR, Raman and spin-labelled ESR techniques. These studies have provided qualitative as well as quantitative information on the intra- and inter-lipid and protein motions (translational and rotational), lipid-protein interactions, membrane associations and temperature and solvent-dependent transitions.

5.2.4 Membrane Proteins

Whereas the lipid bilayer determines the basic structure of any biological membrane, acting as a barrier to most of the water-soluble molecules, the membrane proteins associated with the membrane bilayer are responsible for many specific membrane-associated functions—mass transport, energy and stimuli transduction etc. (Refer to Chapter 15 for further details on membrane transport).

Membrane proteins of the bilayer are peripheral, embedded on the surface of the membrane, and integral, traversing the lipid bilayer (transmembrane proteins). Transmembrane proteins traverse the lipid bilayer and as the segments of the polypeptide chain traversing the hydrophobic lipid bilayer have to be charge-neutralised, all the amino groups of the peptide backbone in the lipid bilayer should, perforce, form hydrogen bonds to the carbonyl groups of the peptide backbone. The favoured conformation satisfying such a requirement is α-helical. Moreover, the amino acid residues of the polypeptide segments in the lipid bilayer should have nonpolar side chains. The transmembrane protein can traverse the bilayer once as a single α-helix (single-pass proteins, e.g. cyclophorin) or as multiple-helices (multi-pass protein, e.g. rhodopsin) (Figs. 5.15a and b).

Membrane proteins are also organised asymmetrically in the lipid bilayers. This asymmetric arrangement of membrane proteins in the membrane is crucial, imparting vectorial functional features to transmembrane protein complexes and is necessary for these complexes to act selectively as 'pumps' and 'gates', etc. (Na^+/K^+–, Ca^{2+}–, H^+–ATPases etc.). The majority of transmembrane proteins are glycosylated. As in the case of glycolipids, the oligosaccharide chains are in the extracellular side of the membrane. Further, asymmetry may also be involved in the case of cysteine residues — they exist as cysteine residues as the cytoplasmic side (reducing environment), but form intra- and inter-chain disulphide bridges in extracellular regions (oxidising environment) (Fig. 5.15a).

Proteins can freely diffuse in the lipid bilayer laterally (translational diffusion) and rotationally (rotational diffusion around an axis perpendicular to the plane of the bilayer). Control of membrane protein mobility is influenced by the presence of integral proteins. Regulation of mobility may also occur through interaction between membrane proteins and proteins located on the membrane surface or in the cytoplasm adjacent to the inner membrane surface (e.g. phagocytosis, pinocytosis and receptor movement etc.).

All eucaryotic cells have carbohydrates on their surfaces, both as oligosaccharides and as polysaccharides, covalently bound to lipids (glycolipids). In addition, many plasma membranes contain integral proteoglycan molecules. The carbohydrates in biological membranes are confined to the noncytosolic surface. Similar to the asymmetric distribution of lipid and protein molecules in the lipid bilayer, the distribution of carbohydrate molecules is also asymmetric in biological membranes. Glycolipids are distinguished from one another by their polar head group, which may consist of one

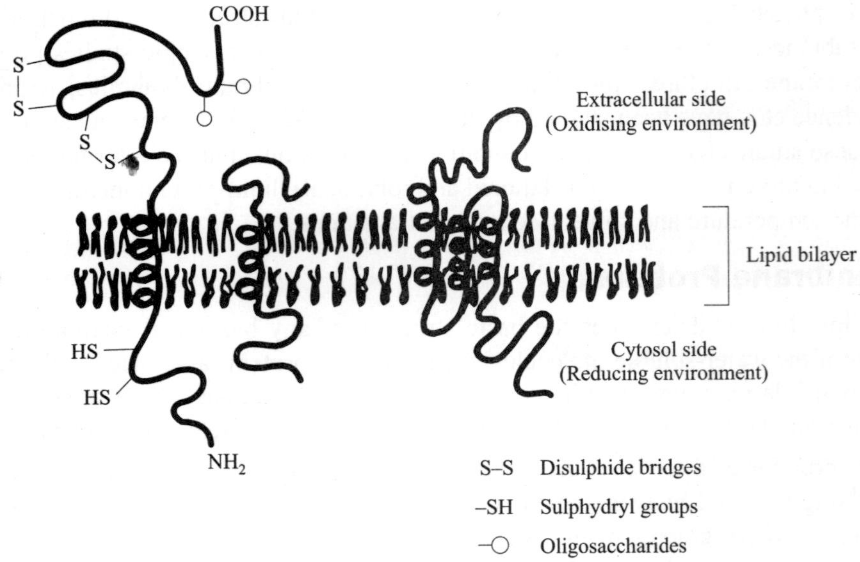

Fig. 5.15a Organisation of transmembrane proteins as single-pass (single α-helical) and multiple-pass (multiple-helical) proteins in biological membranes.
Structural asymmetry exists in the organisation of proteins—e.g. in cysteine-containing proteins, S-S bonds are all on the extracellular and –SH groups on the cytosol region, respectively.

Fig. 5.15b Ribbon diagram of bacterial rhodopsin (7-helix bundle) as an example of multi-pass α-helical transmembrane proteins.*
(*Source:* Chow, K.C. et al. 1 BAD; 7/92).
*Ribbon diagram was generated at the Bioinformatics Centre, Indian Institute of Science, Bangalore, from the PDB coordinates of the indicated source.

or more uncharged sugars. The most complex of the glycolipids are gangliosides, with one or more N-acetylneuraminic acids (NANA), imparting a net negative charge. Gangliosides are most abundant in the plasma membrane of nerve cells. The probable role of glycolipids in cell maybe to serve as receptors for normal signalling between cells.

Suggested Readings

Books

Baum, S.J. & Scaife, C.W.J. "*Chemistry - A Life Sciences Approach*". McMillan/London. (1980).

Ceve, G. & Marsh, D. "*Phospholipid Bilayers: Physical Principles and Models*". Wiley/New York. (1987).

Chapman, D. "*Biomembrane Structure and Function*". Verlagchemie/Berlin. (1984).

Finean, J.B., Coleman, R. & Mitchell, R.H. "*Membranes and Their Cellular Functions*". 2nd Edn. Blackwell Sci/Oxford. (1978).

Gennis, R.B. "*Biomembranes: Molecular Structure and Function.*" Springerverlag/Berlin. (1990).

Harrison, R. & Lunt, G.G. "*Biological Membranes: Their Structure and Function*". 2nd Edn. Blackie & Sons/ Glasgow. (1980).

Hawthorne, J.N. & Ansell, G.B (Eds). "*Phospholipids*". Elsevier/New York. (1982).

Hemming, F.W. & Howthorne, J.N. "*Lipid Analysis*". Bioscientific Pubs/Oxford; UK. (1996).

Jain, M. & Wagner, R. "*Introduction to Biological Membranes*". 2nd Edn. Wiley/New York.

Shapiro, D. "*Chemistry of Sphingolipids*". Herman/Paris. (1969).

Articles

Bloch, K. "*Sterol Structure and Membrane Function*". Crit. Rev. Biochem. 14; 47. (1983).

Bretscher, M.S. "*Membrane Structure: Some General Principles*" Science. 181; 622 (1973).

Bretscher, M.S. "*The Molecules of the Cell Membrane*" Sci. Amer. 253(4); 100 (1985).

Hakimori, S. "*Glycosphingolipids*". Sci. Amer. 254 (4); 44 (1986).

Henderson, R. et al. "*Model for the Structure of Bacteriorhodopsin Based on....*". J. Mol. Biol. 213; 899 (1990).

Israelachvili, J.N., Marcelja, S. & Horn, R.G. "*Physical Principles of Membrane Organisation*". Quart. Rev. Biophys., 13; 121 (1980).

Kjellen, L. & Lindhal, U. "*Proteoglycans—Structure and Interactions*". Annu. Rev. Biochem. 60; 443 (1991).

Lipowsky, R. "*The Conformation of Membranes*". Nature. 349; 475 (1991).

Singer, S.J. & Nicolson, G.L. "*The Fluid Mosaic Model of the Structure of Cell Membrane*" Science. 175; 720 (1972).

Storch, J. & Kleinfeld, A.M. "*Membranes*". TIBS 10; 418 (1985).

Storch, J. & Kleinfeld, A.M. "*The Lipid Structure of Biological Membranes*". Trends Biochem. 10; 418. (1985).

6 Bioinformatics

Synopsis: "Bioinformatics" has emerged as an important trend in modern quantitative biology and medicine, due to the availability of enormous amount of experimental data (due to genome projects and other emerging areas, molecular biology and genetics) and concomitant advances in computer technology to handle storage, retrieval and analysis of complex array of biological systems and cellular and molecular interactions in a systematic and quantitative way (the objectives of Biophysics also!).

6.1 WHAT IS BIOINFORMATICS?

"Bioinformatics" is the generic term coined for the study of, by experimental and computational methods (with greater emphasis towards computational analysis), the information content, its organisation and processing in biological systems (e.g. genes and gene products) and utilization of this "knowledge" towards better diagnostics, therapeutics and design of new drugs and molecular species, tailor-made to desired specifications. It is a "bridge" between biologically related disciplines and extrapolation of information by computational analysis.

Harbinger for the emergence of "bioinformatics" started with simple nomenclatures. It started with "*conformational analysis*" of bimolecules from the experimental quantitative data, obtained (primarily) from X-ray crystallography. Advances in structural and molecular biology and in X-ray crystallography and NMR spectroscopy enabled the extension of this analysis to more complex biomolecules, such as proteins and nucleic acids, and their complexes. The nomenclature was re-christened to "protein-folding problem" to suit the emerging trends. Subsequently, laboratory automation, integration of improved techniques in biological sciences led to accumulation of vast amount of genome sequence information, functional expression and analysis, and other types of experimental data. With biological sciences becoming data-rich, the need for systematic organisation and analysis towards an integrated view of biological sciences necessitated ushering in of computational (in silico) tools for analysis of the experimental data; along with came the need for a better and sought-after nomenclature, "bioinformatics". Notwithstanding the nomenclatures, the basic aims, concepts and procedures are essentially the same—*"conformational analysis"* of biomolecular complexes, i.e., the *"protein-folding problem"*.

6.1.1 Why Bioinformatics?

The considerable "algorithmic" complexity of biological systems requires a vast amount of detailed information at the cellular and molecular levels for their complete description. Availability of three-dimensional (spatial) structural data of biomolecules is vital for understanding their structure-function roles at the molecular level, and to undertake any meaningful molecular design (molecular engineering) tasks. Acquisition of high-thoroughput biological data (e.g. from genomic projects and molecular biology) has been, and is, providing primary structural (sequence) data at a much faster rate than their secondary and tertiary structural data from the existing experimental methods. While single-crystalline state is a prerequisite for initiating X-ray diffraction studies, the size (mass) of the molecule is the limiting constraint for NMR spectroscopic methods. In addition, these experimental methods involve elaborate technical procedures and expertise, and the operational constraints make these techniques for tertiary structure elucidation a challenging and time-consuming task.

As the task of acquiring secondary and tertiary structural data of biological macromolecules by existing experimental methods are still time-consuming, whereas primary structure (sequence) databases are expanding, application of theoretical methodology of "structure prediction"—employing computational tools and algorithms to predict secondary and tertiary structures of biological macromolecules from their primary structure (sequence) data— is the alternative approach, and the major objective of bioinformatics.

In summary, broad objectives of bioinformatics are:

(1) Application of computer software tools for creating computer databases of various kinds.
(2) Development of algorithms to utilise and manage the databases by "knowledge-based" computational analysis in "structure prediction" protocols.
(3) Use of databases and "structure prediction" analysis in the "rational design" of better molecular species/drugs, tailor-made to suit desired requirements (basis of molecular engineering).

6.2 DYNAMICS OF PROTEIN-FOLDING

The first half of the genetic code, namely, how a particular sequence of nucleic acid bases is translated into protein synthesis, has been resolved. Gene sequence is read in triplets of nucleotides (codons) and translated into protein synthesis. That is, there is a direct correlation between a codon (a nucleotide triplet) and the incorporation of a particular amino acid. But, the second half of the genetic code, namely, the principles that govern the folding of a polypeptide (tertiary structure) from its primary structure (linear amino acid sequence) is still a challenging problem. The task of predicting the tertiary structure of a protein, *a priori* from its amino acid sequence data is referred to as the "*protein-folding problem*". Why some sequences fold into ordered segments, such as helix and sheet, while others fold into turns and loops and how these secondary structural elements fold into final tertiary structures are still unresolved issues. Precise knowledge of protein-folding is very important, not only to understand structure and function of the protein but also to establish its phylogenetic status (evolution) and survival probability and its viability in the evolutionary chain. A direct relationship exists between protein folding and genetic diseases at the molecular level. Since the genetic code in the mRNA template is "read" sequentially in triplets of nucleotides (codons) and translated into synthesis of proteins, any defects manifest in nucleotides or shift (frame-shift) in reading will result in synthesis of an altered protein, which is the basis for mutation. While genetic mutations are fundamental to evolution, they are also the reason for genetic (hereditary) diseases. Errors arising out of defective folding, due to mutations at crucial sites and due to chemical and environmental factors can lead to improper functioning of proteins that is manifest in related genetic diseases, some of which can be lethal. Other errors are retained for their selective advantage because mutants are the raw material for evolution. To quote few examples: the characteristic repeating unit of collagen monomers is $(Gly\text{-}X\text{-}Y)_n$, with glycine residues crucial in the formation of collagen helices. So a single-point mutation that results in the replacement of glycine by any other amino acid residue should lead to pathological disorders. Indeed, this is the case in the most common pathological collagen disease, *osteogenesis imperfecta* (OI) or "brittle bone" disease, a clinical syndrome characterised by fragile bones. Defects at any of the levels of collagen formation would result in genetic disorders (Table 6.1a).

Sickle-cell anemia is due to single-point mutation (single-nucleotide polymorphism, SNP) in the 6th position of β-chain of hemoglobin, with the substitution of hydrophobic valine residue in place of glutamic acid, which is acidic (Table 6.1b). This change of a single amino acid residue at the crucial position alters the packing of hemoglobin tetramer in such a way that the deoxygenated protein (dHbS) polymerises and precipitates within the erythrocytes, leading to characteristic sickle shape. Shortening or lengthening of polypeptide chains would lead to host of genetic disorders, β°-thalassemia is an extreme case, where the 17th residue codon (AAG) for lysine mutates to UAG ("stop" codon), leading to the termination of the synthesis of β-chains. Similarly, single-point mutation with substitution of critical threonine residue by lysine in antithrombin results in thrombosis. A single-point mutation in rhodopsin (Hisà Pro) at position 23 leads to retinal degradation and blindness.

Conformation changes may lead to diseases (*conformational diseases*). In cystic fibrosis transmembrane-regulator (CFTR) protein, a single-point mutation, leading to the deletion of the critical phenylalanine residue at position 508, leads to misfolding of the protein and its inability in *protein trafficking* function. The prion diseases— bovine spongiform encephalopathy (BSE) and Cruetzfeld-Jakob disease (CJD)— are transmitted by an induced conformation change, propagated from an abnormal conformer to its normal counterparts.

In many neuro-degenerative diseases, such as Alzheimer's, Parkinson's, Huntington's, agglutination of soluble proteins as fibres (*amyloids*) is related to tri-nucleotide repeat expansion (TNRE), that is, to critical number of particular amino acid residue (e.g. glutamic acid). Therefore, a clear understanding of the principles of protein-folding is, in essence, fundamental to our understanding of the structure-function aspects of proteins, defects and diseases, life processes and evolution.

At first the protein-folding problem appears simple and straightforward. All the structural information necessary for proper three-dimensional folding is inherent in the primary structure data. The molecular forces involved in stabilising protein-folding are also well known—non-bonded interactions (salt bridges, hydrogen bonds and van der Waals forces). All the same, the protein-folding

Table 6.1 Some genetic disorders in proteins: (a) Collagen, (b) Hemoglobin.

(a) Collagen

Disorder	*Defect*	*Manifestations*
Scurvy	Decrease in hydroxyproline	Decrease in growth; capillary fragility
Osteogenesis imperfections	Decrease in type I collagen	Bone deformities
Ehlers-Danlos Syndromes:		
Type III	Decreased cross-linking	Hyper-extensibility of skin and joints.
Type IV	Decrease in type III collagen	Bruised skin.
Type VI	Decreased hydroxylysine	Musculoskeletal deformities.
Type VII	Amino-terminal peptide	Bruised skin; hip dislocation.

(b) Hemoglobin

(i) α-Chain

Species	*α-codon*	*Residue at 142*	*Residues & Disease*
Normal α-chain	UAA	Stop	141 (Normal)
Constant Spring	CAA	Gly	172 (α-Thalassemia)
Icaria	AAA	Lys	172 " "
Seal Rock	GAA	Glu	172 " "
Koya Dora	UCA	Ser	172 " "

(ii) β-Chain

Change at	*From*	*To*	*Disease*
6th-position	Glu (GAA or GAG)	Val (GUA or GUG)	Sickle cell anemia.
6th-position	" "	Lys (AAA or AAG)	Hemoglobin-C.
Residue 145 (McRees Rocks)	Tyr (UAU or UAG)	Stop (UAA or UAG)	144 residues only instead of 146.
Residue 17	Lys (AAG)	Stop (UAG)	β°-Thalassemia (no β-chain).

A = Adenine; G = Guanine; C = Cytosine and U = Uracil.

problem is highly complex, due to the cooperativity of molecular interactions. The protein-folding problem can be analysed under i) structure prediction (theoretical), ii) kinetic (experimental) and iii) phylogenetic (evolution) methods.

6.2.1 Structure Prediction Methods

Prediction of secondary and tertiary structures of proteins from their primary structure data can be attempted at several levels and from various perspectives: incorporating 'knowledge' of protein chemistry and empirical methods.

Influence of solvent and side chains on protein-folding. The parameters that influence protein-folding are solvent, pH and side chains etc. As most polar groups in proteins are bound to the solvent, there appears to be little inherent stability imported to ordered segments in solution. Although intermediate hydrogen bond formation itself may not contribute much to the overall stability of the molecule, if, as a result of hydrophobic interactions, the amide linkages are constrained to exist in the interior of folded regions (in hydrophobic environs), isolated from the solvent, there exists a strong driving force for the formation of ordered structures. That is, a protein assumes a structure in which the polar, nonpolar interactions and the hydrogen bond networks maximise simultaneously.

Proteins and enzymes utilise to their advantage a change of environment to alter their conformations. Structural changes occur after substrate binding to an enzyme; e.g. glucose binds in a cleft between lobes of hexokinase and upon binding the cleft closes and glucose is buried. Similarly, in DNA polymerase I (Pol I), the larger domain of the ***Klenow*** fragment has a cleft ~20 Å wide and ~30 Å deep and upon the binding of the ligand, B-DNA in the cleft, a flexible small domain of the enzyme closes down '**ontop**' of the DNA bound to the cleft, thereby allowing processive nucleotide polymerisation without enzyme dissociation. In carboxypeptidase **A**, a tyrosine side chain moves 12 Å apart after substrate binding. There are numerous examples of structural changes in protein-substrate, protein-protein, protein-nucleic acid interactions. This is a normal occurrence in macromolecular-ligand interactions.

Self associations increase dramatically in solvents of poorer hydrogen bonding capacity. For example, Poly-γ-benzyl-L-glutamic acid adopts a helical conformation in chloroform and dioxane, but exists as a random coil in solvents with hydrogen bonding capacity. Electrostatic factors also play an important role in the conformational stability of molecules. For example, at neutral pH, poly-L-glutamic acid exists as a random coil but is helical in acidic pH, because at neutral pH, the carboxyl groups are neutral and, therefore, helical conformation is feasible. Similarly, polylysine is a random coil at pH ~7.0 (due to its ionised state) and is helical at pH ~12.0 (due to its neutral state). Polyalanine is α-helical at neutral pH, because it is uncharged. Polyisoleucine fails to form an α-helix, even though it is hydrophobic, because of its bulky substituent next to the α-carbon group. That is, for proper conformation of proteins, the size and shape of the residues as well as electrostatic charge are important factors.

Interiors of proteins are entirely apolar, but contain mainly the polar N-H and C=O groups of the backbone. Therefore, α-helices, formed by nonpolar amino acids and existing in interiors of proteins, have residual polarity due to the alignment of N-H and C=O dipoles around the α-helices. The hydrogen bonds in the α-helix point in the same direction *i.e.*, along the helix axis. Therefore, there is a net dipole moment due to partial positive charge at the amino end and partial negative charge at the carboxyl end of the α-helix. As the ends of the α-helices are polar they are almost always found at the surface of the protein molecules.

Most amino acids play more than one structural and functional roles. On the basis of size and shape, glycine is the smallest amino acid with only H-atom for a side chain. This allows glycine to have a greater conformational flexibility and fit in where other residues will be too bulky. Proline lacks a free amide hydrogen group that prevents main-chain hydrogen bonding. It has additional structural constraint on the backbone relative to any other amino acid. While both glycine and proline are helix disrupters, proline is also a sheet-breaker, which is an important factor in globular proteins.

Side chains have an effect on the conformation of polypeptides. Proline is a helix- and sheet-breaker; it is the strongest helix breaker in the backward direction and therefore, usually appears at the trailing end of the α-helix (amino side). There is a correlation between the numbers of proline residues present in a protein and the type of structure likely to occur—for ~3% proline ≈50% α-helix, for ≈8% proline random coil, and above 10% proline content leads to left-handed poly-L-proline type helix. Similarly, several amino acid residues exert an asymmetrical influence between the carboxyl and amino terminal polypeptide segments. For example, aspartic acid residues tend to terminate helices in the backward direction and lysines break helices in the forward direction (on the carboxyl side). Threonine and valine occur frequently at the amino terminal region of helical segments. The preponderance of certain amino acids at termini of helical segments could be understood by invoking the asymmetrical impact of each side chain on the polypeptide backbone.

Several amino acid residues such as **Ala**, **Glu**, **Leu** and **Met** are preferred in α-helices, while other residues such as **Gly**, **Pro**, **Tyr** and **Ser** are poor α-helix formers. Residue type as well as conformation influence helix propagation. Amino acids with small size side chains (**Gly** and **Ala**) can be located either inside or outside the protein interior. Neutral polar residues (**Cys**, **Asn**, **Gin**, **Ser** and **Thr**) are found at the surface of, as well as inside, the protein. Formation of secondary structure largely relies on hydrogen bond networks and, therefore, burying of polar groups inside proteins without loss of free energy, is achieved only if they form hydrogen bonds. **Ser** and **Thr** preferentially hydrogen bond to backbone -NH groups. **Asn** and **Gin** act as donors as well as acceptors of hydrogen bonds. At pK ~10.0 the polar -OH group of tyrosine can dissociate. Therefore, **Tyr** in such environments behaves like a charged side chain (weakly acidic). **Asp** and **Glu** are ionised (negative) at pH = 7.0 and because of the short side chain, the carboxyl group of **Asp** is rigid with respect to the main chain. Positively charged **Lys** and **Arg** have long, flexible side chains and they do not adopt defined and rigid conformations. They also increase the solubility of a protein and take part in salt bridges. **Leu**, **Ile**, **Phe**, **Pro**, **Val** and **Trp**, and to a lesser extent **Met**, are predominantly found inside protein molecules. Short range interactions lead to helix-forming or breaking. So, residues with more than one β-substituents **(Val)** and those with short polar side chains **(Ser)** have helix-breaking propensities. The role of S-S bridges is both structural and functional and S-S bridges are common in proteins that operate in extracellular regions—toxins, hormones, digestive enzymes, immunoglobulins, milk proteins, Iysozymes etc. Very few intracellular proteins possess disulphide bridges, and where they are found, they have functional roles. S-S bridges are found as integral parts of structural motifs in disulphide-containing proteins.

Reverse turns exhibit structural and physicochemical characteristics.

(1) Reverse turns are polar due to the presence of NH and C=O groups; so usually they are situated at molecular surface (boundary separating protein and solvent), in contact with the solvent, water.
(2) Backbone and side chain hydrogen bonds are disrupted in turn regions.
(3) Solvent participation is an important influence in turn stabilization.
(4) Regions adjacent to the reverse turn have many hydrophobic residues.
(5) Turns in a hydrophobic environment (e.g. interior of proteins) are seldom. In such cases, bound water molecules and hydrogen bonds between side chains and the polypeptide backbone are involved in neutralizing the hydrophilicity.

Reverse turns and loops play an important role in determining conformation and specificity of enzyme-active sites and antibody-antigen interactions. Many of the water bridges (solvent) appear to play a crucial role in stabilising and maintaining proper folding of reverse turns and loops, which would otherwise be more flexible.

Protein-folding Rules. Protein-folding rules are relatively straightforward in the case of fibrous proteins. Many of them are helical and in many cases several monomers entwine to form coiled-coil helical cables. In some special case, where only a few amino acid residues form repeating units, the deduction of tertiary structures from their primary structure data becomes simplified. The collagen molecule is one such example. The repeat sequence in collagen monomers is **(Gly-X-Y)**$_n$ (where **X**

and **Y** are generally proline and 4-hydroxyproline, respectively), and glycine in the first position cannot be replaced by any other amino acid residue due to steric restraints. This reduces the folding problem in collagen, essentially, to a 'two amino acid residue' problem. But protein-folding rules, which are simpler in the case of fibrous proteins, cannot be, a priori, extended to predict secondary and tertiary structures of globular proteins from their amino acid sequence data, because (i) all of the twenty amino acids can be involved in the construction of polypeptide chains (no longer a few-residue problem and no repeating units) and (ii) globular proteins are not linear, the direction of backbone folding changes many times and the structures are compact and globular. Therefore, the prediction of protein-folding in globular structures is attempted at several levels, incorporating protein-folding rules, structural dynamics and experimental and other **'knowledge'**.

Secondary Structure Prediction. There are numerous empirical and statistical methods of predicting secondary structures of proteins from amino acid sequence data and a commonly applied one is the Chou-Fasman method. The method is based on the propensities of amino acid residues towards helix forming/breaking, sheet forming/breaking and turns and loops. The frequency, f_α, of a residue occurring in an α-helix in a set of protein structures is

$$f_\alpha = \frac{n_\alpha}{n} \qquad \text{...(6.1a)}$$

where n_α is the number of amino acid residues of a particular type (say Ala) that occurs in α–helices and n is the total number of residues of this kind (Ala) in the total set of proteins. The propensity of a particular amino acid residue occurring in a α-helix is

$$P_\alpha = \frac{f_\alpha}{< f_\alpha >} \qquad \text{...(6.1b)}$$

where $<f_\alpha>$ is the average value of f_α for all twenty residues. From such a propensity chart (Table 6.2) the **Chou-Fasman** rules for secondary structure prediction are:

1. A group of four helix-forming residues (H_α or h_α) out of six contiguous residues will initiate a helix and a group of three β-forming residues (H_β or h_β) out of five contiguous residues will initiate a β-sheet. Any segment that is at least six residues long and ($<P_\alpha>$) > 1.05 is considered to be helical and one with at least three residues long with ($<P_\beta>$) > 1.05 is considered to be a β-sheet.
2. Extend nucleation of the helix and sheet in both directions until sets of tetrapeptides ($<P>$) < 1.0 are encountered.
3. Any segment containing overlapping α and β regions is considered helical if ($<P_\alpha>$) > l($<P_\beta>$) or β-sheet if ($<P_\beta>$) > ($<P_\alpha>$).
4. Reverse turns are located by the criterion ($<P_t>$) > 1.0. Reverse turns can also be predicted from the minimum hydrophobic profiles of the residues along the polypeptide chain.

Whichever procedure is followed, the general features of predictive methods of protein-folding from the sequential comparisons of amino acid residues are to predict regions of ordered segments and turns and loops and formulate further rules to predict how these secondary structural elements fold into a compact tertiary structure. For example, formation of an α-helix in protein folding is

Table 6.2 Chou-Fasman propensity chart for amino acid residues.

No.	Residue	α-helix P_α	Nature	β-sheet P_β	Nature	Reverse turn P_t	Nature
1	Ala	1.39	H	0.79	i	0.66	B
2	Arg	1.00	I	0.94	i	0.95	i
3	Asn	0.78	i	0.66	B	1.56	H
4	Asp	1.06	h	0.66	B	1.46	H
5	Cys	0.95	i	1.07	h	0.19	h
6	Gln	1.17	h	1.00	h	0.98	I
7	Glu	1.44	H	0.51	B	0.74	b
8	Gly	0.63	B	0.87	i	1.56	H
9	His	1.12	h	0.83	i	0.95	i
10	Ile	0.99	I	1.57	H	0.47	B
11	Leu	1.30	H	1.17	h	0.59	B
12	Lys	1.21	h	0.73	b	1.01	I
13	Met	1.32	H	1.01	I	0.60	b
14	Phe	1.11	h	1.23	h	0.60	b
15	Pro	0.55	B	0.62	B	1.52	H
16	Ser	0.72	b	0.94	i	1.43	H
17	Thr	0.78	i	1.33	H	0.46	i
18	Trp	1.03	I	1.24	h	0.96	i
19	Tyr	0.73	b	1.31	H	1.14	h
20	Val	0.97	i	1.64	H	0.50	B

P is propensity for particular structure. H = Strong former, h = former, I = weak former, i = indifferent former, b = breaker and B = strong breaker of secondary structure.

Source: Chou, P.Y. Fasman, G.D. Adv. Enzymol. 47; 45-148. (1978)

Fasman G.D. (Ed). *Prediction of Protein Structure and the Principles of Protein Conformation.* Plenum Press/New York. (1990).

independent of final folding; so an α-helix can be (in general) predicted from the amino acid sequence. α-helix has a periodicity 3.6. So, for α-helices with one face in the protein core and the other exposed to solvent will have conserved hydrophobic residues at positions i, i + 3, i + 4, i + 7 (where i is a residue in a helix) on one face of the helix facing the protein core. Thus, patterns showing such conservation are indicative of α-helical regions. β-strands that are half buried in the protein core will tend to have hydrophobic residues at positions i, i + 2, i + 4, i + 6 … etc., and polar residues at i + 1, i + 3, i + 5, … etc. Conserved glycine or proline suggests a β-turn. Similar attempts can be tried for other secondary structural elements. But, there is a limit for the accuracy of predicting secondary structure. Folding interactions are highly cooperative and treating them as separate entities has no physical meaning and may quite often lead to erroneous conclusions. Knowledge of a few aspects of the structure is not sufficient to predict the whole structure. Tertiary interactions between distant residues in a sequence override the intrinsic conformational tendencies of individual residues to achieve proper folding. That is, during evolution the residues in a protein are chosen, not on the basis of whether they stabilise a particular secondary arrangement, but on the merit of their influence on the overall folding and functions of the protein.

Protein-Taxonomy Procedures. In general, structural information on tertiary folding of proteins can neither be deciphered from their primary structural data nor from the knowledge of the secondary structural elements alone. One of the ways of approaching this complex problem is to classify protein structures by shapes, motifs, subdomains etc., and carry on model-building and other methods based on these shapes and structural motifs and modules. The underlaying premise of these procedures is that proteins with similar sequence will have similar shapes and structures (this need not always be true). In principle and in essence, in the majority of cases the protein-folding problem can be reduced to identifying and characterising common motifs, modules and domains. This method is one or two steps below in the hierarchy of predicting the final structure. Incorporating these additional structural elements, the structural hierarchy in proteins can be organised as:

Primary ⇒ secondary ⇒ motifs & modules ⇒ domains ⇒ tertiary ⇒ quaternary.

The structural basis for this reorganisation is that the three-dimensional folding and packing of proteins rather than their primary structures are conserved during evolution. Even in cases where no obvious sequence homologies may be found, the unknown structures can be considered from the structural motifs, modules and domains. Pattern recognition programmes follow reverse process of sequence analysis. Rather than predict how a sequence will fold, they predict how well a fold will match a sequence. That is, matching of sequence with a given topology rather than search for a topology with a given sequence. Dynamic programming is used to find the best match of the sequence to the pattern of environments found in a given fold.

The number of possible conformations of a polypeptide is too large to be sampled exhaustively. Semiempirical methods (e.g. **Monte Carlo** simulation methods) suggest the possible mechanisms of protein folding. From the available experimental data it appears that protein folding starts by a rapid collapse of the structure from a random-coil state to a random semi-compact globule (molten globule). The folding from semi-compact state(s) to a transition state is a slow process, from which the chain folds rapidly (from the transition state) to the native state.

Overall, the empirical rules that govern the packing that occurs between and among secondary structural elements to form structural motifs, modules, domains and tertiary structure are:

(1) Residues that become buried in the interior of a protein close-pack. That is to say, hydrophobicity is the major factor stabilising protein-protein associations.

(2) Associated secondary structures retain conformations close to the minimum free energy conformations of the isolated secondary structures. That means the conformational angles of the polypeptide backbone and of the side chains lie in the narrowly allowed regions of the Ramachandran map. Therefore, the seconday structures found in proteins interact in a manner that gives the maximum van der Waals energy and induces no appreciable steric strain.

(3) Motifs: modules and topologies play an important role in folding.

(4) Packing interactions stabilise folding and, therefore, tend to be conserved in protein-folding.

(5) In disulphide-containing proteins, S-S bridges are found as integral parts of structural motifs, creating hydrophobic moieties, and also there exists a structural hierarchy of S-S bridges in stabilising structural motifs/moieties which could be of help in improving protein-folding rules in structure prediction.

Protein-taxonomy was initiated by **Rossmann**, **Richardson**, **Chothia**, and **Levitt** and developed further by other groups. The 'combinatorial' method is based on computer modelling by feeding known details of protein chemistry to the computer and arriving at a 'most probable' structure from various 'possible' structures. A similar method, known as the '**knowledge-based**' method, promoted by **Prof. Blundell** is based on identification of analogies in secondary structural elements, motifs etc., between a protein to be modelled and those from known three-dimensional structures. The first step in this method is to find amino acid sequence homology between the test protein and other known proteins. The alignment of amino acid sequences is followed by insertions, deletions and replacements in the test protein and the model is refined either by '**constraints**' procedures (energy minimisation) or on the basis of known analogous three-dimensional motifs, folds, packing and solvent accessibility as 'markers'. Solvent accessibility and hydropathy charts and other information on the individual amino acids are fed into the data bank in computer modelling methods (see Tables 6.3a and 6.3b).

The essence of structure prediction methods is one of pattern matching—matching a sequence with a given topology is more meaningful than matching with a sequence. Therefore, new approaches, such as generating of propensity charts for establishing probabilities of finding each amino acid type in a given environment, followed by dynamical programming to match the sequence to the pattern of environments found in a given fold, and also generating model folds and matching a given sequence to these hypothetical folds, seem to hold greater promise in structure prediction.

Table 6.3a Solvent accessibility of amino acids in proteins.

Residue		*95% buried fraction*[1]	*Relative free energy of the residue in interior to that on the surface relative to Gly*[2]
Als	A	0.38	–0.14
Cys	C	0.40	–0.60
Asp	D	0.15	0.78
Glu	E	0.18	1.15
Phe	F	0.50	–0.60
Gly	G	0.36	0.0
His	H	0.17	0.02
Ile	I	0.60	–0.68
Lys	K	0.03	2.06
Leu	L	0.45	–0.60
Met	M	0.40	–0.65
Asn	N	0.12	0.75
Pro	P	0.18	0.50
Gln	Q	0.07	0.80
Arg	R	0.01	1.40
Ser	S	0.22	0.40
Thr	T	0.23	0.32
Val	V	0.54	–0.55
Trp	W	0.27	–0.40
Tyr	Y	0.15	0.28

(1) From Chothia, C.J. Mol: Biol. 105; 1. (1976).
(2) From Miller, J. et al. *J. Mol. Biol.* 116; 1641 (1987).

Table 6.3b Hydropathy chart of amino acids.

Amino acid	*Single alphabet code*	*Kyte & Doolittle*[1]	*Engelman et al*[2]
Ala	A	1.8	1.6
Cys	C	2.5	2.0
Asp	D	–3.5	–9.2
Glu	E	–3.5	–8.2
Phe	F	2.8	3.7
Gly	G	–0.4	1.0
His	H	–3.2	–3.0
Ile	I	4.5	3.1
Lys	K	–4.0	–8.8
Leu	L	3.8	2.8
Met	M	2.0	3.4
Asn	N	–3.5	–4.8
Pro	P	–1.6	–0.2
Gln	Q	–3.5	–4.0
Arg	R	–4.5	–12.3
Ser	S	–0.8	0.6
Thr	T	–0.7	1.2
Val	V	4.2	2.6
Trp	W	–1.0	1.9
Tyr	Y	–1.3	–0.7

(1) From Kyte J. & Doolittle, R.F.J. *Mol. Biol.* 157; 105 (1982).
(2) From Engelman, D.M., Steitz T.A. & Goldman, A. *Ann. Rev. Biophys. Biophys. Chem.* 15; 321 (1986).

6.2.2 Protein-Folding Kinetics

The protein-folding problem can be treated as a kinetic problem by analysing the pathways of folding. Partially-folded protein intermediates are inherently unstable. The strategy of trapping these intermediates and analysing their structures with the aim of understanding the rules of protein-folding has been followed using biophysical and biochemical methods. **Creighton's** work on bovine pancreatic trypsin inhibitor (BPTI), trapping structural intermediates and analysing S-S bond formation as a way of studying folding pathways, is a method that has been followed by other groups to study folding pathways in other disulphide-containing proteins. General conclusions from these studies are:

(1) The structural elements collapse as a compact unit (*molten globule)* and later reorganize towards the native structure.
(2) The transition state for unfolding has ordered segments largely intact.
(3) Altered regions are exposed to solvent in the transition state.
(4) Folding does not proceed by stepwise acquisition of the native structure.
(5) Partially folded intermediates are inherently unstable.
(6) There is a preferential accumulation of the most stable intermediate in folding pathway, because the productive pathway for refolding leads from this intermediate.

(7) All but rate-limiting steps are readily reversible, so that the initial folding process is rapid.

(8) Inter-conversions of the molten globule state with the fully unfolded state are rapid and non-cooperative, whereas those with the fully folded state are slow and cooperative.

(9) Local sequence alterations and environment can influence the overall structure of proteins.

Folding pathways can also be studied by introducing mutants and analysing the effects of mutations on the kinetics of folding. Engineering mutations in BPTI and other proteins supports the folding model wherein different regions of the polypeptide acquire tertiary structure in a preferred order. Mutations at different sites in a protein have distinct effects on the kinetics and equilibria of different steps in protein-folding pathways. Local sequence alterations and environment can influence the overall structure of proteins.

Protein engineering, or site-directed mutagenesis, is a novel and powerful method which holds a great promise for addressing folding mechanisms, altering chemical behaviour and improving the physical characteristics of proteins. Protein engineering (enzymatic tinkering) brings about changes in the function of the protein by altering amino acid residue(s) at crucial positions. Site-directed mutagenesis techniques have been applied to modify the functions of various proteins/enzymes by rational amino acid replacement (either single mutations cassette mutations) and to study the effects on structural and enzymatic functions of proteins and enzymes. A single amino acid replacement in a protein at a crucial position can have drastic effects on its function without appreciable changes in its folding/conformation. Site-directed mutagenesis, combined with electrophysiological experiments, holds much promise in the detailed analysis of voltage-dependent gatings. These studies emphasise the fact that protein folding is remarkably stable to tinkering but the functional properties of proteins are sensitive to minute changes in geometry (charge, size and environment).

6.2.3 Phylogenetic Methods

Evolution is change in *gene* frequencies resulting, usually, in adaptation to the environment. Principles of evolution are variation, competition and selection. The basic mechanisms for evolution are mutation and recombination. Evolutionary changes, which stem from random mutational events, often alter the function of proteins. Many mutants are fatal and are, therefore, unfit for survival. However, some mutations improve the 'fitness' of the host proteins and they are retained due to their selective advantage (e.g. sickle-cell hemoglobin which confers resistance to malaria). Selection and evolution are governed by the law of natural selection of a genotype on the basis of its phenotypic "fitness". Selection means (i) copying the information of the 'fittest' phenotype for further development and (ii) deleting errors below the threshold level. Self-reproduction in open systems leads to selection, and selection with a finite mutation rate leads, in turn, to evolution.

Evolution and Function. The study of evolutionary aspects of proteins indicates important regions of folding. The basic steps in the evolution of proteins are the modification of side chains, insertion and deletion of amino acid residues, all of which will have bearing on folding. Generally, changes in the interior of proteins are conservative (substitution of apolar side chain by apolar side chain). Amino acid replacements occur mostly on the surface, so that main chain conformations are not drastically affected and more radical substitutions/deletions (replacement of apolar by polar and polar by apolar residues) tend to occur in surface loops and β-bends. The extent of conservative/radical substitutions depend on the relationship between structure and function. Amino acid sequences vary much faster than the chain folds, and packing is conserved in evolution. This chain-fold conservation underscores the fact that folding pathways impose stringent conditions on protein folding. Those proteins which serve a function similar in all organisms manifest strong homology in structure. Comparisons of homologous proteins, therefore, indicate which of the residues in these proteins are invariants (very essential) and which are variants (least significant). The goal of studying protein mutation as way of understanding protein-folding principles is the reconstruction of past events that have given rise to the vast inventory of proteins in existence today. Expression of genetic phenomena in protein structures are effected by several ways. Some of these are:

(i) *Point Mutation*. Changes in homologous proteins can be explained as the result of accumulation of point mutations (by natural drift) (e.g. cytochrome **c**, globins, insulins, and subtilisin). For example, both subtilisins BPN and Carlsberg have the same functional properties and the sequences near the active site are similar, but they differ greatly in their overall amino acid sequences. In ribonucleases the residues around the catalytic site are identical and none of the variant residues are in contact with the substrate. The situation is similar in hydrogenases with the sequences around the reactive residues being identical. A phylogenetic tree of proteases reveals that the fundamental catalytic machinery is virtually identical for all these enzymes, but differences in the substrate-binding regions allow for selectivity of action.

(ii) *Gene Duplication*. Apart from continuous and small evolutionary changes, there are drastic and discontinuous changes caused by gene fusion and duplication (hemoglobins, immunoglobulins etc). The emergence of new proteins/enzymes can be explained by gene duplication. The genetic combination of structural motifs, modules and domains, in various new ways, can result in the rapid development of protein with new function. Evolution of completely new structures by point mutations is comparatively slow. A general mechanism for the evolution of new proteins is gene duplication to form more copies of the polypeptide and the new copies evolve independently by accumulating point mutations, insertions and deletions of amino acids to yield proteins with new or altered functions. Unequal cross-over (in meiosis) is assumed to be responsible for both additions and deletions in the polypeptide chain. After duplication the genes can fuse to yield a longer protein (internal gene duplication) or a protein with different functions. Thus, gene duplication leading to sequence redundancies has been a major driving force in both the elongation of small peptides and in expanding the available repertoire of gene products.

Comparison of amino acid sequences and structural repeats within a protein gives information on internal gene duplication. For example, in ferredoxin the primitive gene coding for 26 amino acids dimerised, and after gene duplication and fusion, a ferredoxin monomer with twice the number of amino acid rcsidues than the initial primitive peptide was created. Similar structural repeats have been observed, indicating internal gene duplication in elongating primitive peptides in rhodanase, troponin **c**, parvalbumin, chymotrypsin etc.

Combination of two dissimilar polypeptides by gene fusion generally results in the formation of a new polypeptide with altered functions. An example is the enzyme lactose synthetase, an enzyme formed by the fusion of α-lactalbumin and transferase. α-lactalbumin shows extensive sequence homology with hen egg-white lysozyme and also the spatial arrangement of the four S-S bridges is similar in both proteins. The conformations of the proteins are also similar, but α-lactalbumin is completely devoid of lysozyme activity, possibly due to point mutations of the catalytically active residue(s). α-lactalbumin is one of the domains (B-protein) of bovine lactose synthetase. Possible evolutionary events in the creation of lactose synthetase are: initially there must have been a duplication of the lysozyme gene and at some later stage one of these protein copies lost its functional activity as lysozyme due to point mutations, but retained the binding affinity for glycosidic compounds. In addition, it developed an affinity for transferase, the other subunit of lactose synthetase (A-protein), present in various animal tissues. Furthermore, α-lactalbumin came under a new set of new genetic controls, being produced only in mammary tissues and being under hormonal control so that its production is low until the onset of lactation. Thus, combination of α-lactalbumin and transferase has resulted in the evolution of a new protein, lactose synthetase, with altered functions. Similarly, in tryptophan synthetase, each chain has distinct catalytic properties, but a new enzymatic function emerges on the combination of the two chains.

Evolutionary Trends in Protein Structure. Analysis of three-dimensional structures of proteins show that proteins with closely related amino acid sequences also have closely similar structures. The converse is not always true, as proteins with very different primary structures can still have closely similar tertiary structures and functions. That is to stress that during evolutionary changes, three-dimensional folding is conserved more strongly than the amino acid sequences. Since the tertiary structure of a protein outlasts changes in the primary structure, sometimes the only surviving-evidence for divergent evolution may lie in the three-dimensional structures of proteins. Furthermore, it is only through the comparison of tertiary structures of proteins, unrelated by primary structures, that convergent evolutionary process can be discerned (e.g. pancreatic proteinases and subtilisins). The three-dimensional structures of molecules (folding) are better guides to similarities and dissimilarities to distantly related proteins than the amino acid sequence comparisons. The role of X-ray diffraction techniques in providing quantitative three-dimensional structural information for these and other biological studies has been immense and central.

(i) *Divergent Evolution.* The production of different species by mutations of genes descended from a common ancestral gene is called divergent evolution. The degree of divergence in the sequence of homologous proteins is, in general, related to the phylogenetic relationship of the species. Mutation rates do not differ greatly in different organisms, but what does differ is the extent of adaptation and selection processes operative on the population of a given species. Large differences in sequences in proteins like hemoglobins may be reflective of evolutionary changes in adapting to different environmental situations—fetal, adult, aquatic life and life at different altitudes etc. Few variations are tolerated in well-adapted proteins, like cytochrome **c** and histones. Numerous examples can be quoted of divergent evolution in proteins. Pancreatic proteinases, trypsin, chymotrypsin, elastase, thrombin and plasmin evolved from a common ancestor. Globin, cytochrome **c** and immunoglobulin families are all examples of divergent evolution.

(ii) *Convergent Evolution.* Convergent evolution is the acquisition of similar functional features by a class of proteins with different conformations and sequence dissimilarities. An example is the similar function of chymotrypsin and subtilisin which are serine proteases. These two enzymes do not have amino acid sequence similarities, and, while chymotrypsin has five disulphide bonds, subtilisin has none. Their three-dimensional structures are also different, but functionally they are similar. These two enzymes evolved from different precursor polypeptide chains, but converged independently to evolve the same type of reaction mechanism. Lysozymes of hen egg-white contain S-S bonds, but phage T4 and T2 lack S-S bridges and plant lysozymes (papaya and ficus) lack sequence homology. Similarly, ribonucleases, pancreatic and bacterial (*Staphylococcus aureus*, *Aspergillus oryzae*) show no structural similarity. In the case of fructose diphosphate aldolases, bacterial, yeast and fungal aldolases require a divalent metal ion for activity, but animal and plant aldolases do not require any metal ion for activity.

Molecular Structure and Evolution

(i) *Chemical Evolution.* Whether chemical evolution of life was via nucleic acids (the RNA-world) or via amino acids (protein-world) is an actively debated issue. Recent discovery of enzymatic properties of RNA does not necessarily mean conclusively that chemical evolution of life processes began with the RNA-world. There must have been simpler molecular systems that preceded RNA and these molecular species (multi-molecular systems) replicated by chemical processes and were able to undergo Darwinian evolution.

Amino acids, simple peptides, nucleic acid bases as well as relevant organic molecules, such as formic acid, acetic acid, succinic acids etc., were present on earth during the prebiotic period. The simple pre RNA-world molecular systems could very well have been molecular aggregates of peptide-peptide, peptide-organic molecule complexes. The specific interactions with characteristic interaction patterns which might have been important in early self-assembly processes of chemical evolution appear to have been predominantly due to hydrogen bonding. Hydrogen bonds between peptide nitrogen atoms and carbonyl oxygen atoms of neighbouring molecules (N-H...O) or through water-mediated hydrogen-bonded water bridges (N-H...O(W)...O) would have held molecules together by head-to-tail sequence as planar straight or zig-zag aggregates. Basic elements of such as aggregation via hydrogen-bonded networks and the important role of water columns in such interactions are indicated in Fig. 6.1.

(ii) *Molecular Evolution.* Evolution can be studied from the molecular perspective, i.e., from the comparative studies of three-dimensional structures of a protein or class of proteins from various organisms and environments. Cytochrome **c** and hemoglobin serve as good examples, because three-dimensional structures at high resolution of these and mutants from several species have been determined by X-ray diffraction studies and their structure-function details have been well documented.

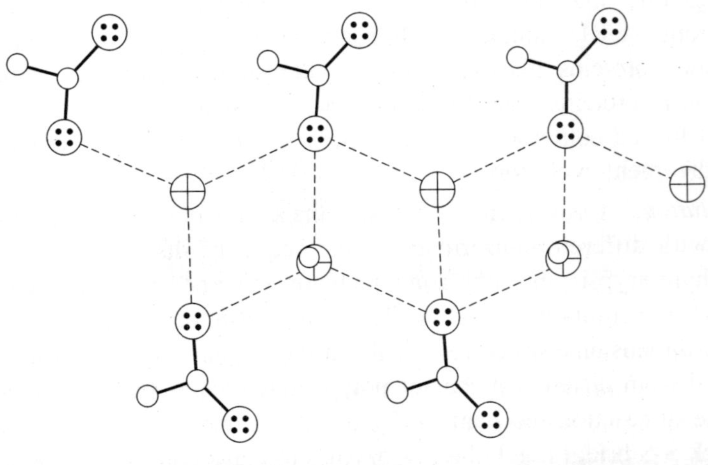

Fig. 6.1a Characteristic interaction pattern of hydrogen-bonded network involving amino and carboxyl groups in L-histidine acetate structure.

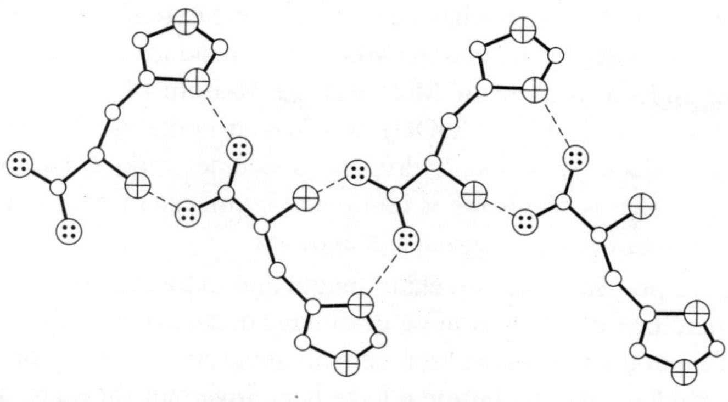

Fig. 6.1b Histidine ribbons formed by hydrogen bonds in L-histidine acetate structure.

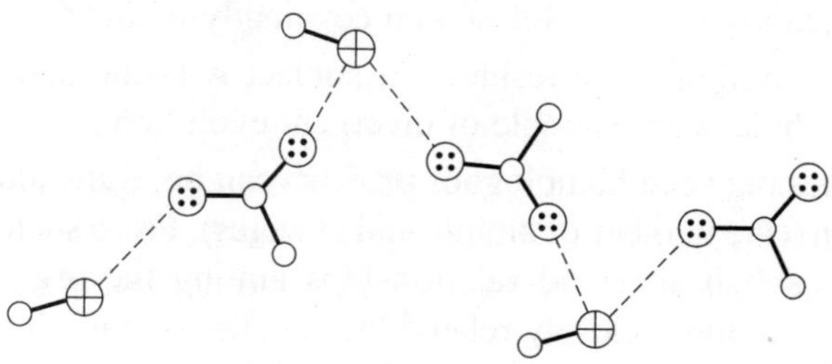

Fig. 6.1c Characteristic interaction pattern via hydrogen bonds in L-histidine acetate structure.

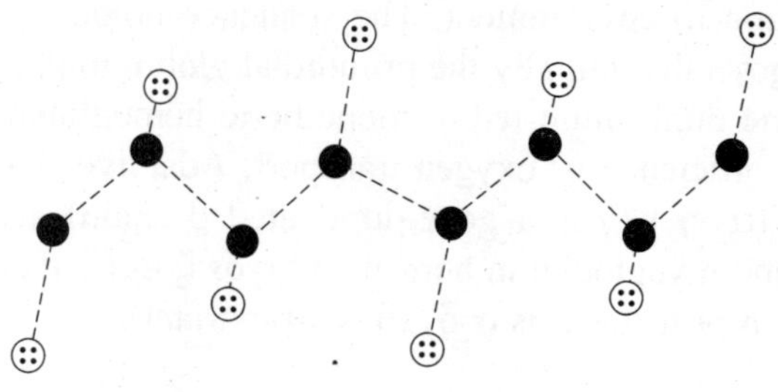

Fig. 6.1d Water column interacting with the molecular aggregate in L-histidine acetate.
O = Carbon; ⊕ = Nitrogen; :: = Oxygen; ● = Water.
Ref. Suresh, S., Sridhar Rao, G. & Vijayan, M.
Int. J. Peptide Protein Res. 43; 139 (1994).
(Photographs : Courtesy of Prof. M. Vijayan, Indian Institute of Science, Bangalore)

Cytochrome **c** is an ancient and well-adapted protein. It is a carrier of electrons from the QH_2-cytochrome **c**-reductase complex to the Cytochrome **c**-oxidase complex in the electron transport chain. It is a heme-containing protein and Fe-heme is enfolded in the heart of a protein chain of 82–134 amino acids. All cytochrome **c** molecules are folded in an identical way, with only the addition or deletion of loops of chain at the surface of the molecule.

Electrons migrate into and out of cytochrome **c** through one exposed edge of the heme. The heme is attached by four bonds, covalently bonded to two cysteine residues and with two coordinate bonds between the iron atom and the S-atom of Met, and the N-atom of His residues. The two polar propionic acid side chains (–CH_2–CH_2–COOH) attached to heme are buried and form hydrogen bonds with Tyr and Trp residues. These two hydrogen-bonded networks are absolutely invariant in all (eukaryotic) cytochromes. That is, the heme is reoriented by means of covalent and hydrogen bonds so that an edge with only small nonpolar groups is exposed.

There is no change in properties due to chain length and extra residues are not strictly essential for protein function. Therefore, chain shortening must have occurred during evolution of this protein. Residues crucial for function or conformation remain invariant in all cytochrome **c** of different species. 26 of the 104 residues of cytochrome **c** have been invariant for more than 1.5 billion years. All cytochrome **c** have invariant **Lys** and **Arg** residues around the heme crevice on one face of the proteins. One of these residues interacts with cytochrome **c**-reductase and the other with Cyt **c**-oxidase. Similarly, heme ligands **His** 18, **Met** 80 and covalently-bound **Cys** residues, **Cys** 14 and **Cys** 17 are invariant. Many hydrophobic residues in contact with the heme are also invariant. The cytochrome **c** family as a whole is an example of divergent evolution.

Evolutionary differences between homologous proteins can be evaluated by counting the number of differences between them (the number of amino acid changes). From such differences phylogenetic trees can be plotted to establish ancestral relationships among the organisms that produced the proteins. In cytochromes **c**, the more closely related the species to man, the fewer the number of the differences in amino acid sequences (chimpanzee = 0; monkey = 1 ; whale = 10; wheat = 35; yeast = 44 out of 104 residues).

The globin family (myoglobin and hemoglobin) is an example of divergent evolution through gene duplication and adaption to environment. The sequence homology between myoglobin and hemoglobin monomers suggests that initially the primordial globin might have functioned simply as oxygen-storage protein. Gene duplication led to monomeric hemoglobin and to tetrameric (α_2 β_2) hemoglobin, with increased efficiency of oxygen transport. Adaptive changes gave rise to different chains—fetal hemoglobin, (α_2 γ_2) (γ is a gene-duplicated β-chain) has oxygen-binding affinity between adult hemoglobin and myoglobin; in human embryos $\zeta_2\varepsilon_2$ (ζ, ε are gene-duplicated variants of α & β, respectively) and in primates it is $\alpha_2\delta_2$ (δ is a β-variant).

6.3 MOLECULAR ENGINEERING

The objective of all molecular protein engineering methods is to produce proteins with improved characteristics. Evolution is nature's method of producing proteins with improved characteristics to suit altered environmental situations, but the timeframe is large and it is basically a trial-and-error process. The rationale of protein engineering is to produce altered proteins on a laboratory time scale.

A rational approach to protein engineering rests on the availability of the three-dimensional structure of the protein that is to be engineered (or its close structural homologue). Two important physical techniques are available to provide the complete tertiary structure of a protein, (i) X-ray diffraction (see Chapter 10 for details), which is applicable to single-crystalline proteins (a very important condition) and (ii) NMR methods (see Chapter 9 for details), which are applicable to proteins of small molecular mass (a limitation). Procurement of pure protein samples in sufficient quantities, for the above methods, can be accomplished either by conventional physico-chemical methods or by genetic methods (gene selection and cloning) (see Chapter 7 for details). The amino acid sequence of the protein can be inferred from its gene sequence (which is a fast and often used approach now) or from amino acid sequencing.

Once the tertiary structure of a protein is available, the task of protein engineering is to evolve a rational design to alter residues/moieties of the protein to achieve an altered protein with the desired characteristics. The general approach is to address the fundamental problems of protein-folding by 'knowledge-based' computer-modelling methods. The crucial problems are: (i) the stability of the folded-state (thermodynamics) and (ii) folding pathways (kinetics). Gibbs free energy, ΔG_{tot} is approximately zero for all polar groups, and for non-polar groups, solvent entropy, ΔS_{sol}, is large and positive and is the single largest contribution to protein stability. That is, both enthalpy and entropy factors are important in protein stability (see Chapter 13 for details).

The strategy of all computer-modelling procedures is to obtain structural and functional features from experimental methods, structure databases, sequence homologies and structure prediction methods, as input data from which several plausible model structures (mutant structures) with altered (desired) functions can be proposed.

The production of these mutants (altered proteins) with altered functions can be achieved by molecular genetic techniques (gene manipulation, alteration, site-directed mutagenesis etc.). The modelled-proteins can be studied for their altered functions and their tertiary structures can also be determined (by difference-Fourier techniques in structure determination procedures). The iterative nature of the protein engineering processes is given schematically in Fig. 6.2.

6.3.1 Site-Directed Mutagenesis

Gross structural features of a protein are not considerably altered by mutation, but mutation(s) at the crucial position(s) leads to drastic changes in the functional properties of the protein. The strategy is to alter amino acid residue(s) at crucial position(s). Site-directed mutagenesis is a powerful strategy to introduce single-point mutation or cassette-mutation (set of selected residues randomly altered), to alter protein functions, to study reaction mechanisms and structure-function relationships.

There has been a heightened interest in site-directed mutagenesis in protein engineering. It has been applied to alter the stability and functions of various proteins and enzymes. A single-residue alteration has been sufficient in several enzymes to alter their substrate-specificity drastically and alteration of few crucial residues can lead to complete loss of function in many proteins/enzymes. Replacement of crucial residues with cysteines can lead to engineered proteins with better thermal stability (such as in lysozyme) and capable of functionally being 'switched on' (on reduction) and 'switched off (on oxidation of S-S bonds). Such 'molecular switch' properties could be useful in various enzyme applications and could increase the shelf-life of enzymes. Several biologically important polypeptides, such as endothelins, several hormones, snake, scorpion and insect venom toxins, are disulphide-containing polypeptides. There exists a structural hierarchy in disulphide bonds, depending on the structural segment of the polypeptide they stabilise. Selective modification of S-S bonds in these proteins may be a way in the future to design more effective and specific antidotes. In addition, these polypeptides, being small size proteins, could be of importance in unravelling the molecular mechanisms underlying pre- and post-synaptic receptor-ligand and protein-protein associations.

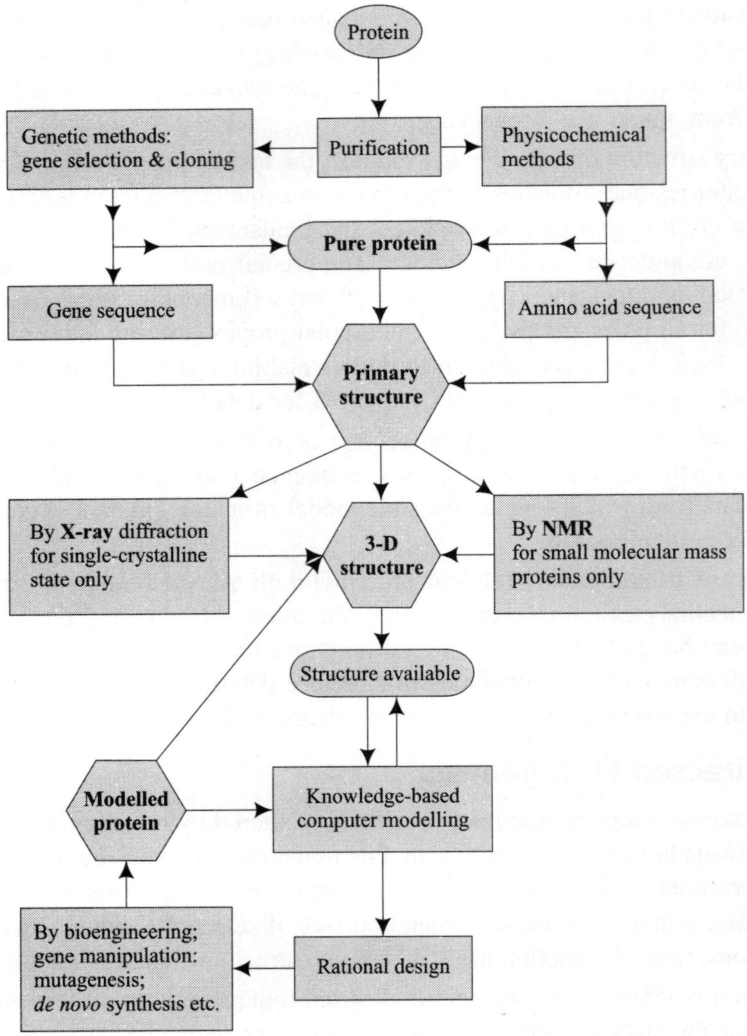

Fig. 6.2 Schematic diagram of the iterative steps in a 'rational approach' to protein engineering.

Site-directed mutagenetic methods, combined with electrophysiological experiments hold much promise to increase our understanding of the structure-function features of voltage-dependent gating proteins.

De novo protein-design is another important approach towards protein engineering. This approach is in its early stages of development and beset with conceptual and experimental problems. However, this is an interesting and challenging approach which might yield rich dividends in the future.

Suggested Reading

Books

Attwood, T.K. & Parry-Smith, D.J. *"Introduction to Bioinformatics"*. Pearson Education/Delhi. (2002).

Davies, K.E. & Reid, A.P. *"Molecular Basis of Inherited Diseases"*. IRL Press/New York. (1988).

Dickerson, R.E. & Geis, I. *"Hemoglobin: Structure, Function, Evolution and Pathology"*. Benjamin-Cummings/ Menlo Park, CA. (1983).

Fasman, G.D. (Ed). *"Prediction of Protein Structure and the Principles of Protein Conformation"*. Plenum Press/New York. (1990).

Lesk, A.M. *"Protein Architecture: A Practical Approach"*. IRL Press/ London. (1991).

Li, W.H. *"Molecular Evolution"*. Sinuaer Associates/ Sunderland, MA. (1997).

Lilley, D.M.J. (Ed). *"DNA-Protein: Structural Interactions"*. IRL Press/ Oxford. (1995).

Narayanan, P. "*Bioinformatics: A Primer*". New Age Intl Pubs/ New Delhi. (2005).

Perutz, M.F. *"Protein Structure: New Approaches to Disease and Therapy"*. Freeman Press/ New York. (1992).

Sternberg, M.J.E., (Ed). "*Protein Structure Prediction– A Practical Approach*". IRL Press/ Oxford, UK. (1996).

Articles

Bajorath, J., Stenkamp, R. & Aruffo, A. *"Knowledge-based Model Building of Proteins: Concepts and Examples"*. Protein Sci., 2; 1798. (1993).

Baker, D. & De Grado, W.F. *"Engineering and Design"*. Curr. Opin. Struct. Biol., 9(4); 485. (1999).

Baldwin, R.L. *"How does Protein Folding Get Started?"* Trends Biochem. Sci., 14; 291. (1989).

Barton, G.J. *"Protein Secondary Structure Prediction"*. Curr. Opin. Struct. Biol., 5(3); 372. (1995).

Baum, J. & Brodsky, B. *"Folding of Peptide Models of Collagen and Misfolding in Disease"*. Curr. Opin. Struct. Biol., 9(1); 122. (1999).

Blundell, T.L., et al. *"Knowledge-based Prediction of Protein Structures and the Design of Novel Molecules"*. Nature, 326; 347. (1987).

Bowie, J.U. & Eisenberg, D. *"Inverted Protein Structure Prediction"*. Curr. Opin. Struct. Biol., 3; 437. (1993).

Carell, R.W. & Gooptu, B. *"Conformation Changes and Disease– Serpins, Prions and Alzheimer's"*. Curr. Opin. Struct. Biol., 8(6); 799. (1998).

Chothia, C. "*Principles that Determine the structure of Proteins*". Annu. Rev. Biochem., 53; 537. (1984).

Chothia, C. & Lesk, A.M. *"The Relation between the Divergence of Sequence and Structure in Proteins"*. EMBO J., 5; 823. (1986).

Chou, P.Y. & Fasman, G.D. *"Prediction of the Secondary Structure of Proteins from Their Amino Acid Sequence"*. Advs. Enzymol., 47; 45. (1978).

Cohen, C. & Perry, D.A.D. "*α-helical Coiled Coils– A Widespread Motif in Proteins*". Trends Biochem. Sci., 11; 245. (1993).

Creighton, T.E. *"Experimental Studies of Protein Folding and Unfolding"*. Prog. Biophys. Mol. Biol., 33; 231. (1978).

Dickerson, R.E. *"Cytochrome C and the Evolution of Energy Metabolism"*. Sci. Amer., 243(3); 136. (1980).

Dobson, C.M. *"Protein Misfolding, Evolution and Disease"*. Trends Biochem. Sci., 24; 329. (1999).

Dobson, C.M. & Karplus, M. *"The Fundamentals of Protein Folding: Bringing Together Theory and Experiment"*. Curr. Opin. Struct. Biol., 9(1); 92. (1999).

Doolittle, R.F. *"Reconstructing History with Amino Acid Sequences"*. Protein Sci., 1; 191. (1992).

Eisenhaber, F., Persson, B. & Argos, P. *"Protein Structure Prediction: Recognition of Primary, Secondary, and Tertiary Structural Features from Amino Acid Sequence"*. Crit. Rev. Biochem. Mol. Biol., 30; 1. (1995).

Englander, S.W. *"In Pursuit of Protein Folding"*. Science, 262; 848. (1993).

Frishman, D. & Argos, P. *"Knowledge-based Protein Secondary Structure Assignment"*. Proteins, 23; 566. (1995).

Harrison, P.M., et al. *"The Prion Folding Problem"*. Curr. Opin. Struct. Biol., 7(1); 53. (1997).

Johnson, M.S. & Overington, J.P. *"A Structural Basis for Sequence Comparison"*. J. Mol. Biol., 233; 716. (1993).

Lee, R.H. *"Protein Model Building Using Structural Homology"*. Nature, 356; 543. (1992).

Levitt, M. & Chothia, C. *"Structural Patterns in Globular Proteins"*. Nature, 261; 552. (1976).

Lupas, A. *"Prediction and Analysis of coiled-coil Structures"*. Methods Enzymol., 266; 513. (1996).

Lüthy, R., Bowie, J.U. & Eisenberg, D. *"Assessment of Protein Models with Three-dimensional Profiles"*. Nature, 356; 83. (1992).

Marrone, T.J., Briggs, J.M. & McCammon, J.A. *"Structure-based Drug Design: Computational Advances"*. Ann. Rev. Pharmacol. Toxicol., 37; 71. (1997).

Orengo, C.A., et al. *"Protein Superfamilies and Domain Superfolds"*. Nature, 372; 631. (1994).

Qain, N. & Sejnowski, T.J. *"Predicting the Secondary Structure of Globular Proteins Using Neural Network Models"*. J. Mol. Biol., 202; 865. (1988).

Richardson, J.S. *"Describing Patterns of Protein Tertiary Structure"*. Methods Enzymol., 115; 349. (1985).

Shao, Z. & Arnold, F.H. *"Engineering New Functions and Altering Existing Functions"*. Curr. Opin. Struct. Biol., 6; 513. (1996).

Sun, Z. & Jiang, B. *"Conformation of Commonly Occurring Super-secondary Structures (Basic Motifs) in Protein Databank"*. J. Prot. Chem., 15; 675. (1996).

Swindells, M.B. & Thornton, J.M. *"Modeling by Homology"*. Curr. Opin. Struct. Biol., 1; 219. (1991).

Vriend, G. & Sander, C. *"Detection of Common Three-dimensional Structures in Proteins"*. Proteins, 11; 52. (1991).

Zaidi, F.N., Nath, U. & Udagaoankar, J.B. *"Multiple Intermediates and Transition States During Protein Unfolding"*. Nature. Struct. Biol., 4; 1016. (1997).

SECTION II

Physical Techniques in Structure Determination

OVERVIEW

Separation (purification) methods are integral parts of characterisation of physicochemical properties (molecular mass, shape and size and structure etc.) of biomolecules. Separation of components to a high degree of purity is essential for some purposes like identification and structure elucidation. Sedimentation (centrifugation), osmosis (filtration), sorption (chromatography and electrophoresis) are employed in cellular, molecular purification and characterisation. Determination of the structure of matter by physical and related techniques is based on various structural and functional parameters. These include: (i) oxidation state, coordination number and electron configuration, (ii) nature of chemical bonds, bond lengths and bond angles, (iii) stereochemistry, configuration and conformation of molecules, (iv) thermodynamic data and (v) reaction kinetics and reaction mechanism parameters. The physical techniques employed, to name a few, are: magnetic susceptibility, electric dipole moment, optical activity, light scattering and spectroscopic methods, such as **UV**, **IR**, **Raman**, **ESR** and **NMR** and **electron microscopy** and **X-ray diffraction**. Some of the various spectroscopic techniques employed to study molecular structure may give complementary structural information and in particular cases one of the techniques may be better suited to tackle the problems than the others. For example, Mössbauer spectroscopy is uniquely suited to study the environment of iron atoms in iron-containing molecules. Similarly, Raman spectroscopy has distinct advantages over infrared spectroscopy to study biomolecules in aqueous media.

Light microscopy is the standard technique in the study of morphological features of microorganisms, depending on the magnification and resolution. Diffraction effects set the resolution limit in light microscopy. Phase-contrast microscopy relies on phase differences to improve the contrast of organisms whose refractive index is about the same as the medium. Electron microscopy (SEM and TEM) is employed in the study of ultra-structure of biological systems at cellular level. 'Optical' abberations and damage of specimens limit the attainable magnification in electron microscopy. Recent advances in microscopies, such as Laser confocal microscopy, Scanning-tunneling microscopy (STM) and Atomic force microscopy (ATM) have enabled the probing of surface details to atomic levels.

Nuclear magnetic resonance (NMR) techniques are employed to study the structure-function dynamics of macromolecule–ligand interactions (e.g. enzyme-substrate interactions) at crucial sites. NMR-radiography (NMR-tomography) is applied in medicine. The development of 2D-NMR has made possible the 3-D structure determination of small macromolecules ($\approx$ 10,000 daltons). X-ray diffraction is a very powerful tool and, to-date, is the only technique to elucidate the three-dimensional architecture of matter at atomic resolution. Advances in data collection and the use of a synchrotron radiation source have increased its potential to address larger and more complex macromolecular structures and structural dynamics; the method has become an indispensable tool in quantitative biology.

Advent of lasers has brought in the coherent optics to the forefront. Lasers are sources of high intensity, monochromatic radiations. Pulsed lasers have enabled the studies in fast physical and chemical reactions. Lasers are employed in variety of applications in diverse fields—scientific, industrial, medical and military. Developments in holography have been mainly due to the realisation of coherent optics. Holographic techniques have been employed in interferometry, information-related fields, visual and entertainment industries.

7 Separation Methods

Synopsis: Diffusion, filtration, sedimentation (centrifugation), and chromatographic and electrophoretic methods are standard methods in purification and physicochemical characterization of biomolecular complexes. Electrophoretic methods in conjunction with blotting techniques and mass spectrometry (MS) find wide applications in molecular biology and genomic studies (gene sequencing).

7.1 DIFFUSION

Movement of matter by a natural process, from a higher to a lower concentration is called diffusion.

7.1.1 Translational Diffusion

It is related to **Brownian motion.** The root-mean-square displacement of a spherical particle, $\overline{x^2}$ in a solution is

$$\overline{x^2} = 2Dt = \frac{2RTt}{N_0 f} \qquad \textbf{Einstein equation} \qquad ...(7.1a)$$

$$f = 6\pi\,\eta r \qquad \textbf{Stokes equation} \qquad ...(7.1b)$$

$$\therefore \qquad D = \frac{RT}{N_0 6\pi\,\eta r} \qquad \textbf{Stokes-Einstein equation} \qquad ...(7.1c)$$

(D = Diffusion coefficient; t = time; R = gas constant; N_0 = Avogadro number, T = Absolute temperature; f = frictional coefficient; r = radius of particle; η = viscosity coefficient).

From thermodynamics, the force, F-acting on a spherical particle is

$$F = \frac{d\mu}{dx} \qquad ...(7.2a)$$

where μ is molecular free energy given by

$$\mu = \mu_0 + \frac{RT}{N_0}\ \ln \mathbf{c} \qquad ...(7.2b)$$

$$F = -\frac{RT}{N_0\mathbf{c}}\frac{d\mathbf{c}}{dx} \qquad ...(7.2c)$$

The opposing force is frictional drag, $F' = f\frac{dx}{dt}$

$$\therefore \qquad \mathbf{c}\frac{dx}{dt} = -\frac{RT}{N_0 f}\cdot\frac{d\mathbf{c}}{dx} \qquad ...(7.2d)$$

Flux, J, is the rate of flow across unit area

$$J = \frac{d\mathbf{n}}{dt} = \mathbf{c}\frac{dx}{dt} \quad \text{by definition} \qquad ...(7.2e)$$

$$J = -D\frac{d\mathbf{c}}{dx} \quad \text{Fick's Ist law of diffusion (mass transport)} \qquad ...(7.2f)$$

For a spherical particle, D is

$$D = \frac{RT}{3N_0\eta}\left(\frac{\rho}{6\pi^2}\right)^{1/3}\frac{1}{M^{1/3}} \qquad ...(7.2g)$$

Therefore, $DM^{1/3}$ should be constant at constant T.

$DM^{1/3}$ is constant if the particle size > size of the solvent molecules.

$DM^{1/2}$ is constant if the particle size < size of the solvent molecules.

Molecular mass determination is shape-dependent.

The diffusion coefficient, D, can be measured in a centrifuge from the optical **Doppler** effect of the laser beam scattered by the moving particles. **Schlieren** optics can be employed in a centrifuge to measure the concentration gradient, $\frac{\partial \mathbf{c}}{\partial x}$ across the boundary. (Figs. 7.1a & b), as the solute behaves like a prism. The refractive index gradient (concentration gradient) can also be measured by the interference method.

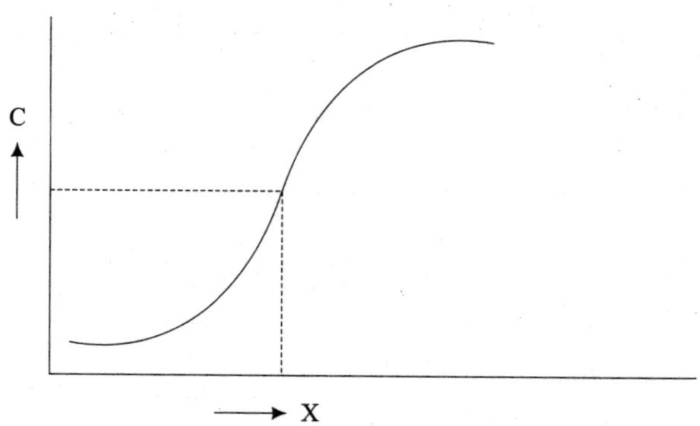

Fig. 7.1a Profile of concentration, **c**, versus distance *x* in a centrifuge.

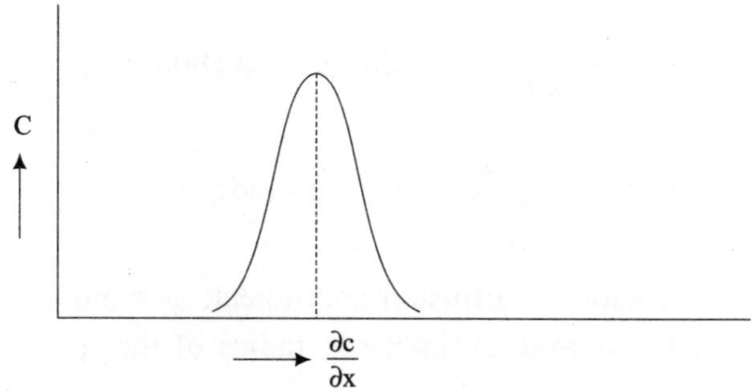

Fig. 7.1b Profile of concentration, **c**, versus concentration gradient, $\frac{\partial c}{\partial x}$ in a centrifuge.

The analogy of transport of matter (Fick's Ist law) can be extended to energy transport (heat transfer) to estimate the diffusion coefficient, *D*.

$$\left.\begin{aligned} J_{\text{energy}} &\propto \frac{dT}{dx} \\ J &= -k\frac{dT}{dx} \end{aligned}\right\} \text{heat transfer} \qquad \text{...(7.2h)}$$

where **k** is the coefficient of thermal conductivity.

The **Einstein-Smoluchowski** equation on **Brownian** motion relates

$$D \sim \frac{1}{3}\lambda\bar{\mathbf{c}}\,;\, D \sim \sqrt{T} \qquad \text{...(7.2i)}$$

where λ is the mean free path of the particle and **c** is concentration.

7.1.2 Rotational Diffusion

Non-spherical molecules undergo a tumbling motion (rotational diffusion)

$$\frac{\partial \Omega}{\partial t} = -\alpha \frac{\partial^2 .\Omega}{\partial \varphi^2} \qquad \text{...(7.3a)}$$

$$\overline{\varphi^2} = 2\alpha t \qquad \text{...(7.3b)}$$

$$\alpha = \frac{RT}{N_0 \zeta} \qquad \text{...(7.3c)}$$

$$\zeta = 8\pi\ \eta\ r^3 \quad \text{Stokes equation} \qquad \text{...(7.3d)}$$

Relaxation time, $\tau = \frac{1}{2\alpha}$

$$\tau = \frac{4\pi N_0 \eta r^3}{RT} \quad \text{Einstein equation} \qquad ...(7.3e)$$

$$\tau = \frac{4\pi N_0 \eta ab^2}{RT} \quad \text{for ellipsoids} \qquad ...(7.3f)$$

(Ω = angular diffusion; α = rotational diffusion coefficient; φ = angle; $\overline{\varphi^2}$ = root-mean-square angular motion; ζ = torque; τ = relaxation time; r = radius of the sphere; a & b are axes of the ellipsoid).

The rotational diffusion coefficient can be evaluated by flow birefringence. Flow birefringence is a method for determining the orientation of elongated molecules in a velocity gradient produced by a mechanical shearing force. A solution between two concentric cylinders is subjected to a shearing force by rotating the cylinders. The light passing through the solution parallel to the axis of the cylinders is observed through crossed-polarisers. Application of external force produces four dark regions (crosses). The angle between the planes of the polarisers and the arms of the cross allows the estimation of the diffusion coefficient and relaxation time. This method is useful for the study of asymmetric rod-like molecules (DNA and fibrous molecules). The depolarisation of fluorescence method can also be applied to estimate the relaxation time and hence the size and shape of macromolecules.

The dielectric relaxation method. Dielectric dispersion (variation of dielectric constant with the frequency of the applied field) is a method that can provide useful information on the rotational mobility of molecular species and the nature of hydration of macromolecules, such as proteins.

The molecular motions of polar molecules in the liquid and gaseous states are random in the absence of an external field and the net average dipole moment is zero. Application of an external field leads to molecular alignment (polarisation) along the field. The permittivity (dielectric constant) is related to the polarisation, *P.*

$$\varepsilon_0 = 1 + \frac{P}{\varepsilon E} \qquad ...(7.4a)$$

(ε_0 is permittivity in vacuum; E = Applied field).

Relaxation time, τ, the time to return to the equilibrium state after the application of a perturbation, is

$$\tau = \frac{3V\eta}{kT} \qquad ...(7.4b)$$

(V = molecule volume; η = viscosity).

The relaxation time, τ, is a measure of the rotational mobility of the molecular species, which can provide information about the shape and size of the molecules. By applying a sinusoidal field to a dielectric sample at a particular frequency, *w,* the permittivity (ε') and the loss factor (ε'') can be determined at that frequency.

$$\varepsilon' = \varepsilon_\infty + \frac{\varepsilon_0 - \varepsilon_\infty}{1 + w^2\tau^2} \qquad \text{...(7.4c)}$$

$$\varepsilon'' = \frac{\varepsilon_0 - \varepsilon_\infty}{1 + w^2\tau^2} \qquad \text{...(7.4d)}$$

(ε_∞ is permittivity at shorter times, due to electronic and atomic polarisations).

By varying the frequency, a frequency spectrum can be obtained. The same is achieved by time-domain spectroscopy (TDS), by applying a pulse to a dielectric material and recording the decay profile (as a function of time). Fourier transformation can convert time-domain decay data to frequency-domain spectra.

Hydration of proteins, membranes and other macromolecules can also be studied by dielectric relaxation method, i.e., the nature of surface-bound water in proteins. The nature of hydration can be either hydrophobic hydration, where water forms clatharite aggregates or hydrophylic hydration, where water molecules are involved in a hydrogen-bond network.

7.2 SEDIMENTATION

Centrifugation is a sedimentation (precipitation) technique, wherein individual constituents of a mixture are separated by an applied centrifugal force (applied gravitation). The constituents are separated based on their sedimentation rates. The rate of sedimentation of a particle is a function of molecular mass, shape and size of the particle.

The rate at which a particle sediments in a centrifuge in a centrifugal field is related to the net force acting on it. The centrifugal force, F_c, acting on a particle of mass, m, at a distance, r, from the revolving axis is

$$F_c = m\,\omega^2\,\mathbf{r} \qquad \text{...(7.5a)}$$

The opposing buoyant force, F_b, is

$$F_b = V\rho\,\omega^2\,r \qquad \text{...(7.5b)}$$

$$\left(V = m\bar{v};\ \bar{v} = \frac{1}{\rho_{\text{solute}}}\right)$$

$$\therefore \qquad F = m\omega^2 r\,(1 - \bar{v}\rho) \qquad \text{...(7.5c)}$$

This force, F, is opposed by the frictional drag, $F' = f\,\frac{dr}{dt}$

Drift speed $\qquad \frac{dr}{dt} = s\omega^2 r$ by definition

At equilibrium

$$m\omega^2\,r(1 - \bar{v}\rho) = fs\,\omega^2\,r \qquad \text{...(7.5d)}$$

but, $\qquad N_0 f = \frac{RT}{D}$

$$\therefore \qquad M = \frac{RT}{(1-\bar{v}\rho)}\frac{s}{D} \qquad ...(7.5e)$$

10^{-13}s is called a **Svedberg.**

(ω = angular velocity; V = volume; $\bar{v}$ = particular specific volume of the particle; ρ = density of the solution; f = frictional coefficient; **s** = sedimentation coefficient; M = molecular mass; N_0 = Avogadro number).

$\bar{v}$ can be determined from the variation of the density of the solution with solute concentration. The sedimentation coefficient, **s**, can be measured in a centrifuge by optical Doppler effect of laser light scattered by the moving particle. The molecular mass, M, can then be calculated if the diffusion coefficient, D, of the molecule is known.

Molecular mass, M, can be determined without the knowledge of the diffusion coefficient, D, by the sedimentation equilibrium method. This is achieved by equating diffusion to sedimentation. At equilibrium the diffusion rate is equal to the sedimentation.

$$\mathbf{c}\,\frac{dr}{dt} = D\,\frac{d\mathbf{c}}{dr} \qquad ...(7.6a)$$

but

$$D = RT/N_0 f$$

From the sedimentation rate,

$$\frac{dr}{dt} = \frac{M\omega^2 r}{N_0 f}(1-\bar{v}\rho)$$

Substituting this in equation (7.6a) and rearranging

$$\frac{d\mathbf{c}}{\mathbf{c}} = \frac{M\omega^2}{RT}(1-\bar{v}\rho)\,r\,dr \qquad ...(7.6b)$$

Considering the concentrations $\mathbf{c}_1$ and $\mathbf{c}_2$ at positions r_1 and r_2 from the axis of rotation

$$\int_{\mathbf{c}_1}^{\mathbf{c}_2} \frac{d\mathbf{c}}{\mathbf{c}} = \frac{M\omega^2}{RT}(1-\bar{v}\rho)\int_{r_1}^{r_2} r\,dr \qquad ...(7.6c)$$

$$M = \frac{2RT}{(1-\bar{v}\rho)\omega^2}\cdot\frac{\ln(\mathbf{c}_2/\mathbf{c}_1)}{(r_2^2 - r_1^2)} \qquad ...(7.6d)$$

Taking the concentrations of a solute at a distance r and at the distance of the minuscus, the plot of **ln** ($\mathbf{c}_r$) as a function of r^2 gives a straight line; from the slope of which molecular mass, M, can be calculated (Fig. 7.2). This method is better as it is independent of the diffusion coefficient.

Density-gradient sedimentation, employing caesium chloride (CsCl) or sucrose gives a better separation and resolution of particles. CsCl or sucrose acts as a 'buffer' to prevent intermixing of neighbouring bands due to thermal and mechanical inhomogeneities. Determination of molecular mass, shape and size of biomolecules is carried out generally by diffusion and sedimentation (centrifugation) methods.

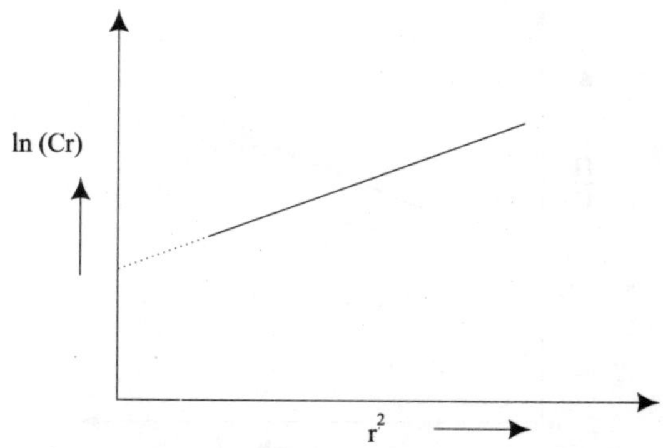

Fig. 7.2 Plot of **ln** (c_r) versus r^2.

7.3 OSMOSIS

Osmosis is a phenomenon in which only the solvent is free to migrate through a barrier (semi-permeable membrane). Osmotic separation methods are membrane-based (barrier) filtration techniques that can be employed for selective separation of constituents from a mixture. Most membrane filtration methods employ semi-permeable membranes for selective transport of constituents.

Molecular mass, *M,* of macromolecules can be estimated by osmotic methods. Osmosis is a colligative property; that is, it is dependent on the concentration of the solute and not on its nature (ideal). By analogy to the gas law

$$PV = RT$$

$$\Pi V = RT \quad \text{(ideal case)}$$

$$\Pi = RT\mathbf{c}/M \qquad \text{...(7.7a)}$$

where $V = M/\mathbf{c}$

A plot of $\Pi/\mathbf{c}$ as a function of **c** should be a straight line from which *M* could be evaluated. (Fig. 7.3a).

For a real situation the van der Waals empirical equation for real gases is invoked.

$$(P + a/V^2)\,(V - b) = RT \qquad \text{...(7.7b)}$$

$$PV + a/V - bRT/V \approx RT; \quad \therefore\ P = \frac{RT}{V}$$

Substituting $P = \Pi$ and $V = M/\mathbf{c}$

$$\Pi/RT\mathbf{c} = 1/M + B\mathbf{c}/M^2 \qquad \text{...(7.7c)}$$

2nd virial coefficient $B = (b - a/RT)$ ...(7.7d)

From the plot of $\Pi/\mathbf{c}$ as a function of **c**, *B* and *M* can be calculated (Fig. 7.3b), from the slope and intercept, respectively.

(P = pressure; Π = osmotic pressure; V = volume; **c** = concentration; M = molecular mass of the particle; R = gas constant; T = abs. temperature; a & b = coefficients; B = 2nd virial coefficient)

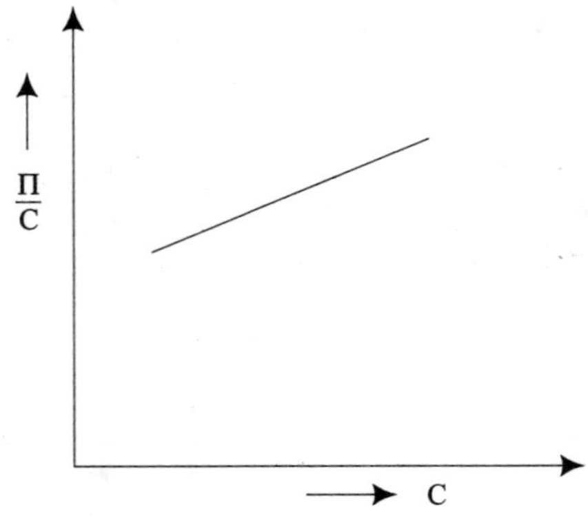

Fig. 7.3a Plot of Π/**c** versus **c**.

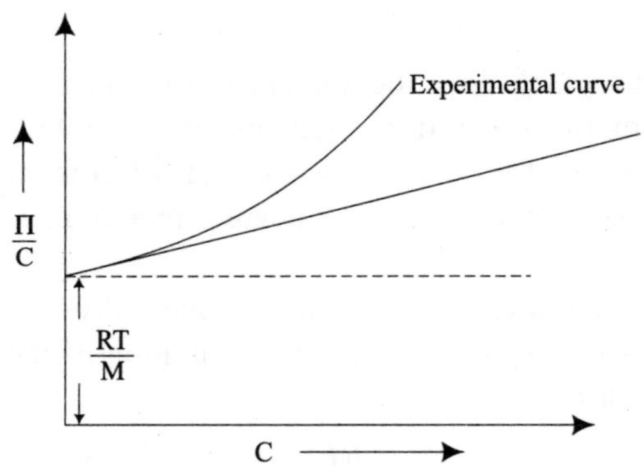

Fig. 7.3b Plot of Π/**c** versus **c**.

$B \approx \frac{1}{2} N_0 \overline{V}_e$, where $\overline{V}_e$ is the excluded volume due to a macromolecule. The excluded volume of the solvent contributes to the excess entropy of the solution and the attractions/repulsions between macromolecules contribute to the excess enthalpy.

Ultra-filtration is separation under hydrostatic pressure. In ultra-filtration, an inert gas at a constant pressure is passed through a filtration cell supported by a membrane disc. Membranes of cellulose nitrate, cellulose acetate and acrylonitrile polymers, with pore diameters in the range 0.001-0.1 μm are used. This technique is used for separation, purification and concentration of high- and low-molecular mass biological macromolecules— proteins, nucleic acids and their complexes, viruses and polysaccharides.

7.4 VISCOSITY

Viscosity of a fluid is an index of its resistance to flow, analogous to inertia in solids. Viscosity measurements can be employed to determine molecular mass, *M*. The following are various viscosities and definitions:

$$\eta_r = \frac{\eta}{\eta_0} \quad \text{...(7.8a)}$$

$$\eta_{sp} = \eta_r - 1 \quad \text{...(7.8b)}$$

$$[\eta] = \mathbf{c} = 0\left(\frac{\eta_{sp}}{\mathbf{c}}\right) \quad \text{...(7.8c)}$$

The flux

$$J = -\eta \frac{dv_z}{dx} \quad \text{(momentum transport)}$$

(η_0 = viscosity of the solvent; η = viscosity of the solute; η_r = relative viscosity; η_{sp} = specific viscosity; $[\eta]$ = intrinsic viscosity; v_z = velocity; $\mathbf{c}$ = concentration).

Intrinsic viscosity is $\eta_{sp}/\mathbf{c}$ at concentration $\mathbf{c} = 0$ (Fig. 7.4).

Viscosity of a dilute solution is

$$\eta = \eta_0 \{ I + [\eta]\, \mathbf{c} + ...\} \quad \text{...(7.9a)}$$

$$\eta = \eta_0 (1 + 2.5\, v) \quad \text{Einstein equation} \quad \text{...(7.9b)}$$

where,

$$v = V_{\text{particle}}/V_{\text{solvent}}$$

$$[\eta] = 2.5\, v/\mathbf{c} \quad \text{...(7.9c)}$$

Intrinsic viscosity, $[\eta]$, is related to molecular mass, M

$$[\eta] = KM^{\alpha} \quad \text{...(7.9d)}$$

where K and α are solvent-dependent constants; α = 1/2 and 2 for globular and random coil macromolecules, respectively.

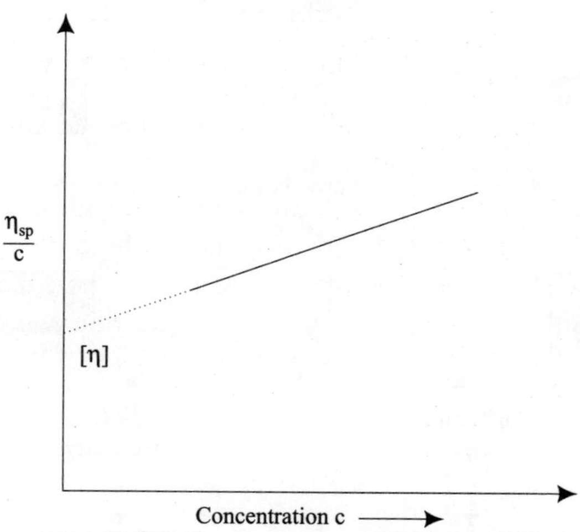

Fig. 7.4 Plot of $\left(\frac{\eta_{sp}}{\mathbf{c}}\right)$ versus concentration, $\mathbf{c}$, to interpolate intrinsic viscosity $[\eta]$.

7.5 CHROMATOGRAPHIC METHODS

Chromatographic and electrophoretic techniques are commonly used in biological sciences in the separation and identification of compounds and in the estimation of molecular mass, shape and size. Their extensive use is due to their simplicity and extensive development.

Chromatographic methods are employed in the separation of compounds based on the differences in (i) adsorption to a solvent (stationary phase), (ii) solubilities, (iii) molecular size exclusion (molecular sieving), (iv) charge, (v) binding affinities and (vi) ionic mobilities etc. In all chromatographic separations, the sample is dissolved in a mobile phase, liquid or gas, and the mobile phase is passed through an immiscible stationary phase.

Chromatographic methods are classified on the basis of the geometry of the system, mode of operation, retention mechanism and phases involved (Table 7.1). Choice of a chromatography depends on the physicochemical and physiological properties of solvents (Fig. 7.5).

Table 7.1 Classification of chromatographic techniques.

Mobile phase	*Stationary phase*	*Classification*
Gas	Liquid	Gas-liquid (partition) chromatography (GLC).
Gas	Solid	Gas-solid (adsorption) chromatography (GSC).
Liquid	Liquid	Liquid-liquid (partition) chromatography (LLC)
Liquid	Solid	Paper chromatography; Thin-layer (TLC); Adsorption; HPLC; Affinity chromatography.
Liquid	Solid matrix	Size-exclusion (molecular-sieve) chromatography.
Liquid	Ion-exchange resin	Ion-exchange chromatography.

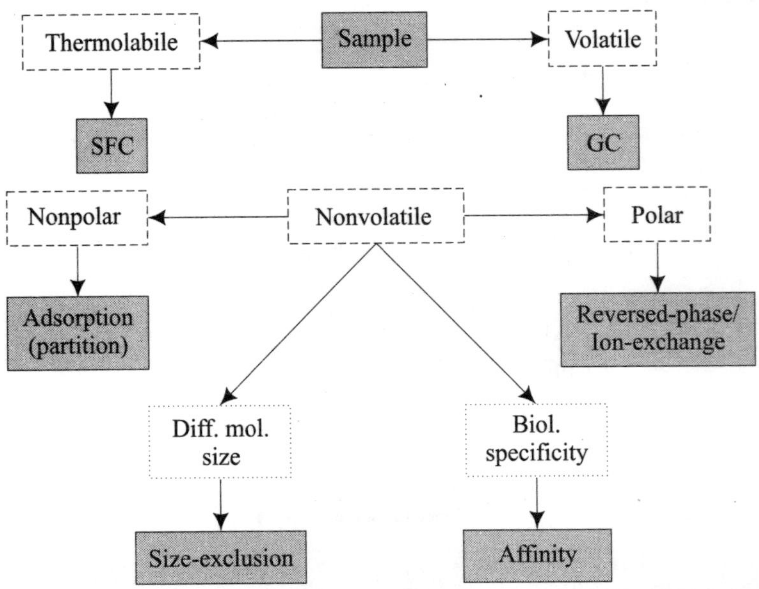

Fig. 7.5 Schematic of chromatographic separation methods.

7.5.1 Thin Layer Chromatography (TLC)

Thin layer chromatography (TLC) is a plane chromatographic method, similar in principle and operational features to paper chromatography. The main difference between them lies in the type of support matrix (stationary phase) material. The stationary phase (sorbent) is a solid, whose particle size can be controlled to improve the resolution. The mobile phase is a solvent or a mixture of solvents, whose selection depends on the sorbent and the type of analytes to be separated.

In this method, the stationary phase (sorbent) is a fine coating of alumina, silica or cellulose nitrate etc., spread uniformly onto a glass or plastic plate. Spots of known and unknown samples are applied at the edge of the sorbent-coated plate and the plate is kept in a closed chamber containing solvent-mixtures (mobile phase) in such a way that the line of spots is just above the absorbent layer. As the mobile solvent phase moves up the plate the samples also migrate upwards, their distances of migration are proportional to their affinities to the solvent and to the components of the solvent mixture. The plate is removed and dried before the solvent front reaches the top edge of the plate. The spots are identified visually due to colour or under ultraviolet light. The relative migration, R_f, is characteristic of the substance of the spot.

$$R_f = \frac{\text{distance of migration of the substance from the origin}}{\text{distance of migration of the solvent from the origin}} \quad \text{...(7.10)}$$

Known compounds spotted on the plate function as controls to identify unknown compounds.

7.5.2 Column Chromatography

The column chromatography system usually consists of a vertical glass column filled with a packing material (stationary phase), equilibrated with solvent. The sample is layered uniformly on top of the column, and eluting buffer from the reservoir, flowing down the column by gravity, carries along the components of the sample according to their solubility. There are several kinds of column chromatographies, classified according to the mobile phase—

1. Liquid chromatography (LC)— mobile phase is liquid.
2. Gas chromatography (GC)— mobile phase is gas.
3. Super-critical fluid chromatography (SFC)— mobile phase is super-critical fluid.

(1) **Liquid Chromatography (LC):** LC separates molecular species of a given sample, based on the difference in solubility. While the mobile phase in LC is liquid, the stationary phase is dependant on the method of separation, accordingly are classified: (i) adsorption (partition), size-exclusion (molecular-sieve), (iii) ion-exchange, and (iv) affinity chromatography.

***(i) Adsorption (partition) chromatography*:** Adsorption (partition) chromatography is the most widely used LC method in biological and medical sciences. In the *normal phase*, the stationary solid phase matrix consists of highly polar a substance such as silica, alumina, and the mobile phase is a *non-polar* liquid. In this technique, the non-polar components are eluted first, and increasing the polarity of the mobile phase decreases the elution time. In the *reversed phase*, the stationary solid phase is a non-polar substance, and the mobile phase is a polar liquid (water, methanol etc). In this method, the most polar components are eluted first, and increasing the polarity of the mobile phase increases the elution time.

In **high performance liquid chroromatography (HPLC),** high pressure pumping of the liquid phase is employed to achieve faster flow rates, higher resolution and minimal sample requirement. The pumping is done at high pressure, pulse free, with a flow rate of 1–10 ml/min. Due to this, HPLC columns are generally of stainless steel tubing, ~100–300 mm long, with inner diameter of 5–10 mm. Common packing materials are partial of silica, alumina, porous polymers and ion-exchange resins.

Two type of partition chromatography are in use, based upon the relative polarities of the mobile and stationary phases. In the normal-phase, the stationary phase is polar (water adsorbed to a stationary silica column) and the mobile phase is nonpolar (hexane). In the reverse-phase, the stationary phase is nonpolar (generally octyl or octyl-decylsiloxane) while the mobile phase is polar, such as water, methanol or acetonitrile. In the normal-phase partition, the least polar constituent is eluted first and in the reverse-phase (in contrast), the most polar constituent is eluted first and in this reverse phase method, increasing the polarity of the mobile-phase increases the elution time (better separation). Generally, the detectors are based upon the detection of UV absorption at 255 and 280 nm, fluorescence, reflective index and light scattering. Diode-array UV Spectrophotometers are versatile and are highly recommended for HPLC partition detection.

Reverse-phase HPLC is widely employed in various fields—biochemistry (separation of peptides, proteins, lipids, sugars etc.) the food industry (additives, toxicants and impurities), environmental and medical fields (pesticides, herbicides, pollutants, drugs, poisons and other hazardous substances).

(ii) Size-exclusion (Molecular-sieve) chromatography: Size-exclusion (also known as molecular-sieve or gel-filtration) chromatography is a molecular separation method, based on the molecular size and shape. The method is particularly applicable to separation of high-molecular mass species (macromolecules). The technique exploits the molecular sieve properties of the porous materials of the packing matrix. The packing columns (matrix) are generally inert, cross-linked glucose polymers in bead form of required pore diameter.

The pore size of the gel depends on the degree of cross-linking of the gel matrix. Molecules of size larger than the largest pore size of the gel beads, that is, above the exclusion limit, cannot penetrate the gel beads (size exclusion), whereas small molecules can penetrate. Therefore, large size molecules pass through the column unhindered and they are eluted first. Smaller size molecules, which enter the pores of the beads, are retained in the column for a longer time and, therefore, these are eluted through the column in the order of decreasing molecular size. That is, the elution time is a function of molecular size and shape and the average pore diameter of matrix beads determines the range of fractions eluted (Fig. 7.6).

The elution volume, V, of a molecule by gel-filtration is

$$V = V_0 + kV_i \quad \text{...(7.11a)}$$

for larger molecules of size larger than average pore size, $k = 0$

$$\therefore \quad V = V_0$$

for small molecules $\boldsymbol{k} \sim 1$ and

$$V = (V_0 + V_i) \quad \text{...(7.11b)}$$

[(V_0 = free volume outside the gel beads (available for all molecular sizes)
V_i = volume inside the pores (available for smaller molecules)].

From a plot of elution volume, V, against molecular mass (m), with molecular mass markers (Fig. 7.7), the elution volume of a species of unknown molecular mass, and thereby its molecular mass can be determined.

There is a variety of molecular sieves available in the market—Sephadex, Sepharose, Sephacryl etc., of various pore diameters for general and specific requirements (Table 7.2).

Gel-filtration is size-exclusion chromatography with a hydrophobic packing, used for separating polar species. Gel-permeation employs hydrophilic packing to separate nonpolar species.

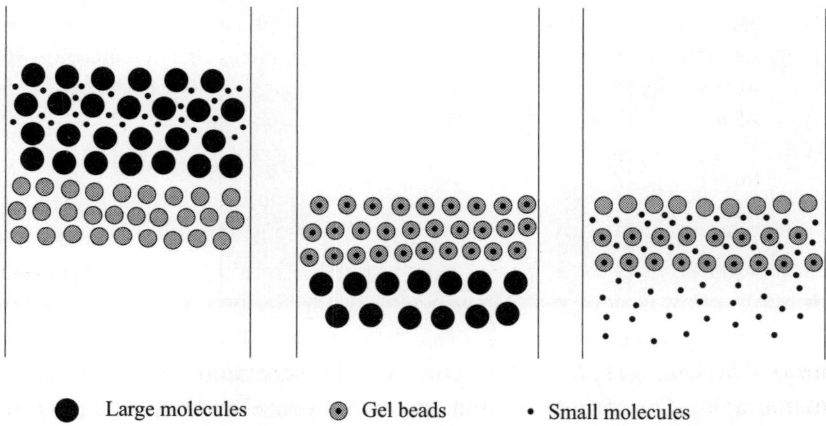

Fig. 7.6 Separation of molecules according to size.

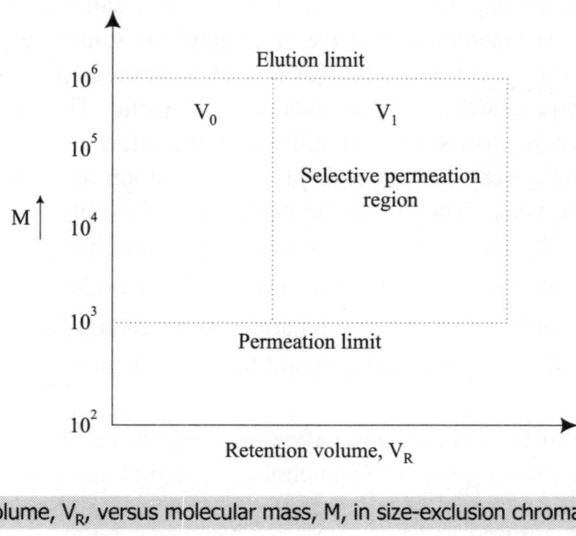

Fig. 7.7 Plot of retention volume, V_R, versus molecular mass, M, in size-exclusion chromatography.

Table 7.2 Molecular sieves used in gel-filtration chromatography .

Type of molecular sieve		*Bead diameter (in dry state)*	*Separation range daltons*
DEXTRAN			
Sephadex	G – 10	40 – 120	~ 700
	G – 25	40 – 200	1,000 – 5,000
	G – 50	20 – 100	1,500 – 30,000
	0 – 100	40 – 100	4,000 – 150,000
	G – 200	10 – 50	5,000 – 800,000
Sephacryl	S – 200	~ 70	5,000 – 200,000
	S – 400		20,000 – 800,000
AGAROSE			
Sepharose	2B		10,000 – 200,000
	4B		50,000 – 400,000
BioGel	A5m		10,000 – 50,000
	A50m		10,000 – 500,000
POLYACRYLAMIDE			
BioGel	P6		1,000 – 6,000
	P30		3,000 – 30,000
	P100		5,000 – 100 000
	P300		5,000 – 300,000

(iii) Ion-Exchange Chromatography. Molecules can be separated based on their charge by ion-exchange chromatography. The sorbent column (stationary phase) consists of a polymer with charged groups (ion-exchanger). A molecular species in a given sample which has opposite charge to that of the charged group of the resin column binds to the column. The bound molecules can be eluted by altering the pH of the buffer or by increasing the salt concentration of the buffer.

Choice of the exchanger depends on the stability of the sample components, their relative molecular mass and other parameters. Generally, if a sample is stable below its isoionic point (pI), giving it a net positive charge, a cation exchanger is useful, whereas if it is stable above its isoionic point, giving it a net negative charge, an anion exchanger is useful. The choice between a strong and a weak exchanger also depends on sample stability and the effect of pH on charge of the sample. Weak electrolytes requiring a very low or high pH for ionization are separated advantageously by strong exchangers and vice versa. The pH of the buffer should be pH = (pI + 1) of the compounds being separated. Separation by ion-exchange chromatography is achieved into stages –

(i) first, adsorption (binding) of the substance(s) to the ion-exchange column, then

(ii) eluting the substance(s) one at a time, separated from each other (Fig. 7.8).

Separations are generally performed under conditions in which one of the ions predominates in both phases.

Again, a wide variety of both anion- and cation-exchangers are available. Aliphatic or aromatic amino groups form anion exchangers, while phenolic, hydroxyl and carboxyl groups form cation-exchangers. Substitution of functional groups into the G-25 and G-50 Sephadex matrices gives eight ion-exchangers, four anion- and four cation-exchangers (Table 7.3a).

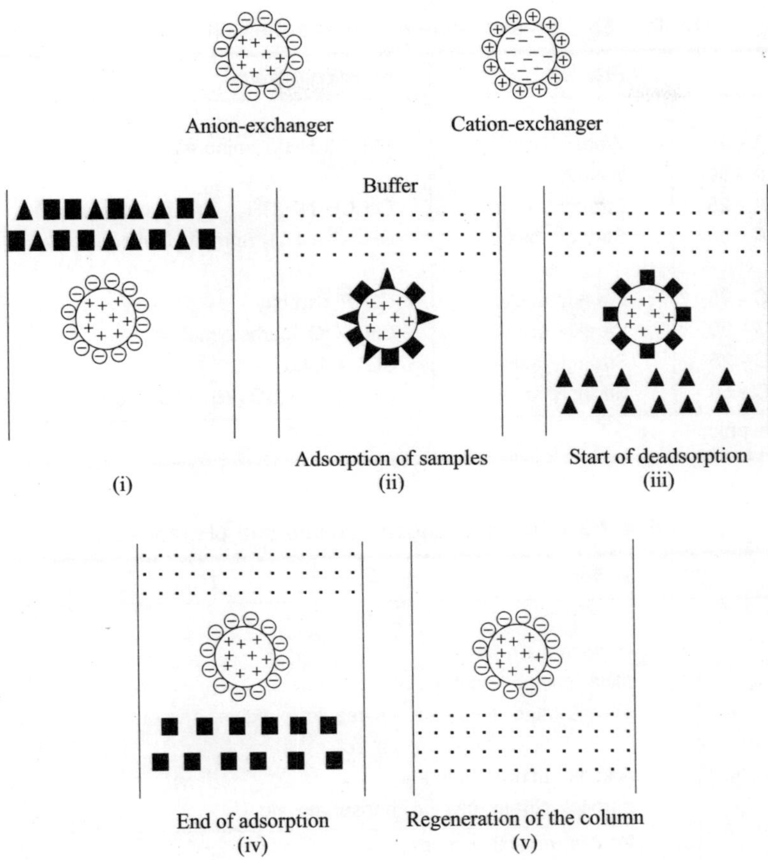

Fig. 7.8 Molecular separation by ion-exchange chromatography.

Proteins are amphoteric polyelectrolytes, with a net positive charge at low pH (below the isoelectric point) and a net negative charge at high pH (above the isoelectric point). Therefore, a protein binds to a cation-exchanger (such as CM- or SP-Sephadex) below its isoelectric point and to an anion-exchanger (DEAE or QAE) above its isoelectric-point. A-25 and C-25 ion-exchangers are intended primarily for peptides and small proteins which are able to penetrate ion-exchange beads (e.g. hormonal proteins, cyto- and neuro-toxins of snake, scorpion and other venoms). Macromolecules of high molecular mass (>200,000 daltons) can be chromatographed from either A- or C-type ion-exchangers, because only functional groups are available for interaction due to lack of penetration of larger molecules into the ion-exchange beads. In such cases, A-25 series are preferable since charge density is greater on the surface of the less-swollen beads and due to greater rigidity of the beads, better flow rates are achievable. In addition, whenever possible, cationic buffers should be used with anionic-exchangers and anionic buffers with cationic-exchangers. Some ion exchangers, preferred buffers and optimal pH are given in Table 7.3b.

(iv) Affinity Chromatography. In affinity chromatography, specific affinity of a molecular species to specific chemical group(s) of the sorbent columns is exploited to bind that particular molecular

Table 7.3a Ion-exchange resins and their characteristics.

Type		*Property*	*Functional group*	*Counter ion*
Anion Exchanger				
DEAE	A - 25	Weakly basic	$CH_2CH_2NH_3^+$ (amino ethyl)	chloride
	A - 50	Weakly basic		chloride
QAE	A - 25	Strongly basic	$CH_2CH_2N^+(CH_3)_3$ (triethylaminoethyl)	chloride
	A - 50	Strongly basic	$CH_2N^+(CH_3)_3$ (trimethylaminoethyl)	chloride
Cation Exchanger				
CM	C - 25	Weakly acidic	COO^- carboxy	sodium
	C - 50	Weakly acidic	CH_2COO^- carboxymethyl	
Sp	C - 25	Strongly acidic	SO_3^- sulpho	sodium
	C - 50	Strongly acidic	$CH_2CH_2CH_2SO_3^-$ (Sulphopropyl)	

A = anionic; C = cationic

Table 7.3b Ion-exchangers, buffers and pH ranges.

Ion-exchanger	*Buffer*	*pH range*
Anion-exchanger		
DEAE QAE	Cations such as: alkyl amines, alcohol; ethyldiamine, ammonia, imidazole, pyridine, Tris etc.	2 – 9
Cation-exchanger		
CM	Anions such as: acetate,	6 – 10
SP	barbitol, citrate, glycine, phosphate etc.	2 – 10
A – 25 or C – 25	for low molecular mass	< 20,000 daltons
A – 50 or C – 50	medium molecular mass	30,000 – 200,000
A – 25 or C – 25	high molecular mass	> 200,000

species to the column containing specific chemical group(s), X, and later removing the bound-molecule by eluting the column with a buffer containing a high concentration of the chemical group(s), X. The necessary and sufficient condition for the use of this method is that the material to be purified should be capable of binding <u>reversibly</u> to specific ligand that is attached to the column matrix.

The matrix column contains beads coated with ligands like antibodies, enzyme substrates and receptors that bind specifically to the molecular species, X (X = antigens, enzymes, receptor proteins). When a mixture containing different molecular species is added to the matrix column, only that molecular species that bind to the ligand are retained by the column matrix. When the sample is passed and washed, all the components except the molecular species, X, bound to the affinity matrix, pass through the column. Later, the bound species, X, are removed by altering the conditions of eluting buffer (e.g., pH) (Fig. 7.9).

This technique has been applied extensively in the purification of antigens, antibodies, enzymes, proteins, hormones, receptors, nucleic acids, carbohydrates, viruses and many other biologically important compounds, and quantified by radiolabeled (RIA and IRMA), enzyme-linked,

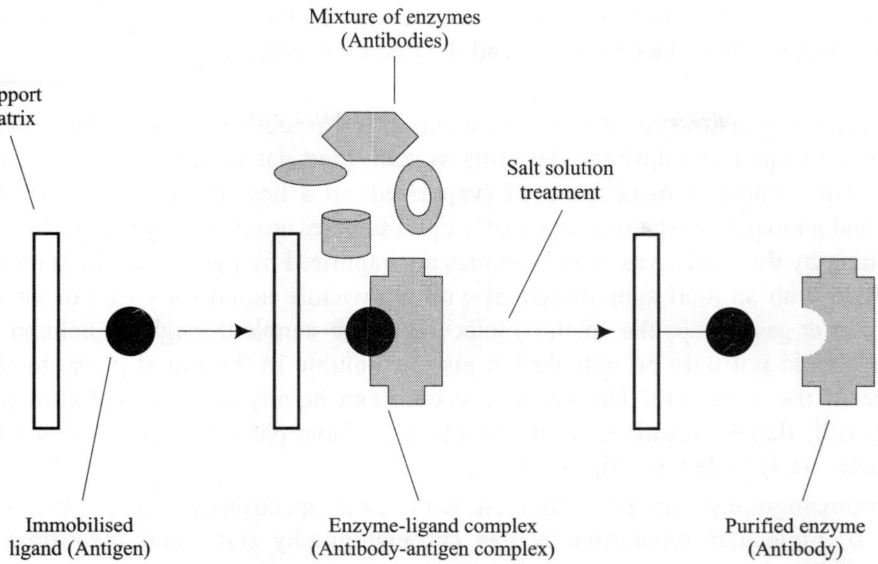

Fig. 7.9 Principle of molecular purification by affinity chromatography.

chemiluminescence and bioluminescence assays. Separation by affinity chromatography and detection by any of the immunoassay methods has a wide range of applications in diverse fields— biology, medicine, clinical chemistry (endocrinology), pharmacology (toxicology), immunology (allergy and oncology), virology, bacteriology and mycology.

Metal-chelate chromatography is an extension of affinity chromatography. In this method, immobilized metal ions, Cu^{2+}, Zn^{2+}, Mo^{2+}, Mn^{2+}, can be utilized to separate many proteins and peptides that bind to these metals. Ligand-exchange chromatography is another version of affinity chromatography, which is based on selective interactions between the analytes and matrix-bound ligands (ligand-ligand interactions). This method is used in the purification of many organic compounds, and peptides, proteins and nucleic acid and their constituents.

For example, soluble carbohydrate-containing molecules can be fractionated by affinity chromatography with wheat germ lectin-Sepharose. This is because of the ability of lectin to bind selectively to carbohydrate moieties containing N-acetyl- glucosaminyl residues. This is useful in the isolation/study of polysaccharides, glycoproteins and cell surfaces. Similarly, various types of affinity columns can be employed to separate different types of molecules, molecular complexes etc. *High-performance liquid chromatography (HPLC)* is the result of attempts to increase the efficiency of the LC. There are several books on separation techniques (practical series), giving detailed information on packing columns, chemicals, buffers, pH ranges, recipes etc., which are quite useful in the proper selection of parameters.

(2) Gas Chromatography (GC). Gas chromatography exploits the relative volatilities of compounds in a given mixture to separate them. Separation of substances in GC depends on their volatility and solubility differences. The basic principle of GC separation technique is—when a mixture of volatile materials, transported by a carrier gas/mobile phase (argon, helium or nitrogen), is passed through a column containing a stationary medium (column matrix) each volatile component will be partitioned between the carrier gas and the adsorbent phase according to its physicochemical properties.

A gas chromatographic system will have a carrier gas (mobile phase), sample injecting system, stationary phase (liquid or solid) and detecting system. Inert gas (helium, argon or nitrogen) is the carrier gas. The sample is made gaseous (vaporised) in a heated column (~ 400°C), which is nonvolatile and nonreactive (stationary phase). Liquid samples are thermally vaporised and are carried into the column by the carrier gas. Solid samples are vaporised by pyrolysis. The stationary columns are tubes filled with an inert support coated with nonvolatile liquid (in GLC) or an adsorbent (in GSC). The carrier gas sweeps the volatile molecules of the sample through the column to a detector. The sample should not only be volatile but also be soluble in the liquid phase to realise proper performance of the technique. The detector system can be any one of the following: a thermal conductivity cell, flame ionisation, electron-capture or flame photometric detectors. Generally, gas chromatograms are recorded on strip-charts.

Gas chromatography can be combined with mass spectrometry (GC–MS) to realise the advantages of molecular separation by gas chromatography (GC) and identification by mass spectrometry (MS).

(3) Supercritical Fluid Chromatography (SFC): Gases above their critical temperature and under high pressures are supercritical fluids with better mobility (which is the case with gases) and solvating properties (which is the case with liquids). The SFC technique exploits this liquid-like solvating property of supercritical fluids to separate polar, thermally unstable, chemically reactive, and non-volatile compounds. GC methods can only be used to separate compounds that can be volatilized. In the case of some substances, which are thermally unstable or cannot be resolved by GC, this technique works better than HPLC.

7.6 ELECTROPHORETIC METHODS

Electrophoretic methods find wider applications in purification and physicochemical characterisation of biomolecules. The technique is based on the differential migration of charged particles in a solution towards the electrodes of opposite polarities under the influence of an applied electric field. The extent of such a separation is time dependent.

The rate of migration of a charged particle under the influence of an electric field is dependent on— (i) the electrophoretic mobility of the charged particle and (ii) the frictional drag in the solution, and other physicochemical parameters. By optimal control of these parameters, the electrophoretic method can be made versatile for the separation of polar molecules such as amino acids, peptides, proteins, carbohydrates, and nucleic acids (DNA, RNA and their components).

Force, **F**, on an ion of charge, q, in an electric field of strength, E, is

$$\mathbf{F} = \mathbf{q}.\, E. \qquad \text{...(7.12a)}$$

The opposing force is the frictional drag, F'

$$\mathbf{F}' = \mathbf{fv}$$

The drift velocity $\mathbf{v} = \mathbf{q}.\, \text{E}/\mathbf{f}$ and electrophoretic mobility of an ion, μ, which is the drift velocity per unit field is

$$\mu = \mathbf{v}/E = V/\mathbf{f} = (ze/\mathbf{f})$$

$$\mu = \frac{\zeta\varepsilon}{6\pi\eta} \qquad \text{...(7.12b)}$$

where $\zeta = \dfrac{\mu}{\varepsilon r}$ by definition

(**f** is frictional coefficient, η is viscosity coefficient; ε is dielectric constant and r is radius of the particle).

Substituting for f and η from the **Stokes-Einstein** equation (Eq. 7.1c) and from the value of diffusion constant, molecular parameters, like mass, M, etc., can be estimated. The values obtained by electrophoresis are not accurate. But, its routine use in biochemical work is because of its simplicity and speed.

The electrophoretic mobility, μ, of a charged particle is a function of—

- The ionisable groups on the surface of the particle.
- The shape and rigidity of the particle.
- Pore size of the separation matrix.
- Physicochemical properties of the buffer medium.

The sign and the magnitude of charge carried by an ionized moiety vary with pH of the buffer, and the shape of the particle may be affected by denaturing agents in the buffer. By optimum selection of controlling parameters, greater flexibility and versatility can be achieved.

7.6.1 Gel Electrophoresis

Electrophoresis is carried out in gels for two reasons—(i) gels suppress convection currents produced by small temperature gradients, a requirement for effective separation and (ii) gels serve as molecular sieves that enhance separation. Molecules that are small compared with the pore size of the gel readily move through the gel, whereas molecules of larger size than the pore size of the gel are almost immobile. Intermediate size molecules move through the gel with varying degrees of mobility.

The pore size of the gel can be varied by varying the polymer concentration so that the molecular separation can depend not only on the charge of the molecule but also on its size and shape. Choice of optimal conditions in electrophoresis depends on (i) gel porosity and degree of cross-linking, (ii) nature, concentration and pH of the buffer medium, and (iii) maximum power applied.

Both horizontal and vertical slab gel setups are in use. Horizontal setup, with agarose gels (they have large pore size) is used for the separation of large-size macromolecules—nucleic acids, plasmids etc. While standard agarose gel electrophoresis techniques can resolve nucleic acids up to 30–50 kb, the limit can be raised, with the application of *pulsed field* techniques, to resolve large size nucleic acids, up to 10 Mb (10,000 kb).

The charge in a nucleic acid is independent of the pH of the medium. That is, the ratio of charge/mass is almost the same for all nucleic acids. Therefore, the electrophoretic separation of nucleic acids is based on the differences in their molecular sizes (masses).

Polyacrylamide gel electrophoresis (PAGE): Polyacrylamide gels are the supporting media of choice for electrophoresis (PAGE), because they are chemically inert and are readily formed by the polymerisation of acrylamide. In addition, the pore size of the gel can be controlled by choosing various concentrations of acrylamide and a cross-linking agent (methylbisacrylamide) at the time of polymerisation. Well-resolved single bands on a PAGE column are generally accepted as a criterion of purity/homogeneity of the separated fractions.

SDS-PAGE: Conventional PAGE, called the native PAGE, is employed for separation of proteins and not for the characterization of molecular mass, shape and size. The electrophoretic mobility of a protein depends simultaneously on its (i) net charge, (ii) its size (molecular mass) and shape and (iii) its structural rigidity. These factors vary with the experimental conditions. In order to establish quantitative relation between one of the parameters and the electrophoretic mobility of a protein, the influence of all other parameters should be suppressed. This is the basis of the *Sodium dodecylsulfate (SDS)-PAGE* separation method.

Sodium dodecylsulphate (SDS), $CH_3(CH_2)_{11}SO_3^-Na^+$, is an anionic detergent and the hydrocarbon chains of SDS can get linked to hydrophobic regions in the interior of (globular) proteins, leaving the ionised sulphate groups, (SO_3^-), jetting out into the solvent medium. This binding of SDS molecules to a protein imparts a large negative charge to the protein and also disturbs its shape and conformation. Large proteins will compete with more SDS molecules than the smaller proteins, and therefore, all proteins, will have the same charge/mass ratio and all will have similar shape. Molecules of different proteins, complexed with SDS and subjected to zone electrophoresis in an unrestrictive support medium and restrictive pore size (e.g. polyacrylamide gel) can be separated entirely on the basis of size. The system can be calibrated with marker proteins of known molecular masses, from which the molecular masses of test proteins can be estimated from the plot of log *M* versus *R* values.

7.6.2 2-D Electrophoresis (2-DE)

2-D electrophoresis (2-DE) (in conjunction with mass spectrometry) has emerged as a method of choice for the large-scale purification and characterisation in genomics/proteomics analysis. The 2-DE technique comprises tandem pair of electrophoretic separations—(i) isoelectric focusing (IEF), which separates proteins according to their isoelectric points (pIs) in the 1st dimension, and (ii) SDS-PAGE in the 2nd dimension for size separation. This method allows resolution of large number of proteins (1,000–2,000) from a single lysate.

IEF relies on the presence of a mixture of ampholytes with wide range of isoelectric points. When an electric field is applied to an ampholyte mixture, those with low pI will migrate towards anode, whereas those with high pI towards cathode. In this process a pH gradient is set up from one end of the gel to the other, as a particular ampholyte comes to rest at a position coincident with its isoelectric point. Proteins, introduced in such a column with a pH gradient, migrate through the column subjected to an electric field until each one reaches a point at which its own isoelectric point exactly equals the pH in the column, and becomes uncharged.

Isoelectric focusing can also be achieved with better efficiency and reproducibility by *immobilized pH gradient (IPG)* strips. This is achieved attaching a set of carrier ampholytes to acrylamide molecules and co-polymerizing into gels to form a pH gradient. By changing the concentration of the different monomers along the strip, pH gradients are covalently immobilized and stabilized into the gel. In this way, IPG strips with various pH gradients (3–12) can be fabricated.

7.6.3 Capillary Zone Electrophoresis (CZE)

Capillary zone electrophoresis (CZE) is an alternative to liquid chromatography (LC) on account of its ability to separate ionisable compounds. Separation of analyte in CZE is based on differences in their electrophoretic mobilities, which can be modified by adding suitable soluble reagents. CZE separations occur entirely in a liquid phase. As no mass transport is involved between mobile and stationary phases, no peak broadening results from this source. Therefore, separation power is much greater than for similar LC. Thus, CZE is more efficient, expeditious and selective than LC.

Suggested Reading

Books

Andrew, A.T. *"Electrophoresis: Theory, Techniques and Biochemical and Clinical Applications"*. Oxford University Press/ Oxford, UK. (1986).

Baugh, P.J. (ed). *"Gas Chromatography: A Practical Approach"*. IRL Press/ Oxford, UK. (1993).

Braithwaite, A., & Smith, F.J. *"Chromatographic Methods"*. Chapman & Hall/ London. (1996).

Clement, R.E. (Ed). *"Gas Chromatography—Biochemical, Biomedical and Clinical Applications"*. Wiley Interscience/ Toronto. (1990).

Dean, P.D.G., Johnson, W.S. & Middle, F.A. *"Affinity Chromatography: A Practical Approach"*. IRL Press/ Oxford, UK. (1985).

Deutscher, M.P. (Ed). *"Guide to Protein Purification"*. Academic Press: New York. (1990).

Dunn, M.J. *"Gel Electrophoresis of Proteins"*. BIOS Scientific Publishers/ Oxford, UK. (1993).

Fried, B. & Sherma, J. *"Thin Layer Chromatography"*, 4th Edn. Marcel Dekker/ New York. (1999).

Gehrke, C.W., Wixom, R.L. & Bayer, E. (Eds). *"Chromatography– A Century of Discovery (1900 - 2000)– The Bridge to the Science /Technology"*. Elsevier: Amsterdam. (2001).

Hames, B.D. & Rickwood, D. (Eds). *"Gel Electrophoresis of Proteins: A Practical Approach"*, 2nd Edn. IRL Press/ New York. (1990).

Harris, E.L.V. & Angal, S. (Eds). *"Protein Purification Methods: A Practical Approach"*. IRL Press/ Oxford, UK. (1991).

Hearn, M.T.W. (Ed). *"HPLC of Proteins, Peptides and Polynucleotides"*. VCH Pubs/ New York. (1991).

Heftman, E. (Ed). *"Chromatography"*. Elsevier/ Amsterdam. (1992).

Melvin, M. *"Electrophoresis"*. John Wiley/ New York. (1987).

Mori, S. & Barth, H.G. *"Size-exclusion Chromatography"*. Springerverlag: Heidelberg. (1999).

Narayanan, P. *"Environmental Pollution: Principles, Analysis and Control"*. CBS Pubs/ New Delhi. (In Print).

Neuberger, A.A. & Van Deenan, L.L.M. (Eds). *"Modern Physical Methods in Biochemistry"*. Elsevier/ New York. (1988).

Rickwood, D. & Hames, B.D. (Eds). *"Gel Electrophoresis of Nucleic Acids: A Practical Approach"*. IRL Press/ New York. (1990).

Scott, R.P.W. *"Techniques and Practice of Chromatography"*. Marcel Dekker/ New York. (1995).

Tinoco, I, (Jr). *"Physical Chemistry: Principles and Applications in Biological Sciences"*, 2nd Edn. Prentice-Hall/ Engelwood, NJ. (1985).

Sewell, P. & Clark, B. *"Chromatographic Separations"*. John Wiley/New York. (1988).

Skoog, D.A., West, D.M. & Holler, J. *"Fundamentals of Analytical Chemistry"*, (5th Edn). Saunders/ Philadelphia, PA. (1988).

Wilson, K. & Walker, J.M. *"Principles and Techniques of Practical Biochemistry"*. Foundation Pubs/ New Delhi. (1994).

Yost, R.S., Ettre, L.S. & Conlon, R.D. *"Practical Liquid Chromatography: An Introduction"*. Perkin-Elmer/ Norwalk, CT. (1980).

Articles

Görg, A. *"Two-dimensional Electrophoresis"*. Nature, 349; 545. (1991).

Lee, M.L. & Markides, K.E. *"Chromatography with Supercritical Fluids"*. Science, 235; 1342. (1987).

8 Spectroscopies

Synopsis: Spectroscopic methods provide a set of versatile analytical tools for determining structural and functional parameters from the analysis of specific lines (wavelengths) and their intensity profiles. A great advantage of these methods is that a large number of elements can be analysed simultaneously with a high degree of sensitivity.

Light scattering (Rayleigh scattering) can be employed to determine hydrodynamic properties of biological molecules in solution. Nephalometry/turbidimetry can be used to study attenuation of light to analyse turbid media.

Optical rotatory dispersion (ORD) and circular dichroism (CD) methods are employed to study ordered regions and conformational changes in chiral macromolecules in solution. Order/disorder, helix/coil transitions can be monitored by CD studies.

Fluorescence and phosphorescence spectroscopies find applications in compounds that contain chromophores, either intrinsic or extrinsic. Fluorescence quenching procedures are employed to study the content and environment of chromophores, such as Trp, Tyr, Phe in proteins, nucleic acid bases in nucleic acids and hemes in conjugated-protein complexes. Because of its sensitivity it finds application in the estimation of drug and toxic constituents etc., in the body fluid by cytofluorimetry.

Ultraviolet (UV) spectroscopy is applicable to the study of electronic transitions σ–π- and **n**-(nonbonded) orbitals. It can be used to identify chromophores, such as C=O, NO_2 etc. It is used in the estimation of protein concentration by monitoring the absorbance of Trp, Tyr side chains at 280 nm. However, UV spectroscopy has limited application in the study of finer structural details.

Vibrational frequencies of molecules can be obtained by infrared (IR) and Raman spectroscopies, which are complementary. They are employed to study polar interactions, polypeptide conformation, helix/coil transitions etc. Raman spectroscopy has more advantages and scope. Time-resolved resonance Raman studies (TRR) can be employed to study structure dynamics of intermediate states.

Electron spin resonance (ESR) spectroscopy is used to study transition metal complexes (e.g. Iron-sulphur proteins) and Mössbauer spectroscopy is useful to study in detail the environment of iron in iron-containing macromolecules.

Atomic absorption (AAS) and atomic emission (AES) spectroscopies are used for the analysis of trace metals.

Mass spectrometry (MS) is increasingly being used in purification, identification and primary structure (sequence) analysis of biological macromolecules.

8.1 LIGHT SCATTERING (RAYLEIGH SCATTERING)

Light scattering can be elastic, in which case there is no frequency shift of the scattered radiation from the frequency of the incident radiation ($\nu_0 = \nu_s$). This scattering is called **Rayleigh** scattering. Inelastic molecular scattering that occurs with the change in the frequency of the scattered radiation ($\nu_0 \neq \nu_s$) is called the **Raman** effect. Rayleigh scattering measurements can provide information on the molecular mass, size and shape (from the diffusion constants) of molecules.

8.1.1 Scattering by Small Molecules (Size of the Particle, $d < \lambda/10$)

Molecules can be considered as point dipoles. The electric field, E, of the incident radiation is homogenous over the entire point mass and there will be no phase difference between the incident and the scattered radiations. The electric field induces a dipole moment, μ, and the induced dipole moment is

$$\mu = \alpha \, . \, E \qquad \text{...(8.1a)}$$

where α is the polarisability of the molecules. The oscillating dipole emits radiation and the ratio of the scattered (I_θ) and the incident (I_0) intensities is given by the **Rayleigh** equation

$$\frac{I_\theta}{I_0} = \frac{16\,\pi^4\,\alpha^2\,N.\sin^2\theta}{\lambda^4\,r^2} \qquad \text{...(8.1b)}$$

The Rayleigh ratio, R_θ, is defined as

$$R_\theta = \frac{I_\theta}{I_0} \cdot \frac{r^2}{2\sin^2\theta}$$

$$R_\theta = \frac{8\pi^4\alpha^2 N}{\lambda^4} \qquad \text{...(8.1c)}$$

(N = number of particles per unit volume; θ = angle of scattering; λ = wavelength of the radiation; r = specimen to detector distance).

The polarisability is related to the refractive index, **n** and the concentration of the particles, **c**

$$\mathbf{n}^2 - 1 = 4\pi\,\mathrm{N}\,\alpha \qquad \text{...(8.2a)}$$

$$\alpha = \frac{\mathbf{c}}{2\pi N} \cdot \left(\frac{d\mathbf{n}}{d\mathbf{c}}\right) \qquad \text{...(8.2b)}$$

$$\mathbf{c}/N = M/N_0 \qquad \text{...(8.2c)}$$

where N = number of particles per unit volume, M = molecular mass and N_0 = Avogadro number.

$$\alpha = \frac{M}{2\pi\,N_0}\left(\frac{dn}{d\mathbf{c}}\right) \qquad \text{...(8.2d)}$$

$$\therefore \quad R_\theta = \frac{2\pi^2}{\lambda^4 N_0}\left(\frac{dn}{d\mathbf{c}}\right) \mathbf{c}.\mathrm{M}. \qquad \text{...(8.2e)}$$

$$\frac{\Omega \mathbf{c}}{R_\theta} = \frac{1}{M} \quad \text{for ideal gases.} \qquad \text{...(8.2f)}$$

where
$$\Omega = \frac{2\pi^2}{\lambda^4 N_0}.\left(\frac{d\mathbf{n}}{d\mathbf{c}}\right)^2$$

The analogy of scattering in gases can be extended to scattering in liquids with the inclusion of the refractive index of the solvent, $\mathbf{n_0}$, and the difference in the polarisabilities of the particles and the medium, α.

$$\mathbf{n}^2 - 1 = \mathbf{n}^2 - \mathbf{n}_0^2 \qquad \text{...(8.2g)}$$

$$\frac{\Omega \mathbf{c}}{R_\theta} = \frac{1}{M} \quad \text{for ideal solutions} \qquad \text{...(8.2h)}$$

where
$$\Omega = \frac{2\pi^2}{\lambda^4 N_0}\mathbf{n}_0^2\left(\frac{d\mathbf{n}}{d\mathbf{c}}\right)^2$$

While dealing with the actual situation, that is with real solutions, the interactions between solute particles have to be taken into consideration. Again, the analogy to the correction terms similar to that applied to real gases, is invoked.

$$PV = RT \qquad \text{ideal gases}$$

$$PV = RT(1 + BP + ...) \qquad \text{real gases} \qquad \text{...(8.2i)}$$

(P = pressure; V = volume; T = absolute temperature; R = gas constant and B is the 2nd virial coefficient).

Similarly

$$\frac{\Omega \mathbf{c}}{R_\theta} = 2M \qquad \text{ideal solutions}$$

$$\frac{\Omega \mathbf{c}}{R_\theta} = \frac{1}{M} + \frac{2B\mathbf{c}}{M^2} + ... \qquad \text{real solutions} \qquad \text{...(8.2j)}$$

Scattering by solvent and solvent + solute (solution) are measured experimentally and the difference is taken. ($d\mathbf{n}/d\mathbf{c}$) is determined by any differential refractometer. The quantity $\Omega\mathbf{c}/R_\theta$ is plotted as function of **c**. The intercept gives molecular mass (1/M). The slope at $\mathbf{c} \to 0$ gives $2B/M^2$ from which the 2nd virial coefficient can be determined (Fig. 8.1).

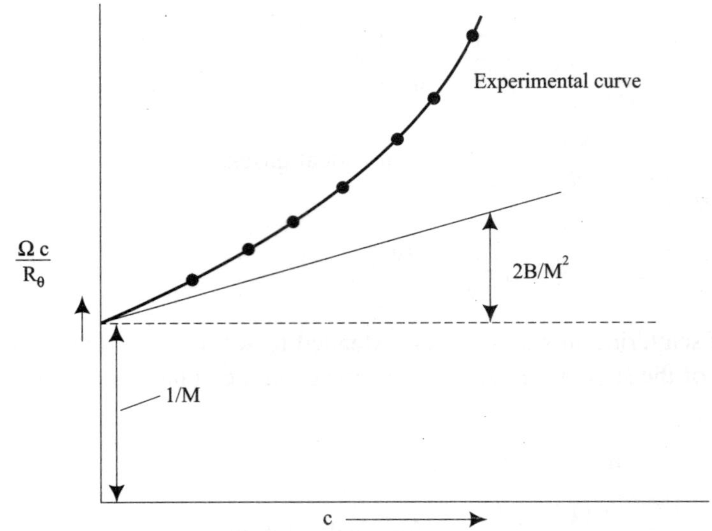

Fig. 8.1 Plot of $\Omega c/R_\theta$ versus **c**.

8.1.2 Scattering by Macromolecules (Size of the Particle, $d > \lambda/10$)

Macromolecules of size $d > \lambda/10$ cannot be considered as point masses. The finite size of the molecule has an influence on the intensity of the scattered radiation due to interference effects and the angular dependence of the scattered intensity should be taken into account. The internal interference of the scattered wave depends on the geometry of the molecule and the direction of the scattered light. The angular dependence of the scattered intensity is accounted for by the P_θ term, which is a measure of distortion from the predicted to the observed values.

$$P_\theta = \frac{I_{\text{obs}}}{I_{\text{Rayleigh}}} \qquad \text{...(8.3a)}$$

The intensity of scattering by large molecules can be computed by considering scattering of pairs of points (atoms), separated by a distance R_{ij} and oriented at an angle, θ, to the incident beam. Summation over all possible orientations of all pairs of atoms (points) of the molecule leads to the **Wierl-type** equation relating the angular-dependence of the intensity of scattered radiation.

$$P_\theta = \frac{1}{N^2} \sum_i \sum_j \frac{\sin \mathbf{s} R_{ij}}{\mathbf{s} R_{ij}} \qquad \text{...(8.3b)}$$

where N is the number of atoms in the macromolecule and

$$\mathbf{s} = \frac{4\pi}{\lambda} \sin(\theta/2) \qquad \text{...(8.3c)}$$

For small molecules

$$R_g = \frac{1}{2N^2} \sum_{ij} R_{ij}^2 \qquad ...(8.3d)$$

For large molecules

$$\frac{\Omega \mathbf{c}}{R_\theta} = \frac{1}{P_\theta}\left(\frac{1}{M} + \frac{2B\mathbf{c}}{M^2}\right) \qquad ....(8.3e)$$

where $$\frac{1}{P_\theta} = \left\{1 + \frac{16\pi^2\, R_g^2 \sin^2(\theta/2)}{3\lambda^2}\right\}$$

(R_g is radius of gyration; R_g^2 is defined as the squares of the mutual distances between all particles). Radiation of gyration is generally independent of molecular mass for globular molecules, but not for non-spherical and rod-shaped molecules.

The molecular parameters, M, B and R_g can be obtained from a plot of $\mathbf{\Omega c}/R_\theta$ as a function of $\sin^2(\theta/2) + \mathbf{kc}$ (**k** is an arbitrary constant), called a **Zimm plot** (Fig. 8.2), from which molecular mass, M, and size and shape of the molecule can be estimated. The experimental points of the Zimm plot are extrapolated to zero angle (X-axis) and zero concentration (Y-axis) according to the equations

$$\left(\frac{\Omega \mathbf{c}}{R_\theta}\right)_{\theta=0} = \frac{1}{M} + \frac{2B\mathbf{c}}{M^2} \qquad ...(8.3f)$$

$$\left(\frac{\Omega \mathbf{c}}{R_\theta}\right)_{c=0} = \frac{1}{M} \cdot \frac{1}{P_\theta} \qquad ...(8.3g)$$

from which M, B and R_g are obtained.

8.1.3 Quasi-elastic Scattering

Spectral width of scattered radiation (quasi-elastic scattering) can also give information on the translational and rotational diffusion motions of molecules in solution. Due to translational and rotational motions of molecules in solution, the width of the scattered line is Doppler-broadened. From the half width of the scattered line (Fig. 8.3) one gets

$$\Delta \nu = \frac{D\vec{\mathbf{k}}^2}{2\pi} \qquad ...(8.4)$$

where $\vec{\mathbf{k}}$ is wave vector $\vec{\mathbf{k}} = \frac{4\pi}{\lambda} \cdot \sin(\theta/2)$ and D is diffusion constant.

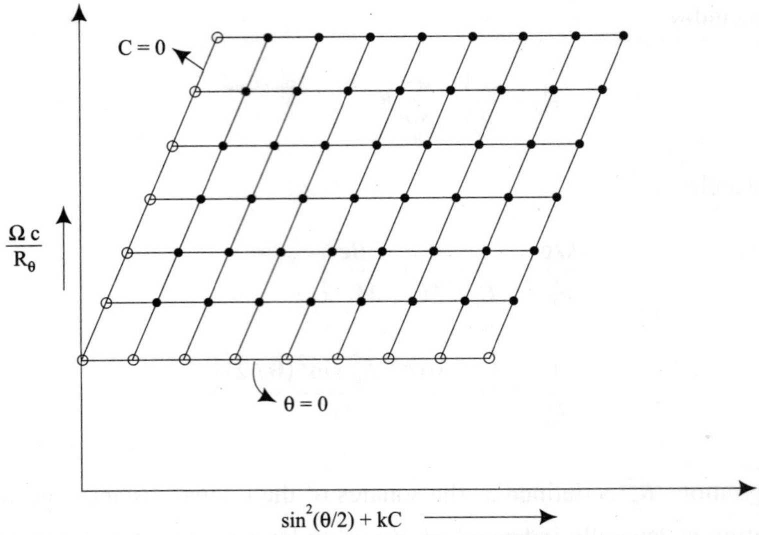

Fig. 8.2 Zimm Plot of light scattering.
• Experimental Points.
⊗ Extrapolated Points.
Zero angle, θ = 0 (X-axis).
Zero conc^n. **c** = 0 (Y-axis).

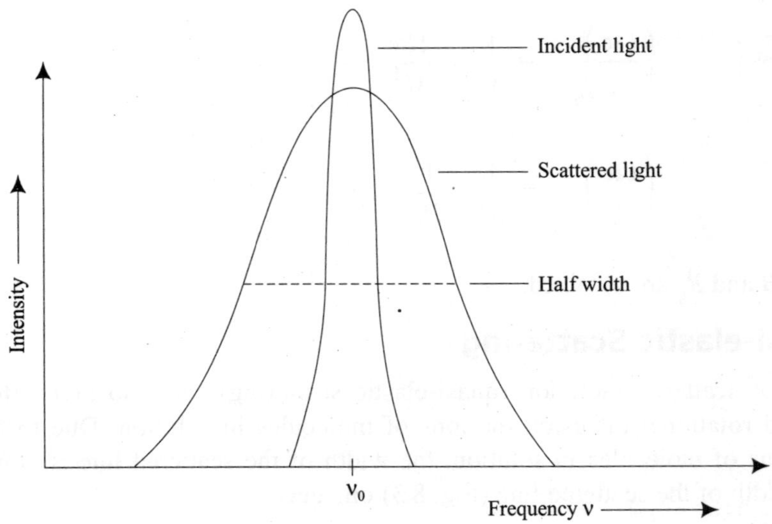

Fig. 8.3 Doppler-broadening of the scattered line.

On the macroscopic scale, diffusion is measured in terms of the change of concentration with time, according to **Fick's** laws. The radius of gyration, R_g, and diffusional coefficients (D_t and D_r) are shape-dependent. Some examples are given in Tables 8.1, 8.2a and 8.2b.

Table 8.1 Relationship between radius of gyration, R_g and shape.

Shape of the molecule	R_g	
Sphere	$\sqrt{3/5}.r$	r = radius
Ellipsoid	$\alpha\sqrt{\frac{2+b^2}{5}}$	a & b axes
Rod-shaped	$l/\sqrt{2}$	l = length
Random coil	$\sqrt{\frac{N}{6}}\,l=\sqrt{\frac{\overline{x^2}}{6}}$	$\sqrt{\overline{x^2}}$ = rms average distance between ends of the coil.

Table 8.2a Molecular mass and radius of gyration, R_g (nm).

Molecule	*Mol. mass (daltons)*	R_g *(nm)**	*Shape*
β-lactoglobulin	36×10^3	—	Globular
Myosin	490×10^3	47	Rod-like
DNA	4×10^6	117	Rod-like
TMV	40×10^6	92	Beam-like

*Data from Stanford, C. Physical Chemistry of Macromolecules. John Wiley/New York (1961).

Table 8.2b Translational, D_t and rotational, D_r, diffusional coefficients.

Molecule	D_t *(nm/s)**	$D_r\,(s^{-1})$	*Shape*
RNase	~ 1000		Globular
BSA	670	...	Globular
Ovalbumin	700	...	Globular
Lysozyme	1100	...	Globular
Immunoglobulin	~ 400		Globular
TMV	40	350	Beam-like
DNA	20	?	Rod-like
Myosin	~ 80		Roe-like

*From Dubin, S.B., Lunacek, J.H. & Benedeic, G.B. Proc. Natl. Acad. Scie., USA. 57:1164 (1967).

Molecular parameters, such as molecular mass, *M*, and molecular size and shape can be determined by other hydrodynamic methods, namely diffusion, sedimentation, osmosis, viscosity and electrophoresis (in some cases in conjunction with light scattering techniques).

8.2 OPTICAL ACTIVITY

Light, an electromagnetic wave, induces a molecule into forced oscillation. The features common to forced oscillation phenomena are : (i) dispersion (in-phase) and (ii) absorption (out-of-phase) effects, which are common to all molecules. However, in the case of chiral molecules, the velocities, and hence the refractive indices of right- and left- circularly polarised components of the plane-polarised light are not equal and, therefore, the chiral molecules (disymmetrical three-dimensional molecules) have the ability to rotate the plane of plane-polarised light. This phenomenon is called **optical activity**. The difference in the refractive indices of left- and right-circularly polarised components, $\mathbf{n_L}$ and $\mathbf{n_R}$ is called *optical rotation* or *circular birefringence*.

$(\mathbf{n_L} - \mathbf{n_R}) = \Delta\mathbf{n}$ Optical rotation or circular birefringence.

The angle of rotation of the plane of polarisation is α and the specific rotation, α_m is

$$\alpha_m = \frac{\alpha V}{ml} \quad \text{...(8.5a)}$$

where α is angle in radians; V = volume (m^3); m = mass (kg); l = length (m).

(Note : Formulas with α measured in degrees, mass in grams etc. are also in use).

Molar optical rotatory power, $[\alpha]_m$ is

$$[\alpha]_m = \alpha_m M \quad \text{...(8.5b)}$$

where M is molecular mass (kg. mol^{-1}).

8.2.1 Optical Rotatory Dispersion (ORD)

Optical rotation varies with wavelength of the radiation. This dependence of optical rotation on wavelength, λ, is called **O**ptical **R**otatory **D**ispersion (ORD).

$$(\mathbf{n_L} - \mathbf{n_R})_\lambda = (\Delta\mathbf{n})_\lambda \qquad \text{ORD,} \quad \text{...(8.6a)}$$

The ORD of a chromophore, far removed from its absorption edge, is given by the **Drude** equation

$$\alpha_m = \frac{\lambda_0^2}{(\lambda^2 - \lambda_0^2)} \quad \text{...(8.6b)}$$

where λ_0 is a constant representing the wavelength of the nearest optically-active absorption band.

If a plot of rotation against wavelength shows maxima and minima, then it is called a **Cotton** effect (see Fig. 8.4a). A Cotton effect is positive if the peak occurs at longer wavelength and is negative if the trough is at the longer wavelength region.

8.2.2 Circular Dichroism (CD)

For optically-active molecules, the intensities of the left- and right-circularly polarised components of the plane-polarised light are not the same, and the difference in the absorbance is **Circular Dichroism** (CD). Circular dichroism changes linearly-polarised light into elliptically-polarised light.

$$\Delta\mathbf{A} = (\mathbf{A_L} - \mathbf{A_R}) \qquad \text{Circular Dichroism (CD)} \quad \text{...(8.7)}$$

The tangent of the ratio of the major/minor axes of the ellipse, θ, is related to the molar ellipticity, $[\theta]$.

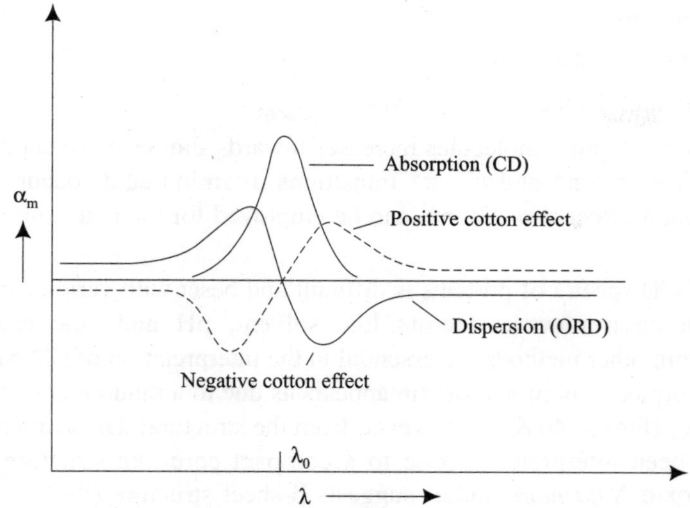

Fig. 8.4a Plot of specific rotation, α_m against wavelength, λ.

$$[\theta] = \tan^{-1} b/a \text{ and}$$

$$[\theta] = 3300\ \Delta A \qquad \text{...(8.8a)}$$

For proteins the mean residue ellipticity [θ] is calculated by

$$[\theta] = \frac{A \times 3300 \times M_w}{\mathbf{c.l}}, \qquad \text{...(8.8b)}$$

where a and b are the axes of the elliptically-polarised light, and M_w is the mean residue weight (~ 100), **c** is the concentration and l is the length of the cell. Circular dichroic data are represented as plots of [θ] against wavelength, λ.

The relationship between ORD and CD is analogous to dispersion and absorption. Whereas ORD is observed at wavelengths far removed from absorption bands, CD is encountered in the region of absorption bands and, therefore, the CD peaks are discrete and have higher resolution. Hence, CD studies have advantages over ORD studies. Both ORD and CD studies can be employed to determine the conformational features of chiral biomolecules in solution, such as helical content, helix-coil transition (denaturation), enzyme-substrate (ligand-ligand) interactions etc. The secondary structure content (helix, sheet and coil) in macromolecules is estimated by the mean residue rotation, according to the **Moffit-Yang** equation:

$$[m']_\lambda = a_0 \frac{\lambda_0^2}{\lambda^2 - \lambda_0^2} + b_0 \frac{\lambda_0^4}{(\lambda^2 - \lambda_0^2)^2} \qquad \text{...(8.9)}$$

where $[m']$ is the mean residue rotation, a_0 is an environment-dependent coefficient and b_0 is a solvent-independent coefficient. The value of b_0 is used as an estimate of the secondary structure. For example, at λ_0 = 212 nm,

$b_0 \approx 0$ for random coil

$b_0 = -630$ for 100% right-handed helix.

Left-handed helix and β-sheet have $+ b_0$ values.

The optical activity of chiral molecules increases towards shorter wavelengths due to $\pi - \pi^*$ and $\sigma - \sigma^*$ transitions. The $\pi - \pi^*$ and $n - \pi^*$ transitions in amino acids occur at 195 and 225 nm, respectively. The Cotton effect at ≈ 230 nm can be employed for the detection and estimation of α-helices in proteins.

Interpretation of CD spectra of proteins is difficult and beset with various uncertainties as CD is influenced by local environmental factors like solvent, pH and side chain orientation etc. Experimental data from other methods are essential in the interpretation of CD data. For example, the CD spectrum of a scorpion venom neurotoxin appears as due to a random coil structure according to the theoretical profiles (Figs. 8.4b & c). However, from the structural data available on other scorpion neurotoxins, it has been interpreted as due to a compact core-like structure. Similarly, the CD spectrum of a cytotoxin *Naja naja* (India) suggests β-sheet structure (Fig. 8.4d). But, the tertiary structures of snake venom cytotoxins are a combination of β-sheet and random coil segments with unique but complex folding patterns.

8.3 ABSORPTION SPECTROSCOPY

Molecular absorption spectroscopy, based upon visible, ultraviolet (UV) and Infrared (IR) radiations, is a widely employed physical technique in physical, biological and medical sciences. In conjunction

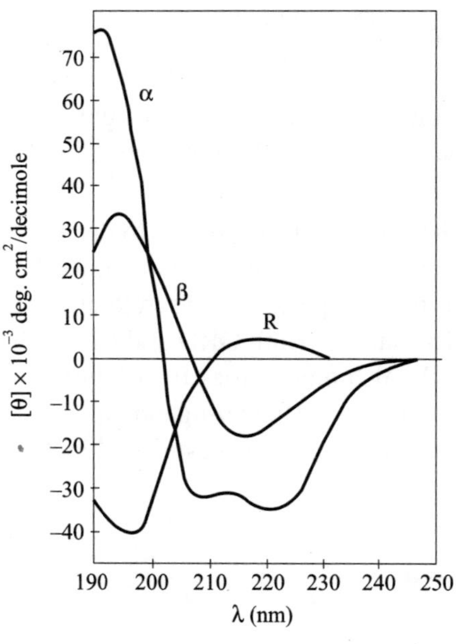

Fig. 8.4b CD spectra of the secondary structures adopted by proteins in solution. For poly *L*-lysine: α = 100% α-helical; β = 100% β-sheet; R = Random coil.
(Source: Greenfield, N. and Fasman G.D. Biochemistry. 8; 4108 (1969)).

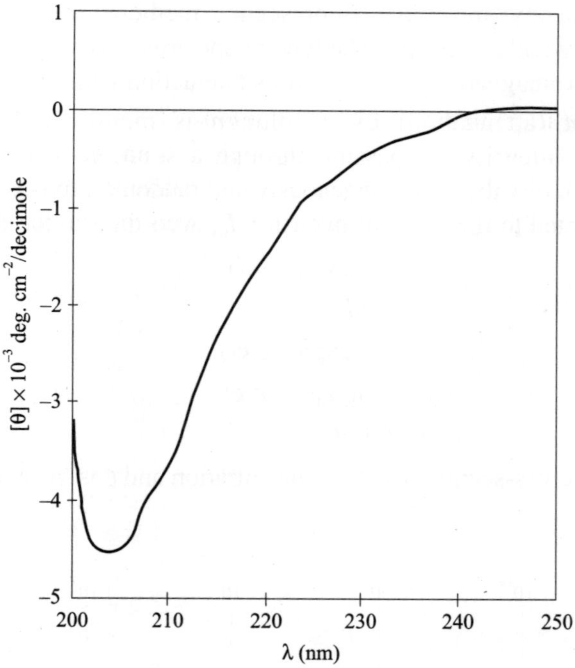

Fig. 8.4c CD spectrum of a scorpion venom toxin.
(Source: Lala, K. & Narayanan, P. Toxicon. 32; 325 (1994))

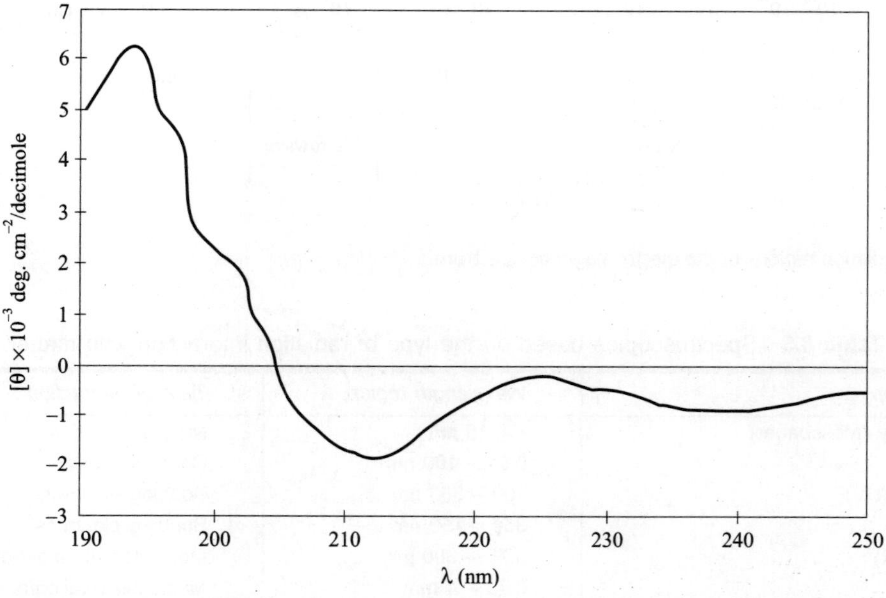

Fig. 8.4d CD spectrum of a snake venom cytotoxin.

with absorption spectroscopy, molecular fluorescence methods are of considerable importance because of their selectivity and sensitivity. The type of spectroscopy which is suitable depends on the energy region of the electromagnetic spectrum and its interaction with matter (Fig. 8.5 and Table 8.3).

Absorbance of light (attenuation) by a solution is monitored by photometric methods. Monochromatic light of intensity, I_0, passing through a solution of length, l, is absorbed by n chromophores (absorbers), distributed homogenously and randomly. In such cases the intensity of the transmitted light, I, is related to the incident intensity, I_0, according to the **Beer-Lambert** law.

$$I = I_0 \exp(-knl) \quad \text{...(8.10a)}$$

Absorbance $A = knl$

$$I = I_0 \exp(-\varepsilon \mathbf{c} l) \quad \text{...(8.10b)}$$

The optical density $OD = \log I_0/I = \varepsilon \mathbf{c} l$ ...(8.10c)

$$\varepsilon = A/\mathbf{c}l$$

where k is the absorption cross-section, **c** is the concentration and ε is the extinction coefficient (molar absorptivity).

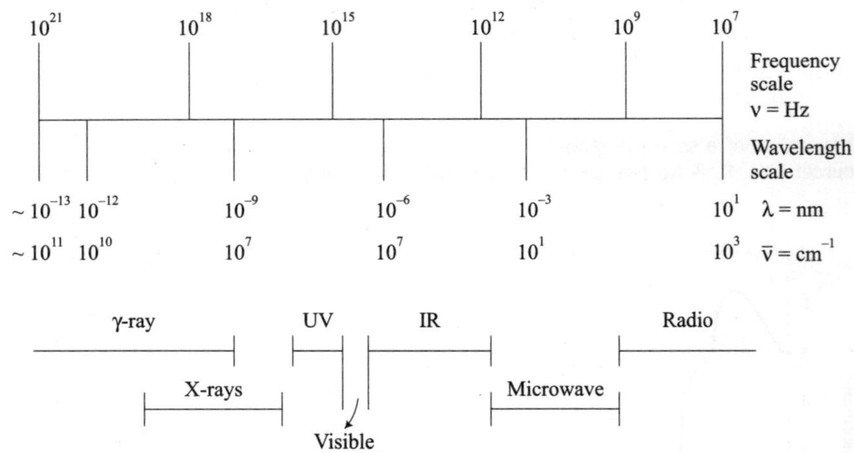

Fig. 8.5 Radiation regions of the electromagnetic spectrum.

Table 8.3 Spectroscopies based on the type of radiation interaction with matter.

Spectroscopy	*Wavelength region, λ*	*Type of Interaction*
Gamma ray (Mössbauer)	< 0.15 nm	Nuclear
X-ray	0.01 – 100 nm	Core electrons
Ultraviolet (UV)	100 – 350 nm	Bonding electrons
Visible	350 – 750 nm	Bonding electrons
Infrared (IR)	0.75 – 300 μm	Molecular vibration/rotation
Microwave	0.75 – 4 mm	Molecular rotation/orientation
Electron **S**pin **R**esonance (ESR)	~ 30 mm	Electron spin
Nuclear **M**agnetic **R**esonance (NMR)	0.5 – 10 m	Nuclear spin

Deviation from the **Beer-Lambert** law occurs due to the asymmetric distribution of the molecules in the medium, association/dissociation, pH changes and other environmental parameters.

8.3.1 Spectrophotometry

Spectrophotometers are commonly used in various kinds of spectroscopies. The general instrumental set up for any spectrophotometry requires — (i) a radiation source, (ii) sample holder, (iii) wavelength selector/monochromator, (iv) detector/recorder system.

Radiation Sources. There are a variety of radiation sources for the different types of spectroscopy and radiation. Both continuous and discrete line sources are available for selection. Continuous radiation sources in the UV region are Argon, Xenon lamps (150–600 nm) and hydrogen/deuterium lamps (180–380 nm by electrical excitation under low pressure). Tungsten filament halogen lamps and Xenon lamps provide continuous radiation in the visible region while the **Nernst glower** provides continuous radiation in the visible and infrared regions (350–20,000 nm). Nichrome wire and globar are continuous radiation sources for infrared studies. (750–40,000 nm). In addition, organic dye lasers provide continuously-tunable radiation in the UV and visible regions.

Hollow cathode lamps provide line spectra in the UV/visible regions (230–700 nm). Laser sources provide line spectra with a high degree of coherence.

A N_2-laser provides an intense line at λ = 337.0 nm, which is very useful for exciting fluorescence in a variety of molecules. Excimer lasers provide spectral lines (193 nm for Ar F; 248 nm for Kr F and 350 nm for Xe F) in the UV region. Gas lasers have spectral lines in the visible region (488 and 515 nm for Ar^+ laser and 633 nm for He-Ne laser). Solid state lasers are employed in infrared studies (Nd : YAG. λ = 1.064 μm; ruby λ = 694 nm).

Sample Holders. The type of sample holder depends on the transmission ranges of the materials. Lithium fluoride (LiF) and fused silica or quartz are employed for UV and can also be employed for the visible regions (150–3,000 nm). Generally glass is used in the visible region (300–500 nm). NaCl and KBr cells can be employed to scan UV to IR regions (150–30,000 nm).

Wavelength Selection. Wavelength selection can be achieved either by absorption filters or by interference filters. Fused silica or quartz prisms in the UV region, glass in the visible and NaCl and KBr prisms in the IR region are employed. Interference filters and gratings provide a wide spectrum, from UV to IR regions.

Absorption filters are generally coloured glass and they are restricted to the visible region of the spectrum (400–700 nm). They have low transmittance (<20%) and wide pass-width (>50 nm) (Fig. 8.6). Selection of the filter depends on the colour absorbed by the sample, as it is the intensity of this colour of the spectrum that varies with concentration. A filter that allows least transmittance of that colour should be selected.

Interference filters provide better transmittance (>50%) and narrower pass width (<10 nm). The interference filter works on the principle of optical interference. It consists of a transparent dielectric strip (magnesium fluoride or calcium fluoride) with both sides coated with semi-transparent silver film. It is sandwiched between two glass plates to protect it from damage. Incident radiation striking the first silver film is partly transmitted and partly reflected. The transmitted rays further undergo transmission/reflection at the second metallic layer. In this way, a particular wavelength is transmitted, excluding other wavelengths by destructive interference. The thickness of the dielectric plate determines the wavelength to be transmitted.

Monochromators. Monochromators are designed for spectral scanning, for proper selection of the wavelength in any spectrophotometry. Any monochromator consists of (i) an entrance slit, (ii) a collimating lens/mirror to produce a parallel beam, (iii) a prism/grating for wavelength dispersion and (iv) a lens/mirror for focussing the radiation onto an exit slit (Fig. 8.7).

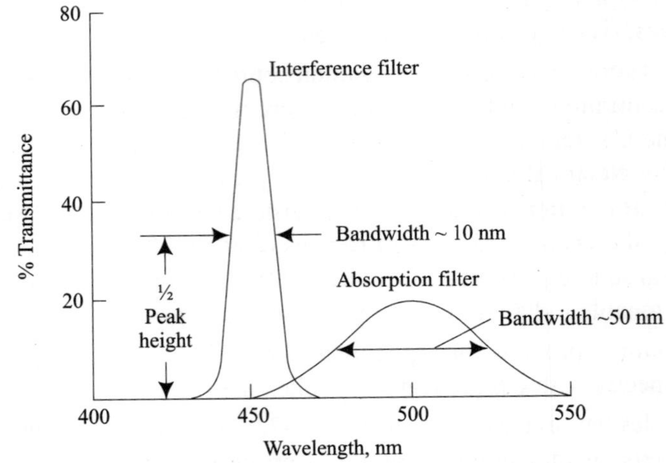

Fig. 8.6 Transmittance and bandwidth features of interference and absorption filters.

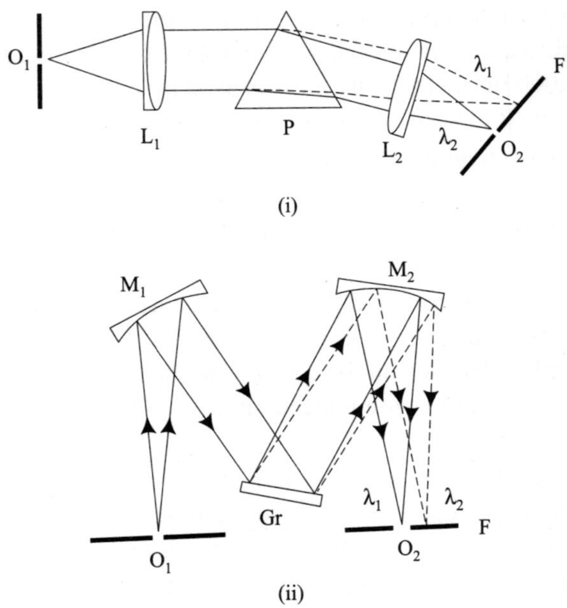

Fig. 8.7 Monochromator set up. (i) Prism type (ii) Grating type.
O_1 & O_2 = entry & exit slits; L_1 & L_2 = Collimating and focussing lenses; P = Prism; F = Focal plane; M_1 & M_2 = Concave mirrors; Gr = Reflection grating; λ_1 & λ_2 = Wavelengths (dispersion).

Grating elements are superior to prisms because of better wavelength separation, linear radiation dispersion and lower cost. Therefore, reflection gratings or holographic gratings are commonly used in all modern spectrophotometers. A holographic grating is an optical grating rather than a conventional mechanical grating. It provides high quality spectra, free from double images (ghosts) etc.

Radiation Detectors. All radiation detectors are either photon or thermal radiation detecting instruments. A transducer detector converts physical/chemical signals to electrical signals which can be amplified and modulated. All photon detectors are of the photoemission or photoconduction type and are employed in the high-energy range, from X-rays to the visible region (<2000 nm). Thermal detectors (thermocouples, bolometer) are employed in the measurement of infrared radiation.

Photon detectors are photo tubes, photomultipliers, photoconductivity tubes and silicon photodiodes. In photo tubes, the cathode is coated with a photoemissive substance (bialkali type), which emits electrons upon irradiation. The photocurrent generated is proportional to the radiation striking the anode. Photomultipliers are extremely sensitive and have a fast response to X-ray, UV and visible radiations.

A photomultiplier tube combines photoemission with amplification of the secondary electron emission. It consists of (i) a photocathode, which converts the incident photons into electrons, (ii) dynodes, which multiply the impinging electrons by three to four times and (iii) an anode which collects the electrons from the last dynode and whose pulse amplitude is proportional to the number of electrons collected (Fig. 8.8)

A thin layer of semi-transparent photocathode film, deposited on the inside of the multiplier tube face (glass or quartz), absorbs the incident radiation and releases electrons (photoelectrons) proportional to the number of incident photons. These primary photoelectrons are accelerated and focussed on the first dynode by the potential difference between the photocathode and the dynode and by the focussing electrodes. The dynode is the multiplier component of the apparatus. The first

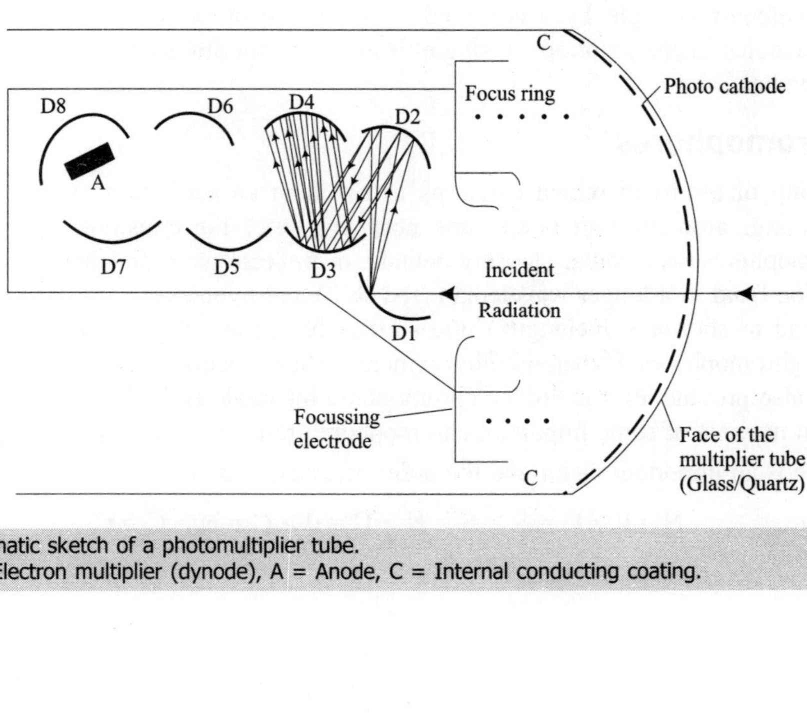

Fig. 8.8 Schematic sketch of a photomultiplier tube.
D = Electron multiplier (dynode), A = Anode, C = Internal conducting coating.

dynode absorbs high-energy electrons (primary photoelectrons) and emits secondary electrons (in the ratio 1:3 or 1:4). These secondary electrons in turn strike the next dynode and so on. In this step-wise process an avalanche of secondary electrons are produced, ultimately collected by the anode. The amplitude of the voltage pulse is proportional to the number of electrons collected by the anode, which is proportional to the number of photons incident on the photocathode.

Photoconductivity detectors are ideal for near-infrared radiation studies (0.75 – 3 μm). A silicon diode detector consists of a reverse-bias *p-n* junction formed on a silicon chip. A large number of such detectors can be assembled in an array on a small silicon chip; with such an arrangement, all wavelengths of dispersed radiation (from a spectrophotometer) can be measured simultaneously. Such multi-channel photon detectors are highly desirable and relevant for high-speed spectral analyses. Charge-transfer devices, made up of an array of tiny semiconductor capacitors, forming an integrated circuit on a small silicon chip, are also in use in multi-channel photon detectors.

Single-beam spectrophotometers are simple and inexpensive and are routinely used in general laboratory work and in colorimetry: More advanced are the double-beam spectrophotometers, designed to annul the variations in the intensity of the light source. This is achieved by 'splitting' the light beam into two parts and directing one of the beams through a reference cuvet (a reference beam) and the other beam through the sample cuvet. The 'splitting' operation can be either temporal or spatial (Fig. 8.9). Diffraction grating monochromators and narrow band pass width (0.5 nm) are other features. Double-beam spectrophotometric methods (difference spectroscopy) can be employed to obtain high precision results and are used to study features such as the influence of solvents, pH, temperature and concentration on the absorption spectra or chromophores.

8.3.2 Colorimetry

Coloured solutions absorb particular colours (wavelengths) of white light and reflect the rest of the visible spectrum. For example, a substance that absorbs yellow light reflects blue (white – yellow = blue) light. This is the principle that is employed in colorimetry. The amount of the absorbence of particular wavelength of light by a coloured solution is a measure of the concentration of the absorbing molecules in the solution. A single-beam spectrophotometer can be employed in most colorimetric studies.

8.3.3 Chromophores

Atoms or group of atoms in which electrons involved in an excitation are localised are called chromophores (e.g., aromatic-ring compounds, porphyrins etc). For transitions in the range of 200 – 800 nm, chromophores must contain loosely-bound n or π-electrons. Bathochromicity, that is shifting of an absorption band to a longer wavelength (red shift) and hypochromicity, that is, shifting of the absorption band to shorter wavelength (blue shift), give details on the structural and functional parameters of chromophores. Hyperchromicity (increase in extinction) and hypochromicity (decrease in extinction) also provide information on chromophore interactions. Details of electronic transitions and absorption maxima of some important chromophore groups are listed in Table 8.4.

Resonance is an important factor and the order of absorption is

$$N = 0 > C = S > N = N > C = 0 > C = N > C = C$$

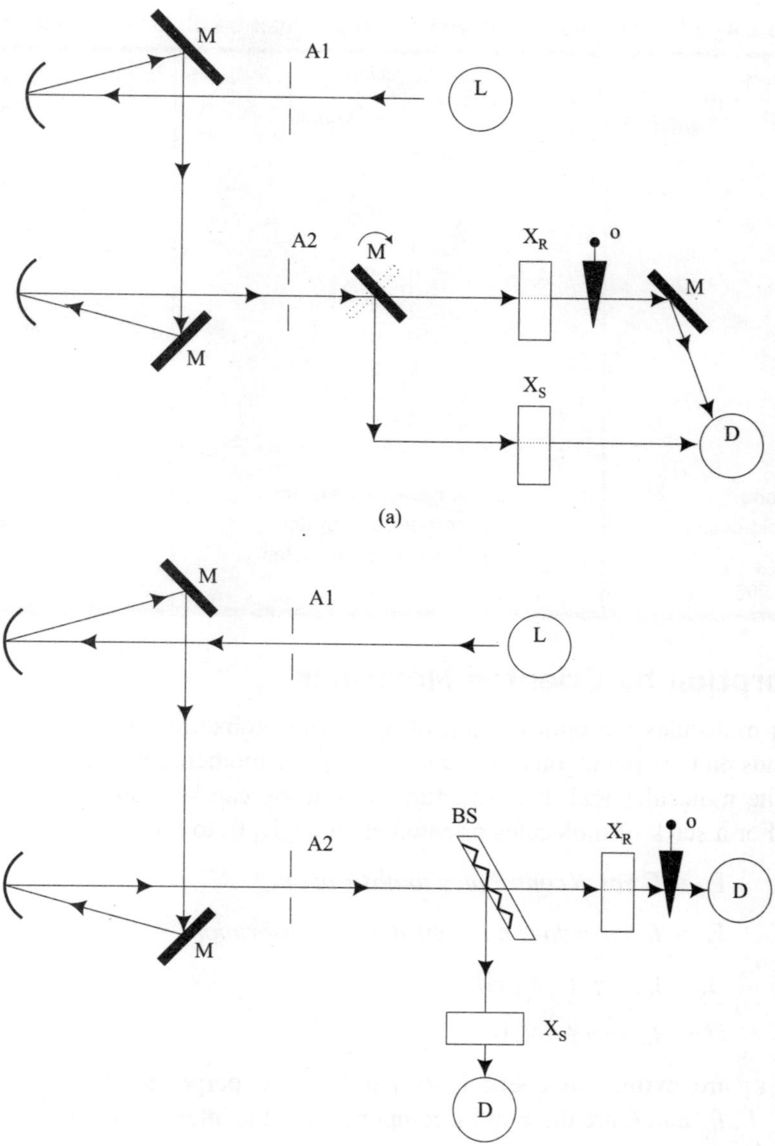

Fig. 8.9 Diagrammatic representation of double-beam spectrophotometers temporal and spatial.
(a) Temporal double-beam spectrophotometer.
(b) Spatial double-beam spectrophotometer.
A_1 & A_2 = entry and exit apertures (slits); L = light source; D = detector; X_R & X_S = reference and sample cuvettes; O = optical wedge; BS = beam splitter; M = mirrors; $\vec{M}$ = rotating mirror (chopper).

Polynuclear hydrocarbons absorb light and the energy is used to promote one of the π-electrons to unused π*-orbitals (the origin of π – π* transitions). For example, naphthalene absorbs light in the blue region and its colour is orange; **orange = white – blue.** Similarly, highly-conjugated compounds, like circumnaphthalene, are black; **black = white – white.**

Table 8.4 Electronic transitions and absorption maxima of some chromophores.

Chromophore	*Transition*	λ_{max} *(nm)*
– COO – R	$n - \pi^*$	205
	$\pi - \pi^*$	165
> C = 0	$n - \pi^*$	280
	$n - \sigma^*$	190
	$\pi - \pi^*$	150
> C = S	$n - \pi^*$	500
	$n - \pi^*$	240
– S – S –	$n - \sigma^*$	250 – 330
> C = C <	$\pi - \pi^*$	190
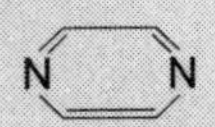	$n - \pi^*$	300
	$\pi - \pi^*$	250
Peptide bond	Small π-electron system	190
Nucleic acid bases	Larger π-electron system	260
β-carotins	Extended π-electron system	400 – 500
Porphyrin ring	" "	> 400

8.3.4 Absorption by Oriented Molecules

Light-absorbing molecules are optically anisotropic. The probability of absorption of light by the molecules depends on the orientation of the transition dipole moment relative to the electric vector, **E**. Orientation of the molecules and their structural anisotropy can be studied by employing linearly-polarised light. For a stack of molecules oriented at an angle, θ, to the electric vector

$$E_x = \boldsymbol{E} \cos\theta \textit{ contributes to absorption}$$

$$E_y = \boldsymbol{E} \sin\theta \textit{ does not contribute to absorption} \qquad \text{...(8.11a)}$$

$$I_x = I_{0x} \exp(-\varepsilon_{11}\, cl) \qquad \text{...(8.11b)}$$

$$I_y = I_{0y} \exp(-\varepsilon_1 cl)$$

where ε_{11} and ε_1 are extinction coefficients parallel and perpendicular to the electric vector, respectively; I_{0x}, I_x, I_{0y} and I_y are the x and y components of the intensities of incident and transmitted lights, respectively.

In the case of molecules aligned parallel to the electric vector

$$I = I_x + I_y = I_{0x} \exp(-\varepsilon_{11}\, cl) + I_{0y} \qquad \text{...(8.11c)}$$

The dichroic ratio, R, is given by

$$R = \left(\frac{\varepsilon_{11} - \varepsilon_1}{\varepsilon_{11} + \varepsilon_1}\right) \qquad \text{...(8.11d)}$$

8.3.5 Action Spectra

Action spectroscopy is used to characterise pigments (photosensors) spectroscopically and kinetically. Any graph representing a photoresponse as a function of λ is called an action spectra. The Beer-Lambert law does not hold good in the case of action spectra. The theory postulated by Mie is more useful. Action spectroscopy is based on the **Einstein-Stark** equivalence— i.e., the amount of transformed pigment is proportional to the number of photons absorbed by the pigment. Action spectroscopy has been employed to study many areas such as, (i) chlorophyll formation and photosynthesis, (ii) photoactivation (e.g., of cytochrome oxidase; nitrate reductase), (iii) photoinactivation and photodimerisation (e.g., of thymine dimerisation in nucleic acids), (iv) rhodopsins and phytochromes etc.

8.3.6 Turbidimetry and Nephelometry

Turbidimetry and nephelometry measure the attenuation of intensity of the transmitted beam and scattered beam, respectively, due to the presence of dispersed particles in a solution. Turbidimetry is similar to colorimetry and both employ similar instruments. Nephelometry is similar to fluorimetry but the scattering is elastic (**Rayleigh** scattering). Beer-Lambert law does not apply in these cases. Transmitted light is measured along the axis of the incident light for turbidimetry, and the scattered light is measured at 90° to the incident light for nephelometry (Tyndall scattering). The intensity ratio, (I_0/I) of the sample is compared with standard suspension(s) of known concentration(s). The instrumental components are similar to those used in spectrophotometry. Mercury vapour lamp with green filter or He-Ne laser would provide monochromatic radiation. Photodiodes or photomultiplier tube (PMT) would serve as detectors.

Turbidimetry and nephelometry are widely employed in biological and clinical studies, to evaluate suspended materials and impurities (in dairy, water-treatment, food, agro- and beverage industries etc.).

8.3.7 X-Ray Absorption Spectroscopy

An X-ray absorption spectrum is simple to interpret and can be used in elemental analyses.

X-ray Absorption. X-ray sources produce both continuous **(Bremsstrahlung)** and line-spectra. Each element has its own characteristic X-ray absorption edges (K, L, M, N... series). Therefore, wavelengths of these absorption edges and their intensities can be used in the quantitative analysis of any element present in a given sample (generally, heavier elements above sodium). Treating the absorption of X-rays through matter as linear, the mass absorption coefficient, μ_m ($\mu_m = \mu/\rho$) is given by

$$\mu_m = KZ^4 \lambda^3 \frac{N_o}{A_m} \qquad ...(8.12)$$

where K is a constant; Z is the atomic number; A_m is atomic mass and N_o is Avogadro number.

In general, standard X-ray sources available in laboratories do not provide a wide spectrum of X-radiation. Therefore, the extrapolation of a single measurement made on the side of the absorption edge cannot be carried out. This has limited the usefulness of X-ray absorption as an analytical method. However, with the availability of synchrotron radiation sources, continuously-variable wide-band X-ray wavelengths are realisable and suitable and improved X-ray spectrophotometers are now available. These developments make this method very versatile and simple to interpret in many elemental analyses (in chemical, clinical and forensic sciences).

X-ray detectors are either gas-filled, scintillation or semiconductor devices. Separation of the signal from noise can be achieved with pulse-height discriminators.

Microradiography. Differences in the X-ray absorbing power of different elements can be employed as a quantitative method of identifying various elements in a specimen. This method can be employed in any laboratory where common X-ray generators are available. A beam of X-rays passing through a specimen is attenuated by the elements in that specimen and transmitted radiation, recorded on a photographic film/plate, is a record of the attenuation characteristics. Heavier atoms attenuate (absorb) X-rays more than lighter atoms and accordingly the photographic film/plate reveals lighter and darker regions, corresponding to the heavier and lighter elements present in the specimen.

8.3.8 Resonance-Ionisation Spectroscopy (RIS)

Resonance-ionisation spectroscopy is an extremely sensitive and highly selective analytical tool. The method is based on resonance absorption of analyte atoms followed by further excitation to ionisation continuum. It exploits high coherence and monochromatic properties of lasers to ionise and eject electrons from selected types of atoms or molecules. The molecular ions (or electrons) are then detected by various means to identify elements or compounds and determine their concentration. Photons from lasers are tuned so that their energies just match allowable transition energies for electrons in a selected atom. The general procedure is to select a wavelength to excite a particular atom from its ground state to an excited state (resonance), while the second photon completes the ionisation process. Ionisation will not occur unless the laser is tuned to the atom.

The combined features of selectivity, sensitivity, and generality make RIS suitable for a wide variety of applications. All the elements in the periodic table, except helium and neon can be detected with RIS. Though mass spectrometry (MS) is well suited for the measurement of isotopes, it has difficulty in resolving the isobars (atomic nuclei with same mass number (A), but different atomic number (Z)). Incorporation of RIS in mass spectrometry solves the isobar problem.

8.4 FLUORESCENCE SPECTROSCOPY

Emission of light from an 'excited' electronic state of a molecule to its ground state is luminescence. Fluorescence is the emission of light that causes a molecular transition from the excited state to the ground state without any change in spin orientation (S_1 to S_0 states and life time 10^{-8} to 10^{-4} sec). Phosphorescence is the transition from a triplet state T_1 (due to change in spin orientation upon excitation from S_1 to T_1 state) to the ground state S_0 ($T_1 \rightarrow S_0$) transition with life time 10^{-4} to seconds). That is, emission in singlet states (no change in spin orientation) leads to fluorescence, and emission from triplet to singlet states (due to change in spin orientation) leads to phosphorescence (Figs. 8.10a, b and c).

In either fluorescence or phosphorescence the frequency of the emitted radiation is less than the frequency of incident radiation. The relationship is

$$\mathbf{v_0 > v_{fluo} > v_{phos}} \qquad \text{(v is frequency).} \qquad ...(8.\ 13\ )$$

Fluorescent chromophores may provide information about molecular environments surrounding the chromophores and thus fluorescence techniques can be employed to study rotational diffusion, collision and reorientation and charge-transfer reactions.

σ-electrons absorb in the ultraviolet (UV), and π-electrons in the near UV and visible regions. Aromatic rings exhibit a strong fluorescence, because of π – π* transitions. Conjugated, fused rings enhance fluorescence. The planarity of the molecule is an important aspect of fluorescent chromophores. *n* – π* and π – π* transitions are very important as they occur in the visible spectral region (200–700 nm).

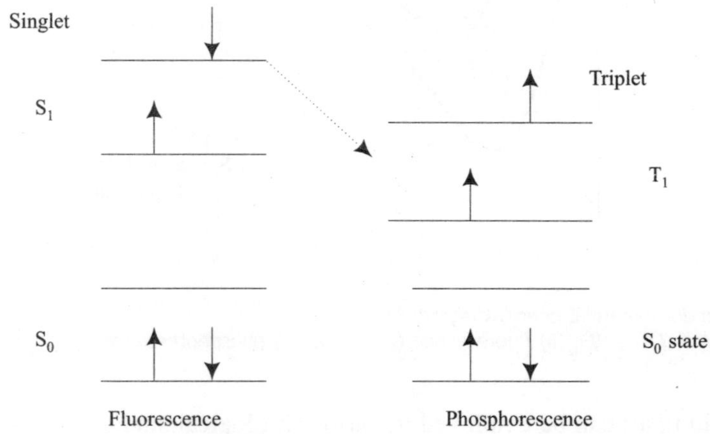

Fig. 8.10a Orientation of spins in fluorescence and phosphorescence.

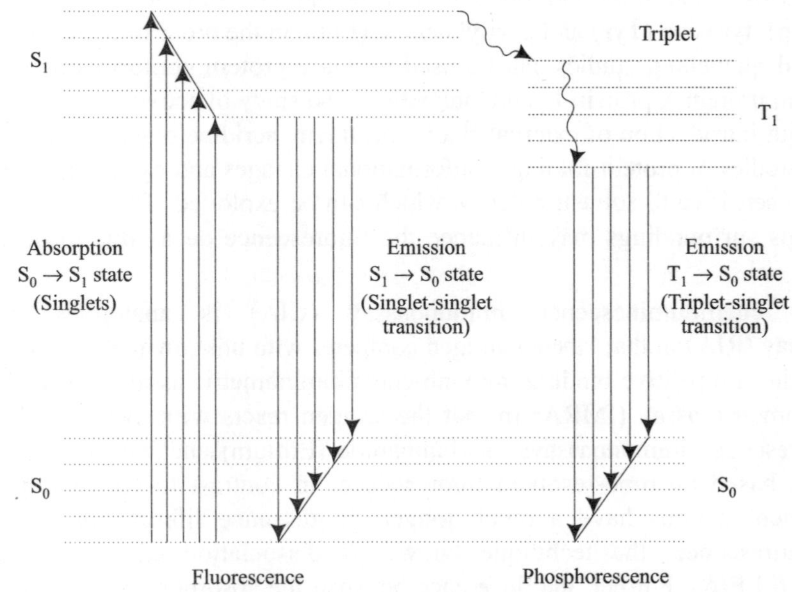

Fig. 8.10b Absorption-emission diagram in fluorescence and phosphorescence.

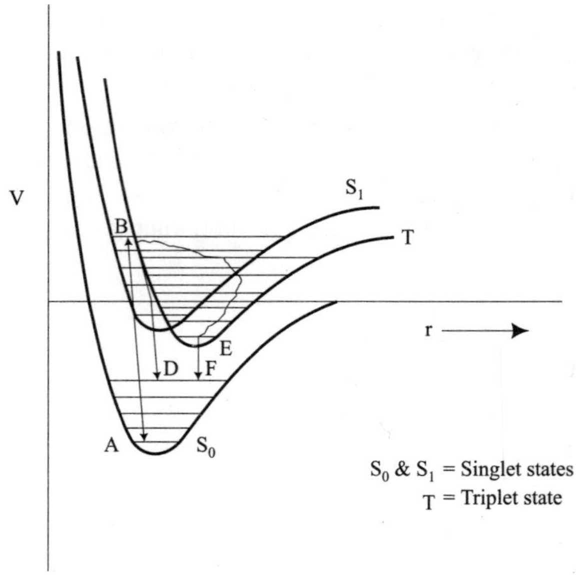

Fig. 8.10c Franck-Condon potential energy diagram for (i) absorption ($A \to B$), (ii) fluorescence ($C \to D$), (iii) phosphorescence ($E \to F$).

Fluorescent techniques can be employed in various biological studies (in histology, immunology and other physiological sciences) and in liquid scintillation counting for assaying radioactive-labelled compounds. Fluorescence can be either intrinsic, due to the presence of chromophores in the molecule, or extrinsic, due to the introduction of chromophores. In proteins, the intrinsic fluors are: tryphtophan **(Trp)**, tyrosine **(Tyr)** and phenylalanine **(Phe)**, in the order of their importance. Intrinsic fluorescence and quenching studies can be used to study protein conformation and folding (e.g., location of **Trp** in proteins), protein-ligand interactions and study of active sites of enzymes. Extrinsic fluorescence, with introduction of external fluors (proflavin, acridine orange, ethidium bromide etc.) can help in the studies of protein-folding, conformational changes and dynamics). Fluorophores with polar groups are sensitive to solvent polarity, which can be exploited. The interaction of an excited molecule with its surroundings may influence the fluorescence decay due to quenching, charge-transfer etc.

While the chemiluminescence immunoassay (CIA) is analogous to conventional radioimmunoassay (RIA) in that labeled antigen competes with unknown labeled antigen for binding to an antibody, the competitive-binding immuno-chemiluminometric assay (CB-ICMA) is analogous to immunoradiometric assay (IMRA) in that the antigen reacts with excess of labeled antibody. Generally, fluorescence immunoassays use lanthanide (Erbium) chelates as fluorophores and the measurement is based on time-resolved fluorescence. In contrast to conventional fluorophores, lanthanide element cations have a much longer fluorescence lifetime and thus suppress the background fluorescence, the technique known as dissociation-enhanced lanthanide fluoro-immunoassay (DELFIA). Further, the difference between the absorption (excitation) maximum and emission maximum (Stokes-shift) is greater than for conventional fluorophores, thus increasing the sensitivity of the measurements.

$$\text{Stokes shift} = \left(\frac{1}{\lambda_{ex}} - \frac{1}{\lambda_{em}}\right) \times 10^7 \quad \text{...(8.14)}$$

Fluorescence activated cell sorting (FACS) is one of the applications of extrinsic fluorescence that is widely used in molecular biology and clinical diagnostics. In FACS, a fluorescent labeled antigen, that is complementary to the required antibody specificity, is introduced into the culture prior to sorting. The labeled antigen binds only to those hybridomas secreting the required antibody and the cell sorter can be programmed to dispense into the clone plate only those cells carrying the fluorescent label.

8.4.1 Fluorescence Quenching

A compound is a fluorescence quencher if it decreases the quantum yield of a fluorophore upon interaction. In static quenching the formation of the complex is non-fluorescent.

$$F + Q \rightleftharpoons FQ. \qquad \text{Static quenching} \quad \text{...(8.15a)}$$

In dynamic quenching, the formation of the quenching occurs in the excited state.

$$F^* + Q \longrightarrow FQ^* \qquad \text{Dynamic quenching} \quad \text{...(8.15b)}$$

(F, F^*, Q, FQ and FQ^* are flourophore, quencher molecules and non-fluorescent and florescent complexes respectively).

Fluorescence quenching relies on the collision of the quencher and the fluorophore, in its excited state. **Stern-Volmer** plots are generally used to represent quenching studies data

$$\frac{I_0}{I} = 1 + K_1 \tau_0 [Q] + K_2 [Q] \qquad \text{Stern–Volmer equation} \quad \text{...(8.15c)}$$

where I and I_0 are intensities in the presence and absence of quenching, respectively; τ_0 is the fluorescence life-time in the absence of quenching; $[Q]$ is the concentration of the quencher; K_1 is the quenching constant and K_2 is the Stern-Volmer quenching constant. The quenching constant, K_1, is given by

$$K_1 = \frac{4\pi RDN_0}{1000} \quad \text{...(8.15d)}$$

where R is the contact radius; $D = (D_F - D_Q)$ is diffusion constant, which is the difference between the diffusion constants of the fluorophore and the quencher, respectively; and N_0 is the Avogadro number.

Many small molecules, e.g., iodides, metal ions and acrylamide can act as collisional fluorescence quenchers.

8.4.2 Fluorescence Energy Transfer

Fluorescence energy transfer is a process by which excited state energy of a donor (D) is transferred to an acceptor (A) due to dipole-dipole coupling interactions. The transfer is radiationless and the rate of transfer, K_T, is

$$K_T = \frac{1}{\tau_0}\left(\frac{R_0}{r}\right)^6 \qquad ...(8.15e)$$

where R_0 is the distance at which 50% energy is transferred, r is the donor-to-acceptor distance. For non-radiative transfer of excitation energy requires some interaction between a donor (D) and acceptor (A) molecule, and it can occur if the emission spectrum of the donor (D) overlaps the absorption spectrum of the acceptor (A). Such transitions are coupled, i.e., they are in resonance. Non-radiative resonance energy transfer (RET) by dipole-dipole interaction is possible at distance up to 8–10 nm (80–100 Å). Such a transfer process provides a tool for determining D-A distances. The donor-to-acceptor distance, r, can be estimated in macro-molecules from

$$r^6 = -\ R_0^6\left(1 - \frac{1}{\varepsilon}\right) \qquad ...(8.15f)$$

where ε is transfer efficiency. ε and R_0 can be determined experimentally.

RET imaging technique is a combination of optical microscopy with resonance energy transfer process, for quantifying the binding of two types of molecules at close proximity. Since this is an imaging-based technique, binding can be localized within the cells. Time-resolved RET imaging allows mapping of D-A distance in more accurate and quantitative way.

8.4.3 Fluorescence Anisotropy

Fluorescent light is generally polarised and fluorescence polarisation method, employing plane-polarised beams of light, can be of help in the study of diffusion anisotropy measurements such as association and dissociation of macromolecules, denaturation, domain motions, orientation of macromolecules (e.g.: DNA in chromosome and chlorophyll in chloroplast), helix-coil transitions, and membrane fluidity and anisotropy. Anisotropy, γ, is defined as

$$\gamma = \left(\frac{I_{11} - I_1}{I_{11} + 2I_1}\right) \text{ for polarised exciting beam} \qquad ...(8.16a)$$

$$\gamma = \left(\frac{I_{11} - I_1}{2I_{11} + I_1}\right) \text{ for exciting beam of natural light.}$$

where I_{11} and I_1 are intensities of parallel and perpendicularly-polarised light, respectively.

The effect of rotational diffusion on anisotropy is given by the **Perrin** equation

$$\frac{\gamma_0}{\gamma} = \left(\frac{1+\tau_0}{\theta}\right) \qquad ...(8.16b)$$

where γ_0 and γ are the anisotropies in the absence and presence of rotational diffusion, respectively, and θ is the correlation time.

Rotational diffusion data can be correlated to the dynamic properties of macromolecules to determine their size and shape as well as structural fluctuations. For this, pulsed-lasers are employed to study time-resolved fluorescent processes. By measuring the time-resolved decays of the parallel and perpendicular components fluorescence, the dynamic anisotropy (time-resolved) γ_t can be calculated

$$\gamma_t = \left\{ \frac{I_{11}(t) - I_1(t)}{I_{11}(t) + 2I_1(t)} \right\} \qquad ...(8.16c)$$

Time-resolved fluorescent studies enable observations to be made on molecular shapes and their changes on ligand binding, molecular or domain mobilities etc. Time-resolved fluorescent techniques are employed in the dynamical studies of membrane fluidity, membrane transport, membrane-protein interactions and conformational changes (as a function of fluorescence decay) (See Chapter 13 for further details on membrane transport).

8.4.4 Flow Cytometry

Fluorescence spectroscopy is more sensitive (about 10^3 times) than absorption spectroscopy. Therefore, it finds ready application in clinical chemistry—analysis of minute quantities of drugs and metabolites in body fluids and in flow cytometry fluorescence immunoassay (FIA) and in various other biological studies. The fluorescence spectrophotometer, used in such studies, is similar in principle to a regular spectrophotometer, consisting of a source of light, and primary filter placed axially in front of the sample tube (the primary filter is to enable the selection of desired wavelength(s) to excite the fluorescent molecules in the sample); a secondary filter followed by detector/recorder system in an axis normal to the axis of the primary filter and sample tube (the secondary filter is to isolate specific spectral lines emitted by the fluorescent molecules of the sample). The filters can be either simple filters or diffraction gratings (Fig. 8.11).

A basic flow cytometer consists of a source of light (low-pressure mercury vapour lamp, xenon arc or tunable dye lasers), a flow cell, optical components to solve different colours (wavelengths) and direct them onto the detector system and a computer system to analyse and display the data. Commonly used fluorophores are fluorescein, *texas red* and coumarin in proteins and propidium iodide, ethidium bromide and acridine orange in nucleic acids. The properties of the fluorophore will depend on its environment. For example, propidium iodide strongly fluoresces when complexed with DNA due to intercalation but is a weak fluorescent in a polar environment. Similarly, acridine orange fluoresces in the green region when intercalated in double-stranded nucleic acids but fluoresces in the red region when complexed with single-stranded nucleic acids. Common applications of flow

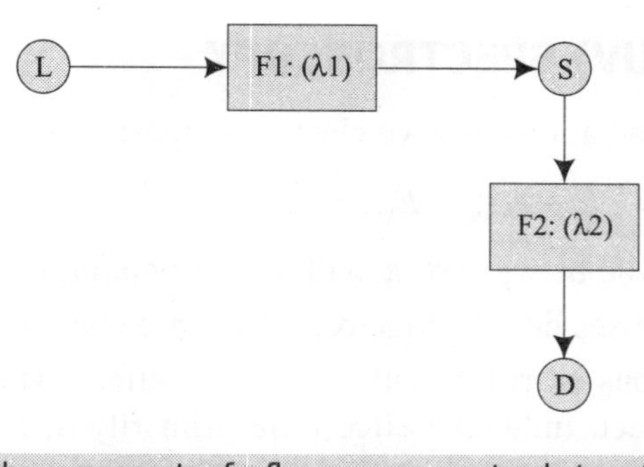

Fig. 8.11 Schematic diagram of the components of a fluorescence spectrophotometer.

cytometry are in the measurement of surface antigens by immunofluorescence labelling using monoclonal antibodies, estimation of protein and nucleic acid content, study of the effect of cytotoxic substances on the cell cycle, chromosomal analysis and sorting (genetics of human diseases), measurement of membrane potential, permeability and intracellular ion concentration and pH etc.

Phosphorescence is even more sensitive than fluorescence (by 10^3 times) because of non-interference of stray incident light in phosphorescence, due to its delayed effect. Therefore, this method can also be employed in biological studies. However, phosphorescence cannot be studied in the liquid state, because of cancelling of phosphorescence due to reorganisation of spins state as a consequence of molecular motions in liquid state.

8.4.5 X-Ray Fluorescence

X-ray spectroscopic methods are widely used for qualitative and quantitative elemental analysis of $Z > 5$. Application of X-ray spectroscopy is due to several advantages it offers. Some of these are:

- Spectra are simple to interpret compared to UV-visible spectra.
- Spectral lines are generally independent of the chemical nature of the element.
- Minimal sample quantity and sample preparation.
- High precision (better than wet-chemical and instrumental methods) and accuracy.
- Nondestructive testing.

X-ray fluorescence is a useful analytical technique employed in clinical and forensic studies. Interpretation of the spectra is quite simple and it can serve as a complementary method to optical emission spectroscopy. In clinical chemistry it is generally employed in the detection of alkaline and rare-earth elements.

The specimen is irradiated by X-rays in an X-ray fluorescence spectrometer and the elements present in the specimen emit their characteristic fluorescent lines, identified by a detector (Fig. 8.12). In the spectrometer, the fluorescent radiation emitted by the specimen upon being irradiated by X-rays, is collimated onto the surface of an analyser (a flat single crystal). Reflected radiations from the analyser crystal are collected by the detector. Rotation of the goniometer by an angle, θ, is coupled to the rotation of the detector by an angle, 2θ, to satisfy **Bragg's** diffraction condition.

As the X-ray fluorescence method is inherently more precise than any wet chemical methods and the spectra are simple to interpret, this method can be conveniently employed for elemental analyses in any clinical/forensic/biological institution.

8.5 ULTRAVIOLET (UV) SPECTROSCOPY

Ultraviolet (UV) and visible radiations involve electronic transitions (of the order of 500 kJ mol^{-1}).

$$E = E_{elec} + E_{vib} + E_{rot} \cdots \qquad ...(8.17)$$

Electronic transitions can be between σ, π, and **n** (non-bonding) orbitals (see Fig. 8.13).

Electronic spectroscopy arises due to inductive, resonance and solvent (environment) effects.

Groups that attract electrons exert a negative inductive effect, while those that donate electrons exert a positive inductive effect. Inductive effects are primarily due to the interaction of σ-bonds (Fig. 8.14). Resonance effects involve π-electron delocalisation.

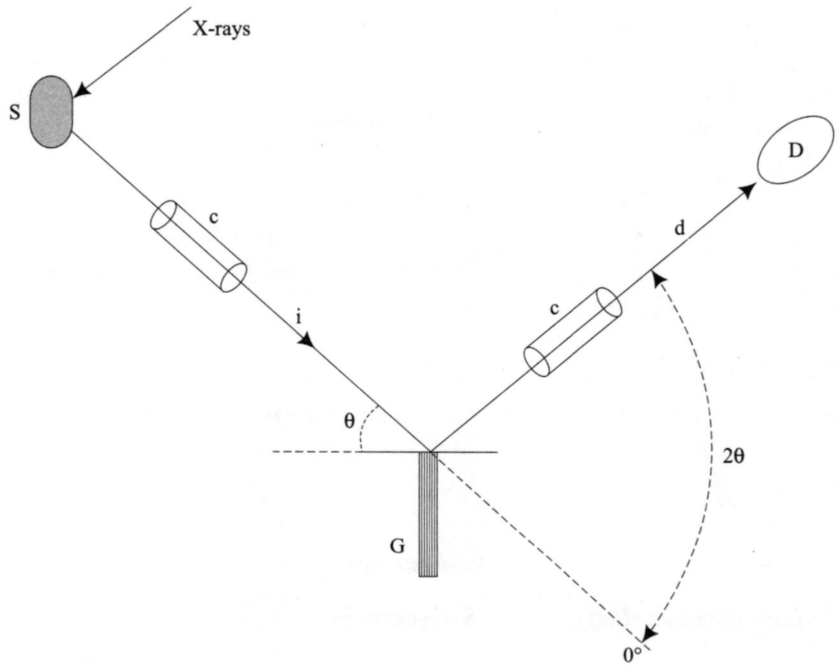

Fig. 8.12 Geometry of X-ray fluorescence spectrometer.
(Angular velocity of the detector = 2 angular velocity of the analiser).
S = sample; A = flat crystal analyzer; G = goniometer head; D = detector; C = collimators, i = incident fluorescent X-ray beam; d = diffracted X-ray beam; θ = angle between the incident beam and the plane of the analyser; 2θ = Angle between the incident and the diffracted beam.

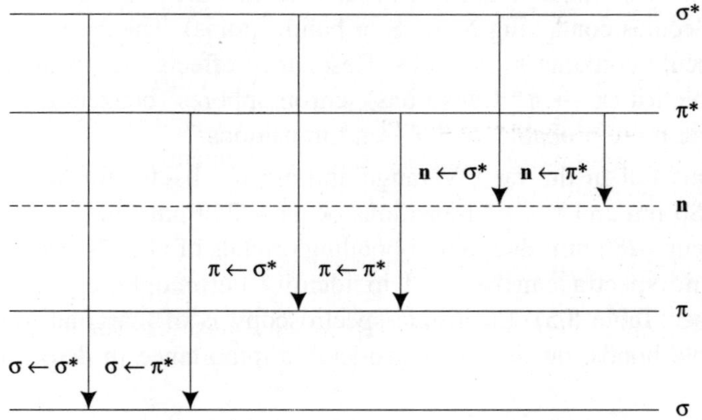

Fig. 8.13 Electronic transitions between σ, π & **n**-orbitals.

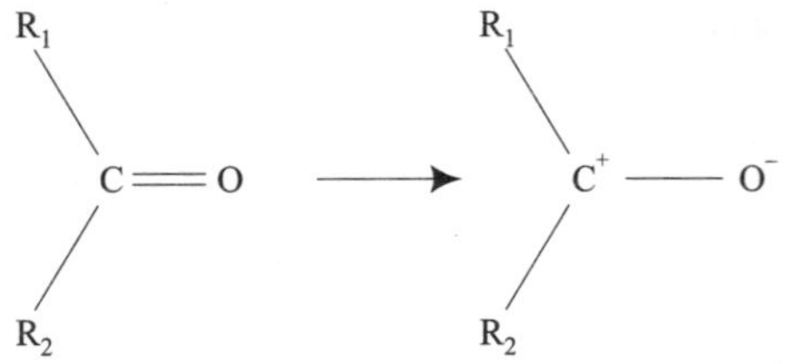

Inductive effect

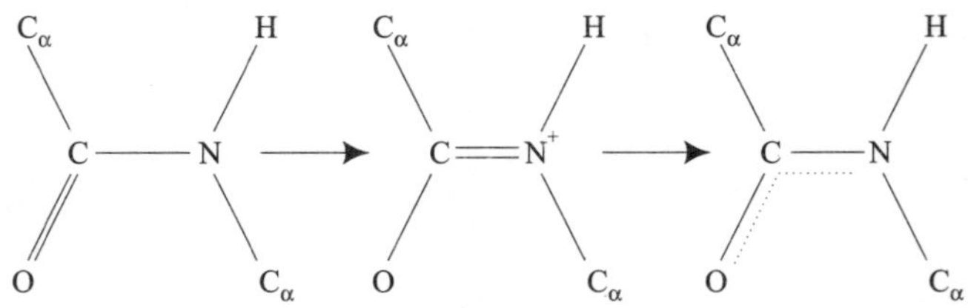

Resonance effect

Fig. 8.14 Inductive and resonance effects due to σ- and π-electrons.

Environmental effects influence spectral lines. On going from less-polar to more-polar solvents, $\pi - \pi^*$ transitions show a shift towards longer wavelengths **(red shift)** due to the dipole moment of the (polar) solvent inducing a dipole moment in the solute. The effect is opposite in $n - \pi^*$ transitions to that of $\pi - \pi^*$ transitions, and larger. Polar solvents cause a shift towards shorter wavelengths in $n - \pi^*$ transitions **(blue shift)**, relative to non-polar solvents (with increasing polarity of the solvent). Hydrogen bonding also induces frequency shifts.

The $n \rightarrow \sigma^*$ and $n \rightarrow \pi^*$ transitions occur in substances, which have at least one free electron pair in the molecule (molecules containing N, O, S or halide atoms). The $n \rightarrow \pi^*$ excitation takes place only when the molecule contains π-electrons. Resonance effects are prominent in molecules that contain aromatic moieties ($\pi \rightarrow \pi^*$ transitions), chromophores (benzenoid rings), porphyrins etc. $\pi \rightarrow \pi^*$ transitions are more probable than $n \rightarrow \pi^*$ transitions.

$\sigma - \sigma^*$ transitions fall in the far UV range and are not useful for structure analysis. $\pi - \pi^*$ transitions occur >180 nm and $\pi - \pi^*$ transitions occur ≈ 260 nm (due to double/triple bonds) and $n - \pi^*$ transitions occur >280 nm (due to non-bonding orbitals of O or **N** to **n**-orbitals, chromophores C=O etc.). Electronic spectra can be used to identity chromophores, such as C=C, NO_2 and phenylalanine etc. (see Table 8.5). Electronic spectroscopy is of marginal importance in detecting isolated-carbon double bonds, but it is of considerable importance in detecting conjugated double bonds.

If a compound does not absorb in the region 200 – 800 nm, it indicates that the compound does not contain conjugated, unsaturated or benzoid groups or aldehyde, keto or nitro groups. Appearance of peaks in the 200–400 nm region generally indicates the presence of unsaturated groups.

Table 8.5 Absorption of functional groups.

Functional group	*Absorption (nm)*
C = C	165 – 185
C = C – C = C	215 – 235
C = O	170 – 190
C = C – C = O	230 – 250

Electronic spectra of the amino acid backbone occur at ≤230 nm. Amino acid side chains with phenyl groups absorb at >250 nm (**His** = 210; **Cys** = 250; **Phe** = 250 and **Tyr** and **Trp** = 280 nm). Absorbence at 280 nm, due to **Tyr** and **Trp** chromophores, is used to monitor protein purification and protein concentration in biological studies. Nucleic acids are monitored by absorbence at 260 nm due to their constituent purine and pyrimidine bases. The phenomenon of hypochromism is employed to monitor the helix-coil transition in nucleic acids (melting studies). The increase in absorbence (extinction coefficient), A, is given by

$$A = \frac{1}{\mathbf{n}_0}\left\{\frac{\mathbf{n}_0^2 + 2}{3}\right\}^2 \qquad \text{...(8.18)}$$

where $\mathbf{n_0}$ is the refractive index.

Polarised light can be used to estimate helical content. The amide bond absorbs at 190 nm, perpendicular to, and at 205 nm, parallel to, the helical axis. The fraction, **f**, of a protein in helical form is

$$\mathbf{f} = \frac{\sum_j \mathbf{n}_j A_j + N_{A_r - A_p}}{N(A_r - A_h)} \qquad \text{...(8.19)}$$

where N = number of peptide bonds; $\mathbf{n_j}$ = number of side-chain chromophores; A_j = absorbence of chromophores and A_r and A_h are absorbence of peptide bonds in random and helical states, respectively.

UV and visible spectroscopies can also be employed to study charge-transfer reactions. However, these spectroscopies do not reveal finer details of structure, as do infrared (IR) and Raman spectroscopies.

8.6 INFRARED (IR) SPECTROSCOPY

The rotational and vibrational frequencies of molecules (equation 8.17) are in the infrared region of the electromagnetic spectrum, and, therefore, infrared spectroscopy (IR) can be employed to study molecular rotations and vibrations in these frequency ranges. As these motions are sensitive to chemical changes, the spectra can be used as monitors of molecular chemistry. The frequencies are expressed in IR and Raman spectroscopies in wave numbers, $\overline{\nu}$ (cm^{-1}).

8.6.1 Infrared Spectra

IR spectroscopy is an absorption phenomenon and to absorb infrared radiations, the molecule should exhibit a change in electric dipole moment, μ, during molecular motion. That is, to observe IR spectrum of a molecule

$$\frac{\partial \mu}{\partial x} \neq 0 \qquad \text{IR active} \qquad ...(8.20)$$

Therefore, polar interactions, such as hydrogen bonding, hydration, proton conduction, conformational changes, order/disorder (denaturation) and metal coordination etc., can be studied by IR spectroscopy.

Rotational Spectra. Molecular rotations occur for diatomic molecules, which can be treated conceptually as dumbbells (Fig. 8.15a).

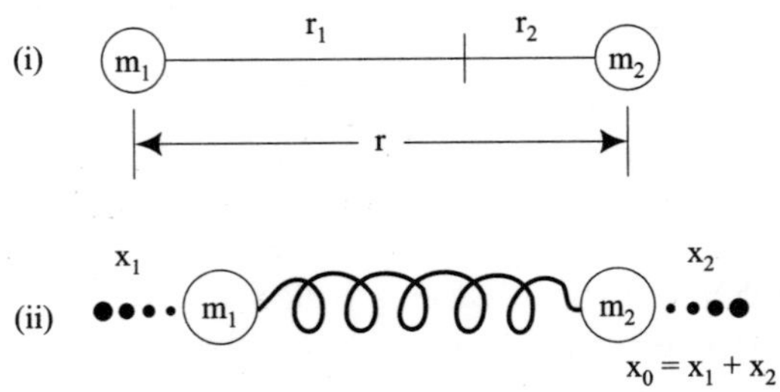

Fig. 8.15a Representation of a diatomic molecule.
(i) As a dumbbell; (ii) as a harmonic oscillator.
m_1 & m_2 are the masses of the atoms and r_1 is and r_2 are the distances from the centre of mass; and r is the distance between the two atoms.

The moment of inertia, I, for a diatomic molecule of masses, m_1 and m_2 at distances r_1 and r_2 from the centre of mass is

$$I = m_1 r_1^2 + m_2 r_2^2 \qquad ...(8.21a)$$

and $$m_1 r_1 = m_2 r_2$$

∴ $$I = \mathbf{m} r^2 \qquad ...(8.21b)$$

where the reduced mass, $\mathbf{m} = (m_1 m_2)/(m_1 + m_2)$ and r is the distance between the two atoms.

The rotational energy, E_r and frequency $\bar{\nu}$, are

$$E_r = \frac{\hbar^2}{2I}\,(J(J+1)) \qquad ...(8.21c)$$

$$\bar{\nu} = 2B\,(J+1) \qquad ...(8.21d)$$

where $\hbar = h/2\pi$ (h is Planck's constant), J is the rotational quantum number and B is a rotational constant. The selection rule is

$$\Delta J = \pm 1 \text{ for rotation spectrum} \quad ...(8.21e)$$

Pure rotational spectra occur in the far infrared and microwave regions and are not so useful in the study of biological structures.

Vibrational Spectra. Vibrational spectra of diatomic molecules can be explained by treating them as harmonic oscillators (elastic springs) (Fig. 8.15a(ii)).

For a harmonic oscillator, the displacement, x, from the mean position, the restoring force; F, and the vibrational frequency, ν, are

$$x = x_0 \sin 2\pi \nu t \quad ...(8.22a)$$

$$F = \mathbf{k} \cdot x = \mathbf{m}\frac{\partial^2 x}{\partial t^2} \quad ...(8.22b)$$

As

$$\bar{\nu} = \frac{\nu}{C}, \bar{\nu} = \frac{1}{2\pi C}\sqrt{\left(\frac{k}{m}\right)} \quad ...(8.22c)$$

where $x_0 = (x_1 + x_2)$ (see Fig. 8.15b), **m** is the reduced mass and **k** is the force constant, which is a measure of bond strength, and C is the velocity of light.

The potential energy, U, of a harmonic oscillator is parabolic,

$$U = \frac{1}{2}\mathbf{k}.x^2 \quad ...(8.22d)$$

and the vibrational energy, E_r is

$$E_r = hC\bar{\nu}\left(\mathbf{v} + \frac{1}{2}\right) \quad ...(8.22e)$$

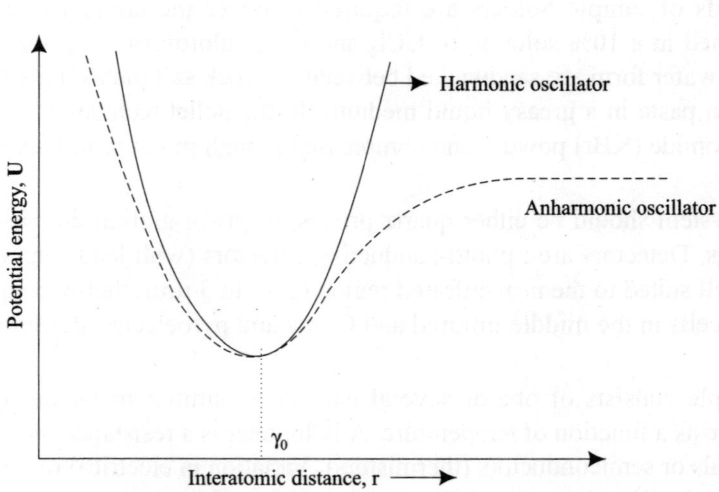

Fig. 8.15b Potential energy curve of (i) harmonic, (ii) anharmonic oscillator.

where, **v** is the vibrational quantum number and the selection rule is

$$\Delta \mathbf{v} = \pm 1 \text{ for vibrational spectrum.} \qquad ...(8.22f)$$

The spectrum of a simple harmonic oscillator consists of a single band at the fundamental vibrational frequency, $\overline{\nu}$, of the molecule. In reality, molecules do not behave like simple harmonic oscillators and the anharmonicity is included by treating them as anharmonic oscillators to interpret the IR vibrational spectra and the potential energy is of the **Morse-potential** form (Fig. 8.15b)

$$U = fx^2 - gx^3 + ... \qquad ...(8.22g)$$

Intra-molecular vibrations are classified under, (i) stretching (↔), (ii) bending (⌒) and (iii) skeletal vibrations.

Homopolar molecules, H_2, Cl_2 and such symmetrical molecules, which do not exhibit change in dipole moment $\left(\frac{\partial \mu}{\partial x} = 0\right)$ while vibrating, are IR inactive.

Molecules like HCl are IR active. Whether the stretching or bending vibrations of a molecule are IR active or IR inactive depends on the mode of stretching and bending. IR spectra are in the frequency range of 4000 to 200 cm^{-1}.

8.6.2 Instrumental Features

Different radiation sources, optical systems and detectors are needed to scan different regions of the infrared spectrum. Radiation sources are the tungsten filament lamp in the near-infrared region, the Nernst glower, globar, and nichrome wire in the near- and middle-, and high-pressure mercury arc lamps in the far-infrared regions. Infrared radiation sources are inert solids heated to temperatures in the range ~2000 K. The Nernst glower is a rod of Zirconium-Yttrium oxides, globar is a silicon carbide (SiC) rod; nichrome and rhodium wire spirals are also sources of IR radiation. The CO_2 laser is also an infrared source ($\lambda \approx 10$ μm).

Different kinds of sample holders are required to cover the entire IR region. For liquids, a spectrum is obtained in a 10% solution of CCl_4 and CS_2, chloroform, acetone or acetonitrile; solid specimens in thin wafer form are sandwiched between flat rock salt plates. Powder specimens can be examined as a thin paste in a greasy liquid medium. In the pellet technique, the specimen is mixed with potassium bromide (KBr) powder and compressed at high pressure to make the KBr plastic and transparent.

The optical system should be either quartz prisms, or prism-grating double-monochromators or diffraction gratings. Detectors are : photo-conducting detectors (with lead sulphide or lead telluride) which are very well suited to the near-infrared region (0.75 to 3 μm), thermocouples. bolometers and Golay pneumatic cells in the middle-infrared and Golay and pyroelectric detectors in the far-infrared regions.

A thermocouple consists of one or several pairs of dissimilar metal junctions that develop a potential difference as a function of temperature. A bolometer is a resistance thermometer constructed from strips of metals or semiconductors (thermistors). Variation in electrical resistance is a function of temperature, Pyroelectric detectors are capacitance-based, constructed from thin wafers of pyroelectric material (barium titanate). These materials are dielectric with special electrical and thermal properties. They exhibit a strong temperature-dependent polarisation and the change in polarisation is related to the temperature change.

8.6.3 Applications

As mentioned earlier, IR spectroscopy can be employed in the study of polar interactions, proton conduction and conformational changes in biomolecules.

Infrared spectroscopy has several advantages—

- The systems can be investigated without changing its composition.
- The IR spectrum is a specific property of the given substance—geometrical isomers and molecules containing different atomic isotopes have each a different spectrum.
- The method is sensitive and selective for the analysis of complex mixtures.
- Continuous analysis can be performed, which lends itself to kinetic investigation.
- Results are quick.
- A very small amount of substance is required.

Some vibrational frequencies are listed in Table 8.6.

In the near-infrared region, stretching vibrations are associated with hydrogen atoms (O–H, N–H etc). Most of the group frequencies (2500 – 1500 cm^{-1}) (C = O, C = C, C = N, N = N etc.) are in the mid-infrared region. Group frequencies help to ascertain whether such groups are present. C–H stretching at ≈ 3000 cm^{-1} indicates the presence of alkane group(s). Peaks at 1380 and 1450 cm^{-1} arise due to C–H bending and a peak at 3200 cm^{-1} is due to hydrogen-bonded OH groups, and the O – H stretch peak at 3600 cm^{-1} is a useful marker for alcohols. The 1300–650 cm^{-1} region is called the 'finger print' region, with single bond stretching and bending frequencies. A band at 1100 cm^{-1} is due to asymmetric stretching of –C–O–C– in ethers. The carbonyl group (C=O) is recognisable from the band in the 1820–1600 cm^{-1} region.

Table 8.6 Some vibrational frequencies.

Type	*Frequency range (cm^{-1})*
C – H stretch	~2850 – 3,000
C – H bend	1350 – 1450
C – C stretch	700 – 1250
C = C stretch	1600 – 1700
C – O	1050 – 1300
C = O stretch	1650 – 1800
C – N	1180 – 1350
C = N stretch	1500 – 1700
N – H stretch	3200 – 3500
N = N stretch	1580 – 1630
Hydrogen bond	3200 – 3600
O – H stretch	3600 – 3650

Hydrogen Bonding. The peptide group $(-\overset{O}{\overset{\|}{C}}-\underset{N}{\underset{|}{N}}-)$ in the polypeptide backbone gives rise to different types of vibrational modes (Fig. 8.16 and Table 8.7a).

Since hydrogen bonding weakens the A–H bond, stretching frequencies are down-shifted and bending frequencies are shifted to higher frequencies upon hydrogen bonding.

The stretching frequency of C=O is at 1700 cm^{-1} and on formation of a hydrogen bond (C=O....H–N), the stretching frequency decreases (~1550 cm^{-1}), while there is increase in the bending frequency (~1620 cm^{-1}). That is, on hydrogen bond formation to C=O of a peptide group, mere is decrease in the amide band I frequency and increase in that of amide band III. These band shifts can be used to determine hydrogen bond formation and its strength in polypeptides and also determine the content of ordered segments, like the α-helix and β-sheet (Table 8.7b). A protein with α-helical structure has a weak amide band I at 1645–1655 cm^{-1} and a strong amide band II at 1264–1300 cm^{-1}. The deuterium exchange method is more useful in the study of hydrogen bonding, and order/disorder phenomena in macromolecules.

Helix-coil Transition. Polarised IR can be effectively used to study helix-coil transitions, order/disorder regions, like the α-helix, β-sheet, reverse turns and the random coil in macromolecules and

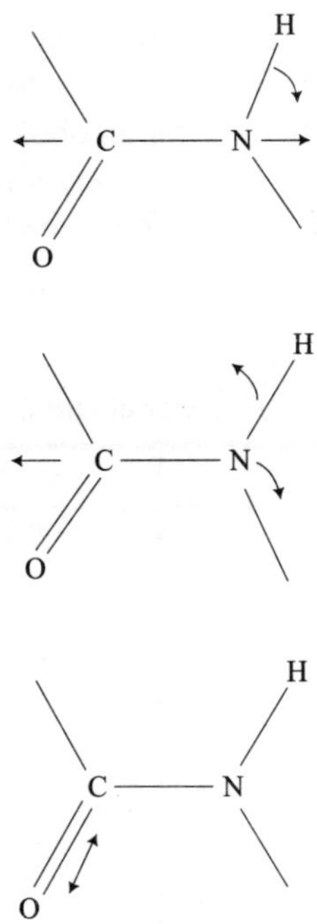

Fig. 8.16 Vibrational modes of the peptide unit $(-\overset{O}{\overset{\|}{C}}-\underset{H}{\underset{|}{N}}-)$.

Table 8.7a Vibrational frequencies of various bond stretching.

Assignment	*Infrared*	*Raman*
Amide band A	3300 cm^{-1}	3300 cm^{-1}
Amide band B	3090	—
Amide band I	1680	1675
Amide band II	1530	1530
Amide band III	1235	1235

Table 8.7b Conformation of polypeptide backbone and vibrational frequencies.

Conformation	*C = O Stretch Amide band I*	*Amide band II*	*N – H Plane bending Amide band III*	*Amide band A*
Random coil	1650 cm^{-1}	1535 cm^{-1}	1245 cm^{-1}	3460 cm^{-1}
α-helix	1650 (II^{l})	1545 (I^{r})	1265 - 1300	3300 (II^{l})
β-sheet	1630 (I^{r})	1530 (II^{l})	1235	3300 (I^{r})

II^{l} = parallel and I^{r} = perpendicular stretch.

also for melting studies. When polarised light is employed, interaction (absorption) is maximum if the electric vector, **E**, is colinear to the oscillating dipole, μ. The dichroic ratio, *R*, ratio of the difference between the parallel and perpendicular absorbences, gives a measure of ordered/disordered regions of macromolecules.

$$R = \frac{|(A_{11} - A_1)|}{|(A_{11} + A_1)|} \quad ...(8.23a)$$

where A_{11} and A_1 are parallel and perpendicular components of absorbence with respect to the incident polarised radiation.

If $R > 1$ the electric vector is parallel to the bond axis

If $R < 1$ the electric vector is perpendicular to the bond axis.

If the transition moment is inclined at an angle, φ, then

$$R = 2\cot^2 \varphi \quad ...(8.23b)$$

For $\varphi = 54.75°$ there is no dichroism $(\because R = 1)$

$< 54.75°$ parallel dichroism $(\because R > 1)$

$> 54.75°$ perpendicular dichroism $(\because R < 1)$

For a right-handed α-helix, the hydrogen bonds are intra-chain and at every fourth residue (A1 and A3). Since there are 3.6 residues per helix turn, the phase angle is, $\varphi = 0$ or $2\pi/3.6$ and the frequency, $\bar{\nu}$ is

$$\bar{\nu}_{(0)} = \bar{\nu}_0 + A_1 + A_3$$

$$\bar{\nu}_{(\varphi)} = \bar{\nu}_0 + A_1 \cos \varphi \,.\, \cos 3\varphi \quad ...(8.24a)$$

For the β-sheet, the inter-chain hydrogen bonds link with a two-fold screw-axis, in either parallel or antiparallel alignment. Therefore, the phase angle is, $\phi = 0$ or π and the frequency, $\bar{\nu}$, is,

$$\bar{\nu}_{(0,\,0)} = \bar{\nu}_0 + A_1 + A_1' \qquad \text{for parallel } \beta\text{-sheet}$$

$$\bar{\nu}_{(\pi,\,0)} = \bar{\nu}_0 - A_1 + A_1' \qquad \text{for antiparallel } \beta\text{-sheet} \qquad \text{...(2.24b)}$$

That is, a parallel β-sheet is IR active and an antiparallel β-sheet is IR inactive. That is to say, in ordered regions, molecular symmetry will make some vibration modes inactive. This can be used to identify different types of ordered regions. In proteins, bands near 1550, 1650 and 3300 cm^{-1} are associated with ordered regions. In addition, structural and conformational changes in macromolecules that occur due to environmental changes, such as those due to solvent, pH, concentration and temperature can be studied by IR spectroscopy.

Ion-ion Interactions. Vibrations of anionic groups change fundamentally with the removal of protons. Thus, from these vibrational bands the degree of dissociation, and hence also the pK_a values of groups in biological systems can be determined by IR studies and information on anion/cation interactions can be obtained from IR vibrational shifts. Tautomeric forms (e.g., nucleic acid bases) can be studied; vibrational bands in the region 1750 – 1400 cm^{-1} give information on base-pairing, while vibrational bands in the region 1300 – 1000 cm^{-1} give information on the structure of the nucleic acid backbone.

Chelation of metals to biomolecules can also be studied by IR spectroscopy. The chelation of a metal changes the polarity of the bonds and changes the frequencies of stretching vibrations of affected bonds. Metal-chelation to nucleic acid bases, electrostatic interactions of polar groups (e.g., amino acid side chains), which vary with the solvent, pH, ionic strength, and temperature are important factors in the study of thermal unfolding and unwinding (melting) of macromolecules, which can be monitored by IR spectroscopy.

Water Hydration. Water hydration, that is, increase of polar environments, usually increases proton transfer. As structurally symmetrical chains of hydrogen-bonded systems show very great proton polarisabilities, such symmetrical hydrogen-bonded chains are particularly effective conductors of positive charge and IR spectroscopy can be employed to study such systems. Such symmetrical hydrogen-bonded systems show continua in the IR spectra. A well-characterised example is that of purple membrane bacteriorhodopsin. The conduction of positive charge within the molecule may proceed via a largely structurally-symmetrical polypeptide chain(s) (consisting of one aspartic acid, six tyrosine and glutamic acid residues). Easily polarisable $O\text{–}H \ldots O^- \rightleftharpoons O^- \ldots H\text{–}Q$ bonds, between these residues are stable even if water is present. Thus, formation of these symmetrical hydrogen bonds, with great proton polarisability, could mediate the uptake of protons at the inner surface of the purple membrane.

Fourier transform (FT-IR) spectroscopy, i.e., interconversion of time-domain spectral data to frequency-domain spectra, has greatly increased the role of IR spectroscopy in data acquisition, data processing and interpretation in molecular structural studies. An FT-IR spectrum of a cytotoxin from the snake venom *Naja naja* (India), with tentative identification of vibrational groups, is shown in Fig. 8.17.

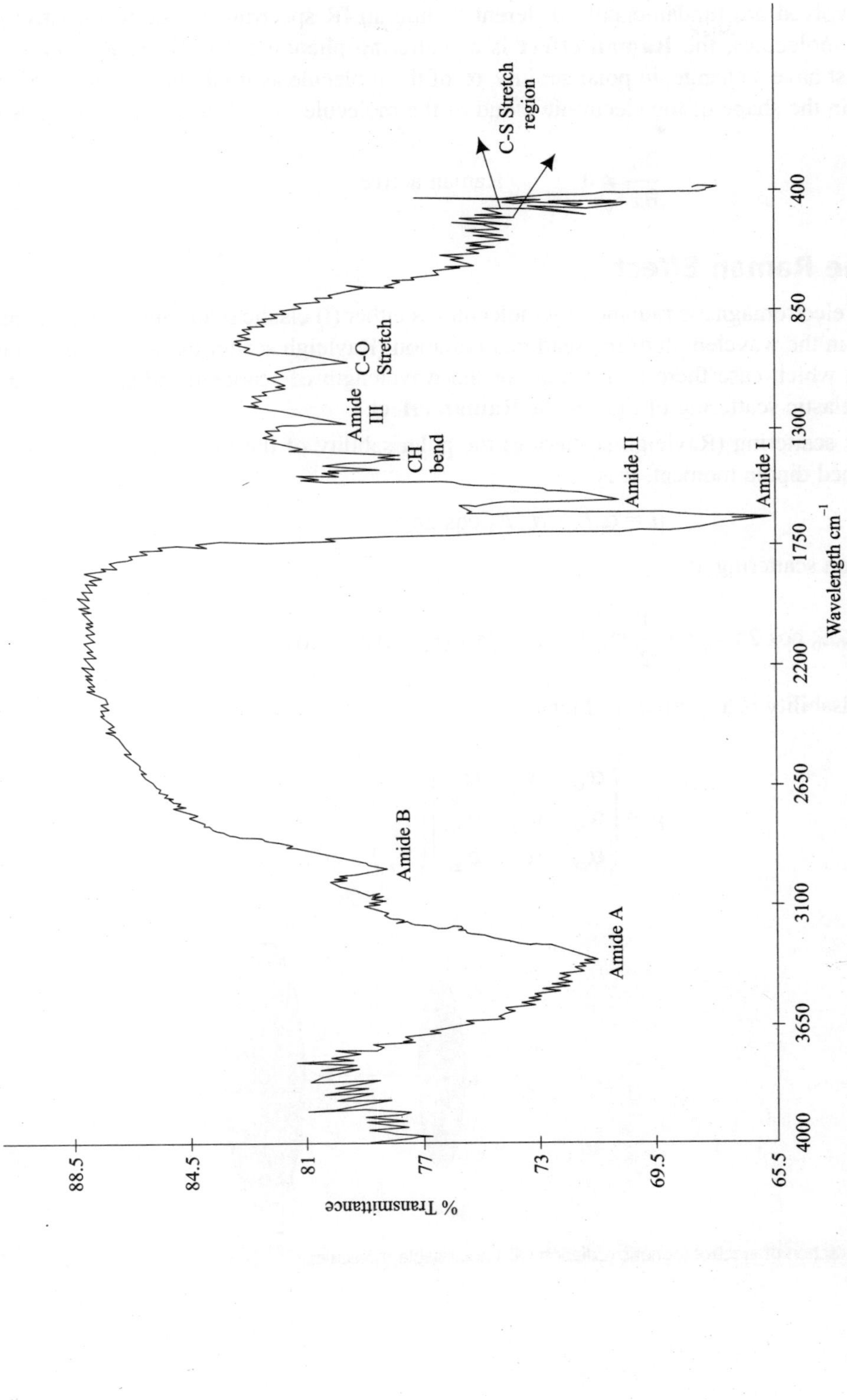

Fig. 8.17 FT-IR spectrum of a cytotoxin from the snake venom *Naja naja* (India).

8.7 RAMAN SPECTROSCOPY

Raman and infrared spectroscopies are complementary in their information content, but the physical principles involved are fundamentally different. While an IR spectrum is due to the absorption of radiation by molecules, the **Raman effect** is a scattering phenomenon. To be Raman active, the molecule must have a change, in polarisability, α, of the molecule as it vibrates. That is, there should be a change in the shape of the electronic cloud of the molecule to be Raman active (Fig. 8.18).

$$\frac{\partial \alpha}{\partial x} \neq 0 \qquad \text{Raman active} \qquad \text{...(8.25)}$$

8.7.1 The Raman Effect

Scattering of electromagnetic radiation by molecules is either (i) elastic scattering, in which case there is no change in the wavelength of the scattered radiation (Rayleigh scattering; $\lambda = \lambda_0$) or (ii) inelastic scattering, in which case there is a change in the wavelength of scattered radiation ($\lambda \neq \lambda_0$). This molecular inelastic scattering of light is the **Raman effect.**

In elastic scattering (Rayleigh scattering) the polarisability of the molecule is isotropic, $\alpha = \alpha_0$ and the induced dipole moment, μ is

$$\mu = \alpha E = \alpha_0 E_0 \cos 2\pi \nu_0 t \qquad \text{...(8.26a)}$$

For Raman scattering, μ is

$$\mu = \alpha_0 E_0 \cos 2\pi \nu_0 t + \frac{1}{2} \alpha_m E_0 \{\cos 2\pi t (\nu_0 - \Delta\nu) + \cos 2\pi t (\nu_0 + \Delta\nu)\} \qquad \text{...(8.26b)}$$

and the polarisability is a symmetrical tensor.

$$\mu = \begin{pmatrix} \alpha_{xx} & \alpha_{xy} & \alpha_{xz} \\ \alpha_{yx} & \alpha_{yy} & \alpha_{yz} \\ \alpha_{zx} & \alpha_{zy} & \alpha_{zz} \end{pmatrix} \begin{pmatrix} E_x \\ E_y \\ E_z \end{pmatrix} \qquad \text{...(8.26c)}$$

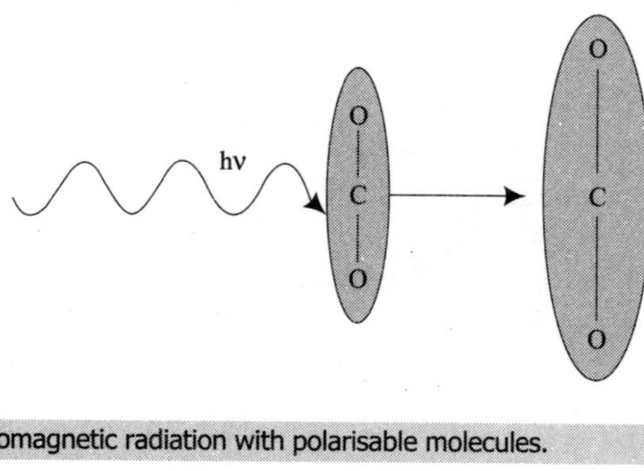

Fig. 8.18 Interaction of electromagnetic radiation with polarisable molecules.

In Raman scattering, the molecule can either (i) absorb a portion of energy from an incident photon ($h\nu_0$) so the scattered radiation has a lower frequency than that of the incident radiation ($\nu_0 - \Delta\nu$); this is called a **Stokes-Raman** spectrum, or (ii) a portion of energy from the molecule in the excited state is transferred to the scattered radiation with an increase in the frequency of the scattered radiation ($\nu_0 + \Delta\nu$; this is known as **AntiStokes-Raman** spectrum (Figs. 8.19a & 8.19b). The probability of molecules being in the ground state is far higher than for molecules in the excited state, and therefore Stokes-Raman lines are stronger: than AntiStokes-Raman lines.

Complementarity of Raman and IR Spectra. The molecular vibrations of bonds (bond stretching) can be either symmetric or nonsymmetric. In Raman scattering, the depolarisation ratio, ρ, which is the ratio of the intensities of scattered radiation perpendicular and parallel to the incident radiation, gives information about the nature of such vibrations, and thereby about molecular structure.

$$\rho = \frac{I_1}{I_{11}} \qquad ...(8.27)$$

where,

I_1 is the intensity of the scattered radiation normal to the incident radiation

I_{11} is the intensity of the scattered radiation parallel to the incident radiation (The parallel component is preponderant).

ρ for non-symmetric vibrations is 3/4

ρ for symmetric vibrations is 0 to <3/4.

In Raman spectroscopy nonpolar bonds with a symmetrical electron cloud distribution produce the greatest polarisability changes. Such bonds are : —C = C—; —C ≡ C—; C ≡ N—; —C = S—; —C — S; —N = N; —N ≡ N— etc.

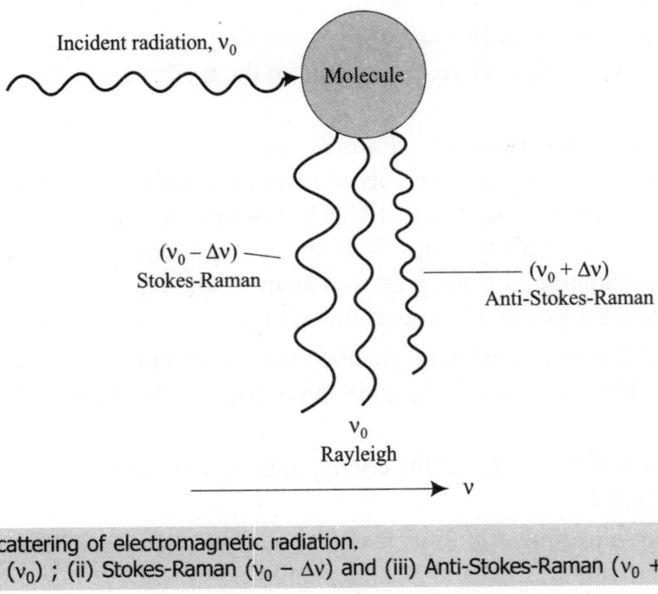

Fig. 8.19a Molecular scattering of electromagnetic radiation.
(i) Rayleigh (ν_0) ; (ii) Stokes-Raman ($\nu_0 - \Delta\nu$) and (iii) Anti-Stokes-Raman ($\nu_0 + \Delta\nu$) scattering.

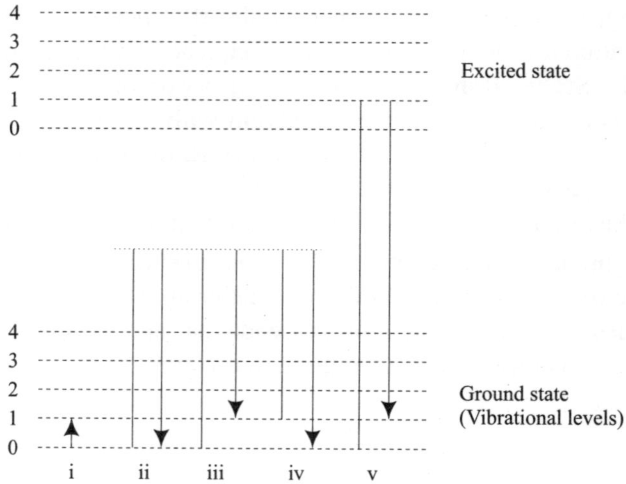

Fig. 8.19b Energy-level diagram of IR and Raman spectra.
(i) Infrared; (ii) Rayleigh; (iii) Stokes-Raman; (iv) Anti-Stokes-Raman and (v) Resonance Raman spectra.

Homopolar diatomic molecules, A-A type (N_2, Cl_2, H_2 molecules etc.) which exhibit polarisability while bond stretching $\left(\frac{\partial \alpha}{\partial x} \neq 0\right)$ but do not exhibit a change in dipole moment $\left(\frac{\partial \mu}{\partial x} = 0\right)$ are Raman active but IR inactive. Stretching vibrations of A-B type diatomic molecules (e.g., HCl) are accompanied by a change in polarisability as well as change in dipole moment and, therefore, give rise to both Raman and IR spectra, which are complementary. Stretching and bending vibrations of more complex molecules are either both Raman and IR active, or Raman active IR inactive or Raman inactive and IR active, depending on the mode of stretching or bending (Fig. 8.20 gives examples).

Advantages of Raman Spectroscopy. Raman spectroscopy has many advantages over IR spectroscopy and with the advent of **time-resolved resonance Raman spectroscopy** (TRRS), Raman spectroscopy has far greater potential than IR spectroscopy to elucidate structural parameters of molecules. Although Raman spectroscopy will never challenge X-ray diffraction as a tool in quantitative structure elucidation, it is the preferred method when qualitative information is sufficient because it is faster and not expensive. Some of the advantages of Raman over IR spectroscopy are:

(1) The Raman effect is a scattering phenomenon in contrast to IR which is an absorption phenomenon. Therefore, the substance need not be transparent (as it should be for IR studies).

(2) Raman spectra can be obtained for compounds in any state (gaseous, liquid, solid, gel or powder/slush form).

(3) Water is not an impediment in Raman studies as it is in IR. Therefore, studies can be carried out in aqueous (natural) media in Raman studies.

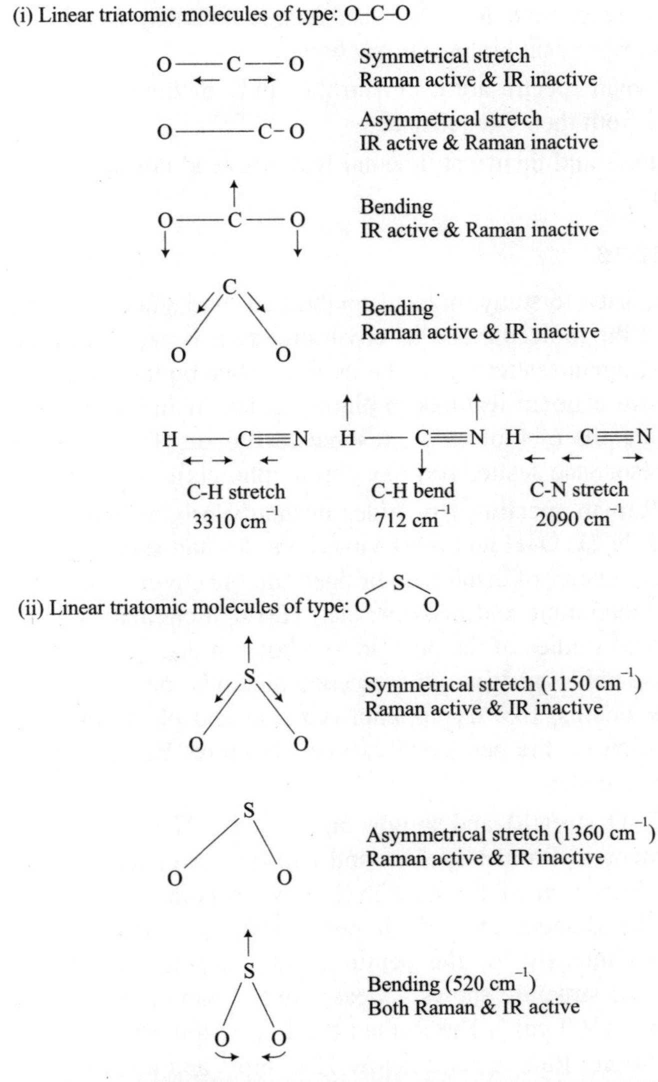

Fig. 8.20 IR/Raman active bond stretching and bending vibrations.

(4) The entire vibrational spectrum, range 4000 to 10 cm^{-1} can be covered in Raman studies (The range is 4000 to 200 cm^{-1} in IR studies).

(5) Accessibility of low-frequency modes which are sensitive to conformational changes (breathing frequencies of molecules).

(6) Raman bands are usually sharper; both polar and nonpolar interactions can be studied.

(7) There is no IR analogy to the resonance Raman effect. Resonance Raman (RR) scattering can be a 'selective' probe to study molecular aspects of chromophores—in hemoglobins, flavins, carotenoids, chlorophylls, rhodopsin etc., and substrate-binding sites in enzymes and metalloproteins.

(8) Time-resolved resonance Raman spectroscopy (TRRS) can be employed to study the dynamics of biochemical/biological reactions.

(9) AntiStokes-Raman spectra are useful in the study of transitions of molecular species (and intermediates) from their excited states.

(10) Radiation sources and their instrumental features need not be special (as compared with IR spectroscopy).

8.7.2 Applications

Raman spectroscopy is used to study molecular vibrations and since the spectrum is sensitive to the conformational state of the molecule and its environment, it is used to elucidate the stereochemical features of molecules. Raman scattering can be used to probe both polar and nonpolar environments, and structural and conformational features in biomolecules. While normal Raman scattering probes the overall structural details of molecules, resonance Raman (RR) spectroscopy focusses on the functional aspects of resonance scatterer(s) like chromophore(s).

Peptides/Proteins. Raman spectra of peptides normally originate from C–C, C=C, C–N, C=N, C–O, C=O, C–H, C–S, N–H, O–H and S–H vibrations. Raman spectra can provide information on the peptide backbone, geometry of disulphide bridges and the environment of some of the side chains (tyrosine, tryptophan, methionine and histidine etc.). Though circular dichroism (CD) is a powerful method in conformational studies of the peptide backbone, it does not provide much information on the side chains and disulphide bridges. Fluorescence methods, on the other hand, are useful in the study of aromatic side-chains; like tryptophan, tyrosine and phenylalanine, but are incapable of providing any information on the peptide backbone structure. Besides, both CD and fluorescence techniques need aqueous media.

Amide band I (C=O stretch) and amide band III (C–N stretch and N–H plane bending) (Fig. 8.16) are prominent in Raman spectra and can be correlated with structural properties of peptides. The $n - \pi^*$ transition of the (–CONH–) group is at 190 nm and the β-helix exhibits hypochromicity and the β-sheet exhibits hyperchromicity. Therefore, for a given excitation wavelength, the Raman intensity of the peptide group depends on the secondary structure. A polypeptide with α-helical structure shows a weak amide I band at 1645 – 1655 cm^{-1} and a strong amide band III at 1265 – 1300 cm^{-1}. The Raman band at ~1400 cm^{-1} is also sensitive to secondary structure. Lack of any strong Raman band below 1275 cm^{-1} and a strong line at ~1310 cm^{-1} in the Raman spectrum is indicative of α-helical conformation. Antiparallel β-sheets can be distinguished by an intense amide band III at 1235 cm^{-1} and a disorder conformation from the band at 1245 cm^{-1}. Amide band III is more sensitive to structural changes than amide band I and the Ramachandran angle, ψ (C_α – C), is the parameter most indicative of α-helix, and β-sheet conformation; with simultaneous monitoring of both amide I and amide III bands, it is possible to estimate α-helix, β-sheet, reverse turn and random coil content in proteins (Table 8.7b). Deuterium exchange (isotopic exchange) is the preferred method followed in these studies.

The disulphide bond gives rise to an intense Raman line near 500–550 cm^{-1} and the conformation of the S–S bridge (*gauche-gauche-gauche* or *gauche-gauche-trans* or *trans-gauche-trans*) can be inferred, depending on the ν_{S-S} stretching frequency of this line. This S–S vibration is very useful as a 'marker' in the study of disulphide-containing peptides and proteins, like hormones, toxins etc. For example, on cleavage of S–S bond, the Raman line at ~510 cm^{-1} vanishes, underscoring the importance of S–S bridges in maintaining 'proper' folding and structural stability in all disulphide-containing proteins. The S–H stretching band at 2550 cm^{-1} can be used as a marker to follow the conversion of –SH to S–S linkage (redox process).

Raman spectra can also provide information about the microenvironment of certain side chains. Raman lines at 1000, 1360 and 830 and 850 and 1260 and 1490 cm^{-1} are indicative of phehylalanine, tryptophan, tyrosine and histidine residues, respectively. Presence or absence of a 1360 cm^{-1} line can be used as and indicator of 'exposed' or 'buried' tryptophans in proteins. Similarly, whether a tyrosine residue is 'buried' (in which case the phenolic group of it is a strong hydrogen bond donor), or 'exposed' (in which case the phenolic group is a donor as well as an acceptor of a hydrogen bond) can be monitored from the intensity changes of Raman lines at 830 and 850 cm^{-1}. Intensity of the 850 cm^{-1} line increases when the tyrosine residue is 'exposed' and the intensity of 830 cm^{-1} increases when the residue is 'buried'. The intensity ratio R_{Tyr} = 850/830 can be used as a sensitive marker to estimate the 'buried' or 'exposed' states of **Tyr** in protein conformation.

The features of acidic amino acid side chains such as **Asp** and **Glu** can also be monitored from the Raman lines; lines at 1700–1750 cm^{-1} correspond to the –COOH moiety and at 1420 cm^{-1} correspond to the ionised $-COO^-$ moiety.

Structural and conformational changes and denaturation of macromolecules (helix/coil transitions, 'melting') can be studied by the Raman method. Also, breathing motions of entire molecules can be monitored from the low-frequency Raman fines. Helix to coil transition of proteins shifts the amide band I to a higher frequency and the amide band III to a lower frequency. Denaturation ('melting') disrupts 'packing' of side chains (hydrogen-bond networks and polar and non-polar associations). All these can be addressed by Raman studies.

Other Biomolecules. Raman spectroscopy can be employed to study the structure (tautomeric forms of nucleic acid bases etc.), conformations of nucleic acid constituent as well as the phosphodiester backbone. The vibrational frequencies of various groups are given in Table 8.8.

The Raman lines at 1575 and 1484 cm^{-1} correspond to purine and lines at 1400–1300 cm^{-1} to pyrimidine bases (1340 for uridine and 1300 cm^{-1} for cytosine). Phosphodiester stretching at 815 cm^{-1} is sensitive to conformational changes, and this line can be used as a marker to determine the forms of nucleic acids (815, 835, 865 and 800 cm^{-1} corresponding to A–, B–, C– forms and random coil, respectively). Raman bands of the ribose residues and phosphate groups in the region 1000–1300 cm^{-1} provide information on the conformation of the ribose-phosphate backbone and chelation of metal ions. Hard metal ions, like Mg^{2+} and Ca^{2+}, tend to bind to the phosphate groups. The phosphodiester and phosphoionic stretching at 815 and 1100 cm^{-1}, respectively, can be used to study conformational changes due to metal coordination and environmental effects. The ratio, I_{850}/I_{1100} can be used to monitor secondary structure (≤0 disordered; >1.5 ordered).

α- or β-anomeric configurations of saccharides and their stereochemistry can be established from their vibrational frequencies in Raman studies (Table 8.9).

Raman spectroscopy is employed in the study of another important class of biological systems, namely the study of structural and conformational aspects of lipids, biological membranes and lipid-protein complexes. The head group and carbonyl group can be monitored via the symmetric C–N and PO_4 vibrational modes at 720 and 1240 cm^{-1} and C=O modes in the 1720–1740 cm^{-1} range. Membrane fluidity is due to rotational isomerism of fatty acid hydrocarbon chains. With changes in

Table 8.8 Chemical groups and vibrational frequencies in nucleic acids.

Group	*Formula*	*Frequency (cm^{-1})*
Carboxyl	—C(=O)—OH	1660 – 1740
Carboxylate	—C(O)(O) (–)	1340 – 1440 1550 – 1690
Ester	—C(=O)—OR	1700 – 1745
Amide	—C(=O)—NH_2	1675
Ketone	C—C(=O)—C	1700 – 1740
Aldehyde	H—C(=O)—C	1710 – 1740
Phosphodiester	O—P(=O)(—O—)—O	815
Phosphoionic	O—P(=O)(—O—)—O	1100

Table 8.9 Vibrational frequencies of α- and β-anomers of saccharides.

Anomer	*Band 1*	*Band 2*	*Band 3 (cm^{-1})*
α-anomer	917	844	765
β-anomer	920	890	745

temperature, phospholipids undergo thermotropic phase transitions, from gel-to-liquid and *vice versa*, which can be detected by C–C and C–H stretching vibrations of *trans*- and *gauche*-conformations. With increase in temperature, intensities of Raman bands reflecting '*trans*' (gel) conformations decrease and intensities of bands representing '*gauche*' (fluid) conformers increase. Therefore, intensity ratios of these Raman bands (I_{1130}/I_{1100} and I_{2890}/I_{2850}), due to C–C and C–H stretching, can be used as 'monitors' to study phase transitions in biological membrane systems. A number of phospho- and sphingo-lipids exhibit liquid-crystalline properties at characteristic temperature ranges which can be studied by Raman spectroscopy. By such studies on *trans/gauche* conformations (gel/fluid transitions) as well as lateral interactions of acyl chains of phospholipids, and the lateral movements of proteins in membrane-protein complexes, the effects of ligands on membranes and protein-lipid interactions (associations) can be probed by the Raman technique. ESR and other techniques can also be employed to study protein-membrane interactions, but Raman study does not require chemical modification and, therefore, is a convenient method to study/monitor membrane system *in situ* due to its non-destructive nature.

One of the Raman spectroscopic techniques that holds a great promise is the Coherent Anti-Stokes–Raman Spectroscopy (CARS). In the CARS method (wave-mixing phenomenon), two strong collinear laser beams of different frequencies ($\nu_1 > \nu_2$) irradiate the sample. If the frequency difference, $\nu_1 - \nu_2$, is equal to the frequency of a Raman-active transition line, ν_R, then the efficiency of the wave mixing is enhanced and Stokes ($\nu_R = 2\nu_2 - \nu_1$) and anti-Stokes lines ($\nu_R = 2\nu_1 - \nu_2$) are produced by wave mixing due to non-linear polarization of the medium. CARS produces intensities $\sim 10^4 - 10^5$ times great. This enhanced signal level can greatly reduce the time necessary to record Raman spectra. Owing to the coherence of the general signals, the divergence of the output beam is small, and good spatial discrimination against the background signal is achievable.

8.7.3 Resonance Raman (RR) Spectroscopy

Biomolecules in general do not give rise to Resonance Raman (RR) spectra. However, if molecules possess chromophore groups, either intrinsically or by extrinsic introduction, and if the energy of the exciting laser beam is coincident with that of an electron transition of the chromophore, then RR scattering occurs. While Raman spectra provide structural information on the entire molecule, RR spectrum provides structural details confined to the chromophores and their environments. In RR spectra the Raman lines are due to the vibrational modes of chromophores and the surrounding environments are selectively enhanced so the spectrum provides a selective probe of the interactions of the chromophore regions. As these sites are generally of functional significance, RR study is very useful in the elucidation of structure-function features of enzyme-catalysed reactions, metal chelation in macromolecules, photochemical reactions in retinal pigment complexes and heme- and flavin-containing molecules. For example, in rhodopsin, the $\pi - \pi^*$ transitions enhance the Raman modes of chromophores. The frequency shift of the C=N bond (**Schiff base** link of the retinal chromophore), due to protonation and deprotonation in the ground and the intermediate state, respectively, is related to the proton pump of the purple bacterial membrane. Similarly, enzyme-catalysed reactions, and hapten-antigen interactions can be studied by RR-labelling by introducing extrinsic chromophore groups, such as the azo group (–N=N–), to substrates. RR spectra of metallo-proteins provide information on ligand-field and charge-transfer transitions, and heme-protein interactions. Such methods can be employed, with the introduction of extrinsic chromophores, to study the antigen-antibody, enzyme-substrate, DNA-drug (intercalation) and drug-dye interactions. Line-shifts and intensity variation in Raman lines can be used as a monitor to study the conformational and stereochemical changes in all these molecular interactions.

Time-resolved Resonance Raman Spectroscopy (TRRS). With advances in picosecond laser technology, it has become possible to undertake the study of fast biochemical/biological reactions by Raman spectroscopy, to investigate reaction mechanisms in detail. Time-resolved Resonance Raman

Spectroscopy (TRRS) provides an ideal method to study the structure, dynamics and functional features, at various intermediate stages, of a host of biologically-important complexes. To name a few: porphyrin-containing protein complexes (hemoglobins, cytochromes, plastocyanins, flavo-proteins etc.), chlorophyll-containing photosynthetic reaction centres, rhodopsin and retinal proteins etc. Higher-order structural features of macromolecules, such as proteins, can be studied by the time-resolved ultraviolet resonance Raman spectroscopy. To highlight the application of TRRS, heme-hemoglobin interactions can be cited as an example. The structures of hemoglobin and several mutants have been determined by X-ray diffraction studies to a high degree of accuracy and the structure-function details have also been well characterised by other biophysical and biochemical methods; therefore, hemoglobin serves a model structure to underscore the detailed structural information that can be obtained from TRR studies.

Hemoglobin (Hb) is the primary oxygen-transport protein in vertebrates. It is a tetrameric protein, with two α- and two β-type polypeptide chains ($\alpha_2\ \beta_2$) (of ~140 amino acid residues each). Each polypeptide contains a heme-prosthetic group with an iron atom coordinated to four nitrogen atoms of macrocylic tetrapyrrole ring (equatorial ligands). The fifth ligand in the axial position is an N-atom of the imidazole ring of the F8-histidine residue of the polypeptide. The sixth (axial) position is occupied by an oxygen molecule, O_2, in the oxy- and is unoccupied in the deoxy-state of hemoglobin. That is, the iron atom is penta-coordinated in the deoxy-state (C.No. = 5) and is hexa-coordinated in the oxy-state (C.No. = 6), respectively.

To understand the mechanism of heme-iron-protein interactions in hemoglobins, the role of the 3d-electrons of the iron atom in the heme must be understood. In the functional forms of hemoglobins (and also in myoglobins), the iron atom is present in the ferrous, Fe (II)-state. The Fe(II) atom has six valence electrons in its 3d-orbitals and on ligation the ligands lift the degeneracy of 3d-orbitals into two $\mathbf{e_g}$* and three $\mathbf{t_{2g}}$ orbitals. Depending on the nature of ligands, Fe(II) exists either in (i) a low-spin state or in (ii) a high-spin state. If the magnitude of the ligand-field is high (strong field), the ligand-field energy, Δ, is greater than electron-pairing energy, E_P ($\Delta > E_P$); E_P is the energy required to force the pairing of two electrons in the same orbital, the valence electrons occupy the low-energy three $\mathbf{t_{2g}}$ orbitals, resulting in the low-spin (S = 0) state of Fe(II). On the other hand, if the magnitude of field splitting is smaller (weak field) than the electron-pairing energy, E_P ($\Delta < E_P$), the 3d-valence electrons occupy all five 3d-orbitals, (both $\mathbf{e_g}$* and $\mathbf{t_{2g}}$) and the resulting state is (S = 2), a high-spin state for Fe(II) (Fig. 8.21a).

The size of the Fe(II) atom varies with the spin-state it exhibits. The size of the Fe(II) atom is smaller in the low-spin state (strong field) as the electron cloud is concentrated to $\mathbf{t_{2g}}$ orbitals, whereas in the high-spin state (low field) the electron cloud is spread out to $\mathbf{t_{2g}}$ as well as $\mathbf{e_g}$* orbitals and, therefore, the size of the Fe(II) atom is larger. The difference in size of the Fe(II) atom with change in its spin-states has a fundamental effect on Fe(II)-heme stereochemistry, and thereby on heme-protein interactions. The central 'hole' of the tetrapyrrole frame is ~2.0 Å, in diameter. The hexa-coordinated Fe(II) atom exists in low-spin state and has a diameter of ~1.9 Å. Therefore, the Fe(II) atom, in the low-spin state, can fit into the cavity of the heme-frame and, therefore, the Fe(II)-heme moiety in oxy-hemoglobin (C.No. = 6) is planar. In contrast, in deoxy-hemoglobin, the penta-coordinated Fe(II) atom is in a high-spin state. As the diameter of Fe(II) atom in the high-spin state is ~2.08 Å, it cannot fit into the tetrapyrrole 'hole'. The penta-coordinated Fe(II) atom is ~0.7 Å above the heme plane and the geometry of heme-Fe(II) moiety is pyramidal in deoxyhemoglobin (Fig. 8.21b). In essence, the binding or removal of the axial ligand of the metal (oxygen molecule in oxy- and deoxy-hemoglobin) has a fundamental effect on the stereochemical and biological properties of heme compounds.

A two-state allosteric model for oxy- and deoxy-hemoglobin quaternary structures proposes that at low oxygen concentration the quaternary structure of hemoglobin (deoxy-hemoglobin) is in a low-affinity, tense, or **T-state**, and at saturated oxygen concentrations, the quaternary structure (oxy-) changes to high-affinity, relaxed, or **R-state**. TRRS has been found versatile to probe intermediate structural changes between these two states; tense- and relaxed-, quaternary structural states of hemoglobin.

Transient Raman studies reveal that for each quaternary structure (T- and R-states) there exist two tertiary intermediate states. That is, there are four spectroscopically detectable configurations, namely, Hb (Td), Hb (Tl); Hb (Rd) and Hb (Rl) (l and d represent ligated and unligated configurations, respectively, in T- and R-quaternary states of hemoglobin). Ligated and unligated tertiary structures

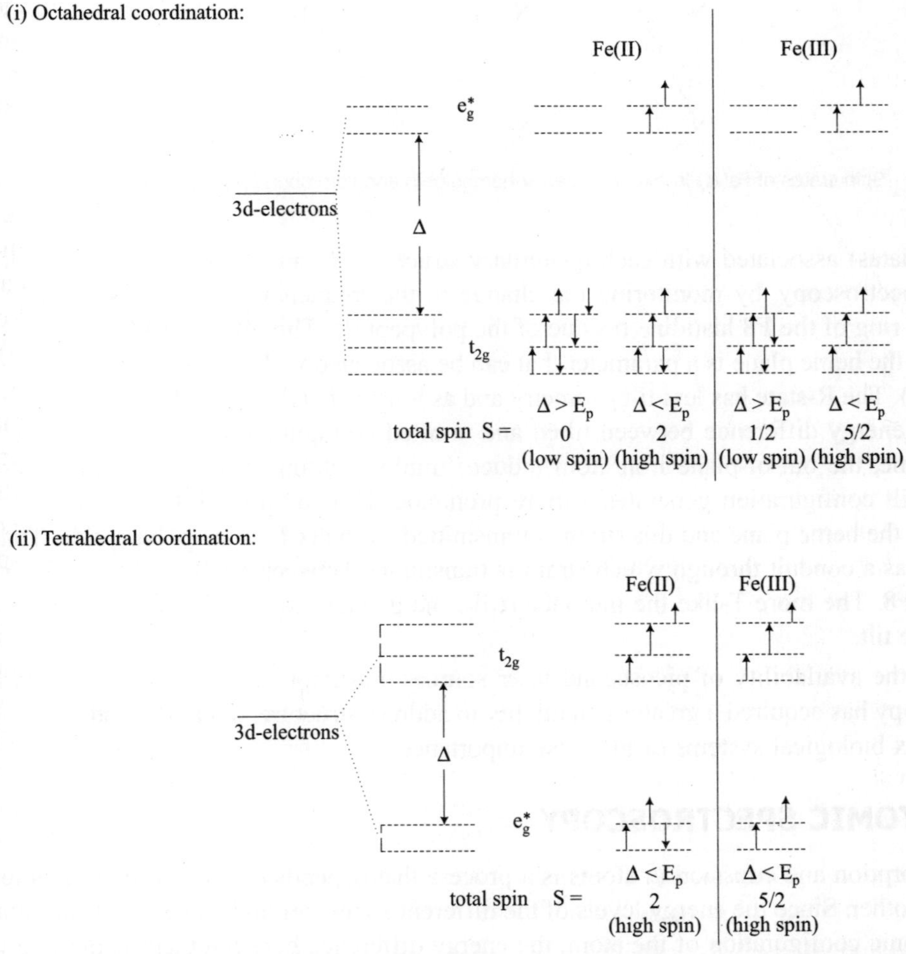

Fig. 8.21a Spin states of Fe (in divalent and trivalent states) in octahedral and tetrahedral coordination due to ligand-field splitting of 3d-electrons.

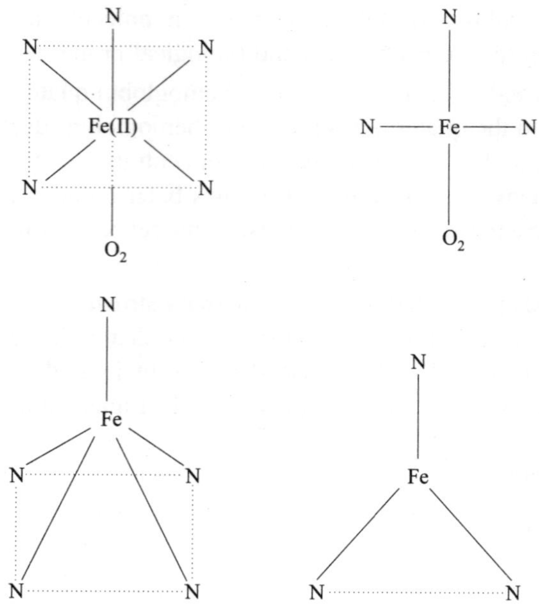

Fig. 8.21b Spin states of Fe(II) in oxy- and deoxy-hemoglobin and corresponding heme-Fe(II) geometries.

(intermediates) associated with each quaternary structure (T- and R-states) can be distinguished by Raman spectroscopy, by monitoring the change in the frequency of the Fe-N bond (ν_{Fe-N}) of the imidazole ring of the **F8** histidine residue of the polypeptide. The tilt of the **F8** histidine residue with respect to the heme plane is a parameter that can be associated with variation in frequency (230 cm^{-1}) (Fig. 8.22). The R-state has less tilt geometry and as long as Fe(II) is out of the heme plane (in deoxy-state) the energy difference between tilted and untilted configurations is small and, therefore, for a deoxy-heme, the out-of-plane iron atom induces minimal strain regardless of quaternary structure. But, the tilt configuration generates a more pronounced torque upon **F8** histidine when the Fe(II) atom is in the heme plane and this strain is transmitted *via* helix **F** to the polypeptide. That is, helix **F** functions as a conduit through which strain is transmitted between Fe(II) and the $\alpha_1\beta_2$ interface, *via* histidine **F8**. The more T-like the interface $\alpha_1\beta_2$, the greater the pull on the histidine and hence the greater the tilt.

With the availability of picosecond laser sources and improvement in instrumentation, Raman spectroscopy has acquired a greater potentiality to address structure, dynamics and functional features of complex biological systems of immense importance.

8.8 ATOMIC SPECTROSCOPY

Light absorption and emission in atoms is a process that depends on excitation of an atom from one state to another. Since the energy levels of the different states (ground and excited states) are fixed by the electronic configuration of the atom, the energy difference between them is discrete (quantised). The wavelengths of light absorbed or emitted are known as 'characteristic wavelengths' because they are representative of a specific atom going through these changes.

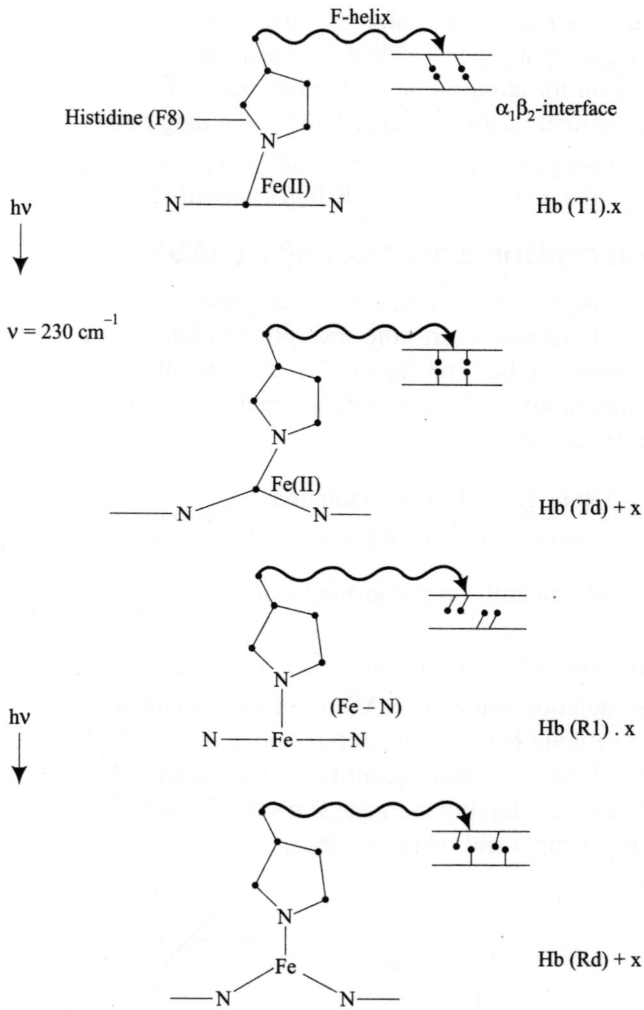

Fig. 8.22 Schematic representation of tilt geometry of histidine **F8** with respect to the heme plane in hemoglobin states—Hb(Tl), Hb(Td), Hb(Rl) and Hb(Rd).
(After Friedman, J.M. Science 228; 1273 (1985)).

Atomic spectroscopy is based upon flame atomisation, that is converting atoms/ions to vapour by heat. The technique is based upon (i) absorption (Atomic Absorption Spectroscopy, AAS), (ii) emission (Atomic Emission Spectroscopy, AES), and (iii) florescence (Atomic Fluorescence Spectroscopy, AFS). Atomic spectroscopy is limited to X-ray, UV and visible regions.

8.8.1 Atomisation

The process of converting a sample to atomic vapour/mist is called atomisation. The quality of data obtained in any atomic spectroscopy depends very much on the quality of atomisation. Atomisers can be continuous or discrete. In a continuous atomiser, a solution of the sample is converted into a mist by a jet of compressed gas by a process called nebulisation. The flow of gas carries the nebulised sample into a heated region for atomisation. A Bunsen burner can be used as a source of flame, but laminar flow burners provide a relatively quite flame and a long path.

Electrothermal atomisers are discrete. The entire sample is atomised in a short period. They accept smaller quantities of samples, and provide higher sensitivity.

8.8.2 Atomic Absorption Spectroscopy (AAS)

Atomic absorption spectroscopy (AAS) relies on the absorption of light by free metallic atoms, when they are dissociated by a flame and records the absorption of light of particular wavelengths by atoms in their ground state. Absorption line spectra are characteristic of the atoms present in the compound. In the AAS what is measured is the percentage fraction of radiation absorbed (A_T), which is proportional to the analyte metal.

$$A_T = \frac{\text{Intensity of absorbed radiation}}{\text{Intensity of incident radiation}} = (\Delta I/I_0) \times 100 \qquad ...(8.27)$$

As the atoms are predominantly in the ground state (>99.1%) at flame temperatures, AAS is a very sensitive technique.

A schematic line-drawing of an atomic absorption spectrophotometer is given Fig. 8.23.

The most common radiation source in AAS is the hollow cathode lamp. It consists of a tungsten anode and a cylindrical cathode sealed in an inert gas tube. The cathode is the metal whose spectrum is desired. In discharge lamps, a small quantity of the metal salt whose spectrum is desired is introduced into a quartz tube containing inert gas and sealed. AAS is a widely used method of atomic spectroscopy because it is simple and relatively cheap.

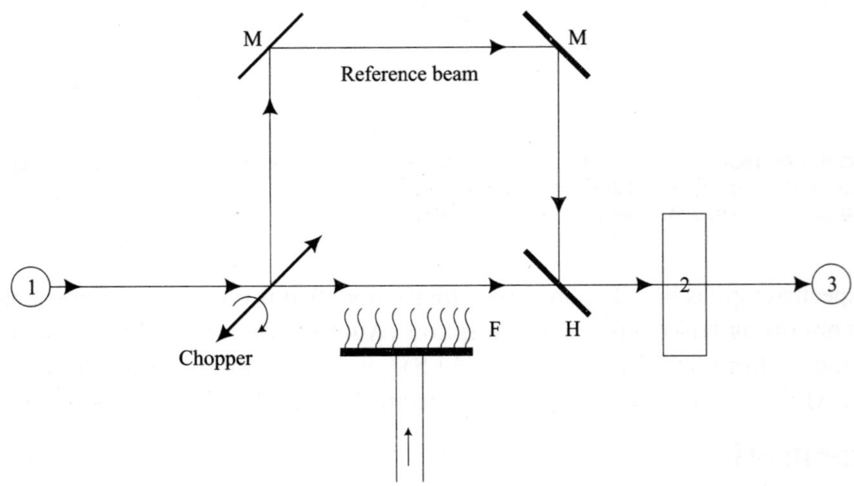

Fig. 8.23 Line-drawing of an atomic absorption spectrophotometer.
1 = Radiation source; 2 = Monochromator; 3 = Detector.
F = Flame; H = Half-silvered mirror; M = Reflecting mirror.

8.8.3 Atomic Emission Spectroscopy (AES)

Atomic Emission Spectroscopy (AES), also known as flame emission spectroscopy or flame photometry, makes use of a flame to dissociate the sample to excited atomic/ionic states. The excited analyte ions serve as the source of radiation (no external source of radiation is required) and the energy of transitions of these ions/atoms from their excited states to their ground states is emitted as spectral lines, and the spectra are characteristic of the elements present in the sample.

AES is a versatile, rapid and sensitive method for the analysis of trace metal constituents in a variety of matrices. Atomic spectroscopic methods are employed in elemental analyses. Their important uses in biological sciences are in clinical chemistry, in the estimation of sodium, potassium and calcium in serum, biological fluids and tissues.

8.9 ELECTRON SPIN RESONANCE (ESR) SPECTROSCOPY

Electron spin resonance (ESR) or electron paramagnetic resonance (EPR) is a branch of absorption spectroscopy that deals with the transitions between magnetic energy levels of electron with unpaired spins. The magnetic energy splitting is created by a magnetic field (**Zeemann** effect) and the transitions are induced by radiation at microwave frequencies. The presence of unpaired electrons in a sample is a condition for ESR. Unpaired electrons are present in free radicals, triplet electronic states, transition metals (like Fe, Co, Ni, Mn, Zn etc.) and rare earth ions. In biological systems, paramagnetic species occur in transition metal complexes, shorted-lived intermediates in electron-transfer processes etc.

8.9.1 ESR Spectra

The basis of ESR is the spin of an electron. A spinning electron behaves like a bar magnet and the associated magnetic moment, μ, is

$$\vec{\mu} = \gamma \hbar \vec{S}$$

$$\vec{\mu} = -\mathbf{g}\, \beta M_s$$

$$\gamma = -\mathbf{g}\, \frac{e}{2mC} \qquad \text{...(8.38)}$$

where g is a splitting factor called **Lande's g**-factor, β is the **Bohr** magneton $\left(\beta = \frac{eh}{2\pi mC}\right)$ and S is the spin quantum number, $\left(M_s = +\frac{1}{2} \text{ or } -\frac{1}{2}\right)$, m is the mass of electron and C is the velocity of light.

In the absence of an external magnetic field, the free electron can exist either in $M_s = +\frac{1}{2}$ or in $M_s = -\frac{1}{2}$ state of equal energy (degenerate states). When an external magnetic field of strength, H, is applied the degeneracy of energy levels is lifted (Fig. 8.24) and the difference in energy between the two levels is

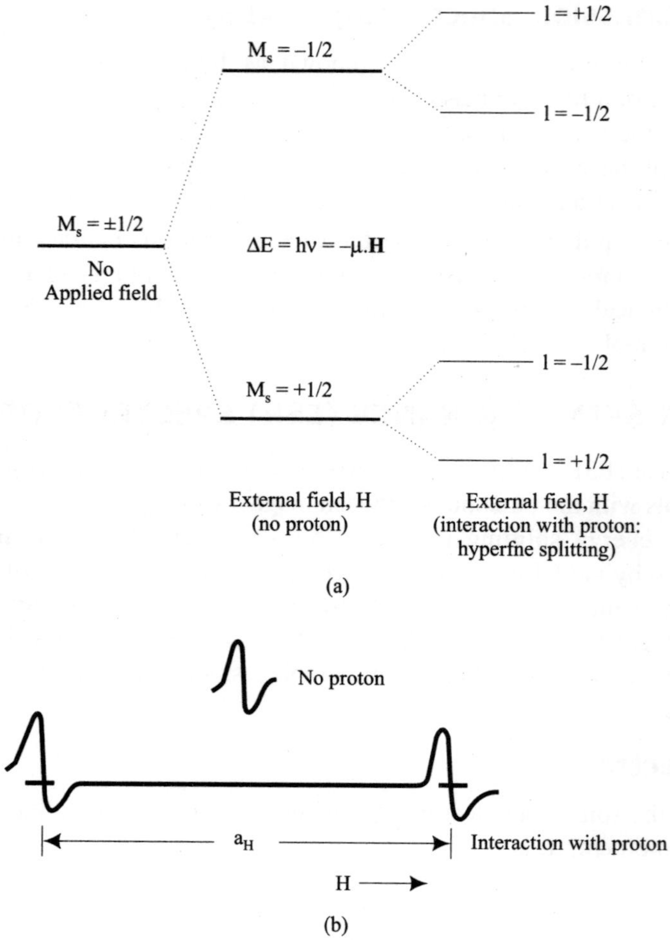

Fig. 8.24 Electron energy levels in a magnetic field, *H*.
(a) Zeemann and hyperfine splitting.
(b) Absorption signal due to interaction without and with proton.

$$\Delta E = h\nu = -\mu \cdot H = \mathbf{g}\beta\, H\, M_s \qquad \text{...(8.29)}$$

When a compound exhibiting the Zeemann effect is irradiated by photons ($h\nu$) in a direction perpendicular to the magnetic field, a magnetic dipolar transition takes place between the split energy states when $\Delta E = h\nu$. This is ESR absorption. An ESR spectrum is a function of the combination of various factors—concentration of the sample, hyperfine splitting and their intensities, **g**-factor, and the relaxation times and line-widths. Hyperfine splitting in ESR (hyperfine Zeemann spectrum) arises due to the interaction between the magnetic moments of the unpaired electrons and the nuclei. Hyperfine splitting in an ESR spectrum can be used as a fingerprint to identify the radicals present in the sample.

The g-factor. The g-factor (Lande's **g**-factor) is a dimensionless quantity.

$$\mathbf{g} = \frac{\text{electron's magnetic moment}}{\text{spin angular momentum}}$$

$$\mathbf{g} \approx 1 + \frac{[J(J+1) + S(S+1) - L(L+1)]}{[2J(J+1)]} \qquad \text{...(8.30)}$$

where S = spin vector; L = angular orbital vector and J is the resultant vector of S and L vectors, and is equal to 2.002319 for the unbound electron. But, the exact value of the **g**-factor reflects the chemical environment surrounding the electron. Different paramagnetic spectra can usually be distinguished quite easily by their **g**-values. For systems which have very little orbital angular momentum, such as free radicals, the **g**-value is the same as for the free electron. However, if the electron has orbital angular momentum (as in metal ions), the **g**-value deviates from the unbound electron value. This deviation is a measure of the additional interactions which the electronic spin system has with its surroundings. **g** is a tensor (direction-dependent) and by choosing principal axes such that off-diagonal terms of the tensor vanish, only three principal **g**-values ($\mathbf{g}_x$, $\mathbf{g}_y$ and $\mathbf{g}_z$ for each unpaired spin are relevant. In perfect cubic symmetry (ignoring **Jahn-Teller** distortions), only one resonance peak is observed. In crystalline state ESR measurements can provide **g**-values based on the orientation of the crystal.

Relaxation Processes. The distribution of the population in spin states, $M_s = +\frac{1}{2}$ and $M_s = -\frac{1}{2}$ has a **Boltzmann** distribution.

$$N_a/N_b = \exp(-\Delta E/kT) \qquad \text{...(8.31)}$$

where N_a and N_b are populations in the upper and lower states and k is the Boltzmann constant. When $h\nu = g\beta H$, transitions between the two states are induced by the applied microwave radiation. There is a greater population in the lower state than in the upper state. The microwave field tends to equalise the populations. However, the interactions of the electrons with the lattice containing them tend to establish and maintain a Boltzmann population. The processes which tend to maintain this population difference are called relaxation processes and the resonant absorption may be viewed as a continued competition between the tendency of the microwave field to equalise the populations and tendency of the spin-lattice interactions to maintain the Boltzmann difference population. The net absorption increases as N_a/N_b decreases, i.e., as temperature, T, decreases.

The line-width of an absorption line is a measure of the interaction between an electron and its environment. The most important is spin-lattice interaction (τ_1). This occurs when the spin energy is shared with the thermal vibrations of surrounding molecules. Crystalline structure produces fields which tend to decouple spin-orbital interaction, so that spin-lattice interaction is reduced. Thus, spin-lattice relaxation time (τ_1; has **Lorentzian** shape) and line-width can give an indication of the crystallinity of the sample. The other interaction is spin-spin interaction that involves a dipolar interaction between an electron and surrounding electrons and nucleic. This field is felt by an unpaired electron, in addition to the applied magnetic field, H. Spin-spin relaxation time (τ_2) tends to broaden the original absorption line (Lorentzian line) into a **Gaussian** shape. The line-width, Δ is

$$\Delta = \frac{1}{\tau_1} + \frac{1}{\tau_2} \qquad \text{...(8.32)}$$

Both spin-lattice and spin-spin relaxations broaden the resonance line-widths and absorption lines are broad (at room temperatures). Therefore, ESR signals are observed at low temperatures.

8.9.2 Spin-Label

Chemical and biological compounds which do not contain an unpaired electron can be studied by extrinsic introduction of a stable free-radical. This spin-label produces a sharp and simple ESR spectrum that gives detailed information concerning the molecular environment of the label. Heterocyclic molecules with a nitroxyl group (whose N-atom is bonded to a tertiary carbon) are excellent spin-labelling compounds for biological macromolecules, such as proteins, nucleic acids and lipids.

8.9.3 ENDOR

Electron nuclear double resonance (ENDOR) is a method for improving the effective resolution of an ESR spectrum. The sample is irradiated simultaneously with a microwave frequency suitable for electron resonance and a radio frequency suitable for nuclear resonance. The ENDOR display is ESR signal height as a function of the swept nuclear radio frequency. Electron double resonance (ELDOR) is a method where the sample is irradiated simultaneously by two microwave frequencies. These methods are useful for separating overlapping multi-radical spectra and for studying various relaxation times.

8.9.4 Interpretation of Spectra

Interpretation of ESR spectra involves **g**-factor, hyperfine splitting and their intensities, relaxation times and line-widths. An ESR spectrum is interpreted by assigning the magnitudes of the coupling constants and counting the lines. The analysis is generally carried out by approximate assignments of coupling constants and comparing the computed spectra with the experimental ones, by computer simulation and modelling methods.

Biological Systems. An important application of ESR in biological systems is in the study of transition metal complexes (Cu- and Fe-proteins). The unpaired electron responsible for the ESR signal in a transition metal is localised in a d-orbital and the pairmg of spins can depend on ligands.

ESR studies are useful in the study of the distortion of metal-ligand coordination of transition metal complexes. Such studies are extensively employed in identifying Fe-S reaction centres in iron-sulphur proteins—ferredoxins (group I), high-potential iron-sulphur proteins (group II), rubredoxins (group IV) and conjugated Fe-S centres (group IV). Electron flow in transition metal proteins, such as copper-containing proteins can be investigated by ESR, stop-flow and rapid-freeze techniques. Such studies indicate that cytochromes **a** and $\mathbf{a}_3$ are in the low- and high-spin states, respectively, in the reduced-state of cytochrome oxidase.

Time-resolved ESR, spin-echo ENDOR and spin-labelling, combined with the latest advances, have enhanced the role of ESR in dynamical studies of biological molecules. Diffusion of spin-labelled molecules through membrane bilayers can be used to study the dynamic aspects of translational and rotational diffusion through membranes, membrane permeability, and accompanying conformational changes of the transport molecules. The dynamics of protein-lipid interactions as a function of transition temperature can be monitored by spin-labelled ESR. Fluorescent spin label probes (and time-resolved methods to monitor fluorescent decay) can be employed to study structural and dynamic features of protein-lipid complexes. Positively-charged hydrophobic spin labels are used for the estimation of trans-membrane potentials. The dynamic behaviour of carbohydrate chains of glycoproteins (e.g., immunoglobulins) can be studied by spin-labelled ESR.

ESR finds application in medical diagnostics. The Fe^{3+}-signal of iron-transferrin, and the Cu^{2+}-signal of caeruloplasmin of blood samples of normal controls and patients can be analysed by ESR methods. Muscular dystrophy and other clinical abnormalities can be studied by ESR methods by monitoring body fluids and cells.

Spin-labelled ESR is employed in radiation biology to elucidate the effects of ionising radiation on molecules of biological relevance— nucleic acids, proteins, polysaccharides and membranes. Irradiation produces free radicals (paramagnetic centres) that can be used as probes in ESR studies to monitor structure-function variations in cells upon irradiation. Enhancement of radiation-damage due to the presence of oxygen radicals can also be monitored by such studies.

8.10 MASS SPECTROMETRY (MS)

Mass spectrometry (MS) is based upon the separation of ionic fragments of a sample according to their mass/charge (m/z) ratios. Mass spectrometry can be employed to determine the molecular mass as well as structural features of molecules.

8.10.1 Mass Spectrometer

Any mass spectrometer has four basic units— (i) an inlet system, (ii) ion source, (iii) electrostatic acceleration system and (iv) detector system.

The sample ($<10^{-6}$ mole), after its inlet in the inlet chamber, is brought to a vapour state by heating as well as by suitable chemical modification and the vapours are led through an orifice (molecular leak) into the ionisation chamber. The sample in vapour form is converted into ionised components in the ionisation chamber by bombarding with electrons, ions or by electrical or thermal ionisation.

The ionisation chamber is the most important part of a mass spectrometer (Fig. 8.25). Its function is to produce ions without mass discrimination and accelerate them across to the mass detector. Some of the types of ionisation in use are: electron-impact ionisation, field ionisation, surface ionisation, chemical ionisation, thermal ionisation and laser microprobes.

Electron-impact ionisation is a commonly, used method. In this method-molecules (at temperatures ~200°C) are led into the ionisation chamber, maintained at high vacuum, through the 'molecular leak' (orifice). An electron gun is located in the chamber perpendicular to the molecular beam and electrons from the gun ionise the molecules in the gaseous phase, producing molecular ions. The electrons after passing through the ion source, are collected by an anode. The intensity of the electron beam is controlled so as to keep the rate of formation of molecular ions constant. Field ionisation occurs when molecules are brought near a sharp metal edge (knife edge) in the presence of a high electric field. Field ionisation and electron-impact ionisation can be combined in the same

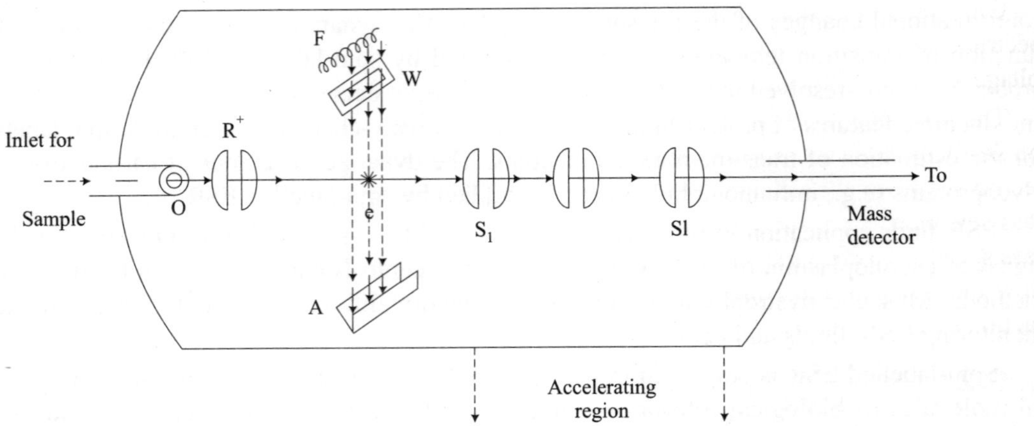

Fig. 8.25 Schematic sketch of an electron-impact ionisation chamber.
O = Orifice (molecular leak); R^+ = Repeller slit; F = Filament of an electron gun; W = Electron slit; e = Electron beam; A = Anode trap; S_1 = First accelerating slit (low negative voltage); S_l = last accelerating slit (at high voltage); * = Ionising region for molecules.

instrument. Spark-source ionisation is advantageous for solids with high-melting temperatures. Surface ionisation is convenient for inorganic materials and finds application in nuclear and geochemistry. In this method, the sample is coated on a tungsten filament and heated until the sample evaporates. For organic and biological specimens the electron-impact process is suitable. In this process, methane gas is introduced in the ionisation chamber (under low pressure). Electron impact produces ionisation of methane and the chemically-reactive methane species react with the sample molecules. Desorption by a laser microprobe vaporises the sample with minimum thermal degradation and these vapours are then ionised in the ionisation chamber. Fast atom bombardment of samples in a glycerol solution by high energetic xenon or argon is applicable to high molecular mass species. The positive ions produced in the ionisation chamber are gently drawn out by a mild electrostatic field maintained between the repeller (charged positive and placed immediately after the 'molecular leak' orifice), and the first accelerating slit (charged negative and placed after the electron-gun system). A strong electrostatic field (500 → 4000 V), maintained between the first and the last accelerating slit, accelerates the ions of masses m_1, m_2, m_3 to the mass detector.

The mass detector resolves the ions of different masses based on the differences in their deflections in a static magnetic field. The Dempster mass spectrometer is one such detector where the ions, accelerated through the ion chamber, are deflected to follow circular trajectories due to the action of the magnetic field applied perpendicular to the ion beam. The m/z ratio of an ion of mass m and charge z is given by

$$\frac{m}{z} = \frac{B^2 r^2 e}{2V} \qquad ...(8.33)$$

where B is the magnetic field strength, r is the radius of the trajectory and V is the accelerating potential.

Only the in-register ions, coinciding with the arc of the detector tube, are recorded. The mass spectrum is obtained by varying at a constant rate either magnetic field strength, H or accelerating voltage, V.

The limit of resolution in a mass spectrometer is generally below 1 part in 500 mass units. With double-focussing mass spectrometers, ions can be focussed both for direction and momentum; the electric sector of the detector provides for energy selection while the magnetic sector provides for mass selection. With such instruments, resolution of 1 part in 2000 mass units can be realised. The mass spectrum can be recorded electronically or photographically. Time-of-flight mass spectrometers allow microsecond recording of events and are employed in studies of fast chemical reactions and reaction mechanisms.

8.10.2 Mass Spectra

A mass spectrum of a sample is a recording of the masses of ion fragments and the relative abundance of these ions of the specimen. At low potential, around 10 eV (the ionisation potential range of most organic compounds) the mass spectrum contains the parent molecule and ion fragments. The mass of the heaviest ion is the molecular mass of the parent substance and mass spectrometry provides one of the most accurate methods of determining molecular mass. In addition, a mass spectrum is a 'finger print' of the compound and its ionic components and the size and structure of a pure compound can be reconstructed from the mass spectrum data of the ion fragments.

Molecular separation can be achieved by gas chromatography (GC) and the resolved species can be introduced into the ionisation chamber for identification by mass spectrometry (MS). This way, the advantages of both gas chromatography and mass spectrometry can be combined by GC-MS.

8.10.3 Mass Spectrometry (MS) in Biology

Until recently, protein sequencing by Edman degradation method was the method of choice for the identification of proteins. The situation has drastically changed with the integration of advanced physical technologies, such as mass spectrometric methods for protein identification and sequence analysis. Incorporation of mass spectrometry in biological studies is made possible with the development of gentle ionisation techniques—matrix-assisted desorption ionization (MALDI), and electrospray ionization (ESI). MALDI and ESI methods permit the efficient transfer of proteins and peptides into gas phase and into the mass spectrometer. Generally, the MS analysis consists of individually excising the spot of interest from a 2-DE gel. The spot is rinsed and subjected to enzymatic (trypsin) digestion. The most commonly used mass analyzers for protein biochemistry applications are time-of-flight (TOF), triple-quadrupole, quadrupole-TOF, and ion-trap instruments.

8.11 MÖSSBAUER SPECTROSCOPY

Mössbauer spectroscopy (nuclear γ-ray spectroscopy) provides a powerful and detailed probe of the electronic state and local structure in the neighbourhood of Mössbauer nuclei, ^{57}Fe, $^{119}S_n$, ^{127}I and ^{129}I. The Mössbauer effect is based on the resonant absorption of nuclear gamma radiation by certain atomic nuclei in solids without energy loss or line-broadening due to the recoil of the nuclei. The recoil of the nucleus is absorbed by the crystal lattice. The γ-ray intensity Z (**v**), recorded behind the absorber as a function of Doppler velocity, **v**, is the Mössbauer spectrum (a plot of resonance γ-ray signal intensity versus relative velocities of the source with respect to the absorber). Basic units of Mössbauer spectroscopy are: a radioactive source (e.g., ^{57}Co), the sample, a Doppler-scanning device and a γ-ray detector. Information that can be obtained includes: (i) isomer shift (chemical shift), (ii) electronic quadrupole splitting and (iii) magnetic hyperfine splitting. A Mössbauer spectrum gives the most complete information when it is observed for paramagnetic ions. In biological sciences, it is employed to study iron-containing molecule (myoglobin, hemoglobin, Fe-S proteins).

From the isomeric shift and quadrupole splitting of the Mössbauer lines, information can be obtained about the nature of the chemical bonding, charge distribution, oxidation state, crystal lattice etc., around the Mössbauer nuclei.

8.11.1 Isomeric Shift

Isomeric shift (displacement of the resonance peak from the zero-Doppler velocity) depends on details of the chemical bond of the Mössbauer nucleus within the molecular matrix. It is a linear function of **s**-electron density and it is also influenced by **p** and **d** electrons. For example, isomeric shift provides information on the oxidation state of iron (Fe(II) or Fe(III)), whether iron is in the high-spin or low-spin configuration and whether the coordinate bond is largely ionic or covalent. High-spin Fe^{2+} can be readily distinguished by quadrupole splitting.

Mössbauer spectra on various hemoglobins give information on the relative distances between the iron atom and its nearest ligands. On binding of oxygen to the iron, one electron is transferred to the oxygen molecule. From Mössbauer studies it appears that the configuration of the iron-oxygen complex is more likely to be $Fe(III)O^-_2$ than $Fe(II)O_2$.

In non-heme molecules, such as iron-sulphur proteins, the Fe is generally in the tetrahedral coordination state as Δ_{tet} will never exceed pairing energy, E_p (see Fig. 8.21a). The **Fe** and **S** atoms of each cluster are located at the corners of a slightly-distorted cube, the corners of which are alternately occupied by **Fe** and **S** atoms. Whether the iron atom in the cluster behaves diamagnetically (if the coupling of spins of two iron atoms in the cluster is antiparallel) or paramagnetically can give information on the coupling of the **Fe-S** clusters.

Mössbauer spectroscopy has limited applications in biology other than in the study of iron-containing molecules.

Suggested Reading

Books

Ackerman, E., Ellis, L.B.M. & Williams, L.E. *"Biophysical Science"* 2nd Edn. Prentice-Hall/New Jersey (1972).

Atkins, P.W. *"Physical Chemistry"* 6th Edn. Oxford University Press/Oxford. (1998).

Brey, W.S. *"Physical Chemistry and its Biological Applications"*, Academic Press/ New York. (1978).

Cantor, C.R. & Schimmel, P.R. *"Biophysical Chemistry"*. Freeman/New York. (1980).

Carey, P.R. *"Biochemical Applications of Raman and Resonance Raman Spectroscopies"*. Academic Press/New York. (1982)

Chang, R. *"Physical Chemistry with Applications to Biological Systems"*. McMillan/New York. (1977).

Clark, R.J.H. & Hester, R.E. (Eds.). *"Advances in Infrared and Raman Spectroscopy"* Vol. I. Heydon & Son/London (1975).

Colthup, N.B. et al., *"Introduction to IR and Raman Spectroscopies"* 3rd Edn. Academic Press/New York (1990).

Cundall, R.B. & Dale, R.E., (Eds.). *"Time-resolved Fluorescence Spectroscopy in Biochemistry and Biology"*. Plenum Press/New York. (1983).

Evans, J.N.S. "*Biomolecular Spectroscopy*". Oxford University Press/Oxford, UK. (1995).

Ferro, J.R. and Nakamoto, K. *"Introduction to Raman Spectroscopy"*. Academic Press/New York. (1994).

Guilbault, G.G. *"Practical Fluorescence"*. Marcel Dekker/New York. (1990).

Hallet, F.R. et al. *"Physics for the Biological Sciences"*. Methusen Pub/Toronto. (1982).

Hoppe, W. et al. (Eds.). *"Biophysics"* Springerverlag/Berlin. (1983).

Hughes, W. *"Aspects of Biophysics"*. John Wiley/New York. (1979).

Lakowicz, J.R. *"Principles of Fluorescence Spectroscopy"*. 2nd Edn. Kluwer-Academic Press/New York. (1999).

Lippard, S.J. & Berg, J.M. *"Principles of Bioinorganic Chemistry"*. University Science Books/Mill Valley/CA. (1994).

Martin, B.R. *"Introduction to Biophysical Chemistry"*. McGraw Hill/New York. (1964).

Moore, W.J. *"Basic Physical Chemistry"*. Prentice-Hall of India/New Delhi (1989).

Neuberger, A. & Van Deenan, L.L.M. (Eds.). *"Modern Physical Methods in Biochemistry"*. Elsevier/ Amsterdam. (1988).

Shanna, A. and Schulmann, S.G. *"Introduction to Fluorescence Spectroscopy"*. John Wiley/New York. (1999).

Taylor, D.L. et al., (Eds.). "*Applications of Fluorescence in the Biomedical Sciences"*. Liss/New York. (1986).

Tinoco, I. Jr. et al. *"Physical Chemistry: Principles and Applications in Biological Sciences"*. 2nd Edn. Prentice-Hall/New Jersey. (1985).

Tu, A.T. *"Raman Spectroscopy in Biology: Principles and Applications"*. John Wiley/New York. (1982).

Valeur, B. & Bronchon, J.C., (Eds.). *"New Trends in Fluorescence Spectroscopy: Applications to Chemical and Life Sciences"*. Springerverlag/Berlin. (2001).

Volkenstein, M.V. *"Molecular Biophysics"*. Academic Press/New York. (1977).

Weiz, B. *"Atomic Absorption Spectroscopy"*. 2nd Edn. VCH Pubs/New York. (1985).

Williams, et al. *"Biophysical Chemistry for the Life Sciences"*. Freeman/New York. (1978).

Articles

Millar, D.P. "Time-resolved Fluorescence Spectroscopy". *Curr. Opin. Struct. Biol.* 6; 637. (1996).

Selvin, P.R. "Fluorescence-resonance Energy Transfer". *Meth. Enzymol.* 246; 300. (1995).

Zumbusch, A., Holtom, G.R. & Xie, X.S. "Three-dimensional Vibrational Imaging by Coherent Anti-Stokes-Raman Scattering". *Phy. Rev. Letts.* 82(20); 41-42. (1999).

9 Nuclear Magnetic Resonance (NMR)

Synopsis: The basic principles of **Nuclear Magnetic Resonance (NMR)** are similar to those of Electron Spin Resonance (ESR). As the name implies, NMR, which is an absorption spectroscopy, is due to nuclear spin of certain atomic nuclei, similar to that of unpaired electron(s), giving rise to paramagnetic resonance. Developments in NMR spectroscopy have made **NMR** an important tool in the elucidation of structure dynamics in a wide variety of chemical compounds. **FT-NMR** and multidimensional **NMR** techniques have extended the range and scope of NMR spectroscopy to elucidate the three-dimensional structural features of macromolecules of molecular mass <20,000 daltons. NMR studies, based on their general applications can be classified into two categories— (i) **NMR** Spectroscopies, which address structural and functional aspects in biochemistry and biology and (ii) biomedical NMR, whose arena is *in vivo* biochemistry and bioenergetics of living systems and NMR imaging of anatomical sections of living systems. *In vivo* **NMR** techniques are very useful in clinical monitoring of metabolic processes at the molecular level and thus in the diagnosis of clinical abnormalities. It is a non-invasive technique.

9.1 ONE-DIMENSIONAL NMR SPECTROSCOPY

Nuclear magnetic resonance (NMR) is the absorption of electromagnetic radiation by magnetic nuclei of molecules in an applied magnetic field.

Nuclear spin giving rise to nuclear spin resonance, analogous to electron spin resonance, is the basis of **Nuclear Magnetic Resonance (NMR)**. The discovery of **chemical shift** arising out of particular nuclei existing in different chemical environments, is the basis of Nuclear Magnetic Resonance Spectroscopies.

9.1.1 Basic Principles

Atomic nuclei possess spin, $I = 0$, half-integral or integral spin, depending on:

(i) nuclei with even mass number and even charge number have spin, $I = 0$ (e.g., ^{12}C, ^{16}O & ^{32}S etc.):

(ii) nuclei with odd mass number have spin, I = half-integral (^{1}H, ^{13}C, ^{31}P etc., have $I = 1/2$ and ^{23}Na, ^{35}Cl, ^{39}K have $I = 3/2$) and

(iii) nuclei with even mass number but odd charge number have spin, I = integral (^{2}D).

Atomic nuclei, such as ^{1}H, ^{13}C etc., which have nuclear spin exhibit magnetic properties and, therefore, when they are subjected to an external magnetic field, B_o, they behave like tiny bar magnets precessing around the axis of the magnetic field, and transitions occur in the radio-wave region.

Magnetic moment, μ, is the moment experienced by an object when it is under the influence of an applied magnetic field, B_o. The magnetic moment for an electron is

$$\mu_e = -g_e\beta S = \gamma\vec{P} = -\frac{\mathbf{g}_e\beta}{h}.\vec{P} \quad ...(9.1)$$

and $$\vec{P} = \hbar\, S$$

where m_e is the mass of an electron, $\vec{P}$ = angular momentum, $\mathbf{g_e}$ = Lande's **g**-factor, β = Bohr magneton and $\hbar = h/2\,\pi$.

Extending this analogy to the nuclear magnetic moment, μ

$$\mu = \frac{Ze}{2m_N}\,\vec{P} = \frac{\mathbf{g_N}\,\mu}{\hbar}\,\vec{P} \quad ...(9.2)$$

where Z = atomic number, m_N = mass of nucleus.

Angular momentum is expressed in terms of nuclear gyromagnetic ratio, γ

$$\mu = \gamma\vec{P} \quad ...(9.3a)$$

where $$\gamma = \frac{\mathbf{g_N}\,\mu}{\hbar}$$

For a nucleus of nuclear spin, I, the magnetic moment is

$$\mu = \gamma\,\hbar\, m_I \quad ...(9.3b)$$

where m_I is the nuclear spin value.

An external magnetic field, B_o (along z-axis) lifts the degeneracy of spin states, m_I, to $2I + 1$ levels. Accordingly, a nucleus of spin m_I = 1/2, is split into two levels under the influence of an external magnetic field (Fig. 9.1); $m_I = \frac{1}{2}$ (↑), called the α-state and m_I = – 1/2 (↓) called the β-state.

The energy of interaction between the magnetic field, B_o and the nuclear magnetic moment, μ is

$$E = -\mu\,.\,B_o$$

$$\therefore \quad E = -\gamma\,\hbar\, m_I B_o \quad ...(9.4)$$

NMR transitions between adjacent spin states are induced by the application of an oscillating magnetic field, B_l (**r.f.** field) perpendicular to the axis (z-axis) of the applied magnetic field, B_o; that is, in the xy-plane.

For resonant absorption, the selection rule is that the nuclear transitions can only occur between adjacent energy levels ($\Delta m_I = \pm 1$).

$$\therefore \quad \Delta E = \gamma\,\hbar B_o \quad ...(9.5a)$$

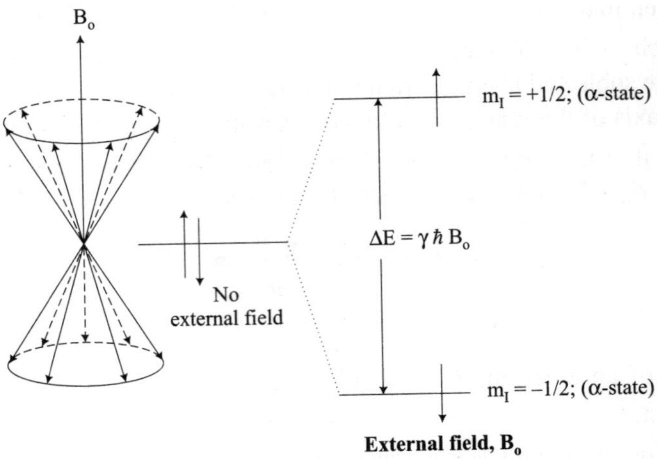

Fig. 9.1 Lifting of the degeneracy of the energy levels of nucleus of spin = 1/2 into +1/2 and –1/2 levels (α and β states) under the influence of external magnetic field, B_o.

and the frequency, ν (**Larmor** frequency) is

$$\nu = \frac{\gamma}{2\pi} B_o \qquad ...(9.5b)$$

Angular frequency $\omega = 2\pi\nu$

The population of nuclei in different energy states have a Boltzmann distribution

$$\frac{n_\beta}{n_\alpha} = \exp(-\Delta E/kT) = \exp(-\gamma \hbar B_o/kT) \qquad ...(9.6)$$

$$\approx 1 - \frac{\gamma B_o}{2\pi kT}$$

where n_β and n_α are the population of nuclei in the spin states –1/2 and +1/2, respectively; k is the Boltzmann constant and T is the absolute temperature. The strength of NMR signals depends not only on the isotope spin, but also on abundance. Some atomic nuclei relevant in NMR are given in Table 9.1.

Table 9.1 Spin properties of some atomic nuclei used in NMR.

Isotope	*Abundance, %*	*Nuclear spin, I*	*Chemical-shift range (ppm)*
^{1}H	99.98	1/2	10
^{13}C	1.1	1/2	250
^{15}N	0.37	1/2	900
^{19}F	100.0	1/2	500
^{23}Na	56.3	3/2	
^{31}P	100.0	1/2	800
^{35}Cl	75.4	3/2	

9.1.2 Chemical Shifts

When a neutral atom (like hydrogen) is in an external magnetic field, B_o, the electrons shield the nucleus from the influence of the applied field. The effective field, B_{eff}, at the nucleus is not B_o, but

$$B_{\text{eff}} = B_o - B_{\text{ind}} = B_o\,(1 - \sigma) \qquad \text{...(9.7)}$$

σ is called the shielding constant. If σ is positive, it is diamagnetic and if it is negative it is paramagnetic ($\sigma = \sigma_d + \sigma_p$).

That is, the phenomenon of "chemical shift" arises due to the shielding of nuclei from the external magnetic field by electrons. The nuclei of an element (say protons) in different chemical environments give rise to distinct spectral lines (chemical shifts), which is the basis of NMR spectroscopy. Chemical shift is an important parameter in structural studies by NMR.

Chemical shifts, δ, of nuclei, are quoted with reference to a standard reference frequency defined as

$$\delta = \frac{\nu_s - \nu_R}{\nu_R} \times 10^6 = (\sigma_R - \sigma_s) \times 10^6\, ppm \qquad \text{...(9.8)}$$

where ν_s and ν_R are resonance frequencies of the sample and the reference compound, respectively. The reference compound is so chosen that it has a single absorption frequency, occurring in a region far away from the signals given by most common chemical groups and the compound should be chemically inert. Tetramethylsilane (TMS), $Si(CH_3)_4$, tetramethylsilyl-propionic acid (TSP) generally fill these criteria and are therefore used as reference compounds in ^{1}H- and ^{13}C-NMR. Triethylphosphate (TEP) and phosphocreatine are used in ^{31}P-NMR studies.

Realisation of the chemical shifts, δ_s, namely, the assignment of resonance frequencies to various nuclei of the NMR-active atomic species of compounds in different chemical environments is the basis of structural characterisation of chemical compounds by NMR spectroscopy. For example, ethanol, CH_3—CH_2—OH, has protons in three different chemical groups — CH_3, CH_2 and OH. A similar case is benzylacetate $—C_6H_5—CH_2—O—\overset{\overset{\large O}{\|}}{C}—CH_3$. As the electro-negative **O** atom pulls away the electron cloud from the **H** atom in the O—H group, the diamagnetic shielding effects are larger in CH_2 and CH_3. One expects at least three resonance lines in the ^{1}H-NMR spectrum of ethanol, characteristic of CH_3, CH_2 and OH groups, which have been experimentally observed (Fig. 9.2). Some of the factors that contribute to the chemical shifts are: atomic and molecular shielding, ring-currents and neighbourhood molecular anisotropy etc.

Atomic Shielding. Application of an external magnetic field, B_o, induces orbital motion in the electrons (Fig. 9.3), generating a secondary magnetic field at the nucleus that opposes the applied magnetic field. This diamagnetic shielding, σ_a is common to all atoms, and lowers the resonance frequency (called the **Lamb** term).

Molecular Shielding. In molecular systems, any particular nucleus is surrounded by other nuclei hindering the motion of the electron cloud of that nucleus. So, for molecules in solution, there exists in addition to diamagnetic shielding, other shielding terms—paramagnetic neighbourhood anisotropy, ring-currents and vent effects etc.

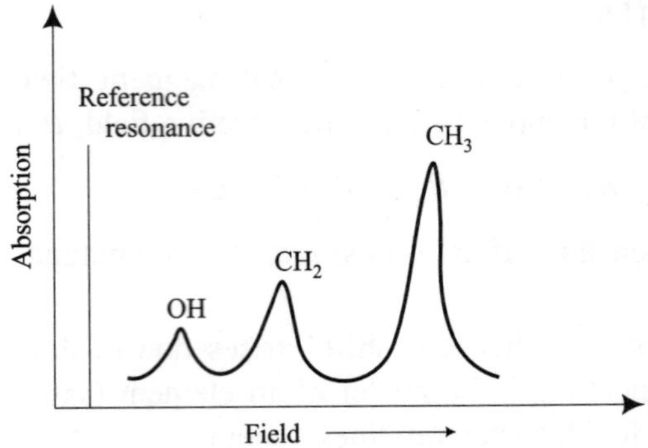

Fig. 9.2 Schematic representation of ^{1}H-NMR spectrum of ethanol, $CH_3 — CH_2 — OH$.

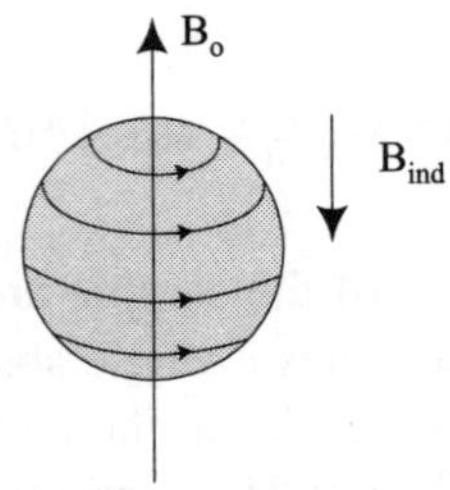

Fig. 9.3 Diamagnetic effect in atoms due to the induced effect of applied magnetic field, B_o.

9.1.3 Relaxation Processes

Nuclear spin acts like a bar magnet and tries to align parallel to the applied magnetic field, B_o (the z-axis is taken as the magnetic field axis by convention). This is resisted by thermal motions and at thermal equilibrium the nuclear spin is precessing around the applied magnetic field with Larmor frequency (Fig. 9.4). The magnetisation vector, $\vec{M}$, has two components—the longitudinal component of magnetisation, M_z, and the transverse component, M_{xy}. Many of the effects that occur in NMR experiments are determined by the behaviour of the magnetisation, $\vec{M}$. At thermal equilibrium, the longitudinal component, $M_z = M_o$ and the transverse component, $M_{xy} = 0$ (Fig. 9.5a). Application of a pulse of strength, B_1 (**r.f.** field), say along the y-axis, perturbs the equilibrium state of magnetisation and $M_z \neq M_o$ and $M_{xy} \neq 0$. $\vec{M}$ is tilted at an angle, θ, to the longitudinal axis and rotates around y-axis (Fig. 9.5b). By choosing the duration of the pulse, *t*, for a given magnetic field, it is possible to turn the magnetisation vector to any desired direction. The time duration, *t*, required to make $\theta = 90°$ is called the 90°-pulse, and to 180° is called the 180°-pulse and so on.

Once the **r.f.** field is switched off, the magnetisation, $\vec{M}$, tries to reach again its initial thermal equilibrium state. The process by which magnetisation tries to regain its thermal equilibrium state is called relaxation, and is characterised by two relaxation components—**spin-lattice** relaxation and **spin-spin** relaxation. These **represent** the return to equilibrium along the z-axis and loss of magnetisation in the x-y plane.

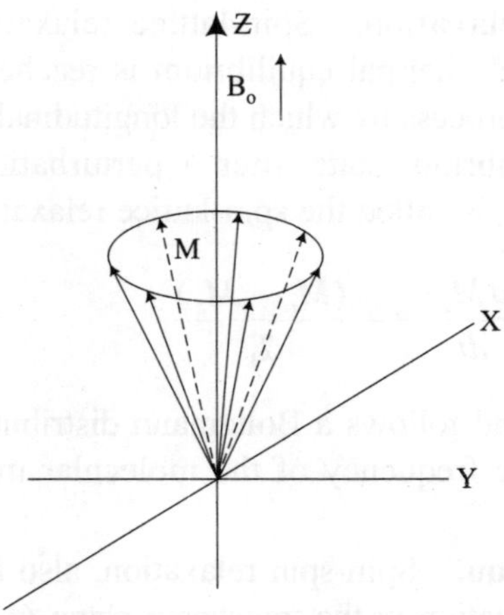

Fig. 9.4 Orientation of magnetisation vector, $\vec{M}$, at thermal equilibrium under the external magnetic field, B_o. $M_z = M_o$; $M_{xy} = 0$.

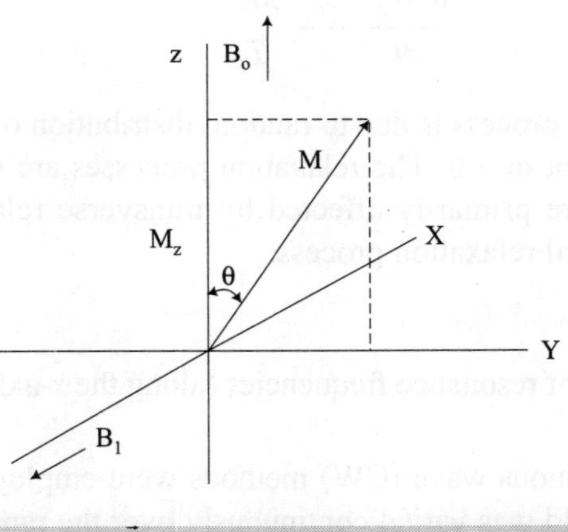

Fig. 9.5a Orientation of magnetisation vector, $\vec{M}$, upon application of transverse **r.f.** field, B_1. $M_z < M_o$; $M_{xy} > 0$.

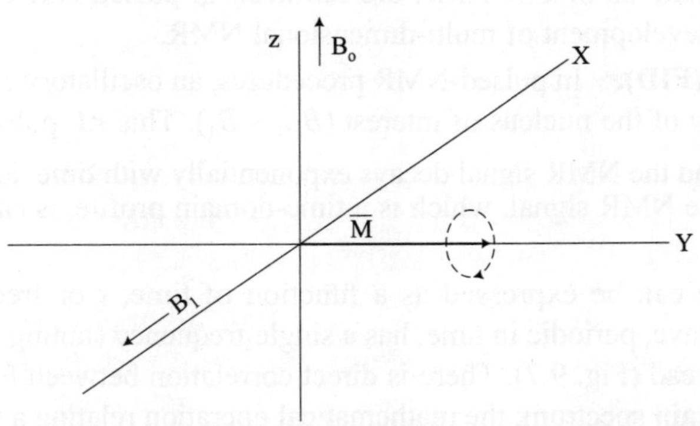

Fig. 9.5b Application of a 90°-**r.f.** field. $\vec{M}$ rotates around *y*-axis. $M_z = 0$; $M_{xy} > 0$.

Spin-Lattice (Longitudinal) Relaxation. Spin-lattice relaxation, also known as longitudinal relaxation, is the process by which thermal equilibrium is reached between a spin system and the lattice (its surroundings). It is the process by which the longitudinal component of magnetisation, M_z' tries to regain its thermal equilibrium state after a perturbation. The time, T_1, to regain this equilibrium state to reach $M_z = M_o$ is called the spin-lattice relaxation (longitudinal relaxation) time.

$$\frac{dM_z}{dt} = -\frac{(M_o - M_z)}{T_1} \qquad \text{...(9.9a)}$$

This is an enthalpy process and follows a Boltzmann distribution. The spin-lattice relaxation is maximum when the characteristic frequency of the molecular motion is the resonance frequency ($\omega \approx \omega_0$).

Spin-Spin (Transverse) Relaxation. Spin-spin relaxation, also known as transverse relaxation, is the process by which the magnetisation in the transverse plane *(x-y* plane), M_{xy}, decays towards the zero value. The time period, T_2, required to reach the state $M_{xy} = 0$ is called the spin-spin or transverse relaxation time.

$$\frac{d\,M_{xy}}{dt} = -\frac{M_{xy}}{T_2} \qquad \text{...(9.9b)}$$

The spin-spin relaxation process is due to random distribution of spin directions in the *x-y* plane. Spin-spin relaxation occurs at $\omega \approx 0$. The relaxation processes are shown schematically in Fig. 9.6. The line-widths of NMR are primarily affected by transverse relaxation and their intensities are influenced by the longitudinal relaxation process.

9.1.4 NMR Spectra

An NMR spectrum is a plot of resonance frequencies (along the *x*-axis) and their intensities (along the *y*-axis).

In the early days, continuous wave (CW) methods were employed. The magnetic field, B_o, was kept constant and the r.f. field was varied continuously over the range of the NMR absorption lines. One of the major disadvantages of this method is that at any given time only one resonance line can be monitored. The limitations of CW-NMR are removed in pulsed-NMR methods and these have paved the way for the development of multi-dimensional NMR.

Free Induction Decay (FID). In pulsed-NMR procedures, an oscillatory r.f. field pulse is applied at the resonance frequency of the nucleus of interest ($B_{\text{eff}} \sim B_1$). This **r.f.** pulse flips the magnetisation, $\vec{M}$, into the *x-y* plane and the NMR signal decays exponentially with time due to relaxation processes. This decay profile of the NMR signal, which is a time-domain profile, is called free induction decay (FID).

FT-NMR. Any wave can be expressed as a function of time, *t* or frequency, ν (= $\omega/2\pi$). For example, a sinusoidal wave, periodic in time, has a single frequency (tuning fork) and a damped wave exhibits a frequency spread (Fig. 9.7). There is direct correlation between FID (**time-domain** signal) and the **frequency-domain** spectrum; the mathematical operation relating a signal in the time-domain and in the frequency-domain is known as a **Fourier transform**.

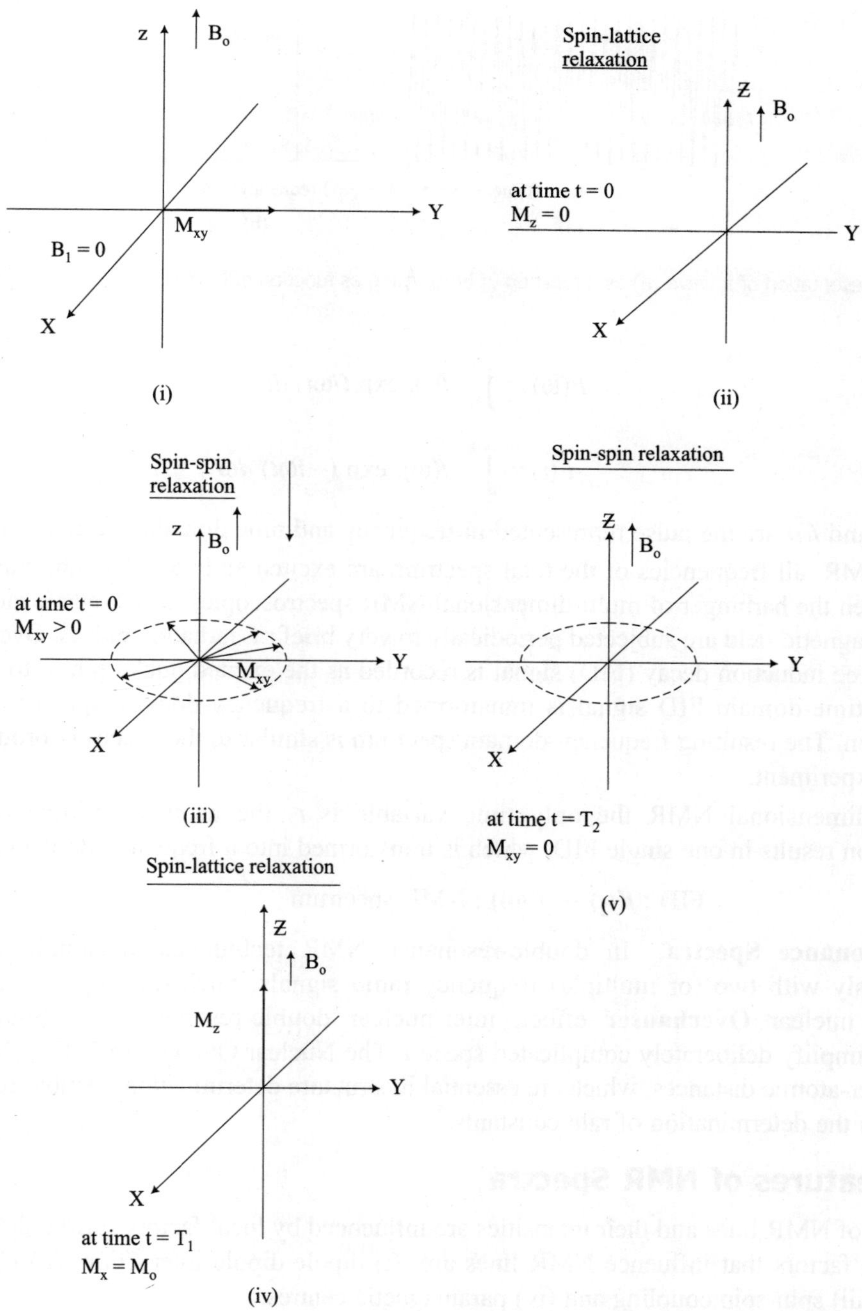

Fig. 9.6 Relaxation processes, after switching off the **r.f.** field, B_1.

(i) Orientation of the magnetisation vector, $\vec{M}$, at the time of switching off of 90°-pulse, $B_1.(B_1 = 0)$; $(t = 0)$

(ii) and (iv) Spin-lattice relaxation (at time $t = 0$ and T_1, respectively)

(iii) and (v) Spin-spin relaxation (at time $t = 0$ and T_2, respectively).

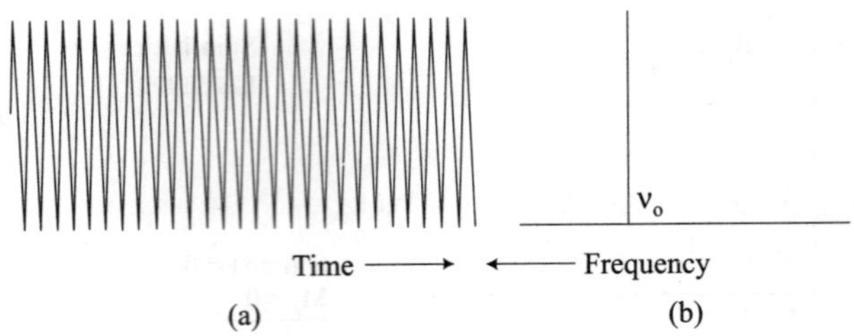

Fig. 9.7 Representation of a wave (a) as a function of time, t; (b) as function of frequency, ν.

$$F(\omega) = \int_{-\infty}^{\infty} f(t).\ \exp\ (i\omega t)\ dt \qquad ...(9.10a)$$

$$F(t) = \int_{-\infty}^{\infty} f(\omega).\ \exp\ (-\ i\omega t)\ d\omega$$

where $F(\omega)$ and $f(t)$ are the pulse represented in frequency and time domains, respectively.

In FT-NMR, all frequencies of the total spectrum are excited and recorded simultaneously. FT-NMR has been the harbinger of multi-dimensional NMR spectroscopies. In FT-NMR measurements, nuclei in a magnetic field are subjected periodically to very brief r.f. radiation pulses. Over a period of time, t, the free induction decay (FID) signal is recorded as the excited nuclei return to their ground states. This time-domain FID signal is transformed to a frequency-domain spectrum by Fourier transformation. The resulting frequency-domain spectrum is similar to the spectrum produced by any CW-NMR experiment.

In one-dimensional NMR the only time variable is t_2 the acquisition time. The Fourier transformation results in one single FID, which is transformed into a frequency-domain spectrum.

$$\text{FID} : f(t_2) \rightarrow f(\omega)) : \text{NMR spectrum} \qquad ...(9.10b)$$

Double-Resonance Spectra. In double-resonance NMR techniques, a sample is irradiated simultaneously with two (or multiple) frequency radio signals. Such techniques comprise spin-decoupling, nuclear **Overhauser** effect, inter-nuclear double-resonance etc. Spin-decoupling procedures simplify deliberately complicated spectra. The Nuclear Overhauser Effect (NOE) is used to obtain inter-atomic distances, which are essential in structure determination. Double-resonance can be applied in the determination of rate constants.

9.1.5 Features of NMR Spectra

The profiles of NMR lines and their intensities are influenced by local factors surrounding the nuclei. Some of the factors that influence NMR lines are: (i) dipole-dipole interaction, (ii) chemical-shift anisotropy, (iii) spin-spin coupling and (iv) paramagnetic centres.

Dipole-Dipole Interaction. The NMR spectrum of a nuclear spin is influenced by the interaction with the magnetic moments of neighbouring spins. An important dipole-dipole that has been found crucial in structure determination of biomolecules is the nuclear Overhauser effect (NOE). In general, the Overhauser effect exists for a system consisting of nuclear spin, I, and electron spins. If two nuclear spins are coupled, which is common in chemical bonds, saturation of one spin of the nucleus influences the intensity of the other NMR line. This interaction is strongly distance-dependent; $\alpha\ \mathbf{r}^{-6}$ (**r** is the distance between interacting nuclear spins). NOE measurements are essential for obtaining bonded and non-bonded inter-atomic distances (<5 Å), and conformational features in the course of structure determination of biological molecules by NMR.

In oriented molecules and in solids, interaction between the magnetic moments of neighbouring spins leads to splitting of NMR lines, given by

$$B_{loc} = \frac{\mu_0}{4\pi} \cdot \frac{\mu_z (3\cos^2\theta - 1)}{\mathbf{r}^3} \quad ...(9.11)$$

where $\mathbf{r}$ = distance between two dipoles;

μ_z = component of the magnetic moment parallel to B_o;

θ = Angle between the vectors $\overline{B}_o$ and $\overline{\mathbf{r}}$.

In liquids, molecular motions lead to $\theta \approx 0$, but not in solids. For $\theta = 55°$, $\cos^2\theta = 1/2$ and $B_{loc} = 0$. This angle is called the '**magic angle**' and many solid-state NMR experiments are performed around the axis of this 'magic angle'.

Chemical Shift Anisotropy. Chemical shift anisotropy is very useful in aromatic systems. In aromatic systems, because of the relatively free movement of delocalised π-electrons, an external magnetic field can induce a ring-current, depending on the orientation of the molecule with respect to the applied field. The current occurs if the plane of the aromatic ring is perpendicular to the applied field, *B*, and does not occur if the plane of the molecule is planar to B_o (Fig. 9.8). For an aromatic ring oriented perpendicular to the magnetic field, protons in the plane of the ring experience an enhanced field, whereas protons above and below the ring experience reduced field. That is, the chemical shifts for such planar molecules (with C=C, C=O, C≡C and C≡N bonds) are highly anisotropic, dependent on their orientations with respect to the field, determined by the chemical shift anisotropy

$$\sigma_{\text{aniso}} = \sigma_{11} - \sigma_{1} \quad ...(9.12a)$$

Chemical shift anisotropy is used in studies of conformation base-stacking in nucleic acids and in membrane fluids, liquid crystals and solids. In general, the total shielding is expressed as the sum of various components,

$$\sigma = \sigma_d + \sigma_P + \sigma_n + \sigma_r + \sigma_s \quad ...(9.12b)$$

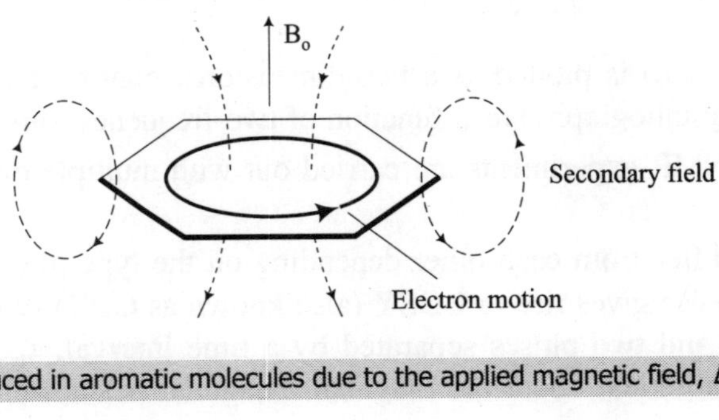

Fig. 9.8 Ring currents produced in aromatic molecules due to the applied magnetic field, B_o.

where σ_d, σ_P, σ_n, σ_r and σ_s are the components due to diamagnetic, paramagnetic, neighbour, ring-current anisotropy and solvent effects.

Spin-Spin Coupling (J-Coupling). Indirect spin-spin interaction (J-coupling) between two nuclear spins is transmitted through bonding electrons. J-coupling is a scalar coupling and is closely related to the covalent bonding in molecules. Knowledge of which nuclei (preferably consecutive) in a molecule are scalar-coupled and angular dependence of such coupling provides valuable information for structure determination and characterisation.

Paramagnetic Relaxation. A nuclear spin of $I = 1/2$ coupled with unpaired electrons predominates nuclear spin relaxation processes. Paramagnetic relaxation procedures are used to obtain inter-atomic distances between paramagnetic centres and the coupled nuclei (in addition to NOE data), and in spin-labelling methods. If paramagnetic centres are not intrinsic to the molecule, then diamagnetic ligands can be replaced by paramagnetic ligands.

9.2 MULTI-DIMENSIONAL NMR SPECTROSCOPY

There is a time lapse between the excitation of a spin system by an applied pulse and its detection. FT-NMR enables the manipulation of the behaviour of nuclear spins by introducing pulses of varying phase, duration or frequency. Multi-dimensional NMR experiments enable information on (i) dipole-dipole (T_1-relaxation) and (ii) spin-spin (J-coupling; T_2-relaxation) interactions. Several 2- and 3-dimensional NMR techniques are in vogue for structure elucidation of macromolecules. Generally used techniques are— **N**uclear **O**verhauser **E**nhancement **S**pectroscopy (NOESY), **Co**rrelated **S**pectroscop**y** (COSY), **S**pin **E**cho-**C**orrelated **S**pectroscopy (SECSY), **D**ouble **Q**uantum-**F**iltered Spectroscopy (DQF-COSY), **T**otal **C**orrelation **S**pectroscopy (TOCSY).

In 2*D*-NMR two time variables are operational: acquisition time, t_2 (similar to 1D-NMR) and evolution period, t_1. In a 2D-NMR experiment, a 90°-pulse produces a transverse magnetisation and after a time interval t_1 (evolution phase), another pulse (mixing phase) is applied and the resulting FID is recorded at time, t_2 ($t_1 < t_2$). Fourier transformation of the FID as a function of time interval, t_2 for a fixed t_1 is similar to an ordinary FT-NMR. The process is repeated for various values of t_1 and a set of FIDs are collected instead of a single FID as in the case of 1D-NMR. The time signal $s(t_1, t_2)$ is transformed into a two-dimensional frequency spectrum, $S(\omega_1, \omega_2)$ by two-dimensional Fourier transform

$$S(\omega_1, \omega_2) = \int_{-\infty}^{\infty} \int_{-\infty}^{\infty} s(t_1, t_2) e^{-i\omega_1 t_1} . e^{-i\omega_2 t_2} dt_2\, dt_1 \quad \text{...(9.13)}$$

The spectrum $S(\omega_1, \omega_2)$ is plotted as a two-dimensional contour map (similar to an electron-density map in X-ray crystallography) as a function of two frequency variables, ω_1 and ω_2.

Multi-dimensional NMR experiments are carried out with multiple evolution times (t_2, t_3, t_4 ...) (Fig. 9.9a).

2-*D* NMR spectra differ from each other depending on the type of pulse applied at the mixing phase. A 90° - mixing pulse gives rise to COSY (also known as COSY-90°); two 90° pulses shifted by 90° to DQF-COSY; and two pulses separated by a time interval, τ_m, to NOESY spectra. The COSY spectrum is symmetrical about the diagonal. Diagonal peaks ($\omega_1 = \omega_2$), called **dispersive** peaks, correspond to the one-dimensional NMR spectrum and the off-diagonal peaks, called **absorptive** peaks ($\omega_1 \neq \omega_2$), provide information about the J-coupling of different nuclei (Fig. 9.9b). One-dimensional and two-dimensional ^{1}H spectra of a protein are shown in Figs. 9.10a and 9.10b.

The DQF-COSY technique provides simplified spectra with certain loss of information, but has several advantages over other COSY methods, such as (i) connectivities between almost degenerate signals can easily be detected, (ii) amide and β-protons in proteins belonging to the same spin system can be distinguished and (iii) glycine amide protons can be distinguished from all others. The complete network of J-coupled spins can be obtained with TOCSY.

Band-selective pulses are useful in selecting a particular region of cross-peaks (zooming-in) by limiting excitation bandwidth.

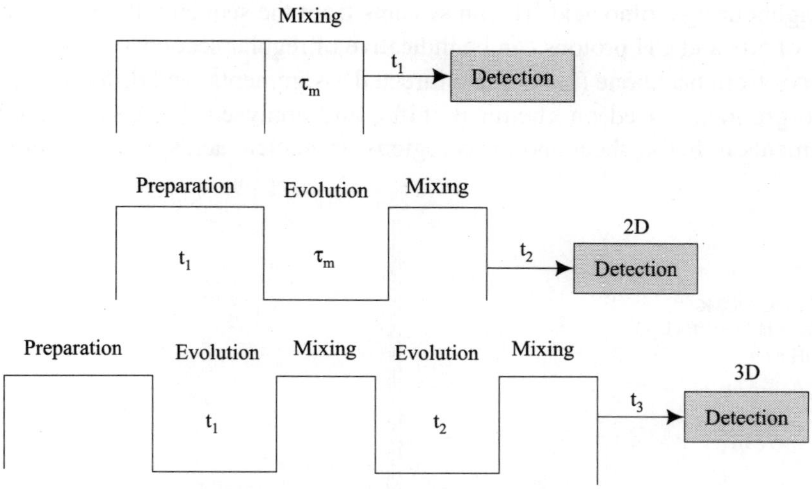

Fig. 9.9a Schematic representation of process in NMR Experiments.

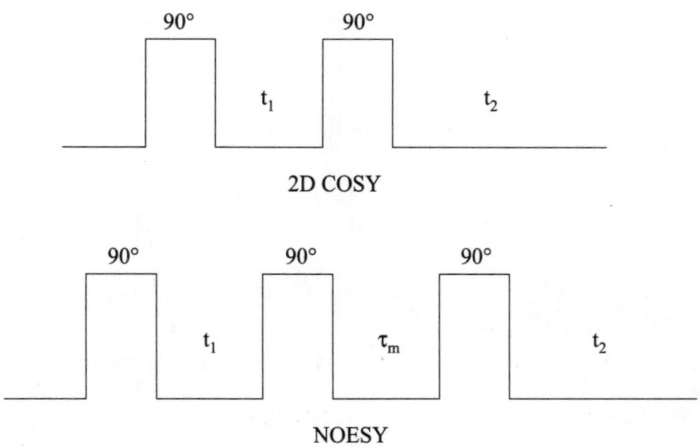

Fig. 9.9b 2-D COSY and NOESY pulse sequences.

9.3 STRUCTURE DETERMINATION

Structure elucidation by NMR spectroscopic methods rests on obtaining information on atomic/ nuclear positions of molecules through measurements of chemical shifts, their nature (e.g., single, double or non-bonded character), and coupling constants.

The general procedure of structure determination followed is—(i) assignment of sequence-specific ^{1}H resonances (from NOESY and COSY), (ii) intra-proton distance information (from NOESY) and (iii) evaluation of J-coupling constants, *Js*, from COSY. Using these data on bonded and non-bonded interactions and incorporating molecular constraints, computer modelling and energy minimisation techniques are employed in structure elucidation.

In proteins, the primary information for identifying the type of amino acid residue associated with each resonance comes from couplings established in DFQ-COSY spectra, and identification of sequentially neighbouring amino acid ^{1}H spin systems from the sequentially NOE connectivities. The chemical shifts of NH and CH protons can be indicative of regular secondary structure. In general, the proton resonances from backbone (main-chain directed assignments) and different types of protons on side-chains are grouped, based on chemical shifts, and analysed. NOESY is used for sequence-specific assignments in helix, sheet and turns regions. In nucleic acids, identification of protons that

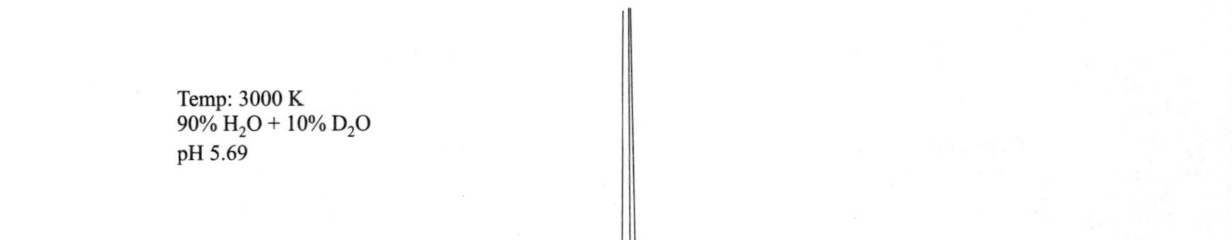

Fig. 9.10a One-dimensional ^{1}H-spectrum of a cytotoxin from the snake venom, *Naja naja* (India).

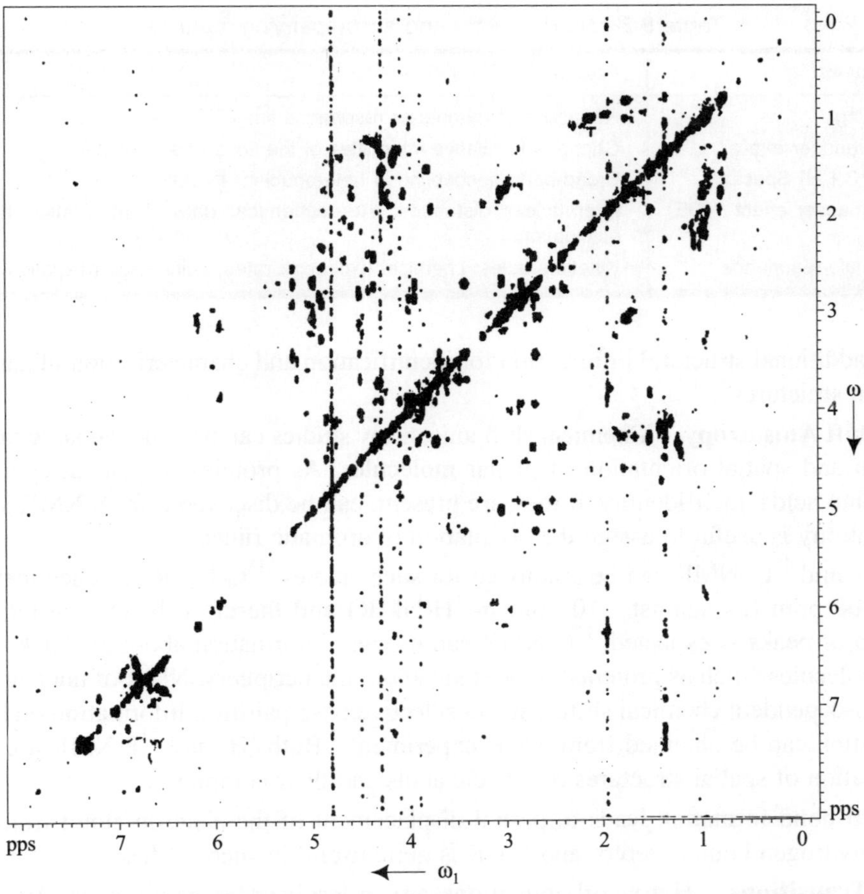

Fig. 9.10b Two-dimensional ^{1}H-COSY spectrum of a cytotoxin from the snake venom *Naja naja* (India). Spectra obtained at the National Facilities for NMR, TIFR; Bombay.

belong to an individual base or sugar is done with COSY. The association of specific bases with sugars is obtained using NOESY spectrum.

9.4 APPLICATIONS

Both 1-*D* and 2-*D* NMR spectroscopies are employed in the structural studies of biological molecules of molecular mass <15,000 daltons. Some of the structural and functional information obtainable from NMR spectroscopies include chemical identification, reaction pathways, substrate binding, conformational changes, molecular associations and secondary and tertiary structures. Some features of NMR spectra are listed in Table 9.2.

9.4.1 Molecular Structure

NMR can be used as an analytical method for chemical identification of molecules. For known compounds, techniques such as gas chromatography and radioimmunoassay may be better but NMR

Table 9.2 NMR spectra and accompanying features.

Type of spectrum	*Features*
Spin-decoupling	Reduction of number of resonance lines
General Overhauser effect	Changes in relative intensities of the lines of a multiplet.
$^{15}N\ \{^{1}H\}$ or $^{13}C\ \{^{1}H\}$ Spectra	Broad-band decoupling in heteronuclear INDOR.
Nuclear Overhauser effect (NOE)	Internuclear distances; stereochemical data; conformation and structure elucidation.
Transient double-resonance	Kinetic studies; chemical exchange rates; spin-lattice relaxation times.

can provide additional structural information for identification and characterisation of partially-known and unknown structures.

Chemical Shift Anisotropy. Chemical shift anisotropy studies can provide valuable information on identification and spatial orientation of planar molecules. As proteins in general contain very few aromatic amino acids, their identity, if they are present, can be discovered from NMR studies. Ring-current anisotropy is useful to assign the orientation of aromatic rings.

Both ^{1}H– and ^{13}C–NMR can be employed for such studies. ^{13}C-NMR has chemical shifts in the range of ~200 ppm (as against <10 ppm in ^{1}H-NMR) and therefore better resolution and non-superposition of peaks is expected. ^{13}C-NMR can provide information about the backbone structure of organic molecules (such as proteins) rather than about the periphery. NMR of nucleic acids exhibit conformation-dependent chemical shifts (due to selective base-pairing). Information on base-stacking and intercalation can be obtained from NMR experiments. Both ^{1}H- and ^{31}P-NMR are employed in the determination of spatial structures of nucleic acids and their complexes.

Hydrogen bond formation leads to partial displacement of the electron cloud from the H-atom towards the hydrogen bond acceptor and NMR is quite useful in such studies.

Helix/Coil Transitions. Helix/coil transitions are order/disorder phenomena. Macromolecular transitions from an ordered state (native conformation) to a disordered state (random coil) can be monitored by ^{1}H-NMR. NMR spectra can indicate whether a macromolecule is in the ordered or disordered state. Most of the resonances of amino acids in ^{1}H-NMR lie between 1-4 ppm and 7-9 ppm for aliphatic and aromatic side chains, respectively, and helix/coil transitions in proteins can be monitored as a 'function of such chemical shifts.

Chemical Exchange. Chemical exchange is generally a time-dependent process. Measurement of such processes as a function of pH changes, ion concentration, solvent etc., can be determined by NMR. In proteins, ionisable side chains play an important role in catalytic process. The mode of action of such an ionisable group depends on its charge under the existing pH value of the solution. At any given pH value of a solution, the charge on an ionisable group is determined from its pK value (equilibrium between protonisation and deprotonisation). From a plot of chemical shift as a function of pH value, the pK values of individual groups can be obtained. This is simpler in the resonance frequency range of the ring proteins (as proteins contain very few aromatic side chains). Histidine, which participates in many catalytic processes, can be identified from NMR spectra in this region.

Selective deuteration of amino acids and interpretation of such spectra is one of the important methods in the assignment of resonant lines in structure determination procedures.

Spin-labelling Spin-labelling by stable isotopes is an important method in *in vivo* NMR as well as in NMR spectroscopy. In these techniques, a naturally-occurring isotope is replaced by a different isotope and its course is monitored continuously by NMR. Typical examples are replacement of ^{12}C by ^{13}C and ^{14}N by ^{15}N. Introduction of paramagnetic centres extrinsically is a procedure that is employed in the measurement of interatomic distances for structure determination. As proteins are diamagnetic (if there are no intrinsic paramagnetic centres), introduction of extrinsic paramagnetic centres is a way of procuring interatomic distances (in addition to NOE data).

Ligand-ligand Interactions. NMR data. combined with X-ray diffraction and other physico-chemical data can be used to interpret the structural arid dynamical features of molecular associations. These include enzyme-substrate conformational features, protein-nucleic acid, protein-heme, and antigen-antibody interactions. NMR is better suited than X-ray crystallography to monitoring conformational changes resulting from mutation (chemical exchange) and ligand binding.

In iron-containing heme proteins, iron is diamagnetic or paramagnetic depending on its oxidation state and on the type of ligand. NMR is a convenient method for identifying such properties. Similarly, physiologically occurring divalent ions, Mg^{2+} and Ca^{2+}, which are not NMR-sensitive, can be replaced by the lanthanide group of metals in proteins/enzymes. For example, Ca^{2+}-binding proteins are studied by NMR methods by replacing Ca^{2+} by the paramagnetic Yb^{3+}ion.

9.4.2 Secondary and Tertiary Structures

^{1}H-NMR chemical shifts are strongly dependent on the character and nature of protein secondary structures. ^{1}H-NMR chemical shifts and α H proton (α-carbon atom proton) of amino acids show an upfield shift, compared to the random coil value, when in a helical configuration and a downfield shift when in a β-strand configuration. Therefore, from *2-D* NMR data, secondary structural segments in proteins can be identified from the α-CH shifts. Three-dimensional structures of proteins of molecular mass $<$15,000 daltons can be attempted from 1-*D* and *2-D* NMR spectral data.

9.4.3 Membrane Structure

Membranes are macromolecules but not polymers and many biological membranes are proteolipids. Also, depending on the composition and temperature, they exist in different physical phase: liquid, gel, liquid-crystalline and solid. Therefore, biological membranes pose challenging problems in structure elucidation. On solid-state NMR, the orientation-dependent terms provide important structural information. There is a correlation between line-splitting and the orientation of a covalent bond in space. Both ^{13}C- and ^{15}N-NMR are preferred in solid-state NMR studies. Phospholipids can also be studied by ^{31}P-NMR. From the relaxation times and line-broadening data, membrane structures in the solid, liquid-cystalline and gel phases can be studied.

9.5 BIOMEDICAL NMR

Studies in biomedical NMR concentrate on the application of NMR to cellular metabolic processes at the molecular level, biochemical energetics of living organisms, and NMR imaging.

9.5.1 *In Vivo* NMR

For continuous monitoring metabolic processes at the molecular level, ^{1}H, ^{13}C, ^{15}N and ^{31}P nuclei are often used in *in vivo* biomedical NMR, for continuous monitoring of metabolic processes at the molecular level. ^{1}H-NMR is the most sensitive method and the spectra are complex. So ^{1}H-NMR is generally employed in high-resolution structure-function studies. ^{13}C-NMR techniques are employed in studies of metabolism, reaction kinetics of enzymes and membrane structure and dynamics. ^{15}N-NMR is useful where nitrogenous compounds are present, such as in nucleic acids. ^{31}P-NMR is useful in the study of bioenergetics, because of the crucial role the phosphate-containing molecules play in the reactions. ^{31}P-NMR spectrum can be a monitor of liver function by observing the changes induced *in vivo* by injection of fructose. ^{31}P-NMR is widely used for intracellular pH measurements and *in vivo* studies of cellular energetics *via* detection of inorganic phosphate (P_i), phosphocreatine (P_{cr}), pH, lipid metabolism and membrane bilayer phospholipids. ^{31}P-NMR has been used to study muscle physiology and pathology. ^{31}P-NMR spectra indicate that in ischemic muscle the level of phosphocreatine is very low and inorganic phosphate very high as compared with normal muscle (Fig. 9.11). *In vivo* NMR procedures are very useful in clinical monitoring of neuromuscular and other disorders due to metabolic deficiencies. *In vivo* NMR studies can monitor enzyme activities under steady-state conditions. Change in line-widths can be correlated with the level of immobilization (substrate binding) and transport of metabolites. Selection of a particular type of NMR depends on the nature of the problem. For example, ^{13}C-NMR is used in monitoring the changes in intracellular concentration of metabolites, anomalies in their uptake by cells and the manifestation of clinical disorders. Similarly, biochemical pathways can be monitored non-invasively by other *in vivo* NMR techniques. ^{19}F-NMR is used in the study of drug metabolism of compounds that contain fluorine.

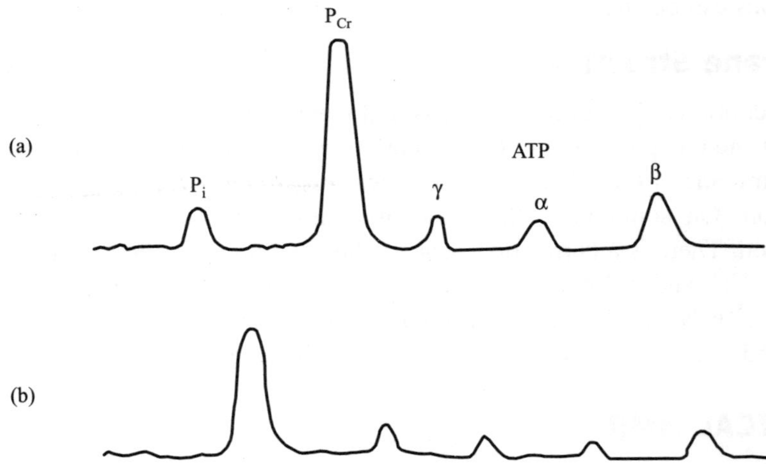

Fig. 9.11 Schematic profiles of ^{31}P-NMR of a muscle cell.
(a) normal muscle cell, (b) diseased muscle cell.
P_i = Inorganic phosphate; P_{Cr} = Phosphocreatine.
ATP = Adenosine triphosphate.

Suggested Reading

Books

Cavanagh, J. et al. "*Protein NMR Spectroscopy: Principles and Practice*". Academic Press/New York. (1996).

Close G.M. & Gronenborn, A.M. (Eds.) *"NMR of Proteins"*, McMillan/London. (1993).

Gadian D.G. "NMR and Its Applications to Living Systems". Oxford Univertsity/Oxford Press (1983).

Govil, G., Khetrapal, C.L. & Saran, A. (Eds.). "*Magnetic Resonance in Biology and Medicine*". Tata McGraw Hill/New Delhi. (1985).

Günther, H. "*NMR Spectroscopy: An Introduction*". John Wiley & Sons/New York. (1980).

Harris, R.K. "*Nuclear Magnetic Resonance Spectroscopy: A Physicochemical View*". Pitman Publishers, London (1983).

Hausser, K.H. & Kalbitzer, H.R. "*NMR in Medicine and Biology: Structure Determination*, Tomography, in vivo spectroscopy. Springer Verlag/Berlin. (1991).

Khetrapal, C.L. & Govil, G. (Eds). "*Magnetic Resonance: Current Trends*". Narosa Publishers/New Delhi. (1991).

Olendorf, W. & Olendorf, W. Jr. "*Basics of Magnetic Resonance Imaging*". Kluwer Acad. Pubs./Boston; MA. (1988).

Sander, J.K.M. & Hunter, B.K. "*Modern NMR Spectroscopy*". Oxford University Press/Oxford, UK. (1993).

"Tomography, In Vivo Spectroscopy". Springer Verlag/Berlin. (1991).

Wennel, J. & Klinowski, J. "*Fundamentals of Nuclear Magnetic Resonance*". Wiley/New York. (1993).

Wüthrich, K. "*NMR of Proteins and Nucleic Acids*". Wiley & Sons/New York. (1986).

Articles

Moonen, C.T.W. "*Functional NMR Imaging*". Science. 250; 53. (1990).

Radda, G.K. "The Use of NMR Spectroscopy for the Understanding of Disease". Science. 233; 640. (1986).

Schulman, R.G. "*NMR Spectroscopy in Living Cells*"; Sci. Amer. 248(1); 76; (1983).

Wishart, D.S., Sykes, B.D. & Richards, P.M. "*The Chemical Shift Index: A Fast and Simple Method of the Assignment of Protein Secondary Structure through NMR Spectroscopy*" Biochemistry. 31; 6147. (1992).

Wüthrich, K. "*Protein Structure Determination in Solution by Nuclear Magnetic Resonance Spectroscopy.*" Science. 243, 45 (1989).

Wright, P.E. "*What Can Two-dimensional NMR Tell us About Proteins*". TIBS. 14; 255 (1989).

10 X-Ray Diffraction

Synopsis: X-ray diffraction is a very powerful and versatile technique and, to-date, is the only quantitative technique available to elucidate the complete three-dimensional structure of matter at molecular and atomic levels and has, therefore, become an indispensible tool in biological sciences. Elucidation of detailed structural features at atomic/molecular levels is possible if the specimen is in a crystalline state and considerable efforts are made by crystallographers to obtain compounds in the single-crystalline state, if feasible. Diffraction data obtainable from semi-crystalline materials (like fibrous molecules, collagen, DNA etc.) are meagre and models have to be proposed and refined to fit diffraction patterns (e.g., the model of the DNA double-helix by **Watson** and **Crick**).

X-ray diffraction data provide information pertaining to the intensities of the diffracted reflections only. The 'phase' information of each of these diffracted spots, which is absolutely essential for structure determination, is not available and, therefore, must be 'somehow' obtained to compute electron-density maps. The solving of 'phase' problem forms a major effort on the part of crystallographers in structure determination by any X-ray diffraction procedure.

X-ray diffraction methods have been and are, applied to determine the structures of simple to immensely-complex molecules — 'small' molecules such as amino acids, nucleic acid components, saccharides, steroids, alkaloids, flavonoids etc., macromolecules such as proteins/enzymes (immunoglobulins, ATPases, photoreaction centres, collagen etc.), nucleic acids (DNA, RNA and DNA-ligand complexes) and viruses etc. Recent advances in the methodology of fast-data collection, processing by area detectors and the availability of intense polychromatic X-rays from synchrotron radiation sources have enabled (i) quick and accurate data collection and proper "exploitation" of structure determination methods and (ii) time-resolved crystallography to unravel intermediary structural states of enzymes and molecules. These developments in technology and methodologies have expanded the scope of X-ray diffraction to address biochemical and structural dynamics in a more quantitative manner.

10.1 DIFFRACTION OF X-RAYS

Diffraction of X-rays was discovered in the year 1912 as an outcome of the speculative discussions **Max von Laue** and **Paul Ewald** had on the properties of X-rays that **Röentgen** had earlier discovered— (i) Do X-rays exhibit wave properties, as they should, like other radiations of the electromagnetic spectrum? (ii) If they do exhibit wave characteristics, could they be employed to probe matter at atomic resolutions, as their wavelengths were estimated to be in the range of ≈ 1 Å (1 Å = 10^{-8} cm = 0.1 nm) and, (iii) could single crystals be employed to perform as natural three-dimensional diffraction gratings, again intuitively speculating that matter in the crystalline, state would have a periodic arrangement of atoms/molecules/groups of molecules in keeping with their well-defined external morphologies? All these assumptions were conclusively proved correct. X-rays do exhibit wave characteristics belonging to the higher-energy region of the electromagnetic spectrum and can be used to probe matter at atomic resolutions and atoms/molecules are arranged periodically in crystals and, therefore, do act as natural diffraction gratings for X-rays. Characteristic X-ray wavelengths, employed in X-ray diffraction studies, are in the range of ≈ 1 Å (e.g., Cu K_α = 1.54178 Å), comparable to interatomic distances in molecules (C – C single bond is 1.54 Å). Therefore, only X-rays of short wavelengths are suitable for unravelling the structure of matter at high resolutions. The method of recording employed by Laue and his colleagues, namely stationary crystal and polychromatic radiation, is called the **Laue** method.

The most common source of X-rays for analytical studies is X-ray tube (*Coolidge* tube). Bombardment of a metal (anticathode), such as Cr, Cu, Mo, Ni etc., with a beam of high-energy electrons produces two distinct profiles of X-ray spectrum, depending on the applied voltage— (i) X-ray continuum (Bremsstrahlung), and (ii) characteristic X-ray spectral lines (Fig. 10.1). For the excitation of characteristic line spectrum for each element, the minimum threshold voltage, V_0, should be above the critical excitation voltage, V_K ($V_0 > V_K$) for that element.

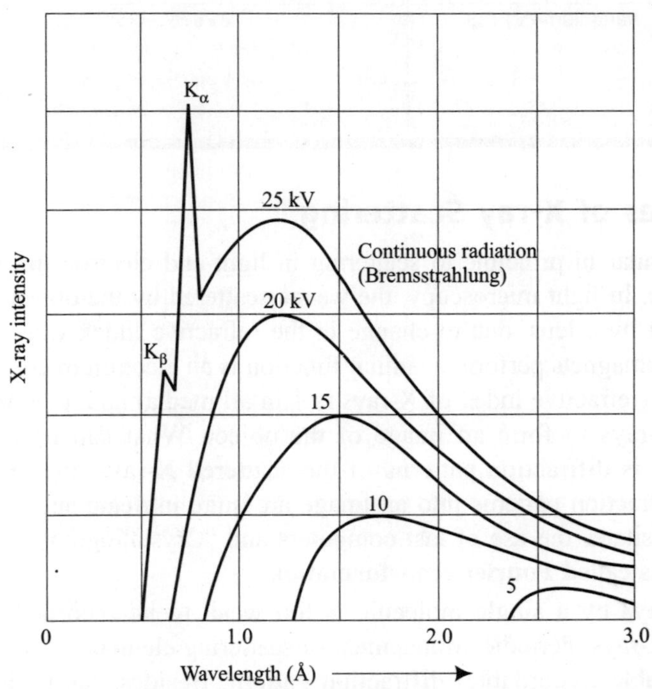

Fig. 10.1 Profiles of X-ray spectra with applied voltage (for Mo target); k_α, k_β are characteristic lines.

Target-filter combination (Table 10.1) is used to obtain monochromatic radiation for diffraction studies. Synchrotron radiation provides a wavelength continuum from which the desired wavelength can be selected.

Table 10.1 Target-filter combination for monochromatic X-ray selection.

Target (anticathode)	*Filter*	*Wavelength of K-absorption edge (Å)*	*Wavelength (Å)*
Molybdenum (Mo) (42)	Zirconium (40)	0.6888	K_α = 0.711 $K_{\alpha 1}$ = 0.70926 $K_{\alpha 2}$ = 0.71354 K_β = 0.63225
Copper (Cu) (29)	Nickel (Ni) (28)	1.4869	K_α = 1.54178 $K_{\alpha 1}$ = 1.5405 $K_{\alpha 2}$ = 1.54434 K_β = 1.39217
Cobalt (Co) (27)	Iron (Fe) (26)	1.7429	K_α = 1.789 $K_{\alpha 1}$ = 1.7889 $K_{\alpha 2}$ = 1.79279 K_β = 1.62073
Iron (Fe) (26)	Manganese (Mn) (25)	1.8954	K_α = 1.937 $K_{\alpha 1}$ = 1.93597 $K_{\alpha 2}$ = 1.93991 K_β = 1.75654
Chromium (Cr) (24)	Vanadium (V) (23)	2.2676	K_α = 2.290 $K_{\alpha 1}$ = 2.28962 $K_{\alpha 2}$ = 2.29352 K_β = 2.08479

10.1.1 Principles of X-ray Scattering

X-ray scattering is similar in principle to scattering in light and electron microscopies, but with a fundamental difference. In light microscopy, the waves scattered by the object are focussed to form an image of the object by a lens, due to change in the refractive index when light passes through different media (electromagnets perform a similar function is an electron microscope). There exist no lenses to focus X-rays (refractive index of X-rays is 1 in all media) and, therefore, it is not possible to focus diffracted X-rays to form an image of the object. What can be recorded in all X-ray diffraction procedures is diffraction patterns of the scattered X-rays; the operations involved in transforming these diffraction patterns into an image are quite intricate and involved. This **"image reconstruction"** necessitates the use of fast computers and "Crystallographers" (to do the job of a lens!). This operation is called **Fourier transformation**.

Scattering of X-rays by a single molecule is too weak to be recorded because of the high penetrating power of X-rays. Periodic arrangement of scattering elements scatter coherently together to produce an observable (recordable) diffraction pattern. Besides, the periodic arrangement of scatterers (atoms/molecules) channels diffraction in a certain direction only, thereby greatly enhancing the contrast (signal to noise ratio) in comparison to diffuse scattering by amorphous material, though the total amount of scattering is same in the both the cases. Matter in single-crystalline state provides more obtainable diffraction data (information content) than in any other state. Due to meagre 'information' obtainable in the case of amorphous and quasi- and poly-crystalline materials, '**powder**' diffraction methods are employed to study very simple structures in higher space groups (such as minerals) and compositions of metals in alloys (metallurgy). Therefore, a substance in the single-crystalline state is a prerequisite for its complete three-dimensional structure determination by X-ray diffraction methods and, naturally, it is very desirable to obtain a compound as single crystals prior to any attempts at in-depth structural studies by X-ray diffraction methods.

10.1.2 The Unit Cell and Space Group

The essential features of a single crystal are that it is composed of a three-dimensional arrangement of atoms/molecules/or clusters of molecules and it can be represented by a basic unit of structure, the unit cell (an imaginary parallelopiped), repeated regularly in a three-dimensional array, called the crystal lattice (an arrangement of points) (Fig. 10.2).

That is, **structure → (Lattice * Motif),** where motif embodies the contents in the unit cell (*symbolises convolution). There are seven types of unit cell and combination of symmetry operations (reflection, glide, rotation and screw operations) gives rise to 230 unique space groups (Table 10.2).

Each reflection of a diffraction pattern can be identified by triple indices, hkl, called **Miller** indices. Indices h, k, and 1 represent the number of times *a, b* and *c* axes are intersected.

A simple and elegant physical interpretation of X-ray diffraction was given by **Lawrence Bragg**, which states that each diffracted spot (hkl) can be thought of as arising from reflection of the incident X-ray beam by a stack of parallel, equally-spaced imaginary planes and the indices, hkl, indicate the orientation of this stack of planes with respect to the unit cell axes. Constructive interference occurs only if Bragg's condition is fulfilled (Fig. 10.3).

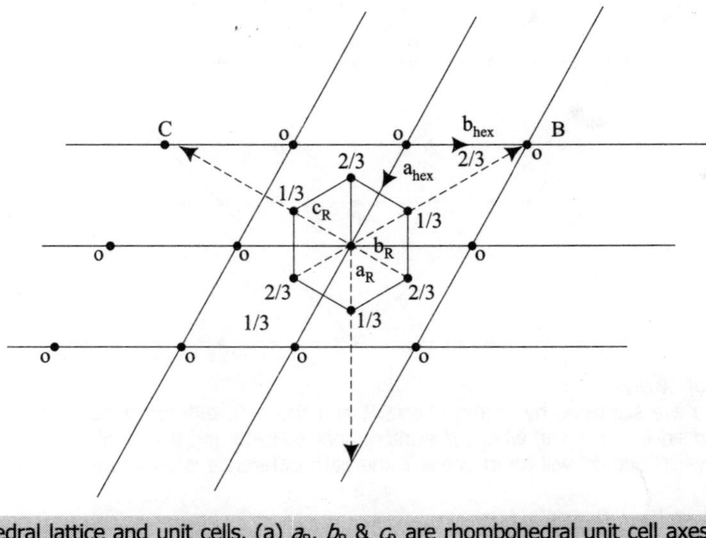

Fig. 10.2 Rhombohedral lattice and unit cells. (a) a_R, b_R & c_R are rhombohedral unit cell axes. (b) a_{hex} and b_{hex} are hexagonal unit cell axes (c_{hex} is perpendicular to the a_{hex}-b_{hex} plane).

Table 10.2 The seven classes of crystal systems.

Crystal systems	*Unit cell parameters*	*Lattice type**	*Lattice symmetry*
Triclinic	$a \neq b \neq c$ & $\alpha \neq \beta \neq \gamma \neq 90°$	*P*	$\overline{1}$
Monoclinic	$a \neq b \neq c$ & $\alpha = \gamma = 90°$; $\beta \neq 90°$	*P*, C	2/m
Orthorhombic	$a \neq \beta \neq c$ & $\alpha = \beta = \gamma = 90°$	*P*, C, *I*, *F*	mmm
Tetragonal	$a = b \neq c$ & $\alpha = \beta = \gamma = 90°$	*P*, *I*	4/mmm
Rhombohedral	$a = b = c$ & $\alpha = \beta = \gamma \neq 90°$ (less than 120°)	*R*	$\overline{3}$m
Hexagonal	$a = b \neq c$ & $\alpha = \beta = 90°$, $\gamma = 120°$	*P*	6/mmm
Cubic	$a = b = c$ & $\alpha = \beta = \gamma = 90°$	*P*, *I*, *F*	m3m

*Lattices: *P* = Primitive; C = C face-centred; *F* = All face-centred; *I* = Body-centred; *R* = Rhombohedral; *m* = mirror plane.
Trigonal system can be expressed as part of a hexagonal system.

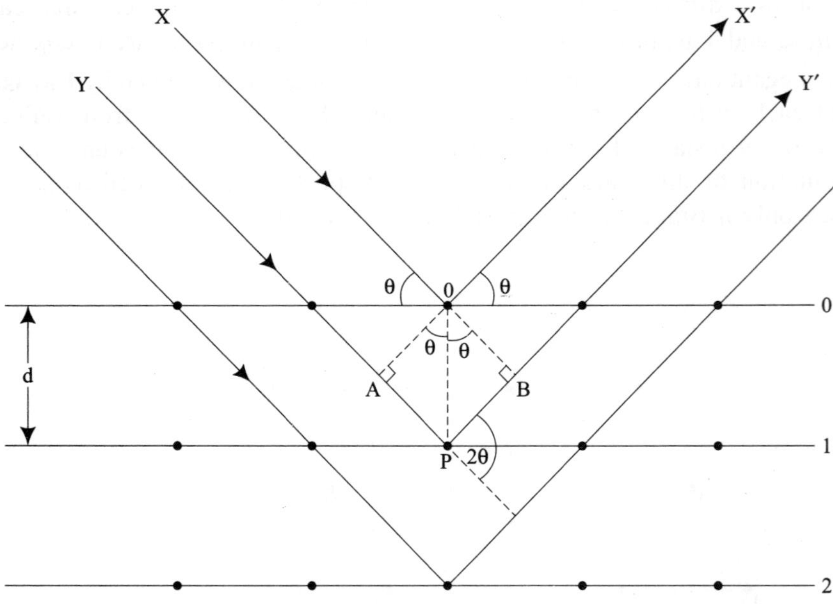

Fig. 10.3 Diffraction of X-rays.
Rays *X* and *Y* are scattered by atoms *O* and *P*, and the path difference between rays *XOX′* and *YOY′* is $AP + PB = d \sin\theta + d \sin\theta$, where *d* is interplanar spacing and θ is angle of incidence.
Scattered rays *X′* and *Y′* will be in phase if the path difference is an integral multiple of the wavelength, λ. That is,
$2d \sin\theta = n\lambda$ → **Bragg's law**
n is an integer, 1, 2, 3...

$$2.d. \sin\theta = n\lambda \qquad \text{Bragg's equation} \qquad ...(10.1)$$

where d is the interplanar spacing, θ is angle of incidence, λ is the wavelength of the radiation and n is an integer. The smallest Bragg reflection gives the largest interplanar spacing, that is to the unit cell dimension (along that axis). Bragg's equation enables one to obtain unit cell parameters, a, b and c (in Å) and angles α, β, γ (in deg) between $\hat{bc}$, $\hat{ca}$ and $\hat{ab}$, respectively. Further, X-ray diffraction photographs by rotation, Weissenberg and precession methods can be used to determine unit cell parameters and space groups (Fig. 10.4). The diffraction pattern can reveal the presence of screw axes, glide planes and centering. The space group of a unit cell can be determined, with few exceptions, from diffraction photographs.

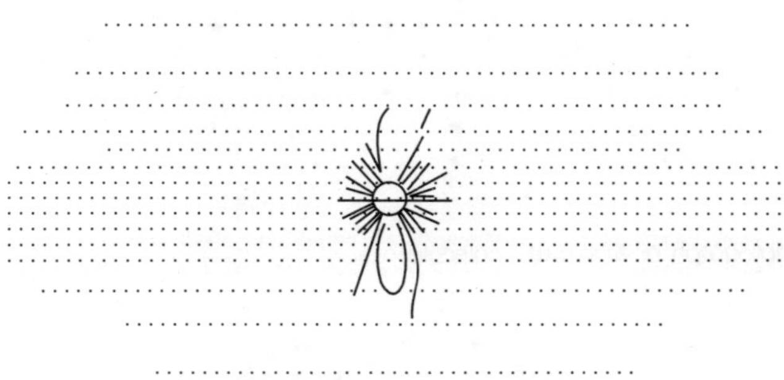

Fig. 10.4a Rotation photograph of a **'small'** molecule.

Fig. 10.4b Weissenberg photograph of a **'small'** molecule.

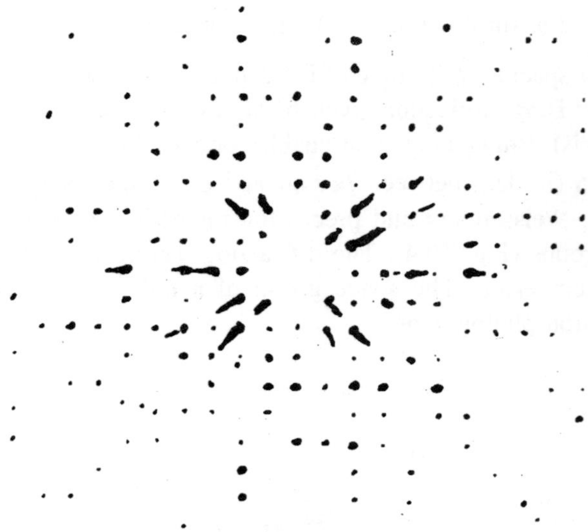

Fig. 10.4c Precession photograph of a **'small'** molecule.

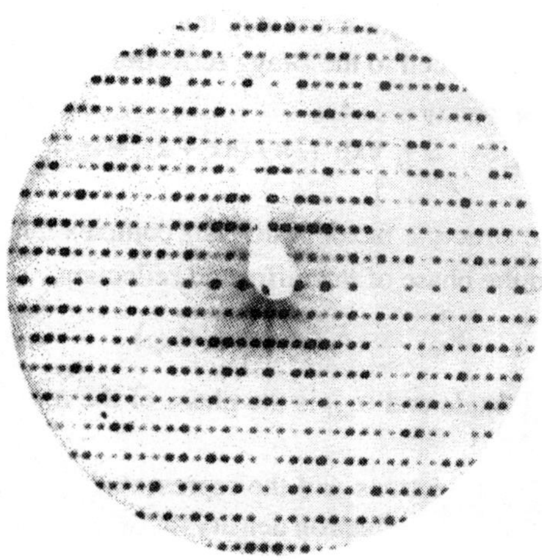

Fig. 10.4d Precession photograph of a single crystalline globular macromolecule (lysozyme at 88% relative humidity).
(Photograph: Courtesy of Prof. M. Vijayan, Indian Institute of Science, Bangalore).

10.1.3 Unit Cell Content (Z)

The number of formula units in the unit cell, Z (generally for **"small"** molecules), can be determined by the formula

$$Z = \frac{Vd_0}{1.66\,F} \qquad ...(10.2)$$

where V is the volume of the unit cell (in $Å^3$), d_0 is the observed density ($g.cm^{-3}$) and F is the formula unit mass. Z should generally be an integer. Presuming deviation from the integral value of Z is due to experimental errors in the density measurements, Z can be set to the nearest integer and calculated density, d_c, can be computed. If d_0 and d_c deviate more than experimentally tolerable values, the error could be due to the noninclusion of the formula mass of the solvent molecules of crystallisation. This is added to the formula mass by adjusting the difference between d_0 and d_c to a minimum.

10.2 STRUCTURE DETERMINATION

Bragg's equation is applicable to the unit cell (external morphology) only, but is inadequate to the determination of structure (i.e., positions of the atoms of the molecule(s) in the unit cell, or in the assymmetric unit, to be precise). For actual structure determination two important parameters fundamental to waves in general, namely, (i) amplitude and (ii) phase, have to be obtained for each reflection, hkl.

10.2.1 Amplitude

All X-ray diffraction methods of data collection (rotation, oscillation, Weissenberg, precession, counter and other methods) provide only the intensity data of the Bragg reflections, hkl, but not the phase information. Whenever the waves are in phase, the intensities of the diffracted spots, hkl, are strong and they are proportional to the square of the amplitude.

10.2.2 Phase Information

The missing information in all X-ray diffraction intensity data is the phase information from reflections, hkl. Resolving the '**phase problem**' is central to structure determination by X-ray diffraction. The mathematical method for transforming a diffraction pattern into an image of the object is called Fourier transform. That is, the

Fourier transform of the diffraction pattern → structure.

The phase difference, α, between a ray reflected by an atom at the origin of the unit cell (0, 0, 0) and by the **j**th atom at position (x_j, y_j, z_j) is

$$\alpha = 2\pi\ (hx_j + ky_j + lz_j) \qquad ...(10.3)$$

If the scattering power of the **j**th atom is f_j, then the sum of the contributions from all the atoms in the unit cell to the Bragg reflection, hkl, is

$$\vec{\mathbf{F}}_{\mathbf{hkl}} = \sum_{j=1}^{All} f_j\ \exp\ (2\pi\ i\ (hx_j + ky_j + lz_j)) \qquad ...(10.4a)$$

where $\vec{\mathbf{F}}_{\mathbf{hkl}}$ is called the structure factor which is a complex quantity representing both the amplitude and the phase of the diffracted reflection.

That is, $$\vec{\mathbf{F}}_{\mathbf{hkl}} = |\ F_{hkl}\ |\ \exp\ (i\ \alpha_{hkl}) \qquad ...(10.4b)$$

where $| F_{hkl} |$ is the amplitude and α_{hkl} is the phase of the reflection, hkl, respectively.

X-rays are scattered by electrons and the representation of the structure is in essence the representation of the electron-density distribution around the atoms of the molecule(s) in the unit cell. Since the electron density in a crystal varies continuously and periodically, the electron density, ρ_{xyz}, at position *xyz* in the unit cell of volume, *V*, is

$$\rho_{xyz} = \frac{1}{V}\sum_h \sum_k \sum_l | F_{hkl} | \exp(i\,\alpha_{hkl}) \,.\, \exp(-2\,\pi i\,(hx + ky + lz)) \qquad ...(10.5)$$

Accordingly, the electron density at any point in the unit cell can be computed from a Fourier series and the coefficients of the terms are structure factors, $\vec{\mathbf{F}}_{\mathbf{hkl}}$, of the diffraction spectra. Both the amplitudes, $| F_{hkl} |$ and phase angles, α_{hkl}, of Bragg reflections, hkl, are required to compute the electron density (from Eq. 10.5). The objective of the phase determining procedures—Patterson, direct, molecular replacement, anomalous dispersion methods etc., is to obtain an initial set of phases for reflections, hkl, to compute the electron density maps of the molecule(s) in the unit cell.

10.3 PHASE DETERMINATION PROCEDURES

Generally, attempted phase determination methods are: (i) the Patterson method (Isomorphous Replacement), (ii) the direct method(s), (iii) molecular replacement and (iv) anomalous dispersion.

10.3.1 Patterson Method

The Patterson method uses the intensities, I_{hkl} (squared amplitudes $|F_{hkl}|^2$), instead of structure factors, $\vec{\mathbf{F}}_{\mathbf{hkl}}$, to calculate Fourier map(s). All the phase angles αhkl are set to zero (or 180°). The Patterson synthesis, when correctly interpreted, provides information pertaining to the inter-atomic vector distances between the atoms in the unit cell. All Patterson functions are centro-symmetric and their space groups are derived from the original space groups by replacing all translational symmetry elements (screws and glides) by corresponding non-translational ones (axes and mirrors) and by adding a centre of symmetry. That is,

Patterson space group = Space group – (screw & glide) + centre of symmetry.

For example, symmetry operations in Patterson maps for

Triclinic space groups	$\equiv P\,\bar{1}$
Monoclinic	$\equiv$ 2/m
Orthorhombic	$\equiv$ mmm

The Patterson synthesis has to be deconvoluted to interpret the structure and obtain the positions of the atoms in the unit cell. The interpretation of Patterson maps is generally not easy or straightforward and also the Patterson space is more crowded than the Fourier space (there are N (N – 1) inter-atomic vectors in the Patterson map for *N* atoms in the structure). Therefore, this method is useful and employable if the structure contains a few heavy atoms and the predominance of these heavy atoms in the scattering process can be exploited to locate the positions of these heavy atoms, based on the electron-density maps computed with the phases derived from the coordinates of these heavy atoms and proceed to interpret the rest of the structure. In the earlier days this was the only method available to determine the structure of any molecule, and atoms with molecular masses above sulphur, phosphorus or potassium were treated as heavy atoms for "small" molecules, but this is not the preferred method to solve structure of these molecules.

Isomorphous replacement procedures, which are based on the Patterson '**heavy atom**' method, are still employed to obtain the initial phase information in macromolecular crystallography. The general procedure followed is to collect intensity data for a single-crystalline macromolecule, say a protein, which is called the native crystal and for its heavy-atom derivatives (the heavy atoms are introduced into the native crystals by soaking them (and monitoring) with heavy-atom (like Hg, Pt, U) solutions. From the differences in the intensities between the derivative(s) and the native diffraction patterns, the contributions to the intensities by the heavy atoms are estimated and used in Patterson syntheses to obtain the positions of the heavy atoms in the unit cell. From the coordinates of these heavy atoms so obtained, phases can be calculated (α_c) and together with the observed amplitudes, ($|F_O|$ s) of the reflections, *hkl,* electron density map(s) can be computed. The vector $\vec{\mathbf{F}}_{\mathbf{pH}}$ is

$$\vec{\mathbf{F}}_{\mathbf{pH}} \approx \vec{\mathbf{F}}_{\mathbf{P}} + \vec{\mathbf{f}}_{\mathbf{H}} \qquad \text{...(10.6a)}$$

where $\vec{\mathbf{F}}_{\mathbf{P}}$ is the structure factor of the native, $\vec{\mathbf{F}}_{\mathbf{pH}}$ of the derivative and $\vec{\mathbf{f}}_{\mathbf{H}}$ of the heavy atom. The structure factor of the heavy atom, $\vec{\mathbf{f}}_{\mathbf{H}}$, cannot be obtained directly from the experimentally available amplitude data, $|F_p|$ and $|F_{pH}|$, without the knowledge of the phases α_P and α_{pH} (Fig. 10.5).

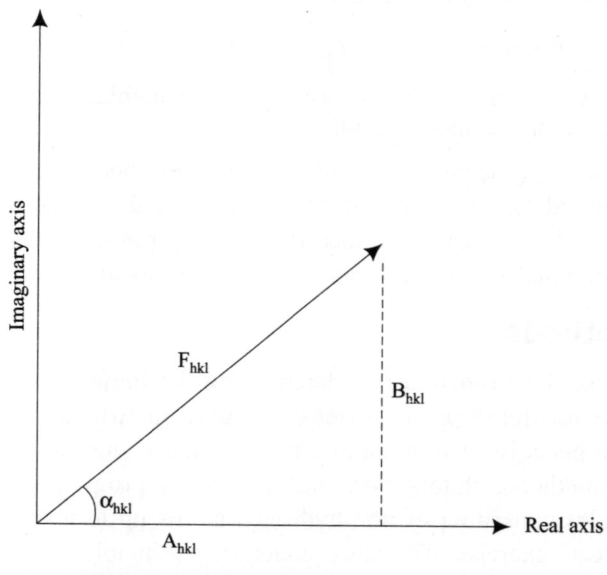

Fig. 10.5a Argand diagram of structure factor, F_{hkl} and phase α_{hkl}.

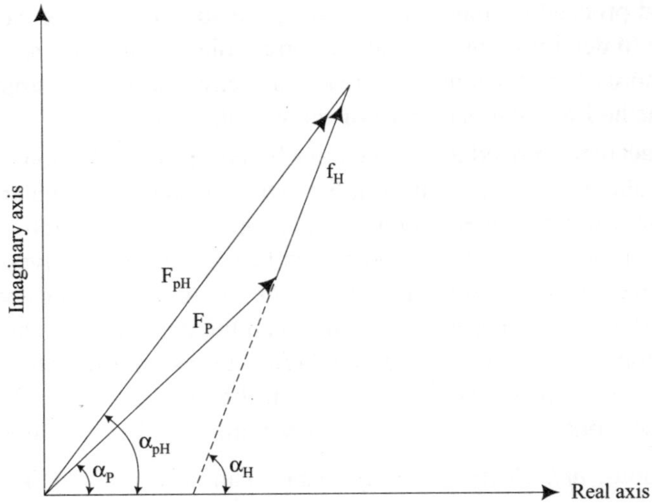

Fig. 10.5b Vector diagram of structure amplitudes and phases of single isomorphous situation. $\mathbf{F_P}$ is structure factor for the native, $\mathbf{F_{pH}}$ for the derivative and $\mathbf{f_H}$ for the heavy atom; and α_P, α_{pH} and α_H, phases, respectively.

In general, the contribution of f_H is small and, therefore, equation (10.6a) can be used to get an estimate of the contribution of the heavy atoms in structure factor calculations. For centro-symmetric reflections, the contributing structure factors, $\vec{F}_p$, $\vec{F}_{pH}$ and $\vec{f}_H$ are real (all the vectors are collinear, and the phases are either 0 to 180°). Then

$$f_H = | \Delta_F | = | F_{pH} - F_P | \quad ...(10.6b)$$

Data from centro-symmetric reflections are used, when available, to locate the position of heavy atoms in the unit cell and solve the phase problem.

In fortuitous situations, the phases obtained from a single heavy-atom derivative, i.e., single isomorphous replacement (SIR), could be sufficient to resolve the phase problem, but, in general, intensity data from several derivatives are needed (multi-isomorphous replacement, MIR) to resolve phase ambiguities, experimental errors and other related problems in solving the phase problem,

10.3.2 Direct Methods

"Direct" methods of phase determination i.e., determining the initial phases of a set of reflections from their internal phase relationships, are extensively used in structure determination of "**small**" molecular compounds, especially of light-atom structures (structures with H, C, N and O atoms). The success of these methods, automation and packaged programmes have made structure determination of molecules consisting of non-hydrogen atoms up to about 80–100, a more-or-less straightforward and "direct" exercise. The basic underlying principles of these methods, which are statistical, are: (i) the scattering powers of the atoms in the structure are about equal (generally the case for molecules containing C, N and O atoms) and randomly distributed (not always the case), (ii) the resolution is such that individual atoms can be located (can be achieved in small molecular structures), (iii) the electron density in the unit cell is always either positive or zero, but never negative and (iv) every atom of the molecule(s) in the unit cell contributes to every part of the transform. That is, different parts of the transform are interrelated. Therefore, mathematical formulations relating amplitudes and phases can be set up. Some procedures are cited below.

Harker-Casper Inequalities. Rewriting the equations (10.4) and (10.5) in a compact form

$$\vec{\mathbf{F}}_{\mathbf{h}} = \sum_j f_j \exp\,(-\,2\pi\, i\,(\underline{h}\,.\,r) \qquad ...(10.7a)$$

$$\rho_r = \frac{1}{V_h}\sum_{h=1}^{\text{All}} \vec{\mathbf{F}}_{\mathbf{h}}\ \exp\,(2\pi\, i\,(\underline{h}\,.\,r) \qquad ...(10.7b)$$

where $\underline{h}$ stands for h, k and l, and r for x, y and z, respectively.

Defining U_h as

$$U_h = \frac{F_h}{\sum_j f_j} = \sum_j n_j \exp\,(2\pi\, i\,(\underline{h}\,.\,r)) \qquad ...(10.8)$$

where $$n_j = f_j \Big/ \sum_j f_j$$

and applying **Scwartz-Cauchy's** inequality for U's, that is,

$$|\sum_j \mathrm{a}_j\ \mathrm{b}_j|^2 \le (\sum_j |a_j|^2)\ (\sum_j |b_j|^2) \qquad ...(10.9a)$$

$$|U_h|^2 \le \sum_j n_j\ \sum_j n_j \cos^2 2\pi\ (\underline{\mathrm{h}}\,.\,x_j) \qquad ...(10.9b)$$

but $$\sum_j n_j = 1;\ \text{and}\ |U_h|^2 \le 1$$

$$\therefore \qquad U_{2h}^2 \ge 2U_h^2 - 1 \qquad ...(10.9c)$$

The bigger the value of U_h, the greater the probability of deducing its sign.

Sayre Equality. Electron density, ρ_r and its square function ρ_r^2 look similar. Therefore, in spite of the differences in form, the relationships found allow some general statements regarding structure factors, namely the phase of $\vec{\mathbf{F}}_{(\mathbf{h}+\mathbf{h}')}$ is related to the phases of $\vec{\mathbf{F}}_{\mathbf{h}}$ and $\vec{\mathbf{F}}_{\mathbf{h}'}$.

$$F_h^2 = \mathrm{S}_h \vec{\mathbf{F}}_{\mathbf{h}} \qquad ...(10.10a)$$

where S_h is a function to take into account the change in atomic shape. Squaring of any function is equivalent to the operation of a self-convoluting array of structure factors.

$$F_h^2 = \sum_{h'} \vec{\mathbf{F}}_{\mathbf{h}'} \cdot \vec{\mathbf{F}}_{(\mathbf{h}-\mathbf{h}')} = S_h \vec{\mathbf{F}}_{\mathbf{h}} \qquad ...(10.10b)$$

For a centro-symmetric case, the phase $\alpha = 0$ or 180°. Then

$$\therefore \qquad \alpha(h) \,.\, \alpha(h') \,\alpha\, (h + h') \approx 1 \qquad ...(10.10c)$$

Hauptman-Karle Method. Phase probabilities are used to solve structures. The normalised structure factor, E_h, is defined as

$$\mathbf{E}_{\mathbf{h}} = \frac{\mathbf{U}_{\mathbf{h}}}{\sqrt{<|U_h|^2>}} \qquad ...(10.11a)$$

where
$$<|U_h|^2> = \frac{1}{N} \sum_h |U_h|^2$$

or

$$E_h^2 = \frac{F_h^2}{\varepsilon \sum_j^N f_{jh}} \qquad ...(10.11b)$$

where ε is space group dependent.

For a centro-symmetric case: In the case of centro-symmetric structures, the planes of atoms either lie on the lattice planes or halfway between them. Therefore, the phase angle, α of the reflection, hkl, is zero or 180°.

$\mathbf{E}_{\mathbf{h}} \approx \sum_{h'} \mathbf{E}_{\mathbf{h}'} \,.\, \mathbf{E}_{\mathbf{h}-\mathbf{h}'}$ and probability, $P_r\,(E_h)$ is

$$P_r\,(E_h) = \frac{1}{2} + \frac{1}{2} \tanh \frac{\sigma_3 |E_h| \sum_{h'} |E_{h'} . E_{h-h'}|}{\sigma_2^{3/2}} \qquad ...(10.12a)$$

where $\sigma_n = \sum_{j=1}^{N} Z_j^n$ and Z is atomic number.

For a non-centrosymmetric case: For non-centrosymmetric structures there is no restriction on the phase angle α, or the positions of the planes of atoms relative to the origin of the unit cell.

$$\tan \alpha_h = \frac{\sum_{h'} |E_{h'}||E_{h-h'}| \sin(\alpha_{h'} + \alpha_{h-h'})}{\sum_{h'} |E_{h'}||E_{h-h'}| \cos(\alpha_{h'} + \alpha_{h-h'})} \qquad ...(10.12b)$$

Hoppe-Zechmeister Method.

$$\mathbf{U}_{\mathbf{h}} = < \mathbf{U}_{\mathbf{h}'} \,.\, \mathbf{U}_{\mathbf{h}-\mathbf{h}'} >_{h'} \qquad ...(10.13a)$$

$$< \alpha_h > = \text{Phase of } \sum_{h'=1}^{m} |U_{h'}| |U_{h-h'}| \exp i (\alpha_{h'} + \alpha_{h-h'}) \qquad ...(10.13b)$$

where α_h is the phase of $< \mathbf{U_{h'}} . \mathbf{U_{h-h'}} >_{h'}$.

The summation of the equation (10.13a), taken over all reflections, *hkl,* is called the **Hughes** formula and is equivalent to squaring the electron density in direct space (which is also the Sayre equation). $m = 1$ is the triple product method.

$$\mathbf{U_{new}} = |U_{obs}| \exp (i, \alpha_m^*) \qquad ...(10.13c)$$

where α_m^* is the phase of the modified electron density, ρ_m^*.

The statistical error ($\Delta \alpha_h$) is given by **Cochran**

$$\Delta\alpha_h = |\alpha_h - <\alpha_h>| = \frac{1}{\pi J_0(X)} \int_0^{\pi} \Delta\alpha_h \exp (2X . \cos \Delta\alpha_h) \, d \, \Delta\alpha_h \qquad ...(10.13d)$$

where $X = 2N | U_h . U_{h'} . U_{h-h'}|$ and $J_0(X)$ is a modified **Bessel** function of the second kind and N is the number of atoms in the unit cell.

The general approach of these direct method procedures is :

(1) Form E's in descending order (i.e., tabulate $E's$. E-list).

(2) Generate Σ_2 relationships.

(3) Form triple product phases ('Pyramid-building' or phase relationships).

That is,

$$\alpha_h = < \alpha_{h'} + \alpha_{h-h'} > \text{ and}$$

$$\alpha_h = \frac{\sum_{h'} |E_h||E_{h-h'}|(\alpha_{h'} + \alpha_{h-h'})}{\sum_{h'} |E_h||E_{h-h'}|}$$

(4) Sort equations on the basis of probability selection criterion).

$$\tan \alpha_h = \frac{\sum_{h'} |E_{h'}||E_{h-h'}| \sin(\alpha_{h'} + \alpha_{h-h'})}{\sum_{h'} |E_{h'}||E_{h-h'}| \cos(\alpha_{h'} + \alpha_{h-h'})}$$

(5) Reduce equations to fewest possible symbols.

(a) Assign origin and enantiomorph (origin of the unit cell can also be temporarily fixed by assigning arbitrary phases to starting reflections and the phases continuously redetermined).

(b) Define unknown symbols (Symbolic addition method of **Hauptman & Karle** and multisolution methods and other variations developed by **Hoppe, Woolfson** and other groups).

(6) Calculate E-Fourier(s) from the multisolution 'pyramid' and locate the positions or atoms of the molecules(s) (partially or fully) in the unit cell.

(7) If the E-Fourier map(s) phase information is correct (to unravel the structure fully or partially), then regular electron density syntheses/refinements iterative procedures are followed to determine the final structure with an acceptability index (R-index).

Further developments in 'direct' methods of phase determination are in the elaboration and automation of various criteria and procedures for reliable 'pyramid' building of phase relationships and in the choice of selecting the 'best' E-Fourier from the multisolution list (MULTAN) to obtain an initial set of phases to solve the structure.

10.3.3 Molecular Replacement

This is a method generally employed in macromolecular crystallography to resolve the phase problem. The strategy of the method is to locate the best orientation and position of a molecule in the unit cell (or a part of it) whose structure has been determined (whose positional coordinates of atoms/molecules are available) and which has structural similarity to the structure to be determined. The known molecule (or a part of it) is rotated around the origin and translated in the unit cell to get the best-fit and the structure amplitudes are phased from the positional coordinates of the best-fit structure.

The methodology is based on the Patterson function of a crystal. In the Patterson function of a crystal there are self-vectors (intra-molecular vectors) as well as cross-vectors (inter-molecular vectors). All self-vectors are contained within a sphere of radius, l, about the origin of the Patterson (l is also the maximum dimension of the molecule). To find the best orientation of the search molecule in the crystal, the orientation that superposes ρ_M upon one of the self-vectors is searched. The correlation function is

$$R(C) = \int \rho_M \, C(U) \, \rho(U) dV \qquad ...(10.14)$$

where $C(U)$ is rotation matrix based on **Eulerian** angles. The integral is evaluated numerically for each value of C and correct orientation is that for which $R(C)$ is maximum. $R(C)$ is large when the peaks and the troughs in ρ_M coincide with the peaks and the troughs in $\rho(U)$, respectively.

Once the orientation of the search molecule in the unit cell (around the origin) has been fixed, then its position in the unit cell is located by the translational search by applying the translational function, $T(X)$

$$T(X) = \int_{\text{unit cell}} \dot{\rho}_{MX} \, (U, X) \, \rho(U) \, dV \qquad ...(10.15)$$

10.3.4 Anomalous Dispersion

The Anomalous dispersion method is being revitalised in recent years in its application to resolve the phase problem in macromolecular crystallography. Even though the feasibility of this method for obtaining the phase information for macromolecular structures was pointed out by **Ramachandran** and colleagues as early as 1965, the method has been applied effectively in the 'exploitation' of phase information to its maximum potentiality only recently, mainly due to technological advances for collecting accurate intensity data and the choice of desired wavelength from a synchrotron radiation source.

All atoms scatter X-rays anomalously, but their contribution to the anomalous scattering depends on their atomic numbers and also whether the wavelength(s) are near to the absorption edge(s) of the anomalous scatterer(s). The atomic scattering factors, f, (tabulated in the International Tables for X-ray Crystallography) are for elastic scattering. To account for anomalous scattering, the atomic scattering factor, f, is represented as a complex number

$$f = f_0 + \Delta f' + i\,\Delta f'' \qquad \text{...(10.16)}$$

where $\Delta f'$ and $\Delta f''$ are real and imaginary parts of dispersion corrections to be applied.

In the case of non-centrosymmetric structures, the effect of anomalous scattering can be determined experimentally by careful measurement of the intensity differences between **Friedel** equivalent reflections, hkl and $\overline{hkl}$. In the case of anomalous scattering, Friedel's condition breaks down and the intensities of the **Friedel** pairs are not equal.

$$I_{hkl} \neq I_{\overline{hkl}} \qquad \text{...(10.17)}$$

(i) **"Small" molecular structures:** A comparison of the magnitudes of the measured and calculated **Bijvoet** differences, B_H, of selected **Friedel** pair reflections provides information on the magnitude of $\Delta f'$ and a comparison of their signs yields the absolute configuration of the molecule.

$$B_H = \frac{F_{hkl} - F_{\overline{hkl}}}{0.5\,(F_{hkl} + F_{\overline{hkl}})} \qquad \text{...(10.18)}$$

The experimental determination of the absolute configuration of "small" molecule structures can be carried out by exploiting the anomalous scattering of anomalous scatterers, like K, Br, P or S, if they are present. This was the procedure that was followed in determining the absolute configuration of the molecule in earlier years and if the anomalous scatterer was not an intrinsic part of the molecule, then an anomalous scatterer was introduced extrinsically (doping) to determine the absolute configuration. But accurate data collection (by diffractometry) has enabled the experimental determination of the absolute configuration of '**light-atom**' structures (containing H, C, N and O only) by delineating the slight differences in the anomalous scattering powers between carbon and oxygen atoms, without any recourse to introducing an anomalous scatterer(s) extrinsically.

(ii) **Macromolecular structures:** As mentioned earlier, the resurgence of this method to resolve the phase problem more effectively has been due to technological developments and the availability of the synchrotron radiation source. The synchrotron radiation source allows a choice of wavelengths nearer to the absorption edges of the anomalous scatterer(s) for optimum use of the dispersion method. Anomalous dispersion methods, like Multiple-wavelength Anomalous Dispersion (MAD) and other procedures being developed can be applied effectively to solve macromolecular structures with heavy atoms (metalloproteins and enzymes).

10.4 STRUCTURE REFINEMENT

Once the approximate coordinates of the atomic positions of the structure (partial or full in the case of small molecules) or the position of the heavy atoms and/or position of the molecule or part of it (in the case of macromolecular structures), in the asymmetric unit cell, are known, then refinements/ electron density syntheses procedures arc employed iteratively to obtain better fit of the calculated structure factors, F_c, with those of the observed structure factors, F_σ, In the least-squares refinement procedures, which are generally employed, the most plausible parameters are those for which the sum of the squares of the differences between observed and calculated values is a minimum. That is,

$$\sum_h \sum_k \sum_l \{F_0(hkl) - F_c(hkl)\}^2 \rightarrow \text{Minimum} \qquad ...(10.19)$$

The goodness-fit of the refinements is monitored by the reliability index, R.

$$R = \frac{|(F_0(hkl)| - |F_c(hkl)|}{|F_0(hkl)|} \qquad ...(10.20)$$

A value of <0.1 for a structure of about ≈ 100 non-hydrogen atoms is considered a well-determined structure.

10.5 STRUCTURE AND FUNCTION

The correlation of structure and function depends, among other things, very much on the quality of the structural information attainable — namely the resolution attainable, the quality of the crystal, clarity of electron density maps and interpretation of molecular features in the electron maps without ambiguity etc.

10.5.1 Small Molecules

Three-dimensional X-ray structure determination of "**small**" molecules (about 50 non-hydrogen atom structures) provides resolution sufficient to resolve individual atomic positions ($\approx$ 0.75 Å) and, therefore, can provide positional coordinates of individual atoms in the unit cell (in the asymmetric unit, to be precise) together with the thermal vibrational parameters of the atoms (Fig. 10.6).

From the atomic coordinates of the molecule, all the essential details of the molecule, namely, the bond lengths and angles, stereochemistry and conformational details of the molecule and thermal motions of the atoms/groups of the molecule, non-bonded interactions, such as ionic (salt bridges), hydrogen bonding, van der Waals and hydrophobic and base-stacking interactions can be calculated and other dynamic and structural aspects, like charge-transfer etc., can also be evaluated. In addition, the absolute configuration of light-atom structures (containing H, C, N, O atoms only) can be experimentally determined by accurate evaluation of Bijvoet differences of a set of selected Friedel pair reflections, without recourse to introducing an anomalous scatterer.

The development of direct methods of phase determination and accurate data collection by diffractometry and fast-computing facilities have made the structure determination of "small" molecules (most chemical and biological compounds belong to this category) less of a challenge. The

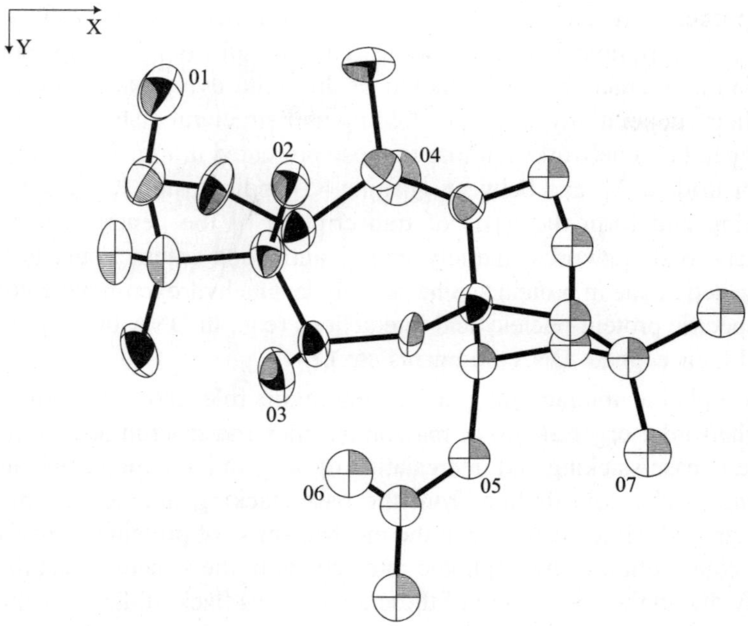

Fig. 10.6 Structure of a **"small"** molecule projected down an axis (z-axis), with well-resolved atomic positions and thermal ellipsoids of individual atoms.

emphasis has been shifted from interpretation of structural data, that can be obtained without too much difficulty or delay (in a well-established laboratory with trained personnel), to functional aspects: understanding biological/biochemical reaction mechanisms at the molecular level.

As an example, the final structural data on a "small" molecule, **pseudouridine**, has been presented (as shown in Figs. 10.7 a, b & c). The molecular formula of pseudouridine is $C_8H_9N_2O_4$. There are two molecules of pseudouridine and a molecule of water of crystallisation in the asymmetric unit belonging to the space group, $P2_1$. The structure had been determined by direct methods and refined to an R-value of 0.037.

Analysis of bond lengths and bond angles of a structure (similar to the values presented for pseudouridine (Fig. 10.7b) can reveal if there are any gross deviations from the expected values and any other features worth explanation (like the deviation from the planarity of the planar groups etc.). Such systematic analyses of data bond lengths and angles in peptide structures led **Pauling** to emphasise the partial double-bond character of the peptide bond due to charge delocalisation, the planarity of the peptide group 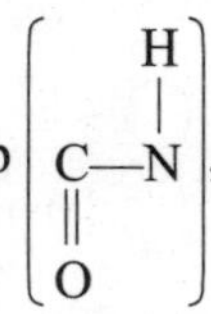

, proposes the structure principles of the polypeptide backbone and predicts the existence of the ordered segments—α-helix and β-pleated sheet, in polypeptides (proteins and enzymes) long before the three-dimensional structure of a protein was elucidated.

Of major importance in understanding the structure-function relationships of biomolecular interactions, such as macromolecular associations, enzyme-substrate binding, protein-nucleic acid and protein-membrane interactions, intercalation of drug and dye molecules etc., are non-bonded interactions and their cooperativity in proper folding and structural stability. For example, parallel oligodentate hydrogen bond networks (similar to those presented in Fig. 10.7c) between adenine (A) and thymine (T) or uridine (U) and between guanine (G) and cytosine (C) bases provide sequence-specific base-pairing and form the basis of transcription of the genetic code. Further, parallel tridentate hydrogen-bond networks, namely codon–anticodon interactions, are involved in the translation of the genetic code in protein synthesis. Polydentate hydrogen-bond patterns play a central role in sequence-specific protein-nucleic acid interactions (e.g., in DNA-binding proteins, restriction endonucleases and their cognate DNA fragments etc.).

Similarly, hydrophobic interactions play an important role and are involved in cooperative interactions in higher-order organisation of macromolecules and macromolecule-ligand interactions. One good example is base-stacking and intercalation of drug and dye molecules in the DNA-planar-molecule interactions. The contribution from the base-stacking interactions in codon-anticodon triplets cannot be ignored while dealing with the mechanism(s) of protein synthesis at the molecular level. Though the contribution of hydrophobic interactions to the structural stability and function is difficult to quantify due to the very nature of these interactions (lack of directionality, non-specificity and weak nature), the cumulative effects are crucial in macromolecular organisation.

Experimental determination of the absolute configuration of 'light-atom' structures is possible without any recourse to the introduction of heavy-atom "markers" from accurate data collection and careful measurement and evaluation of Bijvoet differences I_{hkl} and $I_{\overline{hkl}}$) by exploiting the difference in the strength of anomalous scattering between oxygen and carbon atoms ($\Delta f_0' = 0.02$ with respect to $\Delta f_c'' = 0$). Therefore, the chirality of biomolecules can be experimentally determined in a

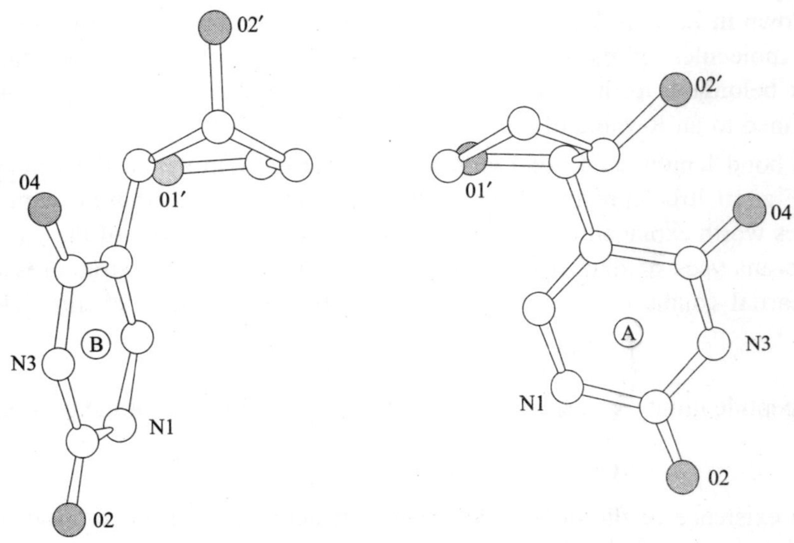

Fig. 10.7a Pseudouridine molecules *A* and *B*, viewed along the normals to the best planes of the sugar rings.

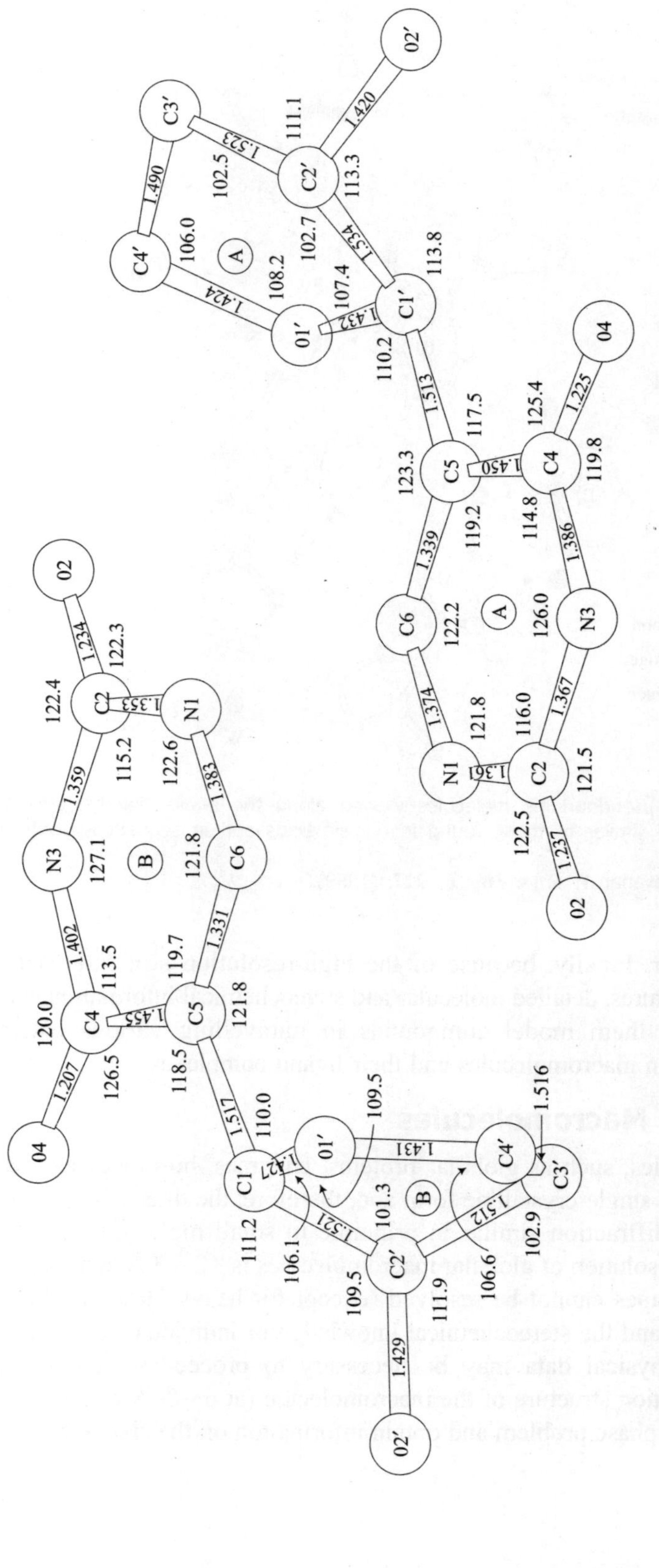

Fig. 10.7b Bond distances (in Å) and bond angles (in o) in pseudouridine molecules *A* & *B*.

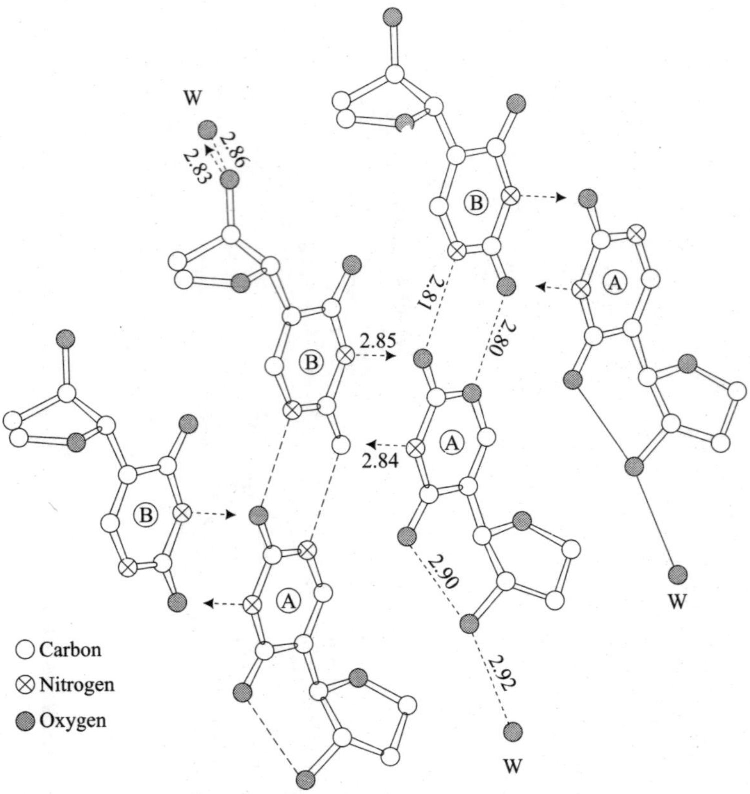

Fig. 10.7c Structure of pseudouridine molecules viewed along the z-axis. Parallel bidentate hydrogen bond networks are similar to those found in nucleic acids and in sequence-specific protein-nucleic acid interactions.
(Source: Narayanan P. *Phys. Edu.* 6; 217. (1989).

straightforward manner. Finally, because of the high-resolution structural data obtainable from "small" molecular structures, detailed molecular and stereochemical information at an atom level from these structures make them model compounds in unravelling various patterns of molecular interactions that occur in macromolecules and their ligand complexes.

10.5.2 Globular Macromolecules

Globular macromolecules, such as globular proteins, enzymes, hormones etc., can be, with effort and luck, crystallised in single-crystalline form and, therefore, the three-dimensional structure can be determined by X-ray diffraction similar in principle to small molecular structure determination. However, the limit of resolution of globular macromolecules is ~2 – 3 Å and therefore, the individual atoms or molecular shapes cannot be resolved (except for heavy atoms in the structure). So, the primary sequence data and the stereochemical knowledge of individual residues and other available biochemical and biophysical data may be necessary to proceed with structure determination. Generally, a low-resolution structure of the macromolecule (at 6 – 5 Å resolution) is attempted as a first step in solving the phase problem and obtain information on the size, shape and other structural features obtainable at such resolutions. High-resolution data are then included to solve the structure by interactive methods of model building, electron-density syntheses and structure refinements. Various 'goodness' tests are applied to monitor these procedures to arrive al a physically- and chemically-acceptable three-dimensional structure.

A reasonably well-determined structure of a macromolecule (Fig. 10.8a & b), even at 2 – 3 Å resolution can provide a host of quantitative as well as qualitative structural information. These include backbone conformation, positions and conformational features, side groups, structural motifs and domains, stereochemical details of substrate-binding sites and details of specific and nonspecific non-bonded interactions. Such information enables one to correlate the functions of the macromolecule with its three-dimensional structure. It enables one not only to propose biochemical/ biological reaction mechanisms at the molecular level but also to get an insight into the nature of molecular diseases due to structural defects, mutation and other causes. Such an insight is the basis of molecular engineering, that is, the altering of specific sites/residues of macromolecules is not only to improve upon existing functions but also to alter the functions to suit particular needs. This outcome of merging areas of structural biophysics and molecular biology has a promising future.

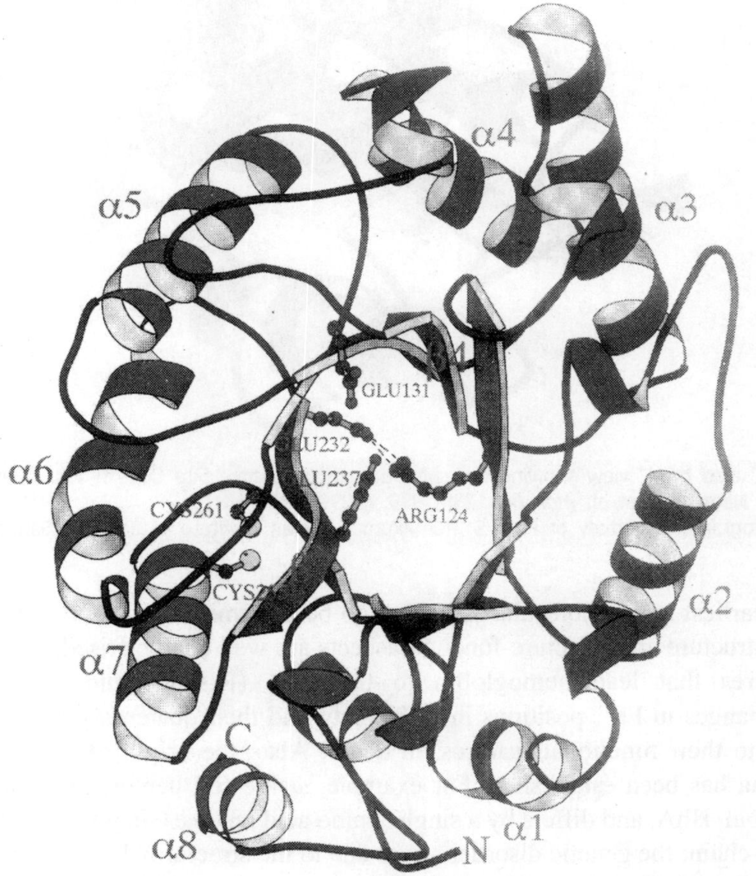

Fig. 10.8a Ribbon diagram of T.A. Xylanase.

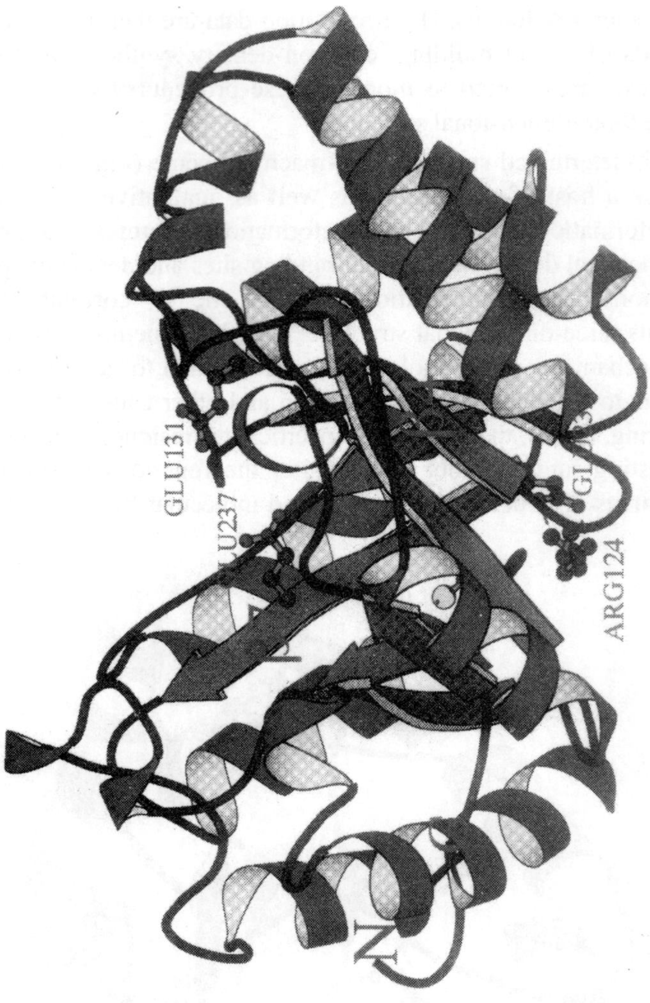

Fig. 10.8b The "salad bowl" view showing the substrate-binding Active Site Cleft in T.A. Xylanase. (Ref: Natesh, R. et al. *Mol. Biol.* 288; 999. (1999)) (Photographs: Courtesy of Prof. S. Ramakumar, Indian Institute of Science, Bangalore, India.)

One of the earliest macromolecular structures to be determined by X-ray diffraction and whose high-resolution structure and structure-function aspects are well characterised, is that of hemoglobin. Structural features that lead hemoglobin to the oxy- (R-state) and deoxy- (T-state) with concommitant changes in Fe^{2+} positions in the moiety and their quaternary structural changes have been correlated to their functional features, in detail. Also, the origin of molecular diseases, like sickle-cell anemia has been established. For example, sickle-cell hemoglobin, HbS, is a mutant of normal hemoglobin, HbA, and differs by a single amino acid residue (single-point mutation) in the 6th position of the β-chain; the genetic disorder arises due to the structural defect by the nonconservative substitution of valine for glutamic acid, which places a nonpolar residue on the surface of HbS and in its deoxy-form HbS polymerises and sickling occurs. Such studies have been and are being carried out for other macromolecules. That is, the molecular-pathology of a macromolecule can be understood by correlating its clinical symptoms, which are external manifestations, to the alternatıons in its molecular structure or subtle variations in topologies of the structural motifs at crucial positions arising from mutation and other causes. Such studies could lead to better 'molecular design' of proteins and other molecules.

Another important field which X-ray crystallography has enriched immensely is immunobiology. It has been solely responsible for unravelling the three-dimensional architecture of immunoglobulins and their constituents (IgG_2, F_{ab}, F_c etc.) and thereby has helped much in our understanding of the basic structure of immunoglobulins: the domain structure of immunoglobulins and the structural motif of these domains, called the '**immunoglobulin**' fold (a sandwich structural motif with two layers of β-strand sheets interconnected by a S-S bridge). It has also opened up new avenues of thinking about mechanisms of immunoglobulin function and histocompatibility and a host of other related problem. The three-dimensional structure of a monoclonal antibody-antigen complex could elucidate the nature of antigen-antibody interactions at molecular level and has brought about the structural understanding of how a monoclonal antibody can recognise the replacement of a single amino acid residue at the antigen-antibody interface and also the importance of proper polypeptide chain folding around the crucial amino acid residue(s).

10.5.3 Fibrous Macromolecules

A fibre is an object with an axial length far greater than its width. It is periodic only along the fibre axis (z-axis) and is random in the other two directions. To this category belong many biological molecules, such as keratin group proteins (keratin of hair, wool and horns), myosin, F-actin of muscle, epidermin of skin, collagen, nucleic acids — DNA and RNA and viruses, like tobacco mosaic virus (TMV), microtubules, flagella and cilia etc. The fibre-diffraction pattern is a rotational average of the diffraction pattern of a repeating unit (azimuthal average) and the data obtainable from fibrous molecules are quite meagre and elucidation of their structures is beset with many difficulties, and is individually challenging, due to the nature of fibre-diffraction in general and due to the uniqueness of the particular fibre (or class of fibres) under consideration.

Many of the fibrous biological macromolecules are helical in form and the repeating units along the fibre (averaged out) are nucleic acid bases in nucleic acids, amino acid side chains in proteins and enzymes and protein subunits in viruses. The general structural parameters that characterise a helix are: radius of the helix, rise per residue and the pitch. Taking into account the cylindrical nature of fibres (helical), the theory of fibre-diffraction is meaningfully represented by carrying out the necessary mathematical formalisms in cylindrical coordinates and the Fourier-Bessel transform of diffraction pattern represents the structure.

$$F(R, \psi, Z) = \Sigma G_n(R, Z).\mathbf{i}^{\mathbf{n}}.\exp\ (\mathbf{in}\ \psi) \quad ...(10.21a)$$

and the triple integral $G\ (R, Z)$ is given by

$$G_n(R, Z) = \iiint \rho(r, \varphi, z).J_n(2\pi\ rR).\exp\ (-\ \mathbf{in}\ \varphi).\exp\ (2\pi\ \mathbf{i}zZ).rdr.d\ \varphi.dz \quad ...(10.21b)$$

where ρ is the electron density, (r, φ, z) and (R, ψ, Z) are real and reciprocal cylindrical coordinates, respectively, and J_n is the **Bessel** function of order n.

The properties of the Bessel functions, $J_n(2\pi\ rR)$, lead to the following :

(1) For the 0th order Bessel function ($n = 0$), there is a maximum at the origin (J_0 = max), and the behaviour is similar to the damped cosine wave.

(2) For higher-order Bessel functions, ($n > 0$), the value of $J_n = 0$ at the origin (minimum at the origin), and further values are analogous to a damped sine wave.

(3) The intensity of any particular value of n depends on the reciprocal lattice coordinate, R, and is independent of the angular coordinate, ψ. That is

$$|F(R, Z)|^2 = [J_n(R)]^2 \quad ...(10.22)$$

(4) Due to the periodicity along the helix axis, the diffraction pattern occurs only on layer lines.

All the X-ray diffraction intensity data available from a fibrous molecule can be recorded on a single photograph consisting of layer lines (Fig. 10.9). Such a diffraction photograph shows the characteristic features:

(i) Mirror symmetry about the helix axis (meridian) due to cylindrical symmetry and also mirror symmetry about the equator due to a two-fold rotation axis perpendicular to the helix axis.

(ii) Characteristic **X** pattern.

(iii) Intensity maxima confined to layer lines only and only a single Bessel function occurs on any layer line.

Due to paucity of diffraction data from semi-crystalline molecules, structure determination of any fibrous macromolecule poses considerable difficulties and is a unique achievement in each case. The strategy of structure determination rests on postulating and refining a structural model that agrees with the X-ray fibre-diffraction data. As the diffraction data are meagre, there exists the possibility of

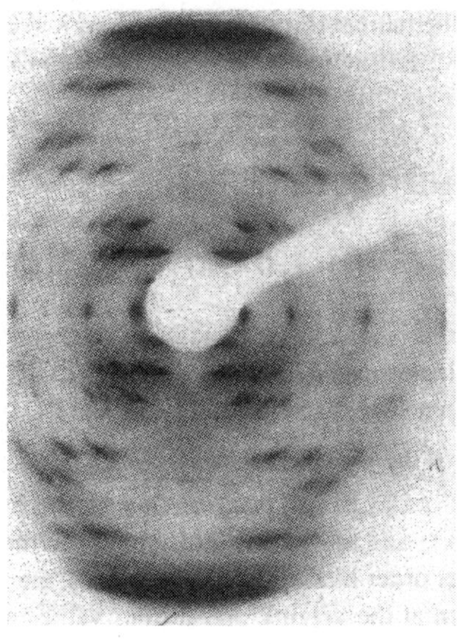

Fig. 10.9 Fibre X-ray diffraction photograph of calf thymus DNA with LiCl, at relative humidity 66%. (Courtesy of Prof. V. Sasisekharan, Indian Institute of Science, Bangalore.)

more than one structural models whose calculated diffraction pattern agrees with the experimental data. A consistent model should not only provide relevant explanations to the existing experimental data but also should stand the test of future experimental observations and also offer better and deeper insight to structure-function features. Some examples are mentioned to underscore this aspect.

Collagen Triple Helix. Collagen is a rod-shaped fibrous protein, comprising three tropo-collagen molecules (monomers), each of ~3000 Å and ~15 Å dimensions forming a left-handed poly-L-proline type II helix and three of these monomers twisted about one another to form a coiled-coil triple helix (Chapter 2 Fig. 2.24b). The amino acid sequence of each tropo-collagen monomer consists of repeating units of three amino acids of general formula- $(Gly\text{-}Pro\text{-}X)_n$ and $Gly\text{-}X'\text{-}HyPr)_n$ (where X and X′ are any amino acid and HyPr is hydroxyproline). There is inter-chain hydrogen bonding between proline and hydroxyproline residues of separate chains, rendering structural stability to the helical form of individual tropo-collagen monomers (for more details see fibrous proteins in Chapter 2).

The triple-helical model of collagen structure, proposed by Ramachandran and his colleagues **(Madras group)**, explains the fibre-diffraction data (2.86 Å meridional, i.e., zero-order Bessel term, and 4.0 and 9.5 Å off-meridional reflections, i.e., Ist-order Bessel terms). However, this initial model from the Madras group could not satisfactorily explain the sequence, **Gly-Pro-HyPr**, because pyrrolidine residues in peptide linkage cannot form such hydrogen bonds and, therefore, one of the hydrogen bonds has to be sacrificed. To overcome this apparent anomaly, **Crick** and **Rich** proposed collagen type I and collagen type II models (poly-glycine models), comprising only one inter-chain hydrogen bond for every three residues, thereby accommodating **Gly-Pro-HyPr** also. But, the revised model of the Madras group (with relaxed van der Waals radii and model-building) can also accommodate **Gly-Pro-GyPr** by sacrificing a hydrogen bond wherever this sequence occurs and with the variations in the packing of helices depending on the sequences. This revised model has been found to be in better agreement with other physico-chemical data, like infrared dichroism, deuterium exchange and melting studies.

Many of the structural features of contraction proteins and the elucidation of the mechanism muscle contraction were based on structural information for X-ray diffraction data, in addition to other physico-chemical data.

DNA Double Helix. The right-handed, double helix model of DNA, proposed by **Watson** and **Crick** was based on X-ray diffraction photographs of B-DNA fibres (similar to the photograph in Fig. 10.9). The 3.4 Å meridional reflection can be accounted for as arising due to base-stacking with a repeat distance of 3.4 Å. The proposed model elegantly and rationally explained all the experimental data available at that time and offered rationale for storage, replication and transcription of genetic information. It was the primary and path-breaking achievement in expanding new vistas in quantitative molecular biology. There were other models proposed for the structure of DNA prior and later to the Watson-Crick hypothesis. However, the Watson-Crick model is still the most consistent, elegant and simple one for the right-handed, double-helical structures of nucleic acids (Refer to Chapter 3 for details on nucleic acids).

Fibre X-ray diffraction methods have been successfully employed, by application of a variety of methodologies and with ingenuity, to unravel (at a molecular level) the structure and functional features of immensely-complex structures, like viruses (e.g. Tobacco mosaic virus) etc. The method has also been extended to study the subunit organisation and structural details of other fibrous macromolecules, such as microtubules and bacterial flagellar filaments. There seems to be no limit to the size of the molecule or to its complexity for structure determination by X-ray diffraction techniques. The flow chart (Fig. 10.10) highlights the salient features of X-ray structure determination of (A) single-crystalline and (B) fibrous materials.

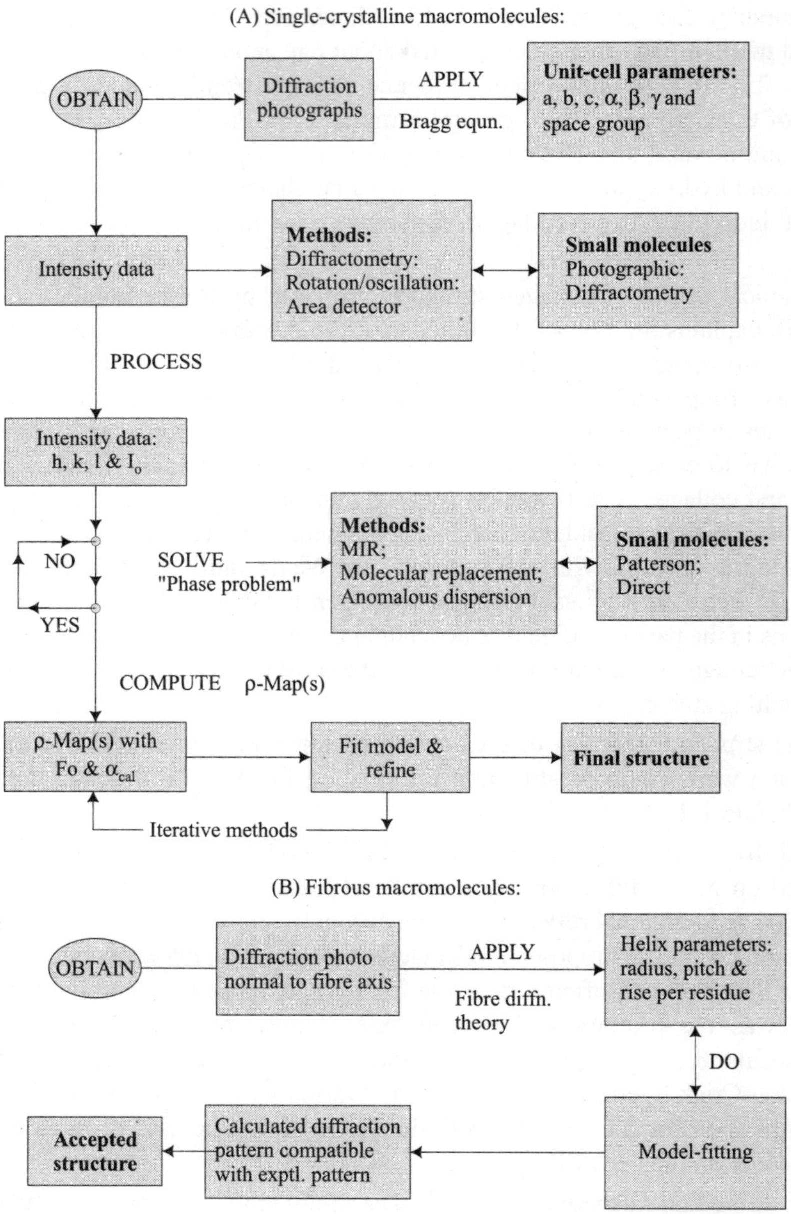

Fig. 10.10 Flow chart : Salient features of X-ray structure determination of macromolecules (and small molecules).

10.6 FUTURE TRENDS

Intensity data collection has become more accurate and fast and with innovative procedures for structure determination, X-ray diffraction techniques are applied not only to elucidate three-dimensional architecture (static picture) but also to address structure-function dynamics. With the emerging trends in this field the central role of X-ray crystallography in biological and medical sciences looks highly promising, exciting and rewarding. Some recent trends are:

10.6.1 Fast Data Collection and Structure Determination

The strategy of intensity data collection of macromolecules by diffractometry, though very accurate, is inefficient because a single diffracted spot is measured at a time. Radiation damage to macromolecules also sets a limitation to this approach to data collection, in addition to the time required to collect a large number of reflections for native as well as for derivative crystals. Diffractometry is still used to collect intensity data for medium range macromolecules. The rotation-oscillation method of data collection allows the recording of a large number of reflections at a single exposure on films with better spatial resolution than diffractometry (Fig. 10.11). But, this advantage is nullified by the time spent in developing and processing of films and digitising the intensity data of recorded spots. Area detectors have been developed to record intensity data of a large number of reflections at a time, electronically, thereby incorporating the advantages of rotation-oscillation. Further improvements to obtain better spatial resolution of spots could make this methodology an ideal method for fast and accurate intensity data collection.

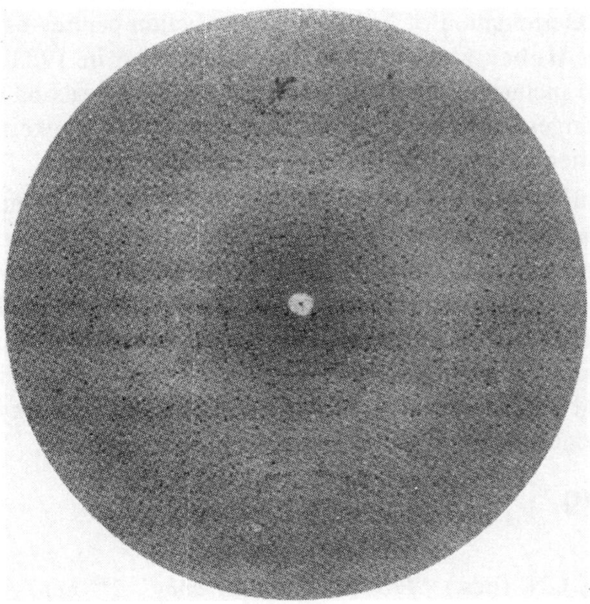

Fig. 10.11 0.5° oscillation photograph of a single-crystalline macromolecule (Sesbania mosaic virus). (Photograph: Courtesy of Prof. M.R.N. Murthy, Indian Institute of Science, Bangalore.)

Availability of X-rays of very high intensity from a synchrotron radiation source enormously reduces the time-span of intensity data collection. The synchrotron radiation source is not only a source for fast data collection because of the intensity but it is also a source of polychromatic X-rays. This allows the selection of desired wavelengths to take optimum advantage of the anomalous scattering of anomalous scatterers in macromolecules to resolve the phase problem by the anomalous dispersion method. Current trends such as Multi-wavelength Anomalous Dispersion (MAD) etc., are indicative of the renewed interest and efforts in this field.

10.6.2 Time-resolved Crystallography

Another important development that has emerged from the availability of the synchrotron X-ray source is 'millisecond macromolecular crystallography'. This development (still in the emerging stage) promises to extend the range of structure dynamics studies of biochemical problems by X-ray crystallography. Because of the high intensity of X-rays from the synchrotron radiation source, the time span of intensity data collection can be reduced to seconds; and under certain orientations of the crystal, a complete intensity data set can be obtained in a single exposure (by the **Laue** method). This technique was applied successfully to study the binding of the ligand, maltoheptose, to glycogen phosphorylase-**b** by taking Laue diffraction sets before, during and after the addition of the ligand, and it was established conclusively that the ligand, maltoheptose, was bound to the glycogen-storage site of the enzyme. More such developments are expected in the coming years.

Both Laue photography and the anomalous dispersion technique are not new methodologies, but with the availability of the right source and concommitant advances in instrumentation, new frontiers of research have been opened up to exploit the full potentiality of X-ray diffraction techniques. For example, the structure determination of bacterial photoreaction centres has been accomplished for which the group led by **Huber** was awarded the Nobel Prize in 1988. However, the structure represents only the static picture of the ground state of the photoreaction centres. Subtle structural (or conformational) changes that occur during charge-separation (excited state) could now be attempted by the application of 'time-resolved' X-ray crystallography.

Detailed structural information on biological molecules is crucial for interpreting their functional features. X-ray diffraction techniques, augmented by NMR and semi-empirical methods of structure prediction and computer modelling can yield the necessary structural information on the structure dynamics and functions of biologically important compounds. In addition, data augmented by recent biochemical methods, such as site-directed mutagenesis, establishing the functional roles of specific groups in a macromolecule, by selectively altering the groups and monitoring the altered functions, has enormously increased the crucial and central role X-ray crystallography will play in the coming years in all fields of biological and medical sciences.

Suggested Reading

Books

Blundell, T.L. & Johnson, L.N. (Eds.). "*Protein Crystallography*". 2nd Edn. Academic Press/New York. (1976).

Drenth, J. "*Principles of Protein Crystallography*". Springerverlag/New York. (1994).

Glusker, J.P. & Trueblood, K.N. "*Crystal Structure Analysis: A Primer*" 2nd Edn. Oxford University Press/ New York. (1985).

Glusker, J.P., Lewis, M. & Rossi, M. "*Crystal Structure Analysis for Chemists and Biologists*". Wiley-VCH/ New York. (1994).

Ladd, M.F.C. and Palmer, R.A. (Eds.). "*Structure Determination by X-ray Crystallography*". Plenum Press/ New York. (1978).

Lattman, E.E. & Amzel, L.M. in "*Methods of Experimental Physics*." Academic Press/New York (1982).

Parthasarathy, S. & Glusker, J.P. (Eds.). "*Aspects of Crystallography in Molecular Biology*". New Age Intl./ New Delhi. (1997).

Stout, G.H. & Jensen, L.H. (Eds.). "*X-ray Structure Determination: A Practical Guide*". McMillan Press/ London. (1968).

Articles

Atwood, D. et al. *"Tunable coherent X-rays"*. Science, 228; 1265. (1985).

Davies, D.R. & Padlan, E.A. *"Antibody-antigen Complexes"*. Am. Rev. Biochem. 59; 439. (1990).

Engel, D.W. et al. *"Absolute Configuration on and the Determination of the Anomalous X-ray Scattering Factors for Light Atoms"*. Nature Physical Sci. 229 (1); 28. (1971).

Hajdu, J. et al. *"Millisecond X-ray Diffraction and the First Electron Density Map from Laue Photographs of a Protein Crystal". Nature.* 329; 178. (1987).

Narayanan, P. *"X-ray Structure Determination of Biomolecules",* Phys. Edu. 6; 217. (1989).

Petsko, G.A. & Ringe, D. *"Fluctuations in Protein Structure from X-ray Diffraction"*. Ann. Rev. Biophys. Bioeng. 13; 331. (1984).

11 Microscopies

Synopsis: The primary objective of any microscopy is the magnification of a microscopic specimen to provide structural detail; the detail available is closely related to the resolving power of the lens system of the microscope. In optical microscopy the resolution limit is set by diffraction effects. Dark-field, phase-contrast, polarising and fluorescence are types of optical microscopy, making use of several forms of interaction of light with matter.

The wave property of electrons and that they can be 'focussed' by magnetic lens systems are the basis for electron microscopy, which enables molecular-level resolutions. Transmission, scanning and scanning-transmission electron microscopic techniques are employed in the ultrastructural studies of biological systems. Specimen preparation is a tedious but a crucial task in electron microscopic studies. Fixing, embedding and sectioning are carried out by normal as well as cryogenic freeze-fracturing methods.

Emerging trends in microscopies, such as confocal microscopy, contrast-enhanced video-microscopy coupled with computer-controlled digital imaging processes and new inventions, such as scanning-tunnelling, atomic-force and interfacial-force microscopies, have improved clarity and resolution and have enabled access to image-manipulation and thereby extending their applications in biological as well as other fields.

Microscopies—optical, fluorescence and electron—are regular features in all biological and medical laboratories and hospitals. Many new developments and inventions in conventional microscopic methods, and emergence of surface-probe microscopies have added better resolution and sophistication to the image analysis.

Combination of spectroscopic parameters (functional) of imaging entities with microscopic (imaging) methods are ushered in *"functional imaging"* microscopy. In addition, emergence of sophisticated imaging methods (tomographies) have brought forward the *biomedical engineering* to the forefront in medical sciences for diagnosis and therapy.

11.1 OPTICAL MICROSCOPY

Light is electromagnetic radiation; each region of the electromagnetic spectrum is characterised by its wavelength and the visible region is in the 4000 – 7000 Å (400 – 700 nm) range. Light in the visible region can be made to refract, due to a change in the velocity of light as it travels from one medium to another with a different refractive index and the purpose of any lens system is to converge/diverge light rays. In optical microscopy, a light source is used to illuminate the object and a lens system focuses the transmitted/scattered light rays from the object, forming its image. The essential purpose of any microscope is to enlarge the image of an object without sacrificing sharpness of the image or finer structural details.

11.1.1 Features of Optical Microscopy

Essential feature of an optical microscope are: Rays of light from an illuminating source are condensed on to the object by a condensor lens (**Köhler** illumination) and the objective lens system collects the light coming from the object (transmitted/scattered) to form a real, inverted and enlarged image of the object (primary image). The eyepiece lens system (or projector lens) focuses the divergent rays of the primary image on the retinal plane/projector screen as a magnified virtual image (Fig. 11.1).

Image Formation. Image formation in a microscope is a two-step process (double Fourier transformation). Abbe's theory of image formation elucidates that during the first stage, a spatial light wave spectrum of the object is created on the focal plane of the objective lens as a result of interference of the radiation emitted from various points of the object (Ist Fourier transformation). During the second stage, another Fourier transformation occurs as a result of the interference of the light radiated from various points of the focal plane of the objective lens (Fig. 11.2).

Magnification. In optical microscopy, the magnification is the sum of the primary (magnification in the primary image plane by the objective lens) and secondary (by the eyepiece lens) magnifications. The magnification power of a microscope is set by the resolving power of the lens system and over magnification beyond the resolution limit does not provide any further information and, therefore, is called "useless" magnification.

Resolving Power. The resolving power of a microscope is the limit of ability of the lens system to distinguish two-point objects, separated above a minimum distance, as distinct and separate identities. This minimum distance is the resolution. Diffraction effects, an inherent wave property, set the resolution limit to optical microscopy. When waves are restrained by any aperture, as in the case of a lens, the diffraction pattern of an object gives rise to imperfections due to the lens itself (series termination effects). The image of a point object has a central diffraction maximum surrounded by

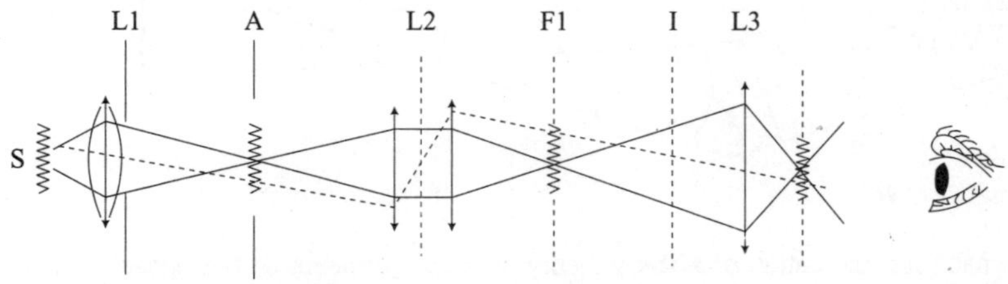

Fig. 11.1a Schematic ray diagram of an optical microscope.
S = source of light; L1 = collector lens; L2 = condensor (objective) lens; L3 = eyepiece lens; A = aperture; F1 = backfocal plane; I = image (primary)

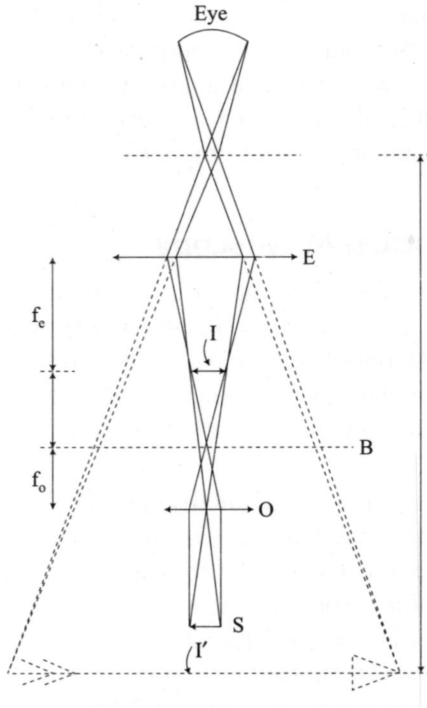

Fig. 11.1b Optics of image formation in optical microscopy.
f_e and f_o = Focal lengths of eyepiece & objective lenses;
E and *O* = eyepiece & objective lenses;
B = backfocal plane; *S* = specimen;
I and *I′* = Intermediate (primary) & final (virtual) images.

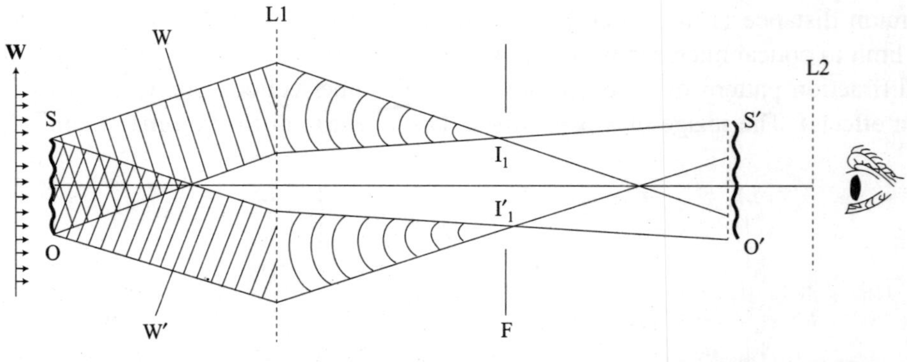

Fig. 11.2 Schematic representation of **Abbe's** theory of image formation in the optical microscope. **W** = incident wave; *OS* = object; *W* and *W′* = scattered waves; L_1 and L_2 = lens planes; I_1 & I_1' = images; *O′S′* = virtual image of the object.

ripples and in a lens such a pattern, called an **Airy pattern**, shows a central intensity maxim (disc) surrounded by haloes (due to ripples) (Fig. 11.3a). When two points of an object are close together, the intensity profiles of the two in the image overlap. The criterion for resolution is that for two point objects the diffraction maxim of the first point (**Airy disc**) is coincident with the first minimum of the diffraction pattern of the second point (**Rayleigh** criterion (Fig. 11.3b). According to this criterion the resolution is given by

$$\text{Resolution} = \frac{0.61\,\lambda}{\text{NA}} \quad ...(11.1a)$$

$$\text{NA (numerical aperture)} = \mathbf{n}\,.\,\sin\alpha \quad ...(11.1b)$$

where λ is wavelength of the light wave, **n** = refractive index of the lens and 2α = angle of the light cone reaching the lens.

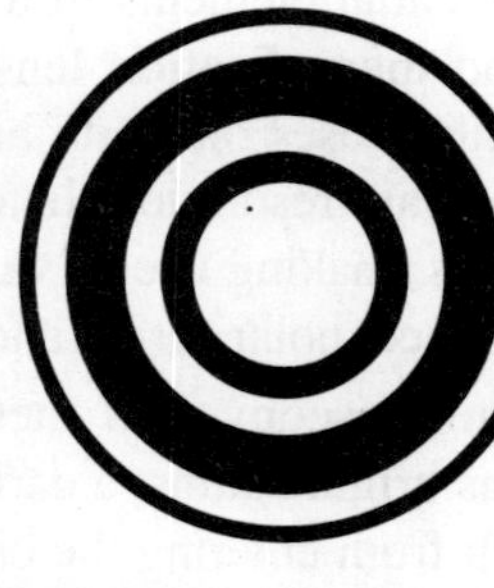

Fig. 11.3a Airy's ring pattern due to diffraction of a point object.

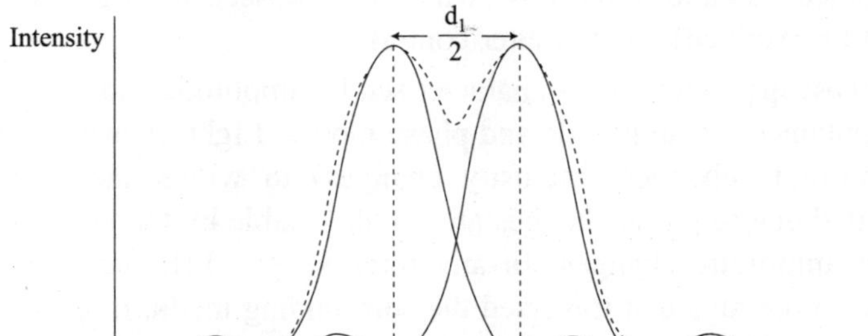

Fig. 11.3b Rayleigh's resolution criterion for resolving images of two-point objects. At the Rayleigh resolution limit, the maximum intensity from one point object coincides with the first minimum of the other; and the resolution limit is $\frac{d_1}{2}$.

The numerical aperture (NA) of the objective lens is a measure of the ability of a lens system to gather light and resolve fine specimen detail at a fixed object distance. Total resolving power of a microscope depends on the numerical apertures of all the lens systems, and by the wavelength of the light, both incident and diffracted. The higher the numerical aperture of the total system, the better the resolution.

Resolution in optical microscopy cannot be lower than 2000 Å (200 nm) due to the diffraction limit and no further improvement is possible. Below the resolution, the image of two point objects appears to be a single object. As resolution is wavelength dependent, one of the ways of increasing the resolving power of a microscope is to look for shorter-wavelength microscopy, the ultimate being X-ray microscopy. This has not been feasible as there is no lens system (yet) available to focus X-rays (the refractive index is 1.0 in all media).

11.1.2 Types of Optical Microscopy

Compound Microscope: The compound microscope, with suitably-designed lens systems to minimise various lens defects—spherical and chromatic aberrations and astigmatism etc., is a routine feature in practically all laboratories. Essential elements of a compound microscope are: condenser lens, specimen stage, and objective and magnification lens system with better optical elements (*apchromatic* lenses). Oil immersion method uses oils with higher refractive index than air between the specimen and the objective lens to increase resolution. In addition, there are other types of optical microscopes for general and specific needs, making use of various forms of interaction of light with matter—absorption/transmission, interference, polarisation and fluorescence.

Dark-Field Microscopy. Dark-field microscopy is a method of detecting objects with better background contrast and objects appear as bright against a dark background. The dark-field condition is achieved by preventing the direct beam from entering the objective lens with the help of a suitable condenser lens system, so that only scattered light enters the microscope objective and contributes to image formation (the direct beam lies outside the aperture of the objective lens) (Fig. 11.4). In dark-field microscopes, objects smaller than those which can be seen in bright-field microscopy can be detected (but cannot be resolved) due to better contrast.

Phase-contrast Microscopy. A wave is characterised by amplitude and phase (in addition to λ) and information can be obtained via amplitude and phase modes. Light transmitted through an absorbing medium shows amplitude changes (intensity changes) to which the human eye is sensitive. Information transmitted due to phase changes are not detectable by the eye, unless they are by some means translated to amplitude changes. In any microscopy, diffracted rays emerging from the specimen interfere with the rays that traversed the surrounding medium. In the case of a transparent specimen, where there is little amplitude change, the resulting intensity of the light beam can be magnified if the light waves passing through the medium are advanced or retarded by an additional wavelength $\left(\frac{1}{4}\lambda\right)$ with respect to the light waves passing through the transparent specimen. This is the underlying principle of phase-contrast microscopy elucidated by **Zernike**.

A phase-contrast microscope should have special features–(i) a condenser which can incorporate an annular diaphragm (or series of diaphragms) and (ii) an annular plate (quarter-wave plate of phase-plate) in the backfocal plane of the objective lens. The arrangement is such that the direct beam emerging through the medium (surrounding the specimen) passes only through the trough

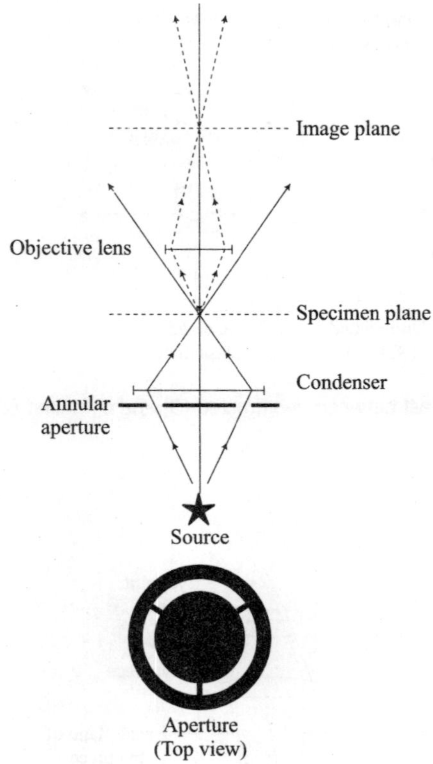

Fig. 11.4 Ray diagram for dark-field microscopy.

of the phase-plate that is coated with light-absorbing material to attenuate its intensity and the light beam that emerges from the specimen passes through the quarter-wave plate to undergo an additional quarter-wave path difference with respect to the direct beam. The interference results thus from the direct beam from the objective and the light retarded or advanced to a total of 1/2 wavelength.

The microscopic setup is to introduce phase difference of half-wavelength between the undeviated (undiffracted; 0th order) light beam passing through the medium (referred to as surround (S) wave), and the diffracted beam (D-wave) so that interference occurs between the S- and D-wavefronts (that is, conversion of phase changes into amplitude changes) (Fig. 11.5).

Phase contrast microscope is useful in observation of living cells (e.g., cell division, chromosome banding etc.).

Polarised-light Microscopy. Ordinary light is unpolarised electromagnetic radiations, with electric vectors vibrating normal to the direction of ray propagation with azimuthal random distribution. Some materials will transmit (or reflect) light with electric vectors oriented in particular directions(s) and then the transmitted (or reflected) light will be polarised. Plane-polarised light has electric vectors oriented in a particular direction normal to the direction of wave propagation. An optically-isotropic material is one where the velocity of light is the same in all directions and an anisotropic material

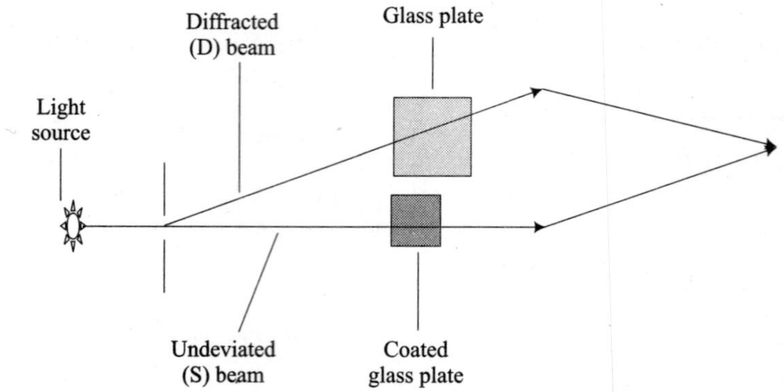

Fig. 11.5a Principle of phase-contrast between undiffracted (S) and diffracted (D) wavefronts.

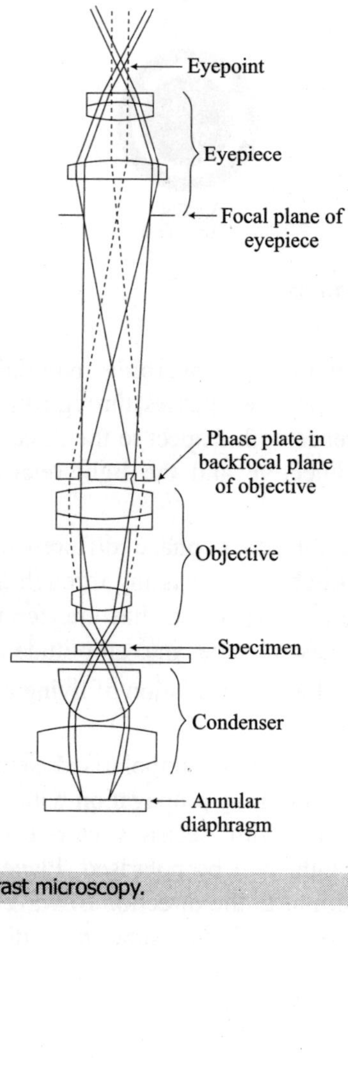

Fig. 11.5b Ray diagram for phase-contrast microscopy.

exhibits optical birefringence—having different velocities of light in different directions. The birefringence of anisotropic materials is utilised in polarised-light microscopy.

A polarising microscope is simply a normal microscope with a polariser-analyser attachment. When polarised light passes through a polarised sheet (film or plate), called polariser, the emerging light is linearly polarised (Fig. 11.6). If the linearly-polarised light is passed through another polariser (called analyser), whose pass-axis is at 90° to the pass-axis of the polariser ("crossed" position), then no light emerges through the analyser and the field of view will be dark. Non-birefringent specimens (glass; uniaxial crystals) placed between the "crossed" polariser-analyser setup will not alter the field of view. In the case of birefringent specimens, their indices of refraction depend on the direction of the incident light beam and the direction of the electric vector. Therefore, their presence between the "crossed" polariser-analyzer setup causes rotation of polarisation vector, and some light passes through the analyser and the field of view will not be dark. Polarised-light microscopy is based on this principle.

One of the polarising sheets, the polariser, is set below the specimen and the other sheet, the analyser, is placed above the objective lens. They can be rotated with respect to each other and, in the crossed position, where the axes of polariser will not be transmitted through the analyser and, therefore, the field of view will be dark. If an isotropic specimen is on the specimen stage at this crossed position the field of view will still be dark, but if it is a birefringent specimen it changes the plane of polarisation and that component will be transmitted through the analyser and observed in the eyepiece. The polarising microscope is employed to study birefringence properties of materials — intrinsic birefringence, for molecules like proteins, nucleic acids etc., form-birefringence (asymmetric arrangement of particles in medium of different refractive index) and strain-birefringence, such as that in the nucleus, connective tissues etc.

Fluorescence Microscopy. Fluorescence microscopy is based on the fluorescence of certain materials, fluorophores, that absorb shorter-wavelength light and emit at longer-wavelengths (see Chapter 8 for more details). Intense fluorescence occurs if the wavelength of the incident light is closer to the peak of the excitation curve of the fluorophore. Essential components of fluorescence microscopy are: a light source; suitable fluorophores; filters to select appropriate excitation (band-pass excitation filters) and to remove those wavelengths which are not effective in creating fluorescence (short and long-pass filters); and fluorescence-detecting setup.

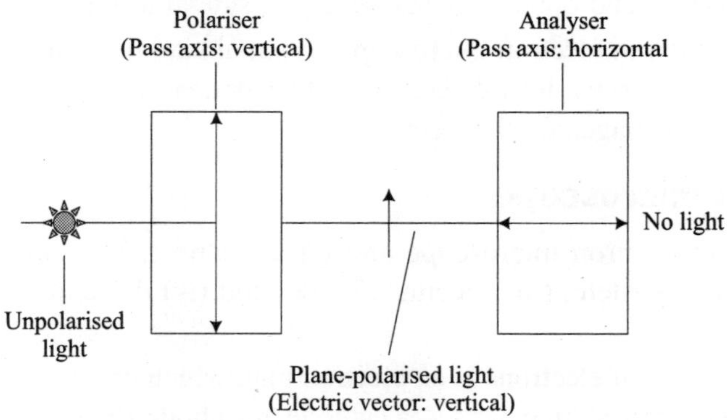

Fig. 11.6 Operational principle of "Crossed" polaroids.

Fluorescence microscopy is an important technique in cytochemical studies and because of the great sensitivity of the technique, the presence of weak fluorophores in the cell (such as vitamin A, riboflavin etc.) can be detected. In immunofluorescence, antigen-antibody precipitation reactions can be observed with the introduction of fluorophores or colloidal gold with silver enrichment. It is an important technique in chromosomal studies (banding techniques). Acridine orange and ethidium bromide are used in intercalation and flow cytometric studies of nucleic acids. In microfluorimetry, a detector, incorporated as an additional component, is used to read/scan the emitted light from the fluorescent specimens. This technique has become a versatile method in various fluorescence microscopic studies.

Fluorescence microscopy is an extremely sensitive imaging technique; further it has become versatile imaging technique that has wide applications in structural and functional characterisation of biological specimens. Many new developments, such as fluorescence lifetime imaging (FLIM), fluorescence resonance energy transfer (FRET; see Chapter 8), "wavelength ratiometric imaging", which combine spectral parameters of imaged species with the imaging techniques, have made the fluorescence microscopy an advanced imaging technique.

11.2 ELECTRON MICROSCOPY

As the resolution attainable in optical microscopy cannot be better than 2000 Å (200 nm), microscopies employing shorter-wavelength sources are the alternative for ultrastructure and molecular structure determination. X-ray microscopy has not been practically feasible due to lack of X-ray lens systems. In addition, X-rays have other debilitating properties that might prove a limiting factor in biological sciences, even if such a microscope were to be developed. Major problems in X-rays are high penetration and radiation damage. Due to high penetration power, the interaction of X-rays with matter is poor and therefore recording time would be exceedingly long. The radiation damage to specimens would seriously curtail its use in the biological sciences. Electrons do have advantages over X-rays in some of these aspects and electron microscopy has become an indispensible tool in ultrastructural studies of biological systems. De Broglie's hypothesis of the wave nature of electrons (matter) was the impetus for the development of electron microscopy. Electron microscopy is essentially similar to optical microscopy in its basic principles, in spite of the differences in mechanisms in image formation and other aspects in these microscopies. Diagrammatic similarities between electron and optical microscopies are shown in Fig. 11.7.

Theoretically, resolution in electron microscopy can attain atomic resolution, but in practice it is never realised for various reasons: lens defects and other optical imperfections, problems in sample preparation and radiation damage, artefacts etc.

11.2.1 Electron Microscope

Basic components of an electron microscope are: (i) the source, i.e., electron gun, (ii) focussing system, i.e., magnetic lens system, (iii) specimen holder and (iv) detecting system—either projector lenses or detectors.

Electron Gun. The source of electrons is an electron gun, which can be either (i) thermionic or (ii) field emission type. It comprises three parts— a cathode, a cathode shield **(Wehnelt cylinder)** and an anode.

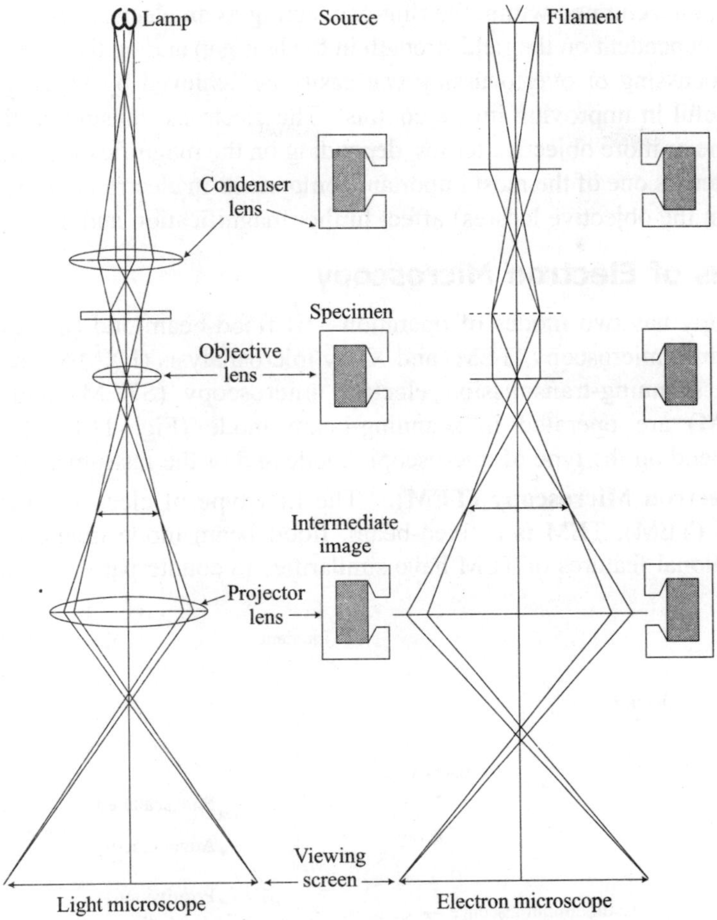

Fig. 11.7 Equivalence of ray diagrams for light and electron microscopies.

(i) Thermionic Gun: In a thermionic emitter gun a V-shaped tungsten filament (cathode) is heated and the emitted electrons are channelled through an aperture towards the anode by a cathode shield between the filament and the anode, maintained at a negative voltage (negative bias) with respect to the filament. The design is similar to a vacuum tube triode and the Wehnelt cylinder functions as a week converging lens.

(ii) Field Emission Gun: In a field emission gun a very high voltage is applied to the cathode, which is a very fine pointed single crystal of tungsten, to 'bleed out' electrons from the tip of the cathode by field emission. The electron beam emerging from the field emission gun has several advantages—the source is of smaller area and brighter and highly coherent, resulting in reduction in chromatic aberration and longer cathode life.

(iii) Electron Lens. Focussing of electrons is achieved by a magnetic lens system, which consists of a coil of wire encased in a soft-iron shroud, allowing only a small ring-shaped gap exposed in the interior side. Passage of current through the coil creates a magnetic field along the axial direction of the electron beam, concentrated within the ring-gap, acting as an electron lens. As the focal length of a magnetic lens is dependent on the field strength in the lens gap and on the velocity of electron beam, focussing, underfocussing or overfocussing can easily be achieved by varying the current through coils, which is useful in improving image contrast. The electrons transmitted through the specimen are focussed by one or more objective lenses, depending on the magnification required. Therefore, the objective lens system is one of the most important components in electron microscopy and any defects or imperfections in the objective lens(es) affect further magnification and other image features.

11.2.2 Types of Electron Microscopy

Electron microscopy has two modes of operation—(i) flood-beam and (ii) scanning-beam modes. Transmission electron microscopy (TEM) and X-ray microanalysis (EXMA) are performed in flood-beam mode and scanning-transmission electron microscopy (STEM) and Scanning Electron Microscopy (SEM) are operated in scanning-beam mode (Fig. 11.8). Detection and image reconstruction depend on the type of microscopic mode and/or the instrumental probe used.

Transmission Electron Microscopy (TEM). The first type of electron microscope built was a transmission type (TEM). TEM is a fixed-beam, flood-beam mode image transmission method. Though the operational features of TEM have similarities to counterparts in optical microscopy in a

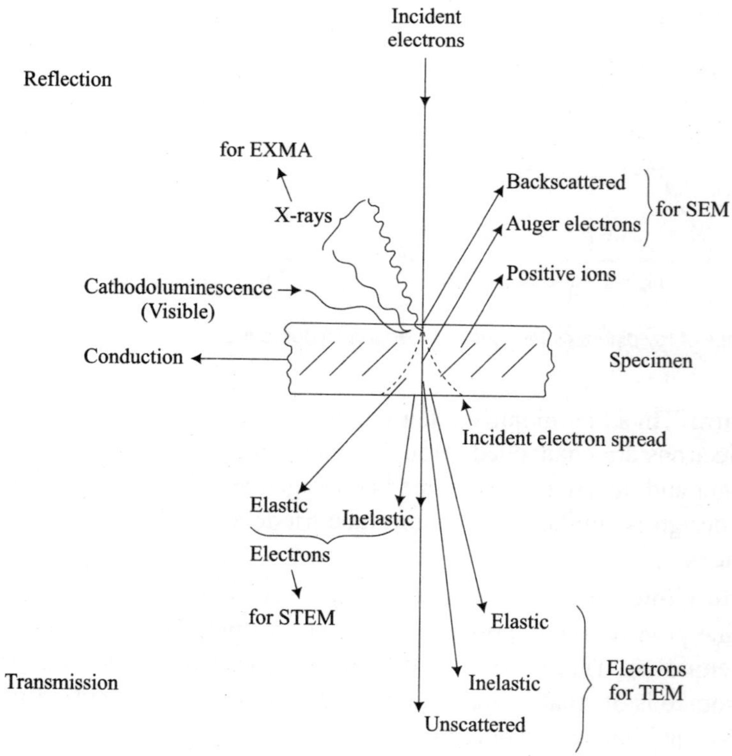

Fig. 11.8 Types of interactions between high-energy electrons and matter.

general way, the mechanisms of image formation and other aspects are different in both cases. The electron beam from an electron gun is focussed to illuminate the maximum area of the specimen (analogous to Köhler illumination) and the transmitted electrons are focussed by objective lenses and the magnified image of the specimen is projected onto a fluorescent screen or recording system (Fig. 11.9).

The mechanism of image formation in optical and electron microscopies is different. While it is due to absorption in optical microscopy, in electron microscopy it is primarily due to scattering of electrons by matter (the specimen). Therefore, the higher the atomic number of an atom the better its scattering power; biological specimens need heavy-atom 'staining' for image formation and contrast.

Since the penetration power of electrons through matter is very low, specimen sections should be ultrathin (<500 Å) for transmission of electrons and, therefore, specimen preparation requirements are stringent, especially in TEM. In TEM, the transmitted electrons are focussed by lens systems and lens defects, such as spherical, chromatic and astigmatic aberrations which are inherent in any lens system, affect the resolution, magnification and clarity of the image. *Whereas the resolution limit is set by diffraction effects in optical microscopy, it is set by lens defects in electron microscopy.* In addition, depth of field is relatively short and it is not possible to view surface topography of the specimen.

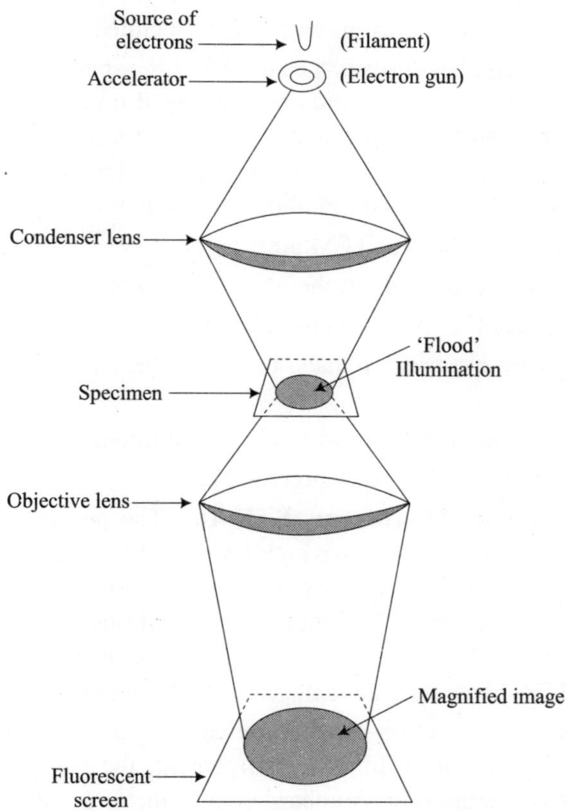

Fig. 11.9 Diagrammatic representation of flood-beam mode TEM.

In microscopy, contrast generally means amplitude contrast, due to intensity changes, which can be detected by the eye. To increase the amplitude contrast the specimens are 'stained' with heavy-atom solutions in electron microscopy. Phase effects due to interference can also partially increase contrast and in electron microscopy this is achieved by slight under-focussing and over-focussing. Such a phase contrast is possible only for ultrathin specimens; thick specimens exhibit only amplitude contrast.

A dark-field condition can also be achieved in TEM to obtain dark-field images. A bright-image is formed if the unscattered electrons of the incident beam interfere with the scattered electrons in image formation. If the unscattered beam is prevented from reaching the objective lens (by tilting of the specimen stage or by displacement of the objective aperture), the image then is formed from only the scattered electrons (dark-field imaging).

Scanning Electron Microscopy (SEM). When electrons of the incident beam strike the specimen several types of interactions occur (as indicated in Fig. 11.8). Most electrons will be transmitted, some of them will be back-scattered at the surface (elastic scattering) and others can interact with the electrons at the surface of the specimen and excite the emission of secondary electrons from the specimen surface. Secondary electron emission is very sensitive to the topography of the specimen and is the type of signal usually used to image the specimen in the SEM. These secondary electrons can provide information on surface features of the specimen while the elastically-scattered electrons provide information on heavy metal atoms present in the specimen.

In SEM, an ultrafine electron beam scans the surface of the specimen, the scattered electrons are collected by detecting counters and the features are displayed on a TV screen or cathode ray tube (CRT) or analysed etc. Non-conducting specimens should be made electrically conducting to avoid build up of charge at the surface. Therefore, non-conducting biological specimens are given a thin coating of conducting material (gold-palladium alloy) by vacuum evaporation.

Some of the advantages of SEM over TEM are:

1. Image formation is achieved without the objective lenses, so lens defects are minimised.
2. Small spot size reduces the spherical aberration.
3. Image formation is by electrical signals, therefore, the contrast, magnification etc., can be manipulated.
4. Good surface topography and depth of field (gives three-dimensional appearance).
5. Elaborate sample preparation is not a prerequisite.

Scanning-Transmission Electron Microscopy (STEM). The penetration power of electrons can be increased by increasing the voltages to ultra-high level. STEM also requires ultra-high vacuum in the electron gun system (field emission gun). As in the case of SEM, a fine electron probe is used to scan in a raster across a thin specimen. The transmitted electrons are collected in detectors placed beneath the specimen and the features are displayed on a TV screen or CRT. Thicker specimens can be examined by STEM. STEM has same advantages as SEM over TEM.

Electron Probe Analysis. The electron microscope can be converted to carryout non-destructive elemental analysis of solid specimens. In this arrangement, the electron beam from the electron microscope is focussed on the target spot. Bombardment of the target spot by a high-energy electron beam gives rise to characteristic X-rays, depending on the nature of the target spot under bombardment. The scattered X-rays are focussed, by a curved crystal, onto an X-ray detector and their energies analysed. Point-by-point microanalysis can be achieved by scanning the specimen in a raster. This method, X-ray microanalysis (EXMA), is useful for specimens with atoms of high atomic mass.

11.2.3 Specimen Preparation

Specimen preparation is an important part in any electron microscopic work. Ultimately, the quality of results cannot be better than the quality of the specimen prepared. The preparatory work includes preservation, embedding and sectioning and other additional features. Chemical fixation and freeze-drying are the common methods.

Fixation. The aim of fixation is to process the specimen in such a way that every detail of cellular ultrastructure is preserved as it was in the living cell. First, the dissected tissue is fixed well before the setting in of postmortem degradation of the tissues. This is done by treating the tissues first with glutaraldehyde followed by 'post fixation' in osmium tetroxide. Glutaraldehyde fixes proteins and some polysaccharides. Osmium tetroxide fixes unsaturated lipids, reacting with the double bonds. It is usually used as a post-fixative after the tissue has been fixed in an aldehyde. The best fixation is carried out with solutions that are slightly hypertonic. The fixed tissues must be 'hardened' to make them withstand the microtome sectioning processes. This is done by embedding the specimen in a hardening media. For additional contrast (in addition to osmium tetroxide), the specimen is treated with heavy-metal solutions (uranyl acetate, phosphotungstic acid etc.).

Embedding. The fixed specimen is subjected to dehydration before the impregnation of resins. The impregnated resins (araldite, epon etc.) are polymerised and hardened (at 60°C) and the hardened epoxy block with the specimen (matrix) is sectioned by an ultramicrotome to obtain the desired thin sections (<500 Å).

11.2.4 Cryo-Electron Microscopy

Some of the problems and shortcomings in electron microscopy, namely problems arising due to chemical fixation, dehydration and staining, can be minimised by cryo-preparation methods. Freezing techniques should ensure minimal freeze-damage and water should be 'sublimated', The freeze-fracturing technique has become routine in membrane research. For this, the sample is rapidly frozen, then fractured under high vacuum and water is 'sublimated'. Fractured specimens are 'shadowed' with heavy-metal staining. The application of cytochemical labelling together with cryo-electron microscopy has helped in the study of various biological processes at the molecular level (with colloidal gold and enzyme markers).

11.3 EMERGING TRENDS IN MICROSCOPY

Several innovative methods in microscopy and related areas, technological advances and new inventions in recent years have greatly enhanced the potential and range of applications of microscopic methods. Some of the latest advances in microscopy are mentioned below.

11.3.1 Differential-Interference Contrast (DIC) Microscopy

Differential-interference contrast (DIC) (Nomarski) microscopy combines the principles of polarised-light microscopy with interference microscopy (Fig. 11.10). A beam of plane-polarised light is split into two beams by a Nomarski prism (called the condenser prism, which is a beam-splitter), oriented

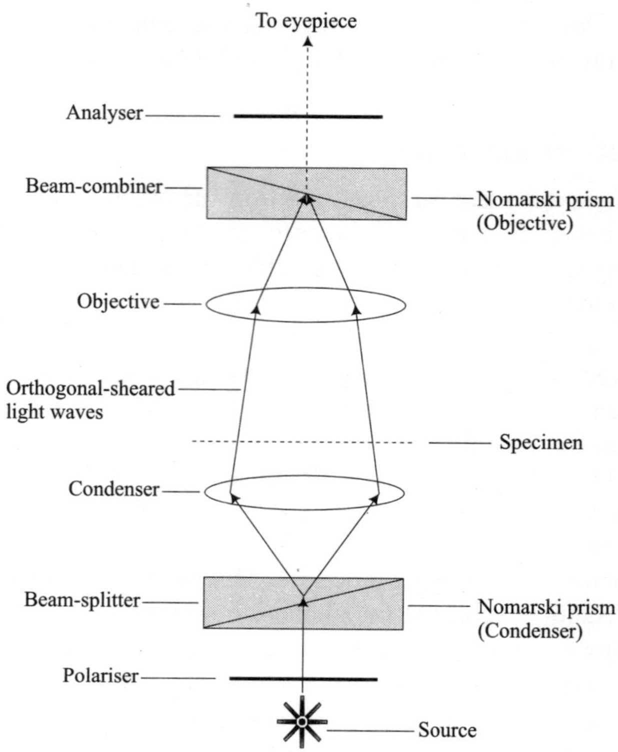

Fig. 11.10 Ray diagram of Differential-Interference Contrast (DIC) microscopy.

with optical axis 45° angle to the polarised light, and placed beneath the condenser. After emerging from the Nomarski prism at the aperture plane, the two coherent beams travel through the specimen before being collected by the lens. The objective Nomarski prism (a beam-combiner), inverted in orientation with respect to the condenser prism, and placed behind the objective lens, recombines the two light beams. Nomarski microscope is superior to phase-contrast microscope with better contrast and resolution.

11.3.2 Confocal Microscopy

In conventional microscopy the images of the out-of-focus planes are blurred and, in addition, the scattered light from the planes reduces the contrast, that is, the information content. Images from conventional microscopy are generally two-dimensional flat pictures. One of the disadvantages in conventional microscopies is the lack of depth of focus. This disadvantage is actually an asset in confocal microscopy. The basic principle of confocal microscopy is—scanning of a specimen plane by plane (optical sectioning). A beam of light is focussed on a single spot on a plane of a specimen and only images on the plane are scanned, filtering out the images from all out-of-focus regions (from) above and below the plane of focus, with pinhole aperture (Fig. 11.11). By this technique information due to blurred images is eliminated, allowing only reflected light rays from the plane of focus to be recorded. A complete three-dimensional image of the specimen can be built by sampling

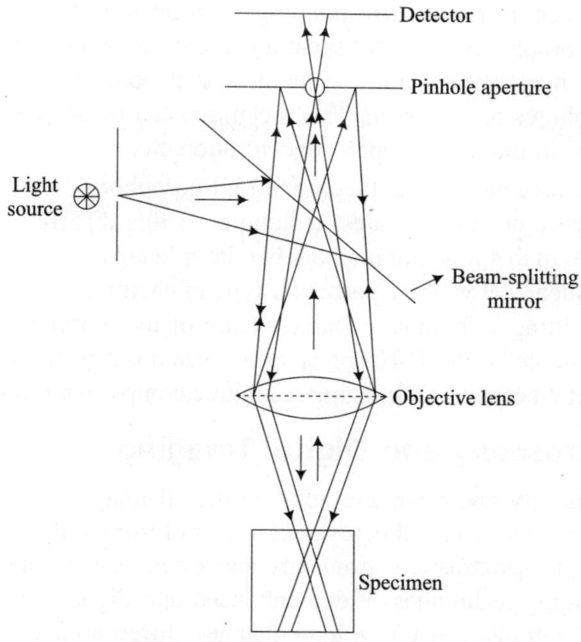

Fig. 11.11 Ray diagram for confocal microscopy.

point by point, section by section, all the volume elements (**voxels**). Computer-controlled automation enables a continuous scan of a plane section of the specimen down an axis and with a step-down mechanism to focus different sections down that axis. The three-dimensional data can be stored and images can be video-reconstructed or manipulated. A confocal microscope, coupled to a computer is, in effect, a three-dimensional image building apparatus. This microscopy has been found very useful in fluorescent specimen studies, by imparting high-detection sensitivities against a dark background.

Two-photon (multi-photon) confocal fluorescence microscopy minimises photo-bleaching and cell damage and provides three-dimensional images of biological specimens in real-time without killing them.

11.3.3 Spectroscopy-Microscopy Imaging

Fluorescence resonance energy transfer (FRET), *Fluorescence lifetime imaging (FLIM)*, *Wavelength-ratiometric imaging* and *Coherent anti-Stokes–Raman scattering (CARS)* are some of the new imaging methods that combine spectroscopy (function) with microscopy (imaging)— functional imaging.

Resonance energy transfer (RET) occurs between two fluorophores when their transition dipoles are approximately oriented and their spectra overlap (when they are within the Forster's distance <10 nm; see Chapter 8). By tagging auto-fluorescent molecules, such as green fluorescent proteins (one protein molecule tagged to a FRET donor and the other to a FRET acceptor), it is possible to map, by fluorescence microscopy, many spatio-temporal dynamics of cellular and molecular species.

Fluorescence lifetime imaging (FLIM) can be used to distinguish between two fluorophores in tissues and permits the spatial imaging and dynamics of biological species. Another fluorescence method, known as "wavelength-ratiometric imaging" can be used to study tissue characteristics. Tissues are doped with fluorophores that have spectra that change in a predictable way. By comparing fluorescence-intensifying images at several wavelengths, it is possible to produce a detailed map of the distribution of fluorophores in the tissue. This technique can be adapted to image local chemical and physical environments in the tissue— pH, concentration etc.

Imaging of interior living cells without tagging those fluorophores, which might interfere with the cell's biochemistry is highly desirable. Latest technique in this direction is coherent anti-Stokes–Raman scattering (CARS). In this imaging method, two laser beams are sent into a cell, at frequencies that differ exactly the frequency at which a particular type of chemical bond in the cell is excited. The excited bond vibrates, emitting a frequency characteristic of its vibrational mode. For example, by tuning C-H vibration in the cells, the C-H bonds would stand out in the resulting image (*chemical microscopy*). CARS might emerge as an imaging technique complement to fluorescence microscopy.

11.3.4 Video-Microscopy and Digital Imaging

Advances in video-microscopy and computer-coupled digital image processing have improved the resolution in optical microscopy to a level equivalent to that of immunofluorescent techniques. This is achieved by improving video contrast electronically and by increasing the signal to noise ratios by digital filtering and averaging techniques. Video-enhanced and digital image processing techniques (such as video-enhanced contrast (VEC), video-enhanced differential interference contrast (VEC-DIC), digital imaging microscopy (DIM) etc.) are increasingly employed to detect even weakly-fluorescing molecules inside living cells; colloid gold and other particles in the specimen under observation; in the studies of structural and assembly dynamics in endocytosis, chromosome banding, micelles, liposomes, membranes, acting bundles and microtubules; also mass transport and other transient phenomena in living cells.

11.3.5 NMR Microscopy

In conventional optical microscopy, specimens have to be sectioned, which is a destructive process and stained, which would mar the structural details and might introduce artifacts. In comparison, NMR microscopy is a non-destructive technique, as there is no physical damage to the specimen. The desired section (plane) of the specimen is sectioned by the magnetic field gradient and that sectioned is scanned and the data on series of sections are collected and processed by image reconstruction procedures. Many of the static (structural) and dynamic (functional) features of cells, tissues and organs—fluid flow in leaves and xylem, blood vessels and other mass transport phenomena—can be studied by NMR microscopy. NMR microscopy can become an additional and perhaps a superior imaging technique that would augment other imaging methods in biology and medicine.

11.3.6 X-ray Microscopy

Soft X-ray microscopy is another imaging method for morphological and chemical analysis of minute structures. It utilises *Fresnel zone plate* gratings as focussing elements. It provides a new method with better contrast for imaging thick-hydrated biological specimens— cells, sub-cellular particles (~10 nm resolution) without requiring elaborate specimen preparation.

11.3.7 Scanning-Probe Microscopies

Some of the recent developments in high-resolution imaging techniques, such as scanning-tunneling, atomic-force and interfacial-force microscopies, do not formally represent conventional microscopies. These are an emerging family of scanning-probe microscopies, which can resolve surface structure at atomic-resolution.

Scanning-Tunneling Microscopy (STM). The principle and the operation of scanning-tunneling microscopy (STM) are quite simple (Fig. 11.12). An extremely sharp conductive tip traces the surface contours to atomic resolution. The tip (Pt-Rh needle) can be moved in three dimensions with an XYZ-piezoelectric translator attached to the needle. The gap between the tip and the surface of the sample is controlled by a voltage applied to the piezoelement. As the X-piezo scans laterally across the surface, the feedback circuit adjusts the voltage to the Z-piezo to raise or lower the tip to keep the tunneling current constant. When the tip of the needle is brought close to the surface, the electrons tunnel across and modify the potential difference. The up and down Z-piezo motion is a function of tunneling current and scanning needle, during motion, follows the contours of the specimen (that is, the electron map is correlated to the tunneling current). Differences in height along the contours are plotted as a function of tunneling current. STM has wide variety of applications in surface structure studies. Its applications in the biological field are limited to specimens with conducting surfaces. Coating with conducting media may result in recording the coated media by STM rather than the specimen.

Atomic-Force Microscopy (AFM). Atomic-Force Microscopy (AFM), which does not require a conducting surface can be employed to study surface structures of biological specimens. In this method, a fine-tipped diamond is attached to a spring in the form of a cantilever. The cantilever is fixed to a small piezo element, called a modulating piezo, which is used to drive the cantilever beam at its resonant frequency. The feedback system prevents the tip from touching and damaging the sample. The small repulsive forces (tracking forces), between the diamond tip and the sample, translated into minute deflections of the cantilever are recorded. Atomic resolutions can be reached by AFM.

The AFM can be used to produce not only spatially-resolved surface maps (topologies) but also to probe the structure function of single, intact biological molecules. These studies include measurement of adhesion and stiffness (*polymer-brush* maps) as well as the force required for

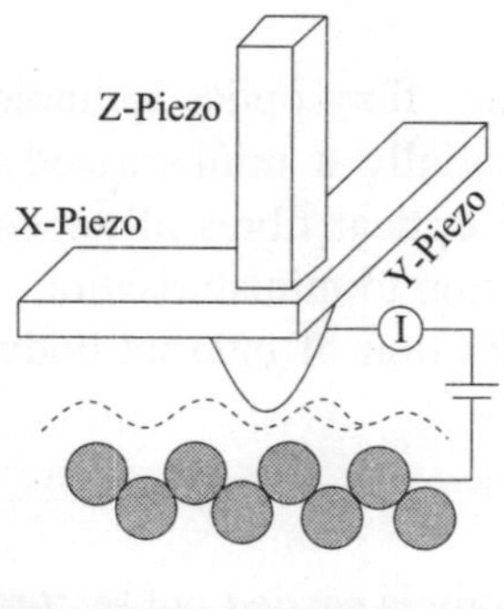

Fig. 11.12 Schematic diagram for scanning–tunneling microscopy.

breaking specific ligand-receptor bonds. By applying specific force (positional force) to specimen and pulling the cantilever tip away from the specimen, minute forces (10^{-9} N) can be measured. That is, the activity of a single molecule (nanoscale technology) can be determined directly by AFM.

Interfacial-Force Microscopy (IFM). Interfacial-force microscopy (IFM) is similar in principle to atomic force microscopy (AFM). IFM is used to study mechanical and adhesive properties of molecules at atomic resolution. In IFM, a piezo tube acts as an XYZ-translator of the sample, and a feedback force sensor incorporates a self-balancing scheme to eliminate the mechanical instability that is inherent in the displacement-type force sensors, such as AFM. The interfacial forces are plotted as a function of separation of the probe from the sample. This microscopy is also useful in surface structure studies of biological specimens.

Near-field Scanning Optical Microscopy (NSOM): In all aperture-based physical probes that confine the waves, diffraction ripples (series termination effects) set the resolution limit. Near-field scanning optical microscopy (NSOM) circumvents this limitation by aperture-less near-field probing rather than beam focussing. In this aperture-less microscope an antenna tip that serves as field-concentrating scatter centre replaces the aperture. An AFM tip scatters radiation onto a detector, while the sample is being scanned in close proximity. Thus, the NSOM has made a sub-micrometer scale microscopy a practical proposition. This method should enable imaging chemical composition at nanometer resolution *(chemical microscopy).*

Recent developments in scanning-probe microscopic techniques hold a great promise in medical and biological sciences.

11.4 DIAGNOSTIC IMAGING

Imaging methods have been a part of diagnostics and therapy in medical fields and hospitals. Recent advances in biomedical engineering and instrumentation have brought in a tremendous progress into medical sciences. These techniques have become regular features in diagnostic and therapeutic protocols, with minimal invasion and damage to body organs, microsurgery and better patient care. For example, laser-based biomedical instrumentation, with new innovations, has found wide applications in many areas in medicine. Another area of progress in imaging methods is the emergence of tomographies. Emergence of tomographies has been due to integration of computers and electronics for imaging body parts.

11.4.1 Endoscopy

A significant contribution of laser-based fibre optics technology in medicine is endoscopy (*endo* = inside; *scopy* = viewing). It is essentially a miniaturised microscopic technique. Lasers allow coherence and small spot size, while optical fibres allow cold-light illumination, advancement in photography and transmission, in addition to miniaturisation of instrumentation. Due to these factors the endoscopic methods allow visualisation of internal body parts, biopsy tests and microsurgical procedures with minimal invasion.

11.4.2 Nuclear Imaging

Nuclear imaging of organs or whole body is carried out by γ-ray emitting radiotracers (radionuclides) and detected by scintillation counters. Distribution of radionuclides spatially over organs or body regions can be imaged (*scintigraphy*) not only to obtain morphological details but also for evaluating organ function. Concentration of organ-specific radionuclides provides areas of activity/abnormalities, providing an early indication of onset of disease/cancer.

Nuclear Tomography: Radionuclides are used in nuclear tomography to measure radioactivity emission by the patient from different orientations. *Single-photon emission computed tomography (SPECT)* permits absolute three-dimensional localisation of radionuclide distribution. The data acquisition consists of a set of multiple sectional digital imaging data (pixels), collected at regular intervals and stored in a computer and then processed by mage reconstruction algorithms to obtain three-dimensional (voxel) images. Though the image resolution is poor (~8 nm), it is simple and inexpensive.

Positron emission tomography (PET) is another nuclear imaging method that employs positron-emitting radionuclides for imaging. The technique is based on the principle that the position of a positron-emitting radionuclide can be precisely determined as annihilation of a positron-electron pair results in emission of two γ-ray photons in diametrically opposite directions (180°). PET is a superior technique with better resolution and sensitivity, but expensive. It can provide *in vivo* quantitative information on the blood flow, metabolite activity, regional tissue biochemistry etc. PET is also used to study brain function. A rapid sequence of brain images is used to analyse the response of brain to various stimuli and pinpoint areas of activity/ abnormality.

11.4.3 X-ray Tomography

Though X-ray radiography (Roentgenography) has been in use as diagnostic tool over a century, and still in use, to produce images in the form of gray shadows, the basic procedure has remained the same— they are two-dimensional projections of three-dimensional objects. The emergence of X-ray tomography, known as *computer-assisted tomography (CAT)*, or simply *computed tomography* (*CT*) has been due to integration of computers to enable the storage of vast amount of digital (pixel) data and processing and manipulation.

The basic operational features are (similar is the case with MRI): a pencil of collimated X-ray beam is directed through an organ of the patient on the examination table and the attenuated beam is detected by an array of detectors on the gantry. At each step a single section (slice) perpendicular to the body axis is scanned as the patient is moved linearly through the gantry (Fig. 11.13), and the attenuation data are stored in a computer. Image reconstruction methods are employed to transform 2-D digital (pixel) data into 3-D images (voxels). CT provides high-resolution, high-clarity morphological details, and three-dimensional localisation of lesions and surrounding structures. Therefore, it has now become a regular method in diagnostic radiography.

11.4.4 Magnetic Resonance Imaging (MRI)

Magnetic resonance imaging (MRI)/NMR Tomography/NMR Zeugmatography is based on the observation that if a constant field gradient is applied along a certain direction the strength of the magnetic field varies (increases or decreases) linearly along this direction. The spin response of a given nucleus along this gradient depends on the spatial position of the nucleus. By introducing magnetic field gradients on a static magnetic field, nuclear spins can be brought into resonance in different parts of a material at different frequencies. This way nuclear magnetisation in a given volume of element can be measured. The procedure is to apply a field gradient on a section of the

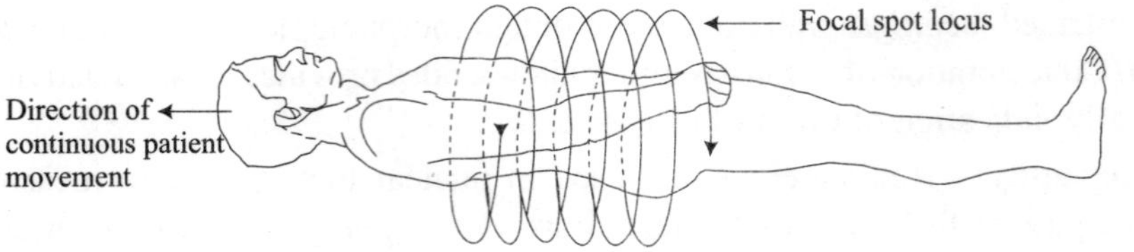

Fig. 11.13 Data acquisition by Chiral X-ray Tomography.

sample and the NMR single is recorded along the axis of the gradient. The gradient (axis) is rotated by 180/*n* times to obtain a spin-density image of the entire plane (similar to X-ray tomography). Both in-plane (pixel), and volume (voxel) imaging NMR methods are now available.

NMR-imaging techniques are nondestructive and are increasingly used in medicine—imaging of anatomical sections (spin-imaging), blood and fluid flow, mass transport and other dynamic features.

Imaging of structures like the skull, skeletal frame and spinal cord is better with NMR imaging than with X-ray tomography (bones hinder X-rays). In addition, the skull, skeletal frame and spinal cord etc., are static (in patients) and imaging is fast and simpler. Diagnosis of infarction, tumour and other damage to the spinal cord and brain and other organs (which might not be readily identifiable by X-ray methods) can be done easily. Contrast in different tissues can be altered by introducing contrast agents. In addition, penetration of these contrast agents through different portions of the cell(s) can be used to monitor the damage to cells like kidney, liver, spleen, and damage to the blood-brain barrier. NMR angiography deals with the imaging of arteries, veins and capillaries to monitor any defects in blood circulation. Paramagnetic contrast agents injected into the blood stream enhance the contrast between blood vessels and the surroundings. NMR signals from tissues are essentially determined by the water content. An important exception is in the case of lipids and this aspect is the basis for the diagnosis of lymph nodes from surrounding fat tissue (mammography). Similarly, in comparison with the MRI of healthy tissues, alterations in the levels of metabolites in the tumour cells can be monitored which would be of importance in cancer biology.

Functional MRI (fMRI): The MRI signal depends not only on the density of tissue, but also on its chemical composition. Therefore, any changes in the chemical composition of that tissue (organ) leading to functional changes can be visualised by dynamic MRI (fMRI). Diagmagnetic/paramagnetic nature of oxy- and deoxy-hemoglobin can be exploited to monitor blood oxygen levels in the brain and brain energy consumption. Thus, the cellular basis of neuronal functions can be correlated to the imaging studies of living organisms by fMRI and PET.

11.4.5 Sonography

Medical sonography relies on ultrasonic sound waves in the 1-15 MHz frequency ranges. Effects of ultrasound in body organs are thermal, cavitation, and shearing and intracellular motion. Various scan modes and Doppler-shift methods can be employed in diagnostic ultrasonic imaging protocols (echocardiography; colour flow Doppler sonography; cephalography). A comparison of tomographies (Table 11.1) illustrates the advantages and limitations of these techniques.

Table 11.1 Comparison of tomographic techniques.

Technique	*Features*	*Advantages*	*Limitations*
Nuclear:			
SPECT	Multi-planar imaging; good signal strength.	Morphological studies; inexpensive.	Attenuation problems; poor resolution; radiation hazard.
PET	Multi-planar imaging; good signal strength; high resolution; high sensitivity.	No attenuation problems; morphological and functional studies.	Positron-emitting nuclides needed; radiation hazard; expensive.
X-ray (CT)	Multi-planar imaging; good resolution; fast acquisition; high sensitivity.	No attenuation problems; morphological features.	Radiation hazard.
MRI	Multi-planar imaging; high resolution.	Morphological and functional studies; no radiation risks.	Attenuation problems; poor sensitivity; expensive.
Sonography	Real-time multi-planar imaging; soft tissue imaging.	Detection of foreign bodies; no radiation hazards; inexpensive.	Poor resolution; surface impedance problems.

Suggested Reading

Books

Bashe, T., et al. (Eds.). "*Single-Molecule Optical Detection, Imaging and Spectroscopy*". VCH/Weinheim. (1996).

Bradbury, S. & Bracegirdle, B. "*Introduction to Light Microscopy*". BIOS Scientific Publishers/Oxford, UK. (1998).

Bradbury, S. & Evennett, P. "*Contrast Techniques in Light Microscopy*". BIOS Scientific Publishers/Oxford, UK. (1996).

Collett, E. "*Polarized Light: Fundamentals and Applications*". Marcel Dekker/New York. (1993).

Cundall, R.B. & Dale, R.E., (Eds.). "*Time-resolved Fluorescence Spectroscopy in Biochemistry and Biology*". Plenum Press/New York. (1983).

Duke, P.J. & Michette, A.G. "*Modern Microscopies: Techniques and Applications*". Plenum Press/New York. (1990).

Glaubert, A.M. (Ed.). "*Practical Methods in Electron Microscopy*". Vols. 1 & 2. Elsevier/New York. (1974).

Goldstein, D. "*Understanding the Light Microscope*". Academic Press/New York. (1999).

Goodhew, P.J. "*Electron Microscopy and Analysis*". Wykeham Science Series/London. (1975).

Herman, B. "*Fluorescence Microscopy*". 2nd Edn. BIOS Scientific Publishers/Oxford, UK. (1998).

Herman, B & Jacobson, K (Eds.). "*Optical Microscopy for Biology*". Wiley/New York. (1990).

Herman, B. & Lemaster, J.J. (Eds.). "*Light Microscopy: Emerging Methods and Applications*". Academic Press/New York. (1993).

Johannessen, J.V. (Ed.). "*Electron Microscopy in Human Medicine*". McGraw Hill/New York. (1978).

Lacy, A.J. (Ed.). "*Light Microscopy Biology in A Practical Approach*". IRL Press/Oxford. (1989).

Lee, R.E. "*Scanning Electron Microscopy and X-ray Microanalysis*". Prentice-Hall/Englewood, NJ. (1993).

Meek, G.A. "*Practical Electron Microscopy for Biologists*". 2nd Edn. John Wiley & Sons/New York. (1976).

Morris, P.G. "*NMR Imaging in Medicine and Biology*". Oxford University Press/Oxford, UK. (1986).

Narayanan, P. "*Clinical Biophysics: Principles and Techniques*". Bhalani Pubs/Mumbai. (2001).

Narayanan, P. "*Biomedical Imaging*". New Age Int Pubs/New Delhi. (in Print).

Partain, C.L., et al., (Eds.). "*Magnetic Resonance Imaging*". Saunders/Philadelphia. (1988).

Ploem, J.S. & Tanke, H.J. "*Introduction to Fluorescence Microscopy*". Oxford University Press/Oxford, UK. (1987).

Powley, J.B. (Ed.). "*Handbook of Biological Confocal Microscopy*". Plenum Press/New York. (1990).

Rawlins, D.J. "*Light Microscopy*". BIOS Scientific Pubs/Oxford, UK. (1992).

Rizzuto, R. & Fasolato, C. (Eds.). "*Imaging Living Cells*". Springerverlag/Berlin. (1998).

Slayter, E.M. "*Light and Electron Microscopy*". 2nd Edn. Cambridge University Press/Cambridge, UK. (1993).

Spencer, M. "*Fundamentals of Light Microscopy*". Cambridge University Press/New York. (1982).

Stark, D.D. & Bradley, W.G. "*Magnetic Resonance Imaging*". Mosby/St. Louis. (1988).

Strocio, J.A. & Cairer, W.J. (Eds.). "*Scanning Tunneling Microscopy*". Academic Press/New York. (1993).

Taylor, D.L., et al., (Eds.). "*Applications of Fluorescence in the Biomedical Sciences*". Liss/New York. (1986).

Valeur, B. & Bronchon, J.C., (Eds.). "*New Trends in Fluorescence Spectroscopy: Applications to Chemical and Life Sciences*". Springerverlag/Berlin. (2001).

Watt, I.M. "*The Principles and Practice of Electron Microscopy*". Cambridge Press/London. (1985).

Weakly, B.S. "*A Beginner's Handbook in Biological Electron Microscopy*". Churchill Livingston/London. (1972).

Wiesendanger, R. "*Scanning Probe Microscopy and Spectroscopy Methods and Applications*". Cambridge University Press/Cambridge, UK. (1994).

Wilson, T. (Ed.). "*Confocal Microscopy*". Academic Press/San Diego, CA. (1990).

Zhu, X. & Ohtsu, M., (Eds.). "*Near-Field Optics: Principles and Applications*". World Scientific Publishing/ Singapore. (2000).

Articles

Betzig, E., Trautmann, J.K. Harris, T.D. "*Breaking the Diffraction Barrier: Optical microscopy at nanometric scale*". Science, 251; 1468. (1991).

Binnig, G. & Rohrer, H. "*The Scanning-tunneling Microscope*". Sci. Amer. 253 (2); 50. (1985).

Binnig, G., Quarte, C.F. & Gerber, C. "*Atomic Force Microscopy*". Phys. Rev. Lett. 56; 930. (1986).

Cheng, J-X., et al. "*Laser Scanning Anti-Stokes-Raman Scattering Microscopy and Applications to Cell Biology*". Biophys. J. 83; 502. (2002).

Dunn, R. "*Near-field scanning optical microscopy*". Chem. Rev. 99; 2891. (1999).

Engel, A., Schoenenberger, C.A. & Miller, D.J. "*High Resolution Imaging of Native Biological Samples Using Scanning-probe Microscopy*". Curr. Opin. Struct. Biol. 7(2); 279. (1997).

Hansma, P.K. et al. "*Scanning-tunneling Microscopy and Atomic Force Microscopy: Applications to Biology and Technology*." Science 242; 209 (1988).

Hounsfield, G.N. "*Computed Medical Imaging*". Science. 210; 22. (1980).

Houston, J.E. & Michalske, T.A. "*The Interfacial Force Microscope*". Nature 256 ; 266. (1992).

Howles, J.E., Kirz, J. & Sayre, D. "*X-ray Microscopes*". Sci. Amer. 264 (2); 42. (1991).

Kirz, J., Jocobso, C. & Howells, M. "*Soft X-ray Microscopes and Their Biological Applications*". Q. Rev. Biophys. 28; 33. (1995).

Knoll, B. & Keilman, F. "*Near-field Probing for Vibrational Absorption for Chemical Microscopy*". Nature, 399; 134. (1999).

Lichtman, J.W. "*Confocal Microscopy*". Sci. Amer. 271 (2); 40. (1994).

Miles, M. "*Scanning Probe Microscopy: Probing the Future*". Science, 277; 1845. (1997).

Millar, D.P. "*Time-resolved Fluorescence Spectroscopy*". Curr. Opin. Struct. Biol. 6; 637. (1996).

Nie, S., Chiu, D.T. & Zare, R.M. "*Probing Individual Molecules with Confocal Fluorescence Microscopy*". Science, 266; 1018. (1994).

Pupples, G.J., et al. "*Studying Single Living Cells and Chromosomes by Confocal Raman Microspectroscopy*". Nature, 347; 301. (1990).

Selvin, P.R. "*Fluorescence Resonance Energy Transfer*". Meth. Enzymol. 246; 300. (1995).

Sereno, M.I. "*Brain Mapping in Animals and Humans*". Curr. Opin. Neurobiol. 8 (2); 188. (1998).

Zumbusch, A, Holtom, G.R & Xie, X.S. "*Three-dimensional Vibrational Imaging by Coherent Anti-Stokes-Raman Scattering*". Phy. Rev. Lett. 82(20); 4142. (1999).

12 Lasers and Holography

Synopsis: Laser is the acronym for Light Amplification by Stimulated Emission of Radiation. Lasers provide highly monochromatic and the least divergent beams of light radiation. Development of laser technology has brought coherent optics to the fore and with it applications of a far-reaching nature. Applications encompass fundamental studies in physics, chemistry, biology, medicine and electroptics as well as a host of applications in industry and defence-oriented technologies.

Holography is a process that records and reproduces the 'phase' information of light waves scattered from the object. Holography records the object wave itself and not the image of the object as is the case with photography. Advances in holography have-been due to the availability of coherent light sources (lasers). Holographic techniques find applications in wide ranging areas—precision measurement, three-dimensional visual perception, performing arts and information-related topics etc. Holography holds much promise in medical and biological fields.

12.1 LASERS

Lasers provide highly monochromatic and coherent light radiation. To understand laser operation certain fundamental properties of light, namely polarisation coherence, stimulated absorption and emission etc., have to be understood.

12.1.1 Photons

Light is electromagnetic radiation and is emitted by various processes—

(i) atomic/molecular transitions from excited to lower energy states,

(ii) deceleration of charged particles, such as electrons (Bremsstrahlung),

(iii) particle annihilation (e.g., electron-positron pair collision) and

(iv) when the velocity of charged particles is faster than the electromagnetic waves in the medium (**Cerenkov** radiation).

In all these processes, radiation of light is emitted as photons. The photon has neither rest mass nor electric charge and the velocity of a photon is equal to the velocity of light in vacuum.

Characteristics of a photon state are defined by the photon energy, E, the direction of its momentum $\vec{\mathbf{P}}$, and its polarisation, α. The energy, E, is

$$E = C\vec{\mathbf{P}} = \mathbf{C}\sqrt{P_x^2 + P_y^2 + P_z^2} \qquad ...(12.1)$$

where C is the velocity of light and P_x, P_y and P_z are components of the momentum vector, $\vec{\mathbf{P}}$ along x, y and z directions, respectively. Any change in any one of the parameters-momentum (P_x, P_y, P_z), and α (polarisation)-leads to a change in the photon state. Two photons are discernible only when they are in different states. The characteristics of a photon state can be described by the characteristics of a plane-polarised monochromatic light wave. That is, a monochromatic plane-polarised light wave is an ensemble of photons existing in the same state.

The behaviour of light is due to the statistical behaviour of photons, i.e., their statistical properties. All fundamental particles in nature can be classified under two distinct categories, according to their statistical properties as the outcome of their spin states. Particles which belong to the first category (with non-integral spin, 1/2, 3/2 etc.) follow one-state-one-particle rule, without exception. That is, once a particle state is occupied by a particle, no other particle of the same type is allowed to occupy the same state (according to **Pauli's** exclusion principle). The particles belonging to this category are called **fermions**, as they obey Fermi-Dirac statistics. As an example, electrons are fermions and they exhibit exclusivity and any energy state cannot be occupied by more than a specific number of electrons (2, 8, 18 ... etc.). Particles with zero or integral spin (0, 1, 2), such as photons, belong to the second category and they obey **Bose-Einstein** statistics and are called **bosons** (in honour of the Indian physicist, **S.N. Bose**). The main characteristic of bosons is that they can populate any state in unlimited numbers. As a matter of utmost importance, the higher the population density of a state, the higher the probability of bosons occupying the same state (birds of the same feather flocking together). This feature of bosons is of fundamental importance in the optical phenomena of photons.

12.1.2 Coherence

A monochromatic (single wavelength), plane-polarised light wave is an ensemble of photons in the same state and is, therefore, ideally coherent. Such a wave is represented by

$$\delta(x, t) = A \cos\left\{\left(\nu t - \frac{x}{\lambda}\right)\right\} \qquad ...(12.2)$$

where δ is displacement from the mean position in the x-direction, A is amplitude of the wave, t is time and ν and λ are frequency and wavelength, respectively. All photons in this (ideal) state have the same energy, E, the same direction of momentum, $\vec{\mathbf{P}}$, and the same polarisation, α. Coherence of a wave can be defined as a measure of its ability to form interference patterns with similar waves. The higher the coherence, the greater the fringe contrast of the interference patterns of the waves. That is, occurrence of interference of waves is an indication of the degree of their coherence. The degree of coherence, expressed by the degree of interference fringe contrast, ξ, is given by

$$\xi = \frac{I_1 - I_2}{I_1 + I_2} \quad \text{...(12.3a)}$$

where

I_1 is the intensity at the centre of a bright interference fringe and

I_2 is the intensity at the centre of a dark interference fringe.

If $\xi = 1$ Maximum coherence (Maximum fringe contrast),

$\xi = 0$ Least coherence (Least fringe contrast).

In actual situations, radiation with $\frac{\Delta\nu}{\nu_0} \ll 1$ is nearly monochromatic and the degree of non-chromaticity, ε, is given by

$$\varepsilon = \frac{\Delta\nu}{\nu_0} \quad \text{...(12.3b)}$$

Lasers provide optical radiations with a high degree of coherence; wavetrains of 100 km long as compared to mm long wavetrains from ordinary light sources. (The length of a wavetrain is an indication of its coherence). The age of optics prior to the invention of lasers can be termed the age of '**incoherent**' optics or '**semi-coherent**' optics.

12.1.3 Absorption/Emission

A deep understanding of the processes of absorption and emission of light and the application of this understanding have been responsible for the advent of lasers.

Spontaneous absorption and emission of light are natural processes. In spontaneous processes, the energy of transition for absorption for two energy levels, $E_{12} = E_1 - E_2$, is the same as for spontaneous emission, $E_{21} = E_2 - E_1$, (that is $E_{12} = E_{21}$). However, these processes occur in two different photon states; direction of momentum, $\vec{\mathbf{P}}$ and polarisation are different. Therefore, spontaneous emission is an incoherent process. In contrast, in stimulated emission, the incident photon may induce the atom/molecule in the upper energy level, E_2, to fall to a lower energy level, E_m (metastable state), thereby inducing the emission of a photon (induced-photon). In this process, the induced-photon and the incident photon will have the same energy, same direction of momentum and identical polarisation and the stimulated wave train is in phase with the initial wave train. That is, both the photons are in the same photon state. If there are many atoms populated in a higher energy level, E_2, than at a lower energy level, E_1, then a photon of frequency, $\nu = E_2 - E_1$, can induce the transition ($E_2 \to E_1$) in many atoms. That is, one primary photon can initiate an avalanche of secondary photons (with the same direction of momentum and identical polarisation). The situation is analogous to a group of birds sitting on a string and falling all at once as the string snaps. This is the phenomenon of stimulated emission of light, postulated by **Einstein** in 1917. The spontaneous and stimulated emission coefficients, A_{21} and B_{21} are related, according to the Einstein equation

$$A_{21} = \frac{8\pi h \nu_{21}^3}{C^3} B_{21} \qquad ...(12.4)$$

The practical realisation of this induced emission, namely the laser, became feasible in the 1960s.

12.1.4 Lasing Action

The important part for laser action is the availability of an active medium (solid, liquid or gas) that contains active centres (atoms/ion/molecules) with a special system of energy levels. Under normal conditions, the lower energy levels, E_1 are more populated than the higher energy levels, E_2 (populations $n_1 > n_2$) and, therefore, absorption process predominates. To achieve laser action, stimulated emission should prevail; that is, higher energy levels (E_2) should be more populated than the lower energy levels, $E_1 (n_2 > n_1)$. This situation is called population inversion (Fig. 12.1). An inverted population for lasing level 1 and 2 is a necessary condition for stimulated emission (**Lasing action**).

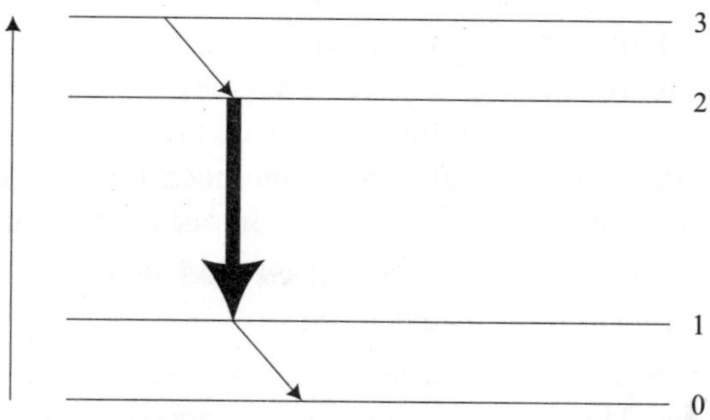

Fig. 12.1 Energy level diagram of active species.
0 = Ground level; 1 & 2 are lasing levels; 3 = pumping level.

12.1.5 Features of a Laser System

The general features of a laser system are given in Fig. 12.2. These include (1) active medium, (2) pumping system (for population invention), (3) resonator (for optical feedback) and (4) intractivity elements (for a high degree monochromaticity).

Active medium. There are a variety of active media (laser materials) available, which can be broadly classified under — (a) solid state, (b) liquid, (c) gas and (d) chemical lasers.

(a) *Solid state lasers:* They are also called doped-insulator lasers, consisting of active media (glass or crystal), doped with ions. The first laser developed was a solid-state ruby laser — a thin rod of crystalline alumina, Al_2O_3, doped with chromium ions, Cr^{3+}. It is a 3-level energy scheme laser emitting at wavelength 694 nm.

A widely used solid-state laser is an yttrium-aluminium garnet laser, Nd:YAG, which corresponds to a 4-level laser scheme. The active medium (lasant), yttrium-aluminium garnet, $Y_3Al_5O_{12}$, is doped with 3% neodymium ions, Nd^{3+}, which act as active centres. The laser operates in the infrared region ($\lambda = 1060$ nm).

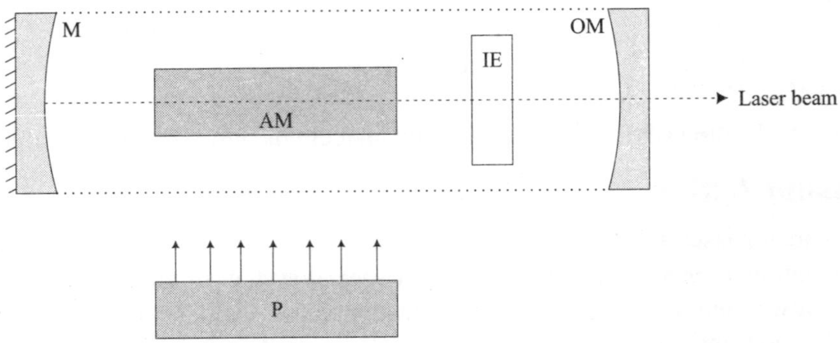

Fig. 12.2 Functional components of a laser.
AM = Active material (ruby rod, gas tube etc.).
P = Pumping system (flash lamps, dc arc-discharge etc.).
M = Total reflecting resonator mirror. OM = Output mirror.
IE = Additional interactivity element (prism, Brewster-angle windows etc.).

These lasers employ light-emitting diodes (LED) and are very cheap and are compact and small. They are, therefore, employed as light sources in miniaturised equipment (fiberoptic systems). Cleavage along the crystalline planes of the diodes produces parallel planes and hence the reflecting property. This eliminates the need for polishing of optical surfaces and external mirrors. Both homo-, and hetero-junction lasers are in use.

Solid-state lasers generally operate in the pulsed mode and they emit radiation with a lower degree of coherence (wave train length ≈100 m).

(b) *Liquid lasers* : The active media in liquid lasers are organic dyes in solution or liquids doped with rare-earth ions. The organic dyes are generally acridines, anthracenes, azines, coumarins and xanthenes. All dyes are optically pumped.

In liquid lasers the concentration of active centres can easily be controlled and the dye laser output can be tuned over a wide frequency range 1350 → 680 nm). Therefore, these lasers are found useful in spectroscopic and chemical reactions and isotope separation (e.g., U^{238} → U^{235}). Also, the laser medium is relatively cheap and easy to handle.

(c) *Gas lasers* : The best coherence is found in gas lasers (wave-train length ≈100 km), due to the homogeneity achievable in gaseous media. Lasers whose lasing action is between rotational transitions emit radiations in the infrared region (orientations of the crystal, complete intensity data set can be obtained (>10 μm); when the lasing action is between vibrational levels the radiation is in the range 5 – 50 μm; and when the transitions are between electronic states, the emitted radiation is in the visible region. Some of the types of gas lasers available are: atomic, ionic, excimer and molecular lasers, depending on the active media. Most gas lasers are excited by means of an electric discharge.

(i) *Atomic Lasers* : The Helium-Neon Laser : The medium consists of a mixture of helium and neon (10:1) and the lasing acts on the atomic transitions between the levels of neon atoms. Helium plays an important role in the excitation of neon atoms, by transferring its energy to Ne atoms by resonant collisions. The He-Ne laser is excited by a stationary glow discharge from a dc current and operates in the continuous wave mode, operating at 633 nm (generally).

(ii) *Ionic Lasers* : Ion lasers are generally noble gas ions — Ar^+, Kr^+. They generally operate in the continuous mode. Whereas Ar^+ and Kr^+ emit light in the blue, green and red regions, a mixture of ions Ar^+ and Kr^+ can emit light in the blue, green, yellow and red regions.

Metal vapour lasers, such as HeCd and HeSe combine the features of atomic and ionic laser systems. They operate in the continuous mode and emit light in the visible regions.

(iii) *Excimer Lasers* : Excimer is a combination of two atoms that survive only in the excited (meta-stable) state. Decay of these excimers produces emission of radiation. The excimers are rare gas atoms (Ar_2^* , Kr_2^* , Xe_2^*), rare gas oxides (ArO^*, KrO^*, XeO^*) and rare gas halides (ArF^*, KrF^*, XeF^*).

(iv) *Molecular Lasers* : CO_2 belongs to the molecular laser category. A CO_2 laser is capable of continuously generating high powers with high efficiency. The active medium of the laser is a gas mixture of carbon dioxide and nitrogen and additives, such as helium (10:50:40) and water vapour. The CO_2 molecules act as active centres, lasing on the transitions between the vibrational levels of the electronic ground state (in the infrared region). The role of nitrogen is similar to the role of helium in the He-Ne laser (excited nitrogen molecules transfer energy to the CO_2 molecules by resonant collision).

Whereas in CO_2 lasers the transitions are between vibrational levels of CO_2 molecules resulting in a laser beam in the infrared region, the transitions in the N_2 laser are between different electronic states and the emitted radiation is in the UV-region (337 nm). The N_2 laser is very useful in photochemical reaction studies as its emitted radiation is in the UV-region.

(d) *Chemical Lasers* : Other lasers such as solid, liquid and gas lasers, are classified with reference to the active medium of the laser system. Chemical lasers are classified with reference to the method of population inversion. Any chemical laser has the general reaction

$$A + BC \rightarrow AB + C + \text{energy} \rightarrow (AB)^* + C \quad ...(12.5)$$

The energy released by the reaction serves to excite the molecule AB, to $(AB)^*$, which serves as the active medium.

Chemical lasers can produce radiant energy directly from the reacting species without the addition of external energy or external energy may be needed to initiate or sustain the reactions.

Types of lasers and their emitted radiations and spectral regions are listed in Table 12.1.

The Pumping System. A laser is triggered by initiating a pumping (excitation) system, which enables the excitation of active centres and build up of inverted population of lasing levels. Both optical and electrical pumping systems are in use. Optical pumping, employing flash tubes and lamps and spark gaps and flames, is the most versatile and is used in solid-state, liquid, gas and semi-conductor active media. Electrical pumping is done generally by means of an electric discharge and is particularly suited to gaseous active media. Pumping processes can be either continuous or pulsed, and pulsed pumping offers a number of advantages over the continuous pumping mode.

Table 12.1 Types of lasers with their spectral lines and spectral regions.

Laser type	*Wavelength λ**	*Spectral region*
Solid State		
Ruby	$\underline{694}$ nm	Red
	1.15 1.39 } μm	Infrared
Nd : YAG	1.06 μm	Infrared
Ga : As	905 nm	Infrared
Liquid State		
Dye Lasers (continuously tunable)	350 → 680 nm	Visible
Gas State		
(i) Atomic/Ionic		
He-Ne	$\underline{633}$ nm	Red
	1.8 1.15 3.39 } μm	Infrared
He-Cd	325 $\underline{442}$ } μm	Ultraviolet Blue
He-Se	460 → 650 nm	Visible
Ne	540 nm	Green
Ar^+	458 477 $\underline{488}$ 515 } nm	Violet-blue Blue Blue-green Green
Kr^+	476 520 568 $\underline{647}$ } nm	Blue Green Yellow-green Red
$Ar^+:Kr^+$	...	Visible
(ii) Molecular		
CO_2	9.6 $\underline{10.6}$ } μm	Infrared
N_2	337 nm	Ultraviolet
(iii) Excimer		
Ar*2 Kr*2 Xe*2 ArF* KrF* XeF*	125 146 172 193 248 351 }	Ultraviolet
ArO^* KrO^* XeO^* }	558 nm region	Yellow-green

*Underlined are intense lines.

Resonator Assembly. Optical feedback is achieved by a resonator assembly (optical cavity). The resonator provides selectivity over photon states. Laser oscillation is initiated by the photons spontaneously emitted by the active species. Due to the spread of these spontaneous photons over different states and hence a wide range of stimulated emissions, coherence cannot be achieved under such conditions. To reduce the spread of emitted radiation, the number of possible photon states has to be restricted by introducing selectivity over photon states. Whereas directional selectivity of the emitted radiation is achieved by obtaining the active medium in the shape of thin long rods/tubes, the selectivity of some photon states while suppressing other unwanted photon states can be realised by an optical resonator, with low losses and high quality factor, $Q = \nu/\Delta\nu$, (which is an essential element of a laser apparatus). A simple optical resonator is a pair of reflecting mirrors, set at the end faces of the rod/tube along the laser beam axis. At least one of the mirrors is made semi-transparent to serve as an output element passing the laser radiation out of the resonator system. The optical resonator cavity provides the desired selectivity of photon states by confining the possible direction of photon propagation, that is lasing action. In addition, the resonator system can suppress undesired transitions, by providing the mirrors whose reflectivity is a function of frequency.

Intractivity Elements. Selectivity of photon states and partial suppression of undesired transitions can be achieved by a resonator cavity system. However, losses due to scattering and absorption by non-active centres and other causes have to be minimised for achieving maximum coherent radiation. This is accomplished by introducing additional intractivity elements, such as additional mirrors, prisms, diffraction gratings and **Brewster-angle** windows etc. These and other additional elements control the lasing action, laser output, switching modes and other characteristics.

One of the ways of suppressing undesired transitions and lasing only desired wavelength(s) is to cover the resonator mirrors with materials which exhibit high reflectivity at the desired wavelength(s) and absorb other wavelengths (e.g., He-Ne laser allows only 630 nm and suppresses 1150 and 3390 nm wavelengths). Selection of required wavelengths from a multi-line emission can also be achieved by introducing in the resonator cavity a dispersive prism, a diffraction grating, or a **Fabry-Ferot** etalon. A configuration with a prism and a rotating mirror selects not only certain wavelengths but also narrows the line-width. Nonlinear optical methods are also used to shift the laser output frequency to other regions of the spectrum, i.e., second harmonic generation (frequency doubling) e.g., Nd line $\lambda = 1.06\ \mu$ into $\lambda = 353$ & 530 nm lines).

Another attachment to increase the coherence of radiation is Brewster-angle windows. The reflected and refracted rays of an incident ray follow the rule

$$\frac{\sin\alpha}{\sin\beta} = \mathbf{n} \qquad \text{...(12.6a)}$$

where α and β are angles of incidence and refraction, respectively, and **n** is the refractive index of the medium. The reflected and refracted rays are partially polarised normal to and in the plane of incidence, respectively. At the Brewster, or polarising angle, satisfying the condition

$$\tan\alpha = \mathbf{n} \qquad \text{...(12.6b)}$$

both reflected and refracted rays are completely polarised. Therefore, placement of output windows at the Brewster angle to the resonator-axis enables the selection of photons of certain polarisation and, secondly, exclusion of losses caused by reflection at the surfaces of the tube windows.

12.1.6 Operation Modes

Lasers can be operated in several modes—continuous lasing, pulsed lasing with controlled losses (Q-switching). The continuous mode is typical for gas lasers and requires continuous steady-state pumping of the active medium (either by steady-state discharge in a gas or continuous optical pumping). Pulsed modes are generally employed in solid-state lasers. A high concentration of light energy (giant pulses) within a short period (pico or femto second) of time can be achieved by a pulsed lasing mode. Time-dependency of laser beams can be realised by controlling losses inside the resonator (*Q*-switching), by introducing an '**optical switch**'. The optical switch in its elementary form is a chopper (a mirror or a prism) that rotates and, therefore, periodically interrupts the light beam. This mechanical *Q*-switching method is rugged and wavelength-independent, though not very efficient.

Efficient *Q*-switching can be achieved by employing electro-optic modulators (electro-optic crystals) as high speed shutters. The birefringence (difference in refractive indices) of an electro-optic crystal can be varied in a controlled manner by the application of controlled voltages across the crystal. Depending on the applied voltage, an electro-optic crystal can serve as a quarter-wave, half-wave or full-wave plate and thereby alter the polarisation of the light beam and control the amount of light transmitted in an optical system.

Acousto-optic modulators (used commonly in neodymium lasers) are based on the diffraction of light by sound waves in a medium. Sound waves are generated by a piezoelectric transducer that is attached to the medium and they create spatially periodic changes in the refractive index of the medium which act as a three-dimensional diffraction grating, modulating incident light.

Q-switching can be realised without any electronic or acoustic methods by using a dye whose absorption band is at the lasing level. In the initial stages the dye absorbs the light but as the population inversion stage is reached the dye is over-saturated ("bleached") and ceases to absorb light and *Q*-switching occurs.

Lasers are operated in continuous mode (CW) as well as in pulsed mode. Pulsed mode operation has many advantages over the continuous mode and is absolutely essential in studying time-dependent events (fast reactions). High-power, short-duration laser pulses can be obtained by '**mode-locking**'. 'Mode-locking' maintains a definite phase relationship of oscillating frequencies and for a pulse (short duration wave) many frequencies are needed and their frequencies maintained in a definite phase relationship. The type of mode-locking can be passive or active. Saturable absorption is employed in a passive mode-lock, by careful placement of the dye in the resonator cavity and adjustment of the dye concentration. This is a very simple, inexpensive and rugged method of obtaining high-power laser pulses. In 'active mode-locking' the duration the shutter is kept open matches the period of the pulse. Because of the extremely short duration of pulses (picosecond range), the peak powers in a mode-locked laser beam can be enormous. With pulsed lasers, problems connected with the continuous mode operation are virtually eliminated. Time-dependent events, such as fast molecular and chemical reactions, time-resolved spectroscopies etc., need pulsed laser sources.

12.1.7 Laser Applications

Laser applications are numerous in divergent fields and are ever-increasing. Lasers are employed in fundamental studies in physical, chemical, biological and medical sciences and have wide-ranging application in industrial, military and information-related areas. The selection of a laser depends on various factors—its spectral characteristics, its power and absorption, thermal conductivity, heat capacity and other characteristics of the material.

Due to a high degree of monochromaticity and intensity concentration, lasers have replaced other sources of light—sunlight, light bulbs, arc lamps and flames etc., in routine and research work. Many of the advances in optics—nonlinear optics and holography—have been due to the availability of highly coherent laser sources. Some of the applications are:

In Science. (a) Lasers are highly monochromatic, coherent and have a concentration of light power within narrow band-widths and hence are ideal light sources in spectroscopic studies. With lasers high resolutions and accuracies are achievable which are unthinkable with ordinary light sources.

(b) Realisation of holography was possible only with the advent of coherent light sources, that is, lasers. A pulsed laser enables the hologram recording of a moving object.

(c) Isotopic separation in nuclear power engineering (e.g., enrichment of ^{235}U from neutral uranium for nuclear fuel and weapons).

(d) Controlled thermonuclear fusion (e.g., **Argus** and **Shiva** plasma systems) for generation of alternate energy sources.

(e) Tunable lasers provide selectivity in exciting or rupturing desired states or chemical bonds. Therefore, desired chemical/biochemical reactions can be initiated, controlled and their reaction kinetics can be studied systemically.

(f) Pico- and femto-second pulsed laser sources have enabled the study of fast reactions, which are common in biology (e.g., photosynthesis, optics of vision), by time-resolved spectroscopy. Time-resolved spectroscopies (IR and Raman) can now address the structural changes and structural intermediates (transient structures) in biomacromolecules by careful analyses of the line-widths and intensity changes of the spectral profiles.

(g) Environmental pollution studies can be monitored by laser technology (concentration of smoke, dust, CO, SO_2 and other parameters etc.). Conventional techniques of pollution monitoring require sample collection and chemical analyses. In contrast, laser techniques, such as **l**ight **d**etection and **r**anging (LIDR), absorption and Raman backscattering, can yield on-the-spot and real-time data without the need for chemical analyses. As pollution data can be obtained on-the spot, environmental control can be better implemented by laser techniques.

(h) Advances in lasers have brought in host of applications in biology and medicine. In addition, single-atom lasers, *monoatomic* condensed matter (*Bose-Einstein* condensate) and *laser-traps (optical tweezers)* have enriched nonotechnology in the study of a single molecule— its position (in nm), force (in nN), and energetics (in μJ)— for exquisitely sensitive measurements on biomolecules at atomic resolution.

In Medicine. Lasers are found to be invaluable in (a) surgery, (b) therapy and (c) ophthalmology.

(a) *Laser Surgery* : The laser beam acts as a highly desirable and unique scalpel that can perform 'bloodless' surgery and also 'weld' blood vessels that are being cut. Besides, the surgery ensures minimum cellular destruction, high sterility and is painless. Ulcers, tumours and nodules are removed by thermal effects and immuno-suppressive effects. Generally, hard lasers are used for surgical procedures on hard tissues, like bone and teeth (Nd- YAG, CO_2 and excimer lasers).

(b) *Laser Therapy* : Low-power lasers (He-Ne) are used to stimulate tissue repair and healing, with minimal edema, scar formation, discomfort and pain. The use of lasers in restorative dentistry is extensive from cavity preparation to photocuring of composite resins. Lasers are used to destroy selectively decayed tooth material, without affecting normal enamel.

(c) *Laser Ophthalmology* : Lasers are used in ophthalmology—in the treatment of eye defects, like retinal detachment, cataract and glaucoma. The green beam of an Ar^+ laser is strongly absorbed by the red blood cells of the retina and consequent thermal energy enables the reattachment of the retina. A pulsed laser beam is used to cure glaucoma by destroying plugs in blood vessels feeding the eye. Excimer lasers are ideally suited to correct corneal defects (by a procedure known as Photo Refractive Keratectomy (PRK)). These, being 'cold' lasers, are free from thermal effects and abrasions are smooth and tissues of $\approx 10^{-3}$ mm thickness can be removed without any damage to underlying tissues while reshaping the curvature of the cornea.

In Industry. Properties of lasers — including very high power concentrated over a small spot ($\approx 10 - 100$ um)— find many applications in industry. Some of these are in (a) welding, (b) cutting and (c) drilling.

(a) *Laser Welding* : Laser welding has many advantages over conventional types of welding—

(i) The welding is contactless; therefore there is no possibility of introducing undesired impurities.

(ii) The welding can be performed in any medium and in accessible regions (very highly desirable in welding in vacuum and at reactive junctions).

(iii) Accurate micro-space welding is ideally suited to precision welding in thermally-sensitive environment, such as in microelectronics.

(b) *Laser Cutting* : Laser cutting has advantages, namely—

(i) A wide range of materials and surfaces can be cut (wood, glass, ceramics, cloth, metals and plastics etc.).

(ii) Precision cutting is possible with minimal mechanical and thermal distortions, which saves material wastage; complicated spatial profile cutting possibilities are widely used in aerospace and other industries.

(c) *Laser Drilling* : Laser drilling is possible due to the evaporation of material heated by powerful laser pulses. High precision drilling can be performed in any direction and there are no broken drills. Drilling of brittle materials (ceramics) can be performed after they are processed (fired) thereby ensuring precise shape of the drilled holes.

In Optoelectronics. The field that deals with information storage, retrieval, processing and transmission is called opto-electronics. A laser beam with high frequency and coherence enormously increases the upper limit of the frequency bands that can be transmitted. That is, lasers have increased the volume of information-handling manifold.

Secondly, electrical signals are unidirectional (a function of time only), but optical signals are functions of spatial as well as time variables (three-dimensional). This multi-dimensional information capacity of coherent radiation in the optical range is useful in image analysis and other multi-channel data processing. Other advantages of laser communication systems are — better directionality and higher efficiencies and immunity to distortion due to electrical and other interferences. Laser communication systems, based on fibre optics, have wide applications in computer networks.

Optical range finders (optical radars) provide more accurate ranging and velocity measurements and are used in aerospace, satellite and other navigational guidance systems—telecommunications, land, sea and air navigation, aircraft landing and laser-guided missile systems etc.

12.2 HOLOGRAPHY

Light is the source of visual cognition; photography and holography are two distinct processes of visual perception of light wave patterns. A light wave, for that matter any wave, comprises i) amplitude and ii) phase components of information and the processes of image transmission are carried by waves *via* amplitude and phase modulations. However, the reception of the information and perception depend on the mode of transmission — either *via* amplitude modulation (recording of intensity changes which is subjective) or *via* phase modulation (which is objective), and mode of perception by the observer.

Photography. Photography and image perception by the retina of the eye (on the plane where the rods and cones are located) are essentially flat-picture recording techniques—of recording a two-dimensional projection of the image of the object and not the 'object' wave. The process through which photography (and our eye) records the information carried by the light waves is through amplitude modulation (intensity fluctuations) and is a subjective process.

Holography. The holographic process records (and reproduces) the '**phase**' as well as '**amplitude**' information of light waves scattered by the object. In essence, the '**object**' wave itself is recorded and subsequently reproduced. Holography is, therefore, an objective process (phase modulation), which is fundamentally different from photography and our perception of objects. Conversion of phase information to amplitude information (our eye can perceive only amplitude modulations) for the recording process is realised by creating a '**standing wave**' (interference pattern) between the reference wave and the scattered wave from the object (object wave). That is, what is recorded in holography (on the photographic plate) is the interference patterns between the reference and the object waves—the hologram. It is the recording of the 'object' wave itself, rather than an image of the object; also the 'object' wave can be manipulated which is impossible in the case of photography. *Holography can be regarded as the optical equivalent of the object.*

Huygen's principle of wave propagation of light directly implies the basic concept of holography. **Gabor's** trials to improve the quality of electron microscopic images, distorted due to the spherical abberation, led to the rediscovery of holography. But real progress in this field was only after the availability of coherent light sources (lasers). Only with a coherent light source can the reference and scattered waves of the light beam form an interference pattern (standing wave) with a high degree of fringe contrast, which is necessary condition for recording holograms.

12.2.1 The Holographic Setup

When a wave field of light reflected by an object is superimposed on a reference wave to create an interference pattern, the bright areas are those where the phases of the scattered wave from the object (object wave) and the reference wave are similar and the dark areas are the regions where their phases differ. This interference pattern (standing wave) is recorded on a photographic plate, which is called a hologram. The intensity at any point in this interference pattern (**hologram**) depends on the phase as well as on the amplitude of the original object wave.

The visual '**reconstruction**' of the object wave is achieved by irradiating the hologram by the reference wave (Fig. 12.3a & b).

The hologram allows the reference wave through only those portions which match the phases of the object and reference waves, thereby reproducing the original object wave. That is, holography is a two-step process of visual reconstruction of the object. In the first step a complex interference pattern (standing wave) is formed and is recorded on a photographic plate (hologram) and in the second step the hologram is illuminated by the reference wave such that part of the transmitted light (with matching phases) is an exact replica of the original object wave. That is, recording a hologram '**freezes**' the object wave, while the reconstruction '**unfreezes**' it. Since the human eye is adapted to record images due to intensity fluctuations only, the holographic plate, which is a record of interference pattern (phase fluctuations), does not look like a conventional photographic negative.

Gabor's method of holography was an on-axis set-up, with the holographic plate kept on axis of the source and the observer for recording. As the true and conjugate images, O and O′ are in the same line, the quality of image reproduction in this set-up is very poor due to mutual interference of the real and virtual images. In addition, only transparent objects could be used that form no shadows. These disadvantages can be eliminated by the off-axis set-up of **Leith** and **Upatnieks** (Fig. 12.4).

In this arrangement, the photographic plate is shifted to the off-axis position, P_2 (from the on-axis position, P_1). A monochromatic light beam is split into two components, one of them is directed towards the object and the other (reference beam) is directed towards a recording medium

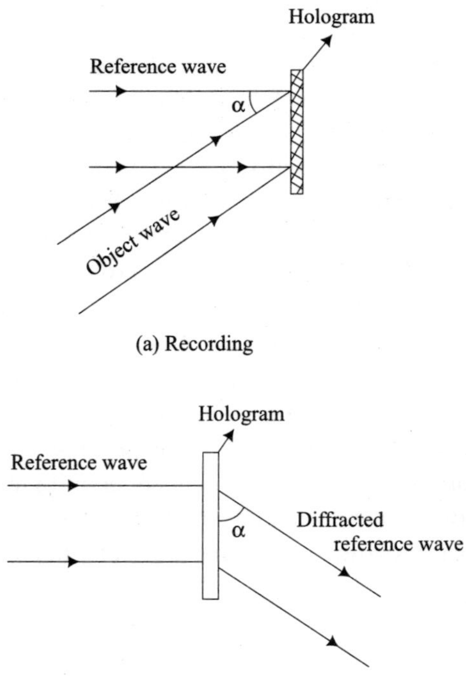

(b) Reconstruction

Fig. 12.3 Hologram recording and reconstruction.

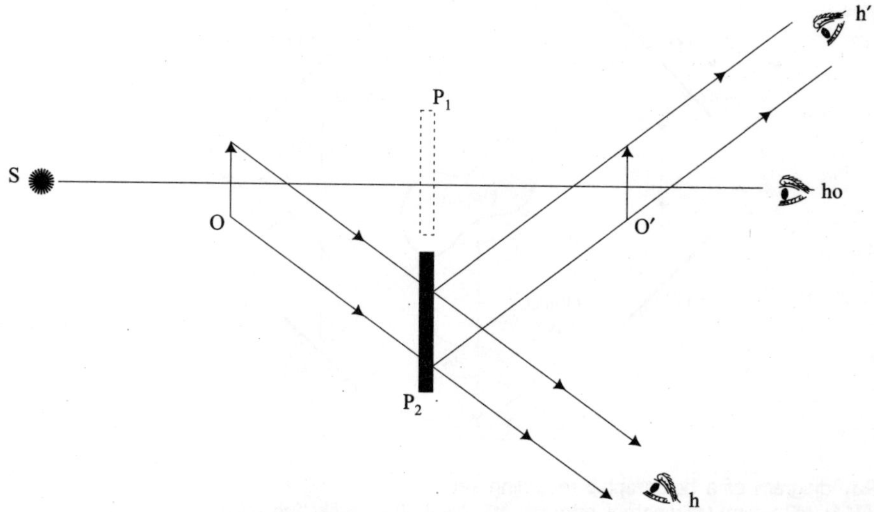

Fig. 12.4 Off-axis set-up of Leith & Upatnieks.
S = Laser source; *P* = photographic plate; *O* and *O′* = real and conjugate images, respectively; *h* and *h′* = observing positions of *O* and *O′*.

(photographic plate). The component beam, scattered by the object, constitutes the object wave which is allowed to form an interference pattern with the reference beam and the interference pattern is recorded on the same photographic plate. As mentioned earlier, as the object wave and the reference wave are mutually coherent they will form an interference pattern (standing wave) at the recording medium with high-fringe contrast (Fig. 12.3a). Though images O and O′ remain in the same positions in an off-axis set-up also, they are viewed at spatially-separated positions, *h* and *h′*, thereby eliminating the interference of images (Fig. 12.4).

Holograms. Holograms can be recorded with parallel, divergent or convergent reference beams. Holograms recorded with diffuse illumination of the objects have many advantages. Diffuse light causes each point of the object to radiate a spherical wave and, therefore, the information concerning each point of the object is spread out over the entire hologram. Each point of the hologram thereby contains information about the whole object. A ray diagram of a holographic (transmission) recording is shown in Fig. 12.5.

(i) When the object and the reference waves are plane waves, their interference pattern is **Young's** interference strips, as obtained with a diffraction grating.

(ii) A **Frauenhofer** hologram is realised when the object is far away from the recording medium. The object is placed in the focal plane of a lens so that each object point gives rise to a parallel beam of light incident on the photographic plate (that is, both reference and object waves are effectively at infinity).

(iii) A hologram of a point object is the **Fresnel** zone plate. Which is a system of alternating opaque and transparent rings. The Fresnel zone plate is a planar equivalent of a lens and, therefore, possesses the focussing properties of a lens.

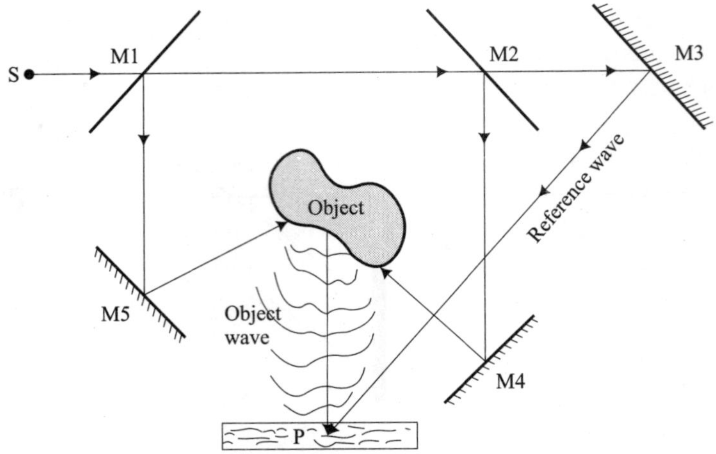

Fig. 12.5 Ray diagram of a holographic recording set-up.
M1 & *M2* : semi-transparent mirrors; *M3*, *M4* & *M5* : reflecting mirrors;
S = laser source; *P* = photographic plate.

(iv) A volume hologram is one where the thickness of the photographic emulsion is greater than the interference fringe spacing. All holograms recorded on photographic plates are actually volume holograms. Volume holograms find applications in colour holography and computer technology.

12.2.2 Applications

Holographic techniques find applications in wide-ranging areas: visual imaging, precision measurements (interferometry), information-related technologies and movie and entertainment arts.

Visual Perception. Our brain is more accustomed to grasp information in the form of visual images than through analytical data. So, visual observation of defects, cracks etc. of objects gives a better insight into the defects. Also, holography can give faithful reproduction of brightness gradations of objects which is not possible in photography — glittering of a diamond or the gleam of an eye etc.

Another of the many advantages of holography in visual perception is its ability to overcome a basic limitation of microscopy. Biological specimens in aqueous media and microscopic objects (such as bacteria) tend to drift around and the objects under observation move out of focus under the microscope. Holograms can 'freeze' these **Brownian** motions while preserving all three-dimensional information. The 'reconstructed' images, recorded at successive foci can be viewed at leisure.

Holographic techniques have been used to make optical elements, such as diffraction gratings and Fresnel's zone plates (lenses) etc. They are produced by optical method and have better line shape dimensions and perfection than can be obtained by mechanical means, and therefore are free from defects like double images (ghosts) and image-blurring etc. In a holographic grating, two identical laser beams are focussed at a suitable angle upon a photoresist-coated glass plate. Sensitised photoresist regions, corresponding to interference fringes, are dissolved, leaving the grooved structure, which can be coated with a reflecting substance to produce a reflection grating. A similar procedure is used to produce Fresnel zone plates. Holographic diffraction gratings and Fresnel zone plates have become common optical elements in modern spectrophotometers and optical instruments, because of their better performance and relatively low cost of production.

Holographic interferometry. Interferometry is the technique of making accurate measurements of physical objects employing optical interference methods (e.g., Fabry-Ferot interferometer). Conventional interferometry is applicable only to highly polished surfaces and to simple shapes. Holographic interferometry does not suffer from this limitation and measurements can be made on arbitrary shapes. Abrasions, objects under stress and deformation and other such material defects (stress and fatigue of materials) under hostile and corrosive environments, thermal expansion and diaphragm oscillations etc., can be studied by this method. Fluid flow dynamics (blood flow in capillaries and wind-tunnel experiments) and many other phenomena of mass transport can be studied by phase hologram interferometry.

Double-exposure holography— superposition of two holographs, one from the object in the initial state and the other when the object is subjected to stress or deformation, on a single photographic plate, is suitable for the study of transient phenomena such as objects under impact or change.

Information-related Areas. A hologram recorded on a photographic plate is impervious to damage and decay. This is because each point of the object scatters light in all directions and as a result, in the course of recording, a hologram the radiation from every point of the object is scattered to all the points of the hologram. When 'reconstructed', an image of the entire object can be seen through every tiny portion of the hologram. Wear and tear of specific portions of a hologram will not erase specific portions of the image, rather only loss of finer details of the image as a whole entity. This property, i.e., information redundancy of holography has a great advantage over photography in the preservation of recorded information. While a hologram recorded on a photographic plate cannot be erased nor can a new hologram be recorded in its place (read only memory), holograms recorded on magnetic tapes, on thermoplastic and photochromatic materials can be erased and new holograms be recorded in their places (read-write memory). In essence, holography offers a very reliable and flexible method of data storage, which can be exploited in computer and entertainment technologies.

Holographic information storage needs less space and is useful in libraries, documentation centres, credit card systems and video industries.

Holography can be applied in pattern recognition and associative retrieval of information and in encoding and decoding of information (cryptography). In addition, more importantly, the stored data (object wave) can be manipulated/modulated selectively for visual reconstruction of models of required or imagined objects or shapes. These computer-synthesised holograms, referred to as 'digital holography', offer immense possibilities in many fields—computer model-building, object visualisation and perception in aero and automative industries and animated forms and objects in performing arts.

Film and Entertainment (Performing Arts). Volume holograms enable colour holography which holds great potential in film, television and other performing arts. The set-up in colour holography is to record at least three separate holograms, each with a different colour on a single photographic plate. Each colour forms its own set of interference patterns (each with different fringe spacings) in the photographic emulsion (Fig. 12.6,). White light is used for 'reconstruction' and each set of Bragg planes filters out the colour with which it was formed and a full-colour reproduction of the object results. Holographic cinema is a distinct possibility and 'live' three dimensional holographic images with natural colours would be more realistic and entertaining than possible with present-day movie and entertainment technology.

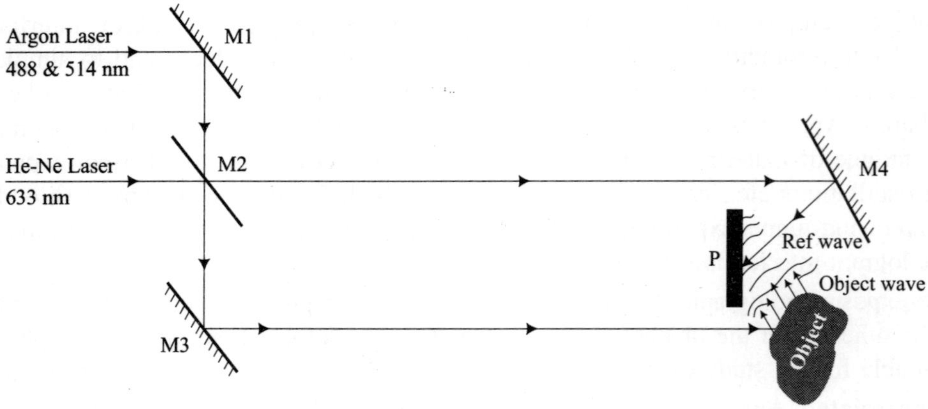

Fig. 12.6 Setup for colour holography.
M1, M2, M3 and M4 are mirrors; P is a photographic plate; M2 is a transmitting mirror.

Suggested Reading

Chopra, S. & Chawla, H.M. (Eds.). "*Lasers in Chemical and Biological Science*". Wiley Eastern/New Delhi. (1992).

Denisyuk, Yu. "*Fundamentals of Holography*", Mir Publications/Moscow. (1984).

Hariharan, P. "*Optical Holography: Principles, Techniques and Applications*", Camblidge University Press/ Cambridge. (1986).

Koch, W.E. "*Lasers and Holography*". Double Day/New York. (1969).

O'Shea, D.C., Callen, W.R. & Rhodes, W.T. "*An Introduction to Lasers and Their Applications*", Adison-Wesley Pub./London. (1979).

Smith, H.M. "*Principles of Holography*", John Wiley/New York. (1969).

Soroko, L.M. "*Holography and Coherent Optics*". Plenum Press/New York. (1980).

Tarasov, L.V. "*Laser Age in Optics*". Mir Publications/Moscow. (1984).

Tarasov, L.V. "*Laser Physics and Applications*". Mir Publications/Moscow. (1986).

SECTION III

Bioenergetics

OVERVIEW

All physico-chemical processes (chemical, biochemical, biological, evolution etc.) are invariably coupled to energy. Thermodynamics deals with the energy processes of a reacting system. Reactions in a 'closed' system can be addressed by reversible thermodynamics, but those in an 'open' system (life processes such as autocatalysis, evolution etc.) can be understood from the concepts of irreversible thermodynamics.

Concepts of bioenergetics (thermodynamics) are applicable in many chemical reactions that occur in living systems. Photosynthetic reactions in plant and bacterial systems and oxidative phosphorylation reactions in mitochondria are examples of areas involving bioenergetics. In photosynthesis the light energy (photon) is utilised by the chloroplast cells to synthesise organic compounds (energy storage) from CO_2 and H_2O. Photosystems PSI (P 700) and PSII (P680) are involved in photo-synthetic reactions. ATP molecules are the main source of energy in many biological processes which involve active transport and ATP is synthesised in mitochondria by oxidative phosphorylation. The energy required to synthesis ATP is realised by utilising the energy released from the redox reactions of the electron-transport chain (electrochemical proton gradient). The electron-transport system of mitochondria is coupled with the phosphorylating system and the proton-translocating enzyme complex, FoF_1-ATPase, catalyses the synthesis as well as the hydrolysis of ATP.

13 Thermodynamics

Synopsis: Thermodynamics deals with the energy processes of any reacting system. The laws of thermodynamics are universal, independent of molecular models. No physical, chemical or biological processes violate these laws. An understanding of the basic concepts of thermodynamics is absolutely essential to comprehend and appreciate problems in bioenergetics.

Thermodynamical processes are either (i) reversible or (ii) irreversible. Generally, **closed** systems (chemical reactions) are reversible and **open** systems (biological processes) are irreversible and nonlinear. The First Law of Thermodynamics is a universal statement of the Law of Conservation of Energy ($dU = \delta Q + \delta W$), and the Second Law postulates the direction in which a system changes spontaneously ($TdS \geq \delta Q$). **Enthalpy, H,** is heat content at constant pressure and is invoked while dealing with the problems concerned with the heats of chemical reactions (thermochemistry). Entropy, S, of a system is indicative of its state of disorder and spontaneous processes always lead to increase in entropy of the system. Entropy and statistical probability are closely related to the information content of a system in a specific state. **Gibbs free energy, G,** is "useful" work done by the system at its expense. A reaction is spontaneous if and only if $\Delta G < 0$.

Irreversible thermodynamics deals with "**open**" systems and tries to explain various biological processes, such as mass transport, membrane permeability; muscle and nerve actions. Non-equilibrium, unsteady states are relevant in biology and have been shown to underlie growth (autocatalysis) and evolution.

13.1 REVERSIBLE THERMODYNAMICS

A system is either **isolated**, **closed** or **open**, depending on the conditions (Fig. 13.1). A system is **isolated** if $Q = 0$, $W = 0$ and $n_x = 0$. That is, neither energy nor matter can pass through the system. A **closed** system is one where $Q \neq 0$, $W \neq 0$ and $n_x \neq 0$. That is, only energy can pass through the system but not matter. Generally, chemical systems are closed systems. In an **open** system $Q \neq 0$, $W \neq 0$ and $n_x \neq 0$. That is, both energy and matter pass through the system. Biological processes are open systems, isobaric (pressure constant), isothermal (temperature constant), irreversible and non-linear.

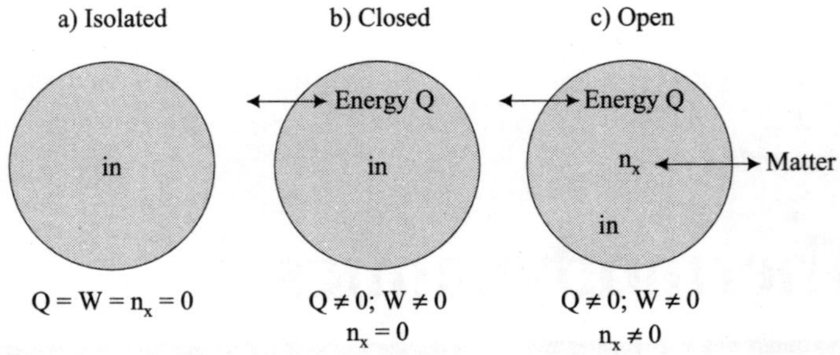

Fig. 13.1 Thermodynamical systems.

13.1.1 The First Law of Thermodynamics

For **closed** systems

$$U = Q + W \qquad \text{Ist Law of Thermodynamics} \qquad ...(13.1)$$

$$dU = \delta Q + \delta W$$

According to the convention, U = internal energy, Q = quantity of heat absorbed **by** the system, W = work done **on** the system and n_x is matter transported.

The first law of thermodynamics is a statement of the law of conservation of energy. Energy can neither be created nor destroyed. Total energy of the universe is constant, but conversion of energy from one form to the other is feasible. The law does not state which energy processes are spontaneous nor does it specify the state of a system under given conditions.

Enthalpy. Enthalpy, H, is heat content at constant pressure, P. The internal energy, U, is altered by exchange of work, W, and heat, Q, with the surroundings.

$$dU = \delta Q + \delta W \text{ (Ist Law), but}$$

$$W = -PV \text{ (Pressure–Volume work)} \qquad ...(13.2a)$$

Enthalpy, H is defined as

$$H = U + PV \qquad ...(13.2b)$$

For an isobaric case (P = 0)

$$(Q)_P = U \pm P \Delta V = \Delta H \qquad ...(13.2c)$$

But, $PV = nRT$, where n is the number of particles, R is the gas constant and T is the absolute temperature.

$$\therefore \qquad \Delta H = \Delta U + P \Delta V = \Delta U + \Delta nRT \qquad ...(13.2d)$$

If $\Delta H > 0$ the process is **endothermic** (heat is absorbed)

$\Delta H < 0$ the process is **exothermic** (heat is released).

13.1.2 The Second Law of Thermodynamics (Entropy, S)

The useful work obtainable depends not only on ΔU, or ΔH, but also on entropy, S, a state function. Entropy is a state of disorder and all spontaneous processes (all irreversible processes) occur in the direction that results in increase in the total entropy of the system. Entropy is an index of exhaustion.

$$T\,\Delta S \geq \Delta Q \qquad \text{IInd Law of Thermodynamics} \qquad ...(13.3a)$$

$$\Delta S \geq 0 \qquad ...(13.3b)$$

"Die Energie der Welt ist Konstant; die Entropie der Welt strebt einem Maximum zu" (Clausius).

As any spontaneous process always results in increase in entropy, such a state must be more probable than other states. Accordingly, statistical probability and entropy must be closely related. Such a relation, as suggested by **Boltzmann,** is

$$S = K_B \ln Z_i \qquad ...(13.4)$$

where K_B is the Boltzmann constant and Z is the statistical probability of the ith state. S cannot be negative. A substance has high entropy if many states are available to the molecules of the substance.

13.1.3 The Third Law of Thermodynamics

$$S_T = S_0 \int_0^T \frac{C_P}{T}\,dT \qquad ...(13.5a)$$

where C_P is heat capacity. At absolute temperature (T = 0), the entropy of perfect crystalline substances becomes zero.

$$S_0 = 0 \qquad \text{IIIrd Law of Thermodynamics} \qquad ...(13.5b)$$

The third law of thermodynamics allows calculation of absolute entropies of substances.

Summing up the three laws of thermodynamics —

Ist Law : Statement of conservation of energy.

IInd Law : Statement of the direction of the spontaneous processes.

IIIrd Law : Relates to experimental approach to the absolute zero temperature.

13.1.4 Thermal Analysis

Thermometric methods (thermogravimetric analysis, TGA; differential thermal analysis, DTA; and differential scanning calorimetry, DSC) provide useful experimental data on thermodynamic parameters, such as heat capacity (enthalpy), kinetics of chemical reactions, structural intermediates, phase transmissions and final products. Thermometric methods rely on the rate at which substances undergo physical and chemical changes (thermodynamical changes) as they are heated or cooled.

Thermogravimetric Analysis (TGA). In thermogravimetric analysis (TGA), the weight of a sample is measured continuously as the temperature of the sample is linearly increased with time. The device consists of a balance, a furnace, an inert gas unit to flush out decomposed samples and a data display setup. The loss of weight of the sample, ΔW, as it is heated (obtainable from a thermogram which is a plot of weight as a function of time) provides the change in weight of the sample associated with thermodynamic transitions.

Differential Thermal Analysis (DTA). In differential thermal analysis (DTA), the differential temperature, ΔT, between a sample as a function of sample temperature, T, while the sample and the reference material are heated under the same conditions. A graph of ΔT against T, at which the whole sample changes (denaturation) is a differential thermogram, which provides information on the nature of the transitions (endo- and exo-thermic).

A DTA instrument consists of two identical and symmetrically-arranged crucibles in a furnace. Each crucible is provided with a thermocouple and the sample is placed in one of the crucibles and the reference material (aluminium oxide, silicon carbide or glass beads) in the other. As the temperature of the furnace is increased linearly with time, the difference in the two thermocouple readings provides ΔT, while the reading of the third thermocouple, connected to the furnace, provides the temperature, T.

Differential scanning calorimetry (DSC). In differential scanning calorimetry (DSC) the heat flow into the sample and the reference material are measured as a function of temperature as they are heated. In DSC, the differences in temperatures are monitored. The sample and the reference material are heated individually and the average temperature and the temperature difference are continuously recorded as the thermogram is a plot of temperature difference versus average temperature.

TGA, DTA and DSC analyses provide useful information on physical and chemical changes of substances due to thermal energy. They can be employed to study or monitor melting and decomposition characteristics of polymers, phase-transition studies of condensed matter (solid-to-liquid, liquid-crystalline-to-liquid etc. e.g., DNA exhibits a liquid-crystalline phase that is dependent on temperature, concentration, nature of solvent etc.), testing of purity of biomolecules, drugs, environmental pollutants, membrane fluidity and mass transport etc., as a function of temperature.

Thermograms show the heat capacity (enthalpy) of the substance as a function of temperature. Broader peaks of endothermic phase transitions indicate temperature-dependent phase transitions or depolarisation of noncrystalline materials, while sharp endotherms indicate phase transitions of crystalline states. (Endothermic processes are: fusion, melting, absorption, desorption, dehydration and decomposition). Similarly, broader peaks of exothermic changes are indicative of chemical reactions, polymerisation while sharp exothermic peaks suggest molecular associations, rearrangements and crystallisation. (Exothermic processes are: adsorption, crystallisation, polymerisation and oxidation in air). Thermograms can provide valuable and easy-to-monitor data on polymer substances, such as melting-point, phase transitions (helix/coil transitions), polymerisation, crystallisation, oxidation, denaturation, chain-branching etc. Random and branched-chain polymers of low molecular mass show broad endothermic peaks while being ordered, and unbranched polymers of high molecular mass give rise to sharp endothermic peaks. Similarly, exothermic peaks and their characteristics provide valuable information on the molecular rearrangements and annealing features of polymer substances.

As an example, in a DTA thermogram of a polymer (Fig. 13.2), the first transition peak (dip) is due to transition of the polymer from amorphous state to glassy, rubber-like flexible state (T_g). This transition occurs without any change in enthalpy ($\Delta H = 0$). At the characteristic temperature, T_c, many amorphous polymers undergo a microcrystalline-phase transition, which is exothermic. The

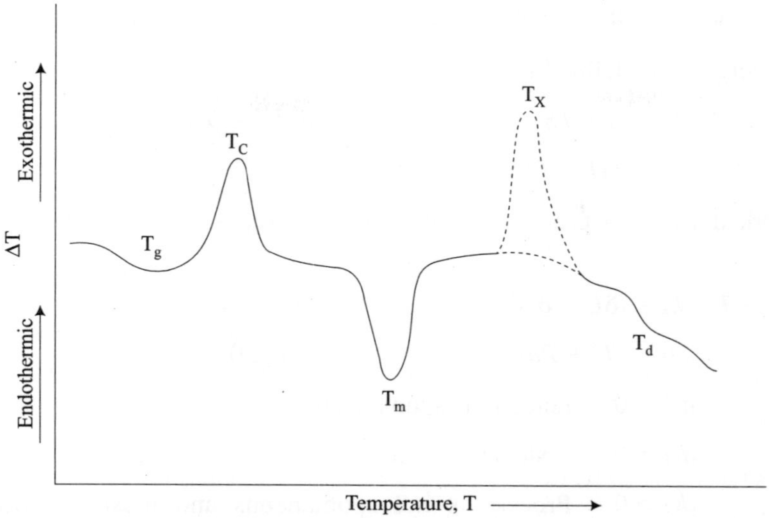

Fig. 13.2 A schematic DTA thermogram of a polymeric material.
$T_g \rightarrow$ glass transition temperature
$T_c \rightarrow$ crystallisation temperature
$T_m \rightarrow$ melting temperature
$T_x \rightarrow$ oxidation (in the presence of oxygen) temperature
$T_d \rightarrow$ decomposition temperature

melting temperature, T_m, of a polymer is an endothermic process; and if the polymer is heated further, it gives rise to an exothermic oxidation peak, T_x, if the sample is heated in the presence of oxygen, or it leads to decomposition, (T_d) (in the absence of oxygen).

13.1.5 Free Energy

'Free energy' is a property of the system at whose expense the system does work. The state functions dealing with free energy are :

(i) Helmoltz free energy, A, at constant volume, and (ii) Gibbs free energy, G, at constant pressure.

Helmoltz Free Energy, A. The maximum work, W_{max} is

$$W_{max} = \Delta U - \Delta Q_{rev} \qquad \text{...(13.6a)}$$

$$= \Delta U - T\Delta S$$

Helmoltz free energy, A, is defined as

$$A = U - TS \qquad \text{...(13.6b)}$$

Helmoltz free energy is not useful in biology, because it is a process at constant volume, whereas in biological systems reactions occur at constant pressure.

Gibbs Free Energy, G. Like enthalpy, it describes processes at constant pressure. All the maximum work (Helmoltz free energy) may not be 'useful' work. The actual amount of work available, 'useful' to a system, is W_{useful}

$$W_{useful} = W_{max} - W_{PV} \quad \text{...(13.7a)}$$

Gibbs free energy, G is defined as

$$G = H - TS \qquad \therefore\ G = U + PV - TS \quad \text{...(13.7b)}$$

$$\Delta G = \Delta H - T\Delta S \quad \text{...(13.7c)}$$

'Useful' work done, in a process at constant pressure, is at the expense of the Gibbs free energy, G.

$$T\,.\,dS \geq (\delta U - \delta W) \quad \text{...(13.8a)}$$

$$\therefore\ 0 \leq dU + PdV - TdS \qquad \therefore\ dG \leq 0 \quad \text{...(13.8b)}$$

for $dG < 0$ reaction is spontaneous

$dG = 0$ system is in equilibrium ...(13.8c)

$dG > 0$ Process is not spontaneous and must be coupled with other reactions so that the resultant $dG < 0$ for the reaction to proceed.

Free-energy change, ΔG, is a manifestation of two different entities, ΔH and ΔS. The relative predominance of these entities will influence the course of the reaction (Table 13.1).

The concept of Gibbs free energy is central to various energy-related areas in chemical, biochemical and biological reactions, as shown in the schematic diagram (Fig. 13.3).

Examples: $\Delta G = \Delta G^{\circ} + RT \ln X$

$\Delta G = \Delta H - T\Delta S$ — Enthalpy-entropy reactions

$= RT \ln \Pi$ — Pressure reactions. (Π = hydrostatic pressure)

$= -RT \ln K$ — Chemical reaction (K = equilibrium constant)

$= -RT \ln (C2/C1)$ — Chemical reaction (C = concentration).

$= -n\mathcal{F}\varepsilon$ — Redox reaction

$\Delta G = n\mathcal{F}\varepsilon = RT \ln (C2/C1)$

For Na^+ ions $\varepsilon = (RT/n\ \mathcal{F}) \ln \{[Na^+]_e/[Na^+]_i\}$ **Nernst** equation.

Table 13.1 Influence of ΔH and ΔS on the free energy of the reaction.

ΔH	ΔS	($\Delta G = \Delta H - T\Delta S$)
+	+	Temperature-dependent reaction. Entropically favoured. Spontaneous at higher temperature ($T > \Delta H/\Delta S$)
+	–	Unfavourable enthalpically and entropically. Reaction does not occur at all temperatures.
–	+	Favourable enthalpically and entropically. Reaction is spontaneous and temperature-independent.
–	–	Enthalpically favourable but entropically unfavourable. Reaction is temperature-dependent. Spontaneous at low temperatures ($T < \Delta H/\Delta S$).

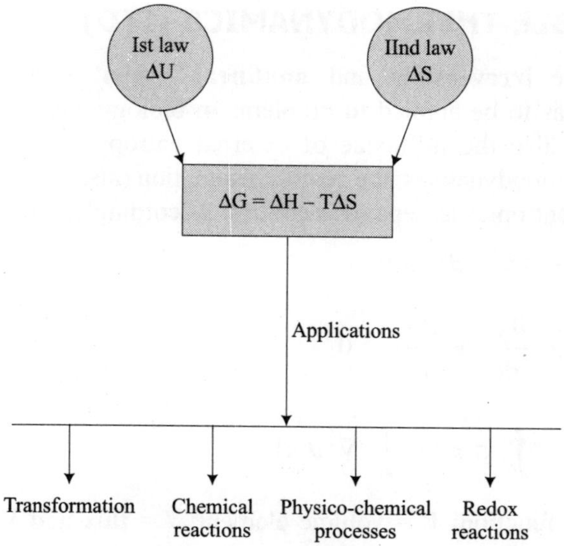

Fig. 13.3 Applications of ΔG.

where $\mathcal{F}$ = Faraday constant; ε = electromotive force; $[Na^+]_i$ and $[Na^+]_e$ are concentrations of Na^+ ions *in* and *out* of the system.

Chemical Potential. The molecular Gibbs free energy of a pure substance is given a new label, chemical potential, μ.

$$G = \mu_A n_A + \mu_B n_B + \mu_C n_C + \ldots$$

$$G = \sum_i \mu_i n_i \quad \text{...(13.9a)}$$

$$\mu_i = \left(\frac{\partial G}{\partial n_i}\right)_{T, P, n_j} \quad \text{...(13.9b)}$$

where n_i = number of particles and at low concentrations, C_i; $n_i \approx C_i$

$$\mu_i = \mu_i^\circ + RT \ln (C_i) \quad \text{.... (13.9c)}$$

Macromolecular folding. The total free energy, ΔG_{tot} of a macromolecular folding in solution is

$$\Delta G_{tot} = \Delta H_{ch} + \Delta H_{sol} - T(\Delta S_{ch} + \Delta S_{sol})$$

$$\Delta G = -RT(\ln K + \ln C) \quad \text{...(13.10)}$$

where ch and sol stand for chain and solution; and K and C are equilibrium constant and molar concentration respectively.

Both energy (enthalpy) and entropy factors are important for overall macromolecular stability. For polar side chains $\Delta G_{tot} \approx 0$; and for nonpolar groups ΔS is positive and far greater and, therefore, would be the single largest contributor to structural stability in these regions.

13.2 IRREVERSIBLE THERMODYNAMICS (ITD)

Biological processes are **irreversible** and **nonlinear 'open'** system. As such, irreversible thermodynamics (ITD) has to be applied to problems in biology, taking into account not only the internal entropy, S_i, but also the influence of external entropy, S_e, surrounding the system. In irreversible, non-linear thermodynamics, the central assumption (also an approximation) is to consider the internal and external entropies as separable entities. Accordingly, the total entropy change, dS, is

$$dS = dS_i + dS_e \geq 0 \quad \text{...(13.11a)}$$

$$\left(\frac{ds}{dt}\right) = \frac{\partial s_i}{\partial t} + \frac{\partial s_e}{\partial t} \geq 0 \quad \text{...(11.11b)}$$

$$= \int_V \sigma\, dV - \int_V \nabla . J\, dV \quad \text{...(13.11c)}$$

where T σ = dissipation function; V = volume element; J = flux and ∇ = **Laplace** operator (for divergence).

Table 13.2 Thermodynamical processes in "Closed" and "Open" systems.

"Closed" system	*"Open" system*
Reversible process	Irreversible process
No exchange of matter	Exchange of matter
Rate of production is not controlled	Rate of production of a component is coupled to its rate of utilisation
Rate of entropy production is maximum	Rate of entropy production is minimum
Entropy increases ($\Delta S \geq 0$)	$\Delta S > 0$ (generally);
	$\Delta S < 0$ is also possible (microreversibility)

13.2.1 Isolated State, $dS_e = 0$

Most successful applications of irreversible thermodynamics have been achieved in the areas of transport, energy transduction and membrane permeability, by addressing these problems at near-equilibrium (linear) state. Then $dS_e \approx 0$. In such situation, dS is

$$dS = dS_i > 0 \quad \text{...(13.12)}$$

In a chemical reaction at near-equilibrium

$$T\, dS_i = -\, dG = \sum_k \mu_k\, dn_k \quad \text{..(13.13a)}$$

$$= -\, d\,\xi \sum_k \mu_k\, \nu_k \quad \text{...(3.13b)}$$

$$T\, dS_i = a_k\, d\xi = C_k\, d\xi \quad \text{... (13.13c)}$$

where μ_k = chemical potential; n_k = number of particles and dn_k = change in molar concentration; a_k = affinity; C_k (concentration) at low concentrations; ν_k = stoichiometric coefficient; $d\,\xi$ = extent of reaction.

Rate of change of entropy, dS_i/dt is

$$\frac{dS_i}{dt} = \int_V \sigma \, dV \geq 0 \text{ from Eq.} \qquad \text{...(13.11b)}$$

The dissipation function $T\sigma$ is

$$T\sigma = -\frac{d\xi}{dt} \sum_k \mu_k \nu_k = \sum_k a_k V_k \qquad \text{...(13.14)}$$

For a single reaction, $T\sigma_1 < 0$, $(a_1 V_1 < 0)$ is impossible. If this reaction is coupled to another reaction $a_2 V_2$ whereby

$$(\sigma_1 + \sigma_2) = a_1 V_1 + a_2 V_2 > 0 \text{ the reaction is possible.} \qquad \text{...(13.15)}$$

That is, the total sum of the internal entropies in any reaction(s) must always be greater than zero for such reaction(s) to proceed.

Local entropy production, σ, can be expressed in terms of conjugate fluxes, *J*, and forces, *X*.

$$\sigma = \sum_k J_k X_k \geq 0 \qquad \text{...(13.16a)}$$

$$\frac{dS_i}{dt} = \int_V \sigma \, dv = \int_V \sum_k J_k X_k \, dV \qquad \text{...(13.16b)}$$

All fluxes, *J*, are interrelated to forces, *X*, acting.

"For every flow there is a conjugate force"—(Onsager).

Example:

Transport of mass, *m*	driven by density gradient, ρ
Transport of heat, *Q*	driven by temperature gradient, *T*
Diffusion, *D*	driven by concentration gradient, *C*
Chemical reaction,	driven by chemical affinity, **a**
	driven by potential difference, ψ

Solution to one flow ⇒ one force cases is simple.

For example, **Fick's laws** of diffusion.

$$J_k = -D_k (dC/dx) \qquad \textbf{Fick's 1st law}$$

In general, flows are not exclusively driven by their conjugate forces. The flows, *J*, are interrelated to their forces, *X*, by coupling terms $L_{kk'}$ (which are tensors).

$$J_k = \sum_{k'=1}^{n} L_{kk'} X_{k'} \qquad \text{...(13.17a)}$$

$$J_k = \begin{pmatrix} L_{11} & L_{12} & L_{13} & \cdots \\ L_{21} & L_{21} & L_{23} & \cdots \\ L_{31} & L_{32} & L_{33} & \cdots \\ \cdots & \cdots & \cdots & \end{pmatrix} X_{k'} \qquad \text{...(13.17b)}$$

The individual fluxes, J_1, J_2, J_3 etc., are

$$\begin{aligned} J_1 &= L_{11} X_1 + L_{12} X_2 + L_{13} X_3 + \ldots \\ J_2 &= L_{21} X_1 + L_{22} X_2 + L_{23} X_3 + \ldots \\ J_3 &= L_{31} X_1 + L_{32} X_2 + L_{33} X_3 + \ldots \text{ etc.} \end{aligned} \quad \ldots(13.17c)$$

At near-equilibrium, the linear approximation holds good and, therefore, the coupling factors are symmetrical tensors.

$$L_{kk'} = L_{k'k} \quad \textbf{(Onsager theorem)} \quad \ldots(13.18)$$

Also, the principle of microscopic reversibility, that is, the decrease in the internal entropy due to fluctions at the near-equilibrium state holds

$$\delta S_i \leq 0 \quad \textbf{(microscopic reversibility)} \quad \ldots(13.19)$$

Irreversible thermodynamics (ITD) at near-equilibrium state can be applied to explore the mechanism of active transport of Na^+/K^+ ions across the cell membrane with concommitant hydrolysis of ATP to drive the uphill transport.

General features of active transport of Na^+/K^+ in the cell are:

(i) Most cells have a high concentration of K^+ ions and low concentration of Na^+ ions inside the cell.

(ii) Na^+ must be inside and K^+ must be outside the cell for active transport and to activate ATPase. That is, asymmetric distribution of cations is coupled to the function of the enzyme.

(iii) ATP is not hydrolysed unless Na^+/K^+ are transported.

(iv) The free energy required for the active transport is provided by the hydrolysis of ATP (Fig. 13.4).

$$ATP^{4-} + H_2O \xrightarrow{Mg^{2+}Na^+/K^+} ADP^{3-} + P_i^{2-} + H^+ \quad \ldots(13.20)$$

The chemical potential,

$$\Delta\mu = RT \ln (C2/C1) + zF\psi \quad \ldots(13.21a)$$

where C_1 and C_2 are concentrations, z is charge, F is Faraday constant and ψ is potential difference.

In the resting state, K^+ ions permeate freely through the cell membrane.

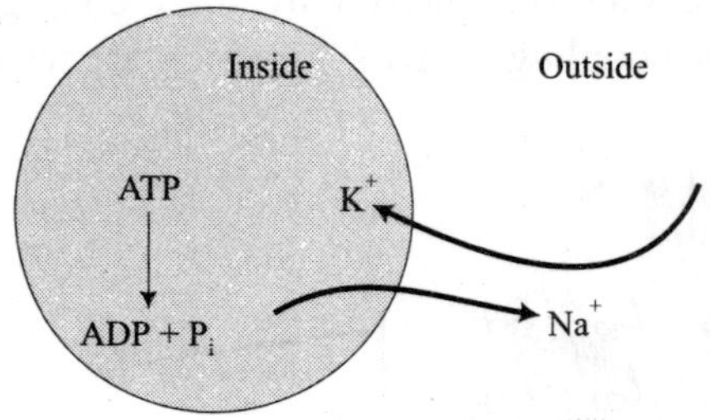

Fig. 13.4 Mechanism of active transport of Na^+/K^+ in cells.

$$\Delta\mu_K^{\ +} = 0 = RT \ln \left\{\frac{[K^+]_i}{[K^+]_e}\right\} + z\mathcal{F}\,\psi_K^+ \qquad \text{...(13.21b)}$$

$$\psi_K^+ \approx -\,65 \text{ mV}$$

But, $$\Delta\mu_{Na,K} \neq 0$$

$$\therefore \qquad \Delta\mu_{Na,K} = RT.\ln\left\{\frac{[K^+]_i/[Na^+]_i}{[K^+]_e/[Na^+]_e}\right\} \qquad \text{...(13.21c)}$$

(*i* and *e* represent the concentrations of respective species *inside* and *outside* the cell). Energy should be expended, even at equilibrium, to keep the observed imbalance (asymmetricity) in cation concentration and the coupling between the hydrolysis of ATP and the Na^+/K^+ flux must be finite. All these inferences have been supported by experimental results.

13.2.2 Open Systems, $dS_e \neq 0$

Far from equilibrium (non-linear states), (equation 13.17a)

$J_k = \sum_{k'=1}^{n} L_{kk'} X_{k'}$ is a poor approximation and coupling factors are not symmetric. That is, for non-linear states

$$L_{kk'} \neq L_{k'k} \qquad \text{...(13.22)}$$

Again, in non-equilibrium, steady state

$$dS = dS_i + dS_e = 0 \qquad \text{...(13.23a)}$$

$$dS_e = -\,dS_i < 0 \qquad \text{...(13 .23b)}$$

The steady state is maintained by increasing the entropy of the surrounding medium.

"An organism feeds with negative entropy" (Schrödinger).

Study of **open** systems **far from equilibrium** provides insight into many biological and biochemical processes — active transport, biological rhythms, information and memory, autocatalysis and evolution etc.

Volterra Systems. For closed systems, the equilibrium values depend only on the equilibrium constants and not on the rate constants. In open systems, on the contrary, individual rate constants play a decisive role in reaction processes. Slight differences in the reactants available can become exaggerated and produce clear-cut differences. In open systems, a change in the equilibrium state changes the organisational set-up and consequently changes the balance of forces. Dissipation tries to bring the system back to equilibrium. Dynamical systems cease to change when and only when they reach equilibrium. Since non-equilibrium is a perpetual state of change in any organisation, the dynamical aspects of such a system (Volterra system) are explainable on the assumption that the initial state of such a system must have been of '**unstable equilibrium**'. Any perturbation in such a system can lead to an asymmetric process, that is introduction of polarity which is very common in biological processes.

Active transport (flow against the concentration gradient), positive-feedback and the law of mass action can be explained on the basis of the **Volterra** system. For any oscillatory system, (i) the reactants must occur far from equilibrium, (ii) reactants must have an autocatalytic step and (iii) the system must be able to exist in at least two steady states (bioavailabling). As an example, the possibility of non-linearities in a kinetic system coupled with oscillatory variation of constraints in the proton pump could have given evolutionary advantage.

Biological Oscillations. An unharmonic oscillator (e.g., damped motion of a pendulum) is represented by the equation

$$\frac{d^2x}{dt^2} + \mu \frac{dx}{dt} + kx = 0 \qquad \text{...(13.24a)}$$

where μ and k are the friction coefficient and restitution constant, respectively.

Situations where friction is a function of the variable, x, is represented by the **van der Pol** equation

$$\mu = -\varepsilon(1 - x^2) \qquad \text{...(13.24b)}$$

$$\therefore \quad \frac{d^2x}{dt^2} + \varepsilon(x^2 - 1)\frac{dx}{dt} + kx = 0 \quad \text{where } \varepsilon > 0 \qquad \text{...(13.24c)}$$

The **van der Pol** oscillator is a relaxation oscillator and the trajectory is called the **'limit cycle'**; any perturbations to the oscillator make the oscillator move back to the 'limit cycle': (linear systems do not have a 'limit cycle') (bistability). Oscillations in a bistable system (X, Y) occur due to the influence of a third substance, Z, which can react with X to produce Y, and with Y to produce X. The substance as a result leaps from one state to the other.

$$[X] + [Z] \rightarrow [Y] \qquad \text{...(13.25)}$$

$$[Y] + [Z] \rightarrow [X]$$

Chemical and biological oscillators are similar to electrical oscillators, with autocatalysis playing the role of positive-feedback. Belousov-Zhabotinski reactions, and autocatalysis (in biology) are examples of oscillator systems. In autocatalysis, the reactants, intermediates and products will vary periodically. All biological oscillators can be influenced (perturbed) by external agents (ions, light, chemicals, stimuli, electromagnetic fields etc.) and biological oscillators (e.g., cAMP oscillations in adenylate cyclase function, glycolytic oscillations in yeast-cell cultures) and circadian rhythms show characteristics typical of van der Pol oscillators.

13.3 INFORMATION AND TRANSMISSION

Any functioning system must have a source of information, and any system that deals with information is either (i) informed system, an information gathering system with at least one input, (ii) informing system, a communicating system with at least one output, and/ or (iii) information system, an interacting system with both inputs and outputs. Living organisms are functioning information systems.

In biology, the information source is DNA and codons are bits of information communicated to the ribosome, the receiver. The haircells of the organ of Corti act as transmitters and the hearing vortex is the receiver. The central problem in any communication process (signal transmission) is the information content available in any message. Any message is mixed with noise (random signals) and the content of information in any message is dependent on the 'signal-to-noise' ratio. This is related to the entropy, S, the measure of disorder (chaos).

A message transmitted in a language of Z alphabets with N sequential words is given by **Shannon's** equation (on information theory)

$$S = -KN \sum_{j=1}^{Z} P_j \ln P_j \qquad \textbf{Shannon} \text{ equation} \qquad \text{...(13.26)}$$

where P_j is the probability of the appearance of a given word and K is a normalisation factor. Extension of Shannon's information theory to multi-directional communication systems can be applied to formulate the behavioural patterns and group psychology of lower-level organisms.

But the amount of information and content of information alone cannot shed much light on the nature of biological processes. The instructive value of the information, the qualities which determine the function of the message have to be known. These are adaptive processes—information value that biological molecules possess in their processes of natural selection. An organism can change its structure by mutation, and the selective value, according to **Manfred Eigen**, is a kinetic parameter characterising the autocatalytic synthesis of that particular molecular species at the expense of other similar species of less value, which is the basis of the 'survival of the fittest'. This process occurs in an open stationary system that is far from equilibrium. That is, evolution is a process of evolving and selecting informationally-meaningful structures *via* non-linear and dissipative processes operating in the **far-from-equilibrium "chaotic"** regions. Two prerequisites of information are: **structural randomness** and **asymmetry**.

13.3.1 Information Storage and Memory

The processes of storage of information (memory) in machines and living systems are entirely different. In mechanistic storage processes, information is stored in stable parts and is retrieved. Increase in entropy (increase in noise) in such processes is equivalent to diminution of information; that is, the effect of noise is to reduce the information available. Mechanistic processes can give a good approximation to the short-term structural features and functions of organisms (short-term memory)—neurons as relays, signal transduction as a rectifier and muscle contraction as an amplifier etc. In such mechanistic processes, the operation of scanning and sorting, i.e., "mechanistic recognition" by machines, computers and robots, would be faster and superior with "routine" and "repetitive" information.

Long-time-memory processes: Long-time-memory elements in organisms are essentially non-structured and complex, and these cannot be deduced from the mechanistic structural memory principles.

Long-time-memory processes of organisms can be explained by "*biotonic*" principles. Biotonic principles are relationships among structured and semi-structured properties as functions of time.

$$\text{Biotic behavior} \rightarrow \text{Non-storage of information} \quad ...(13.27)$$

Storage of information and retrieval in the living organisms are, in general, realised by structural modifications of chemical molecules. Structural variations (e.g.; mutations) are linked to information in organisms, which can be related to metabolic processes. This continuous process of "stirring" (*manthan*) is dynamic (chaotic), which is absolutely essential for the survival of living organisms.

The situation is comparable to the problems encountered in physics. Whereas simple physical laws can explain celestial orbits and predict the positions of the heavenly bodies at any given time with a high degree of reliability, they cannot predict positions of atoms and molecules and their motions in strongly-interacting ensembles.

Noise, as understood in communication networks, cannot increase the amount of information present in any system; it can only degrade it. If one equates 'chaos' with noise (turbulence), it might appear paradoxical that 'chaos' or dynamical instability is a necessary condition for evolutionary characteristics in organisms. Additional, transient and dynamic information, such as 'memory' is the outcome of such turbulent processes, These apparently paradoxical situations arise because of our perception of the concept of entropy, based on classical physics entities, developed to deal with macroscopic ensembles. Unless and until the very concept of entropy in the regions of dynamic instability and also the microscopic features of entropy are understood and defined (which might be impossible), our understanding of the approach to biological phenomena will always be approximate and sometimes self-contradictory. The situation is similar to the concepts of classical physics being an approximation to quantum physics phenomena and yet correct at the macroscopic scale. For example, sub-atomic particle trajectories, energy levels and transitions can never be explained, starting from the foundations of classical physics.

But, what is the general function which represents entropy at a position far from equilibrium? The **Liouville** equation

$$\frac{i\,\partial\rho}{\partial t} = L\,\rho \quad ...(13.28)$$

where ρ is the density of state operator and L is the **Liouville operator**, which is invariant under time reversal and therefore it can only describe reversible processes (in actuality). Irreversibility turns out to be closely associated with dynamical instability. One must, therefore, study the behaviour of a large number of many-body systems at the thermodynamic limit, to underscore the concept of entropy.

Non-linear, non-equilibrium and non-steady states are very relevant to the very fundamentals of biology. Such systems seem to be the basis of evolution, growth, ontogeny etc.

Extending the principle of minimum entropy production, σ, to non-linear states (according to **Glansdorff** and **Prigogine** of the **Brussels' school),** the rate of change of minimum entropy production at a steady and constant flow is

$$\frac{d\sigma}{dt} = \frac{d}{dt}(\sum_k J_k X_k) = \left(\sum_k J_k \frac{dX_k}{dt}\right) + \left(\sum_k X_k \frac{dJ_k}{dt}\right) \quad ...(13.29a)$$

$$= \left(\frac{d\sigma_X}{dt}\right)_J + \left(\frac{d\sigma_J}{dt}\right)_X$$

1st term 2nd term

Under constant forces, the first term vanishes; and under constant flux, the second term vanishes.

At constant flux (J = constant),

$$d\,\sigma_x = d\,\sigma_J = \frac{1}{2}\,d\,\sigma \qquad ...(13.29b)$$

which is the **'universal evolution criterion'** at constant flux. A universal evolution criterion at constant force, namely for $\left(\frac{d\sigma_J}{dt}\right)_X$ does not exist. Therefore, the total entropy production, $\left(\frac{d\sigma}{dt}\right)$, does not fulfill any universal criterion.

Nalinē dhalagatha salilam taralam;

Thadva j ē vitham athisayachapalam.

(From Dhwadasha Panjarika Sthothram by **Sankaracharya**).

"Life is very fickle, as fickle as water (drop) on a lotus leaf".

Life is like a bubble on ever-changing (water) currents of a flowing river.

Everything changes. Nothing is omnipotent. (Buddhist Weltanschaung).

Suggested Reading

Any textbook on thermodynamics.

Ackerman, E. Ellis, L.B.M. & Williams, L.E. "*Biophysical Science*". Prentice-Hall/New Jersey (1979).

Chance, B. et al. (Eds.) "*Biological and Biochemical Oscillators*". Academic Press/New York. (1973).

Hoppe, W. et al. (Eds.) "*Biophysics*". Springerverlag/Berlin. (1983).

Klotz, I.M. "*Introduction to Biomolecular Energetics*". Academic Press/New York (1986).

Marrkus, M., Muller, S.C. & Nicols, G. (Eds.) "*From Chemical to Biological Organization*". Springerverlag/ Berlin. (1988).

Martin, R.B. "*Introduction to Biophysical Chemistry*" McGraw Hill/New York (1964).

Van Wylen, G.J. & Sonntag, R.E. "*Fundamentals of Classical Thermodynamics*". 2nd Edn. John Wiley & Sons/New York (1978).

Volkenstein, M.V. "*Molecular Biophysics*". Academic Press/New York (1977).

14 Photo- and Chemo-Bioenergetics

Synopsis: Photosynthesis in bacterial and plant systems and oxidative phosphorylation in mitochondria are the most important sources of energy for cellular processes in all living systems. In Photosynthesis, light energy from the sun is absorbed by chlorophyll molecules and transferred to reaction centres, which are complexes of integral membrane proteins and cofactors. Light energy is converted into chemical energy at the reaction centres by inducing the transport of electrons from a weak electron-donor, i.e. H_2O, to an electron-acceptor $NADP^+$ to form NADPH and ATP by noncyclic-photophosphorylation. NADPH and ATP, generated in the 'light' reactions of photosynthesis are utilised to "fix" atmospheric CO_2 into carbohydrates (dark reactions). In essence, photosynthesis is combining H_2O and CO_2 to synthesise organic compounds and release O_2 in the process. Light quanta, photosynthetic units, H_2O and CO_2 are essential in the photosynthesis of carbohydrates. Molecular oxygen, generated in the photosynthetic process, is utilised by living cells for the synthesis of ATP by aerobic respiration (oxidative phosphorylation) for various cellular reactions. Liberation of H_2O and CO_2 by heterotrophs completes the energy flow.

The synthesis of ATP by oxidative phosphorylation is carried out in mitochondria. In the mitochondrial matrix, food molecules (carbohydrates, proteins and fats) are converted into NADH or $FADH_2$ via the **Krebs cycle**; these are the starting products for the electron-processes that are carried out in the inner membrane. The inner membrane of mitochondria contains the major electron-transport and energy-transduction complexes. The electron-transport chain (respiratory chain) can be functionally compartmentalised into four enzyme-cofactor complexes. Ubiquinone and cytochromes act as redox-electron carriers, working vectorially from negative redox potential towards the more positive side, to transport electrons from electron-donors to reduce the ultimate electron-acceptor, O_2 to H_2O. The electrochemical proton-gradient, generated by proton translocation across the inner membrane, is utilised in the synthesis of ATP. The chemiosmotic hypothesis proposed by **Mitchell** envisages a direct coupling between the electron transfer by the respiratory chain and ATP synthesis by oxidative phosphorylation.

14.1 PHOTO-BIOENERGETICS

All living organisms depend, ultimately, on photosynthesis for their existence. In photosynthesis, light energy from the sun is used to transport electrons from water to "fix" atmospheric carbon dioxide, to produce organic compounds and molecular oxygen, which are used by living organisms on this planet. The general reaction process is

$$CO_2 + H_2O \xrightarrow{\text{Photon } h\nu} (CH_2O) + O_2 \qquad ...(14.1)$$

Photosynthesis in green plants and algae occurs in chloroplasts. The energy-transduction proteins are in thylakoid membrane system (inner-membranes of mitochondria and bacteria contain these proteins (Fig. 14.1a).

14.1.1 The Chloroplast

The chloroplast is a specialised photosynthetic organelle (≈ 5 μm long) compartmentallised into three main spatial segments — inter-membrane space, stroma and thylakoids (Fig. 14.1b).

Stroma: The stroma is a gel-like fluid matrix surrounded by the inner-membrane of the chloroplast. It contains many soluble enzymes and is the region where CO_2 "fixation" and synthesis of sugars take place.

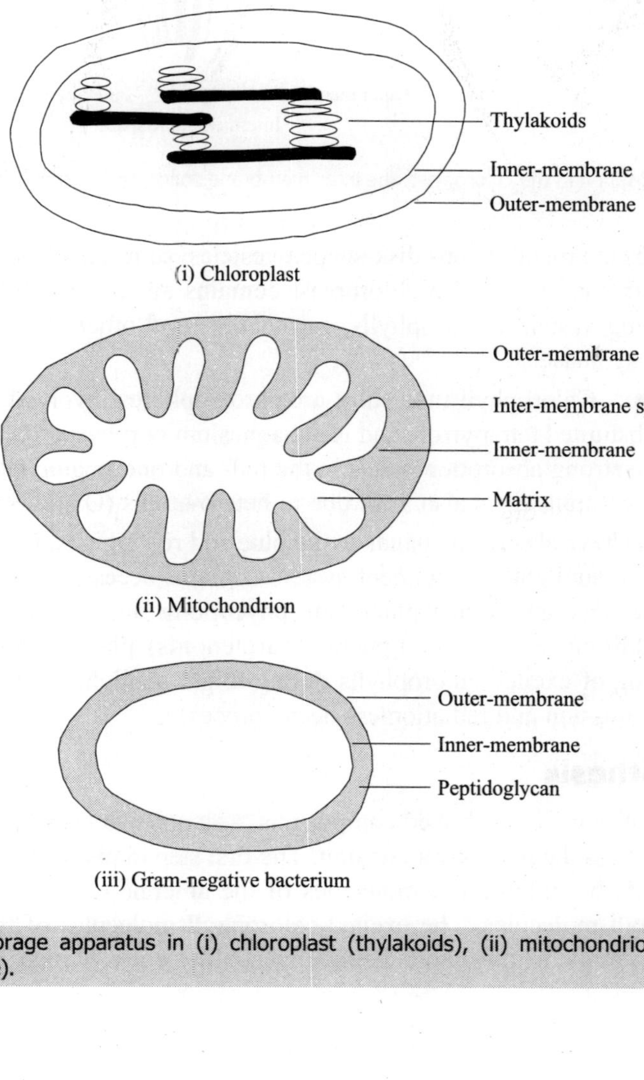

Fig. 14.1a Energy-storage apparatus in (i) chloroplast (thylakoids), (ii) mitochondrion and (iii) bacteria (inner membrane).

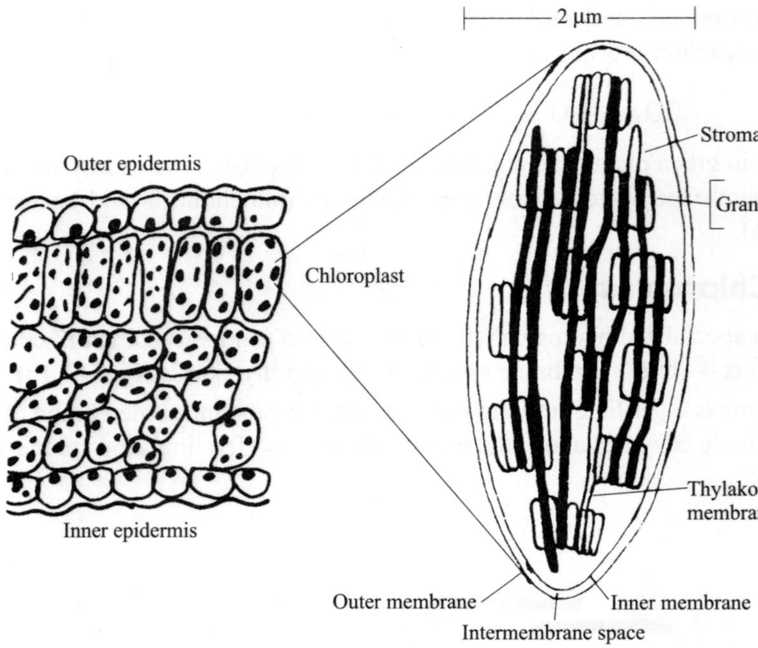

Fig. 14.1b Compartmentallisation in the chloroplast; the inter-membrane space, the stroma and the thylakoid.

Thylakoids: Thylakoids are membranous disk-shaped vesicles, surrounded by the stroma. A pile of these thylakoids is called a granum and a chloroplast contains about 10 – 100 grana. Thylakoids contain the light-absorbing system (chlorophylls, carotenoids and other chromophores), electron-transport chain and ATP synthase.

Chromophore Pigments: Chlorophyll molecules are chromophores involved in the absorption of light. Chlorophyll is a substituted tetrapyrrole and is a magnesium porphyrin. It contains a conjugated π-electron system and the strong absorption bands in the red- and blue-region (**Soret bands**) are due to $\pi - \pi^*$ transitions. $n - \pi^*$ transitions also arise due to hetero-atoms (O and N).

Chlorophylls (a & b) have absorption bands in the blue and red regions. For improvement of the spectral exploitation of the sunlight, the pigment system contains accessory pigments—chlorophyll and carotenoids in green algae and higher plants and phycobilins and carotenoids in red and blue algae (Fig. 14.2). In addition, accessory pigments (carotenoids) play a crucial protective role, preventing the overloading of excited chlorophylls during bright sunlight by dissipating the excited energy via fluorescence emission and radiationless decay processes.

14.1.2 Photosynthesis

A photosystem consists of two closely-linked components—an antenna complex and photochemical reaction centres, referred to as the photosynthetic unit. The first step in photosynthesis begins with the absorption of light quanta by chlorophyll molecules of the antenna complex that contains about ≈99.5% of the chlorophyll molecules. The excited-chlorophyll molecules of the antenna complex transfer their absorbed energy to the photochemical reaction centre, which is a transmembrane protein-pigment-cofactors complex, by resonance energy transfer and/or by electron transfer to a nearby electron-acceptor molecule(s). The antenna complex is important for light harvesting, but the photochemical transduction processes of converting the absorbed light energy to chemical energy take place only at the photo-chemical reaction centres (≈ 0.5% of the chlorophyll molecules are reaction centres which contain "special" chlorophyll pairs to trap excitation energy on a picosecond scale from the excited state to an acceptor. From the functional point of view, the photosynthetic unit can be considered as a lattice, with the antenna complex acting as a "funnel", collecting light energy and transferring it to a specific site, i.e., to the photochemical reaction centre which in turn acts as an effective energy trap. The photochemical reaction centre converts "low-energy" electrons (from the weak electron-donor H_2O) to "high-energy" electrons, creating strong electron-donors.

Electron transfer over a long distance, in a controlled manner, is possible due to the unique organisation of individual components so as to enable unidirectional electron-transfer reactions,

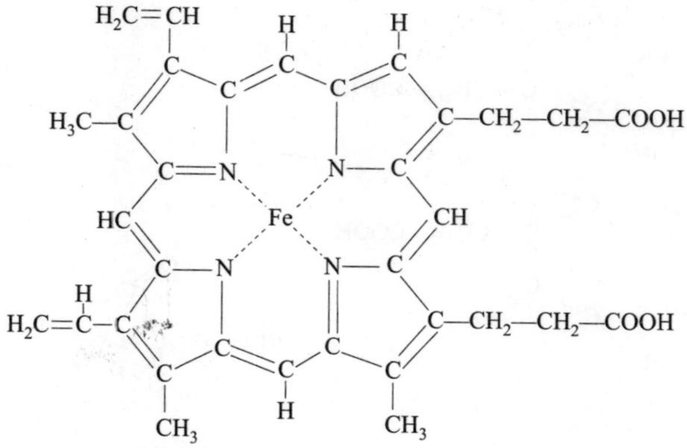

Heme of Myoglobin/Hemoglobin

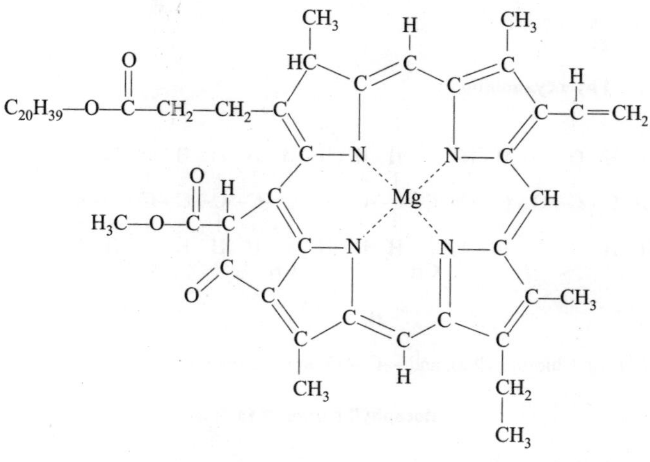

Heme of Chlorophyll a

Fig. 14.2a Several porphyrin ring structures that occur in biomolecules.

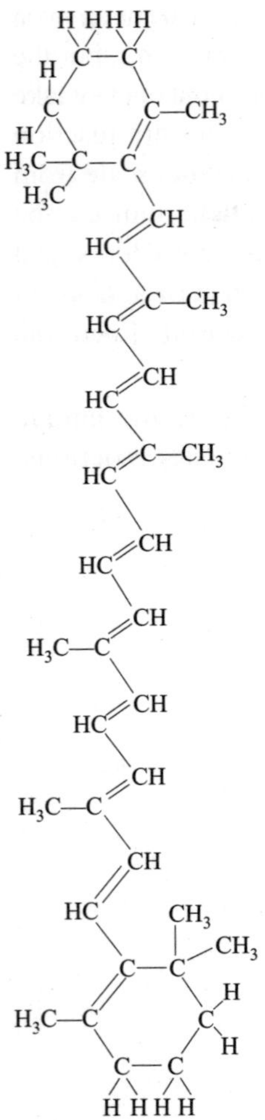

β-Carotene

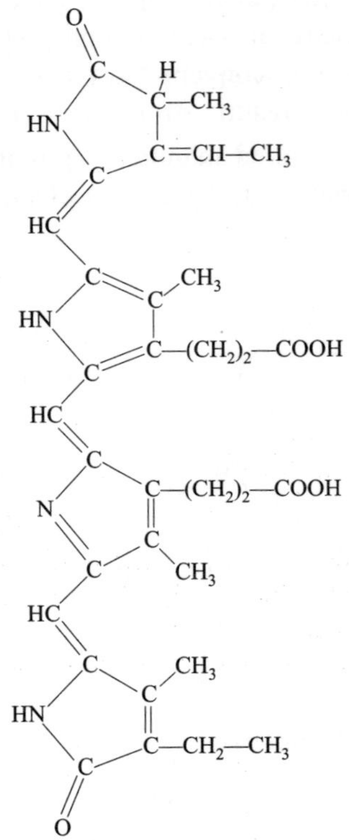

Phycocyanobilin

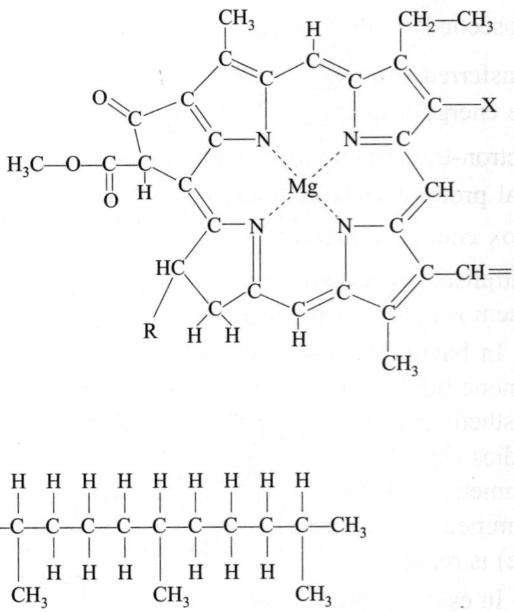

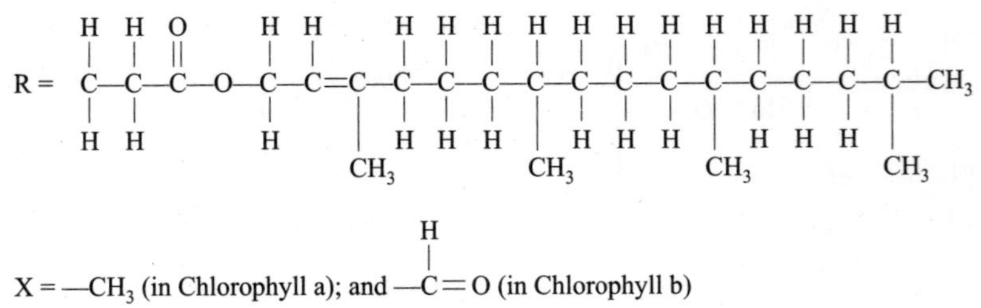

Chlorophyll pigments (a & b)

Fig. 14.2b Structural formula of chlorophylls, carotene and phycobilin.

giving the characteristics of a molecular transistor. The bacterial reaction centre contains three protein subunits—light (L), medium (M) and heavy (H), and the cofactors—two bacteriochlorophylls, two bacteriopheophytins, two quinones and one non-heme high-spin Fe^{2+}. The reaction centres are anisotropically arranged within the thylakoid membrane. The chromophoric groups in the reaction centre are assembled in such a way that a pair of chlorophyll molecules form a closely-associated "special pair", forming a unique spatial organisation in a predominantly hydrophobic environment with the nearby aromatic side chains of the protein complex taking part in the electron-transfer processes. The vectorial electron-transfer processes lead to (i) energy migration within the system, (ii) photochemical reactions leading to charge separation and (iii) electrochemical potential gradient formation.

The photosynthetic reactions that lead to charge separation start with the absorption of light quanta by the light-harvesting pigment complex (LHC), that is, by the antenna. The absorbed light quanta by the antennae migrate toward the reaction centre, i.e., toward the 'special **Chl pair**' in a picosecond time scale *via* excitation and resonance energy transfer processes. An electron of the 'special pair' of chlorophyll molecules absorbs the photon energy and this excited-electron moves to a neighbouring prosthetic group, pheophytin, **Ph**, in ~3 picoseconds with 'special pair' **Chl**-complex rendered positive. The electron is subsequently transferred to a quinone acceptor, Q_A (in 200 picoseconds) which is near the outer surface of the membrane, reducing Q_A to Q_A^-. The electron is ,transferred from Q_A^- to the symmetrically-related quinone molecule, Q_B (ox) in milliseconds. The free energy trapping of the reaction centre in the excited state relies on fast picosecond time scale electron-transfer processes. The oxidation of Q_A^- and transfer to Q_B, forming Q_B^{*-} is coupled with local protein conformational changes which occur on a millisecond time scale, thereby converting redox energy to membrane potential, $\Delta\tilde{\mu}_{H^+}$. In the course of these reactions, the "special pair" is neutralised by a cytochrome molecule that acts as an electron donor to the 'special pair' and the system is ready for the recurrence of the process of photo-induced charge transfer.

In bacterial photosynthetic reaction centres, the light (L) and medium (M) polypeptides are quinone-binding proteins. They are transmembrane proteins with five-pass α-helical segments. The prosthetic groups are: four bacteriochlorophylls, two bacteriopheophytins and one Fe-atom. Structural studies (by **Huber's** group) show that **BPhe** and **BChl** are arranged in an approximate two-fold symmetry with the Fe-atom on the two-fold symmetry axis (Fig. 14.3). However, there is functional asymmetry in the reaction centre processes in that there is only one, Q_A on the L-side (**BPhe** on the L-side) is reduced.

In essence, photosynthesis combines H_2O and CO_2 to synthesise organic compounds and in that process releases O_2. The conversion of photon energy to chemical energy is realised only at the reaction centres. In general,

$$\underset{\text{(Hydrogen donor)}}{\underset{\downarrow}{2H_2D}} + \underset{\text{(Hydrogen acceptor)}}{\underset{\downarrow}{CO_2}} \xrightarrow[\text{light}]{h\nu} \underset{\text{(Reduced acceptor)}}{(CH_2O)} + \underset{\text{(Dehydrogenase donor)}}{2D} + H_2O \qquad ...(14.2)$$

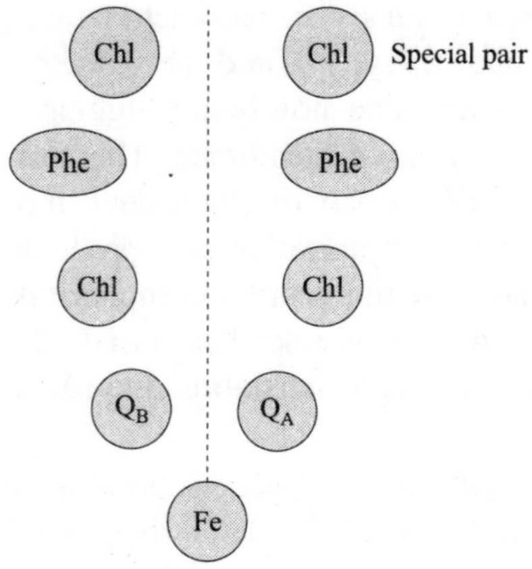

Fig. 14.3 Structural arrangement of the bacterial photosynthetic reaction centre with pseudo two-fold symmetry (through the Fe-atom).

Except for some photosynthetic bacteria, which use H_2S, water is used as the electron donor in photosynthesis. Upon photo-excitation of chlorophylls, the charge transfer in the photosynthetic unit between an electron donor (D) and an acceptor (A) is

$$D.A \xrightarrow{h\nu} D^*.A \longrightarrow D^+.A^- \qquad ...(14.3)$$

The primary donor of electrons is the chlorophyll dimer (special pair) and acceptors of electrons are quinones and other intermediate acceptors. Donor (D) and acceptors (A) of electrons are spatially separated and uniquely assembled.

14.1.3 Photosynthetic Reactions

Photosynthetic reactions can be grouped under (i) light reactions, which use light energy to produce NADPH and ATP and (ii) dark reactions, which utilise NADPH, ATP and CO_2 to synthesise organic compounds (carbohydrates). Light-reactions occur in the thylakoid membranes and the dark-reactions in the stroma (Fig. 14.4).

Light Reactions: Light quanta excite electrons of chlorophyll molecules in the photosynthetic system to higher energy states and this leads to (i) oxidation of H_2O to convert $NADP^+$ to NADPH and liberation of O_2 and (ii) the transport of H^+ across the thylakoid membrane and the resulting proton-motive force drives the synthesis of ATP in the stroma. There are two photosystems, PSI and PSII involved in the photosynthetic processes. Both PSI and PSII have absorption bands below 680 nm, but only PSI has absorption above 680 nm. PSI and PSII are also identified as P700 and P680 pigments, respectively.

(a) Photosystem I (PSI) (P700): Absorption of light quanta leads to the formation of NADPH, without the liberation of oxygen. When the reaction centre, P700, is excited it transfers an electron to bound-ferredoxin (P430, a 11.6 kilodalton Fe-S protein). The bound-ferredoxin, P430, in turn transfers the electron to soluble-ferredoxin, which in turn transfers it to $NADP^+$ to convert it to NADPH (Fig. 14.5). Non-oxygen evolving photosynthetic bacteria contain photosystem I only.

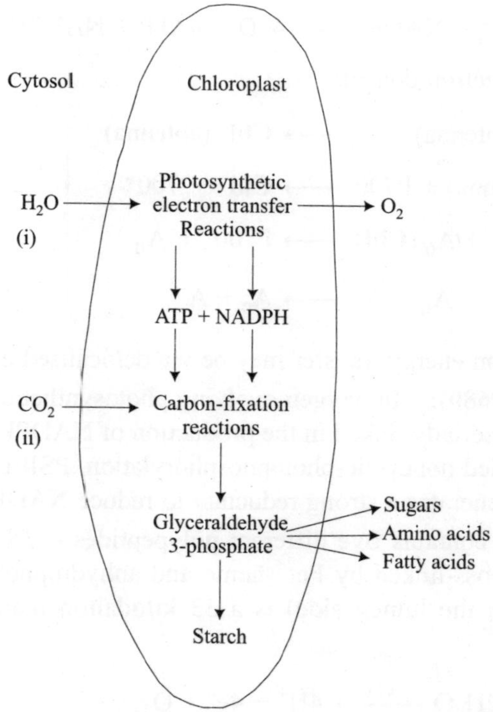

Fig. 14.4 Functional separation of photosynthetic reactions in chloroplast: (i) Photosynthetic electron-transfer (ii) Carbon fixation.

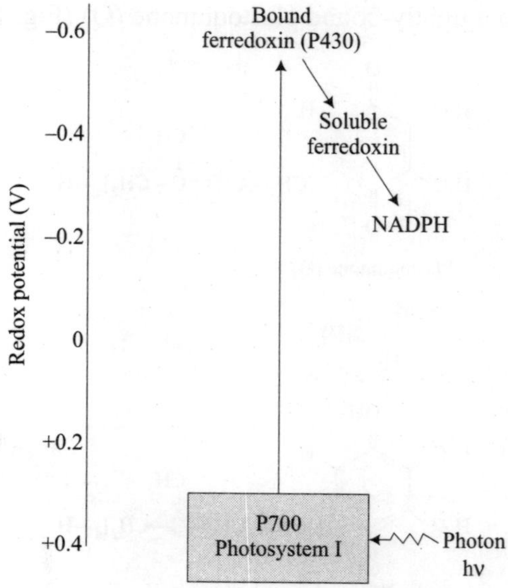

Fig. 14.5 Formation of NADPH by photosystem I (P700).

$$H_2D + nADP + nP_i + NADP^+ \xrightarrow{h\nu} D + nATP + NADPH + H^+ \quad ...(14.4a)$$

(where $n = 0$ to 2 and D is electron donor).

$$\left.\begin{aligned} \textbf{Chl} \text{ (antenna)} + h\nu &\longrightarrow \textbf{Chl}^* \text{ (antenna)} \\ \textbf{Chl}^* \text{ (antenna)} + P700 &\longrightarrow \textbf{Chl} + P700^* \\ P700^* + A_0 \textbf{ (Chl)} &\longrightarrow P700^+ + A_0^- \\ A_0^- + A_1 &\longrightarrow A_0 + A_1^- \end{aligned}\right\} \quad ...(14.4b)$$

In photosystem I, the excitation energy transfer may be via delocalised excitons.

(b) Photosystem II (PSII) (P680): In oxygen-evolving photosynthetic systems (cyanobacteria and plants) both PSI and PSII are serially linked in the production of NADPH and ATP and the release of O_2 by a two-step process, called noncyclic-photophosphorylation. PSII generates a strong oxidant to oxidise H_2O to O_2 and PSI generates a strong reductant to reduce $NADP^+$ to NADPH.

The PSII reaction centre contains five different polypeptides— *D*l and *D*2, (9 + 4) kilodalton dimeric cytochrome **b**-559 cross-linked by Fe^{2+}-heme and anhydrophobic polypeptide. The water-splitting extrinsic protein (in the lumen side) is a 33 kilodalton mangano-protein complex that catalyses the reaction

$$2H_2O \xrightarrow{4h\nu} 4H^+ + 4e^- + O_2 \quad ...(14.5)$$

Four Mn^{2+} ions are thought to be necessary for PSII function.

Formation of one molecule of O_2 requires at least eight electrons.

Absorption of light raises PSII to the excited state and an electron from the excited reaction centre of P680 is transferred to a tightly-bound plastoquinone (Q) (Fig. 14.6).

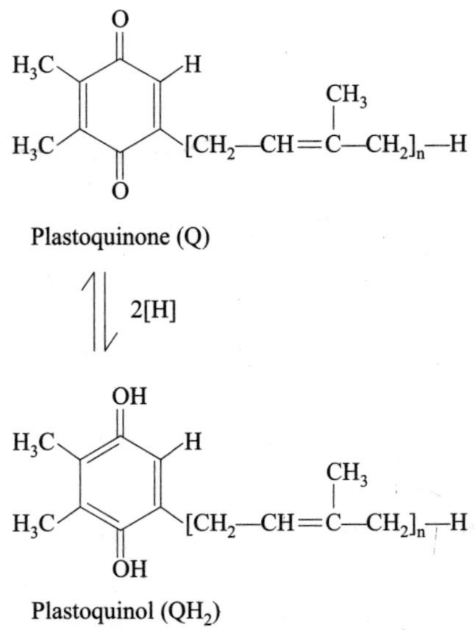

Fig. 14.6 Structural formulae of plastoquinone — Q and QH_2.

The electron is transferred from Q to a pool of mobile plastoquinones and subsequently to cytochrome **b**559 (**b**$_6$ – **f** complex). The final electron-carrier from PSII to PSI is a plastocyanin (a 10.5 kilodalton, copper-containing peripheral membrane protein). PSI is the acceptor of electrons from PSII. By means of two-step electron-energising processes, catalysed by each photosystem (Z-scheme, Fig. 14.7a), absorption of light quanta transfers electrons from H_2O (redox potential = + 0.8 V) to the final acceptor, P430. In effect, light energy causes electrons to flow from H_2O to NADPH (redox pot. = – 0.32 V). The segregation of electron transport components requires the carrier role of quinone (Fig. 14.7b).

The processes of energy migration and transformation occurring within the photosynthetic systems, P680 and P700, are schematically summarised in Fig. 14.8.

The electron-transport process creates an electochemical proton gradient across the thylakoid membrane (H^+ is pumped out of the stroma into the thylakoid space) and the proton-motive force that drives the synthesis of ATP from ADP and P_i by ATP synthase (CF_0F_1-ATPase).

$$2H_2O + 2NADP^+ + 3ADP^{3-} + 3HPO^{2-} \xrightarrow{4h\nu} 2H^+ + O_2 + 2NADPH + 3ATP^{4-} \quad ...(14.6)$$

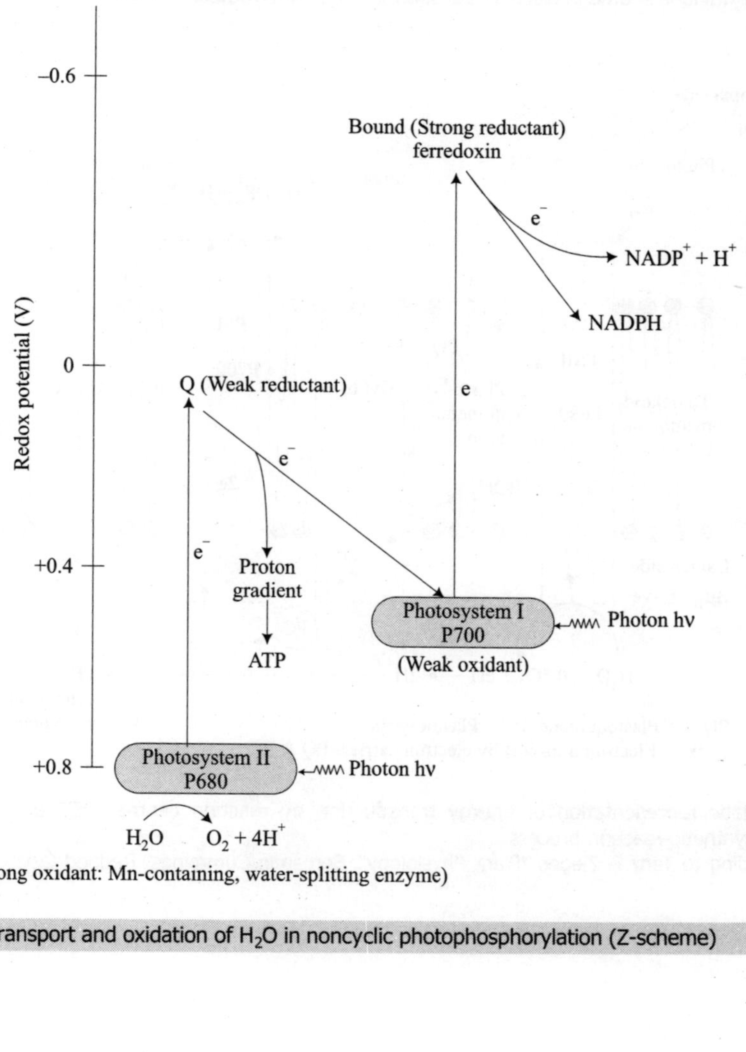

Fig. 14.7a Electron transport and oxidation of H_2O in noncyclic photophosphorylation (Z-scheme)

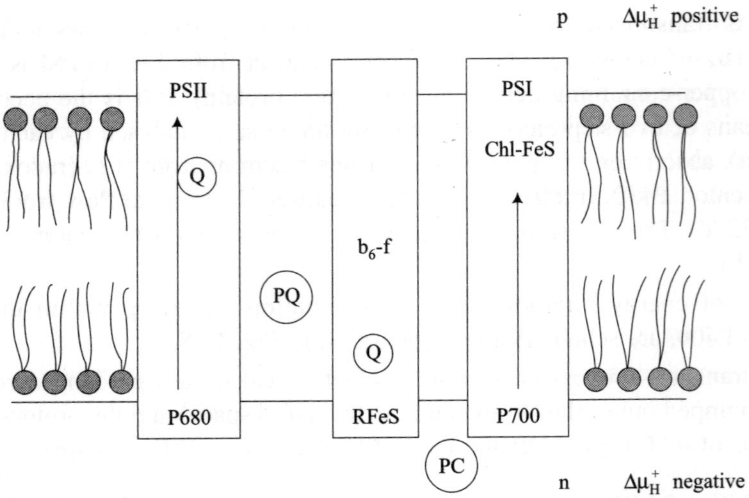

Fig. 14.7b Mobile quinone shuttle in electron transport in the photosynthetic system.

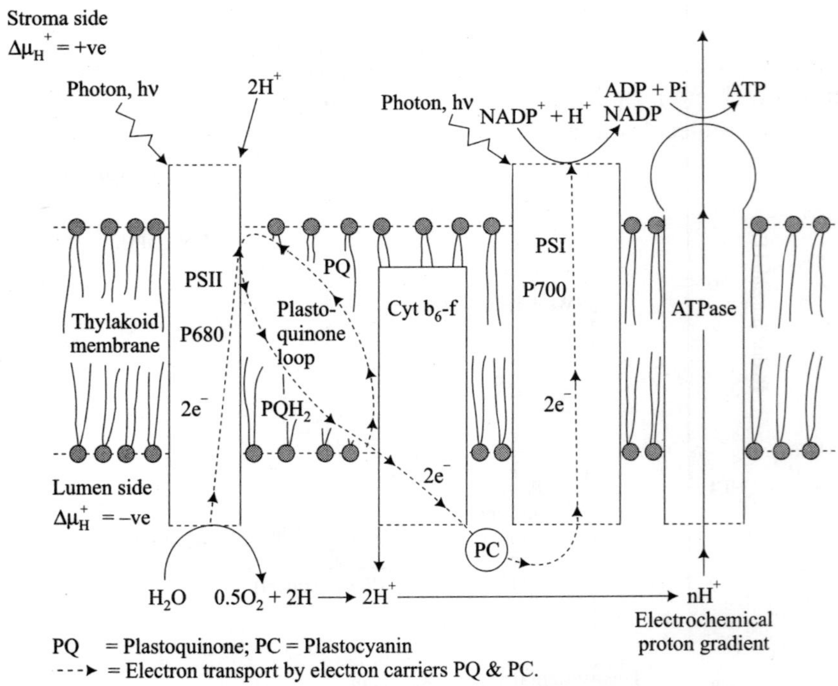

Fig. 14.8 Schematic representation of energy transduction by reaction centres PSII and PSI in the noncyclic photosynthetic-reaction process.
(According to Tanz & Zieger, "Plant Physiology", Benjamin/Cummings, Redood City, CA. (1991).

Chloroplasts can also synthesise ATP only without producing NADPH by cyclic-photophosphorylation. Switching of photosystem I (only) into cyclic mode (electrons are not removed from the transport system by reduction of a metabolite) leads to the transfer of electrons from ferredoxin back to oxidised P700 through electron carriers, cytochrome **b**563, cytochrome **b**552 and plastocyanin, to accomplish H^+-translocation and synthesis of ATP. Photosystem II does not participate in the cyclic-photophosphorylation mode and O_2 is not liberated from H_2O during this process (Fig. 14.9).

$$H_2D + n\text{ADP} + nP_i + \text{NADP}^+ + 2h\nu \longrightarrow n\text{ATP} + \text{NADPH} + H^+ + D \qquad ...(14.7)$$

Dark Reactions. The NADPH and ATP produced by the "light" reactions (photosynthetic electron-transfer reactions), for which sunlight is essential, are the starting chemicals for "fixing" CO_2 to synthesise carbohydrates in the stroma, for which sunlight is not required. These CO_2 "fixing" reactions are called dark reactions, which are **Hill** reactions. In the carbon-fixation cycle, three molecules of ATP and two molecules of NADPH are utilised for each CO_2 molecule that is "fixed". The metabolic pathway by which plants "fix" CO_2 to synthesise carbohydrates is called the **Calvin-Benson** cycle. The cycle generates glyceraldehyde-3-phosphate (GAP) from three molecules of CO_2, expending nine ATP and six NADPH molecules.

$$3CO_2 + 9\text{ATP} + 6\text{NADPH} + H_2O \longrightarrow \text{GAP} + 6\text{NADP}^+ + 8P_i + 9\text{ADP} \qquad ...(14.8)$$

GAP, the primary carbohydrate of photosynthesis, is the precursor of many other organic compounds (sugars, starch etc.). Both energy (ATP) and reducing power (NADPH) are essential for the formation of organic molecules from CO_2 and H_2O.

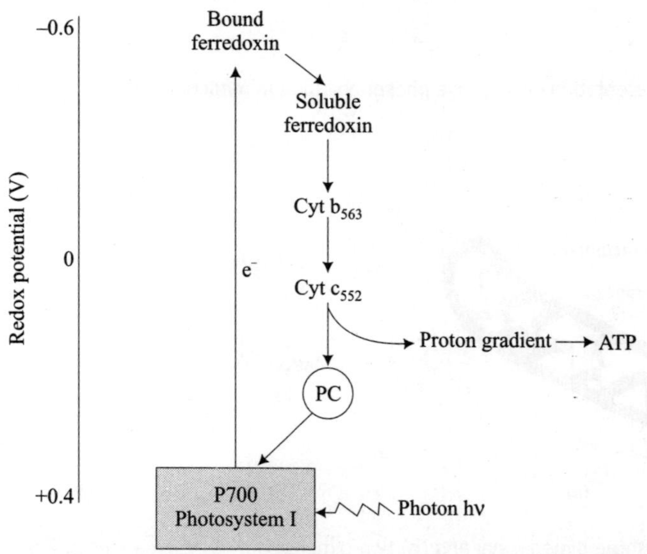

Fig. 14.9 Production of ATP (only) in cyclic photophosphorylation mode by PSI.

14.2 CHEMO-BIOENERGETICS (OXIDATIVE PHOSPHORYLATION)

Aerobic respiration (oxidative phosphorylation) is carried out in mitochondrion, where the foodstuffs (carbohydrates, proteins and fats) are converted into energy-providing molecules, ATP, *via* the **Krebs cycle** (citric acid cycle), and the electron-transport chain coupled to ATP synthesis by oxidative phosphorylation. The primary function of mitochondria is to supply the cell with ATP molecules during oxidative phosphorylation—36 molecules of ATP for every molecule of glucose oxidised. The general scheme of the reactions that takes place in mitochondria is as represented in Fig. 14.10.

14.2.1 Mitochondrial Structure

The mitochondrion consists of an outer membrane, an inner membrane, separated from the outer membrane by an intermembrane region and a matrix region, surrounding by an inner membrane and inner-membrane particles (knobs) protruding into the matrix region (Fig. 14.11).

The bioenergetic reactions in the mitochondrion take place in the matrix region and in the inner membrane surrounding it. Most of the soluble enzymes of the mitochondrion are found in the matrix, including the enzymes required for electron-transport (respiratory-chain enzyme complexes) and oxidative phosphorylation (F_0F_1-ATPase). The cluster of knobs on inner membranes, protruding into the matrix are F_1-particles of F_0F_1-ATPase. The location of enzymes in the matrix and inner membrane of the mitochondrion is given in Fig. 14.12.

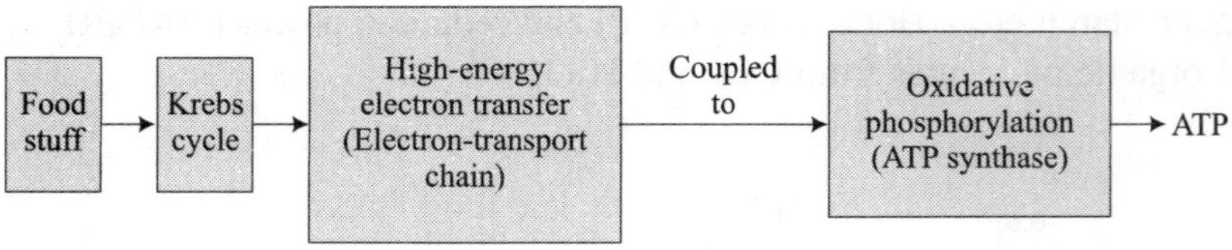

Fig. 14.10 Schematic representation of oxidative phosphorylation in mitochondrion.

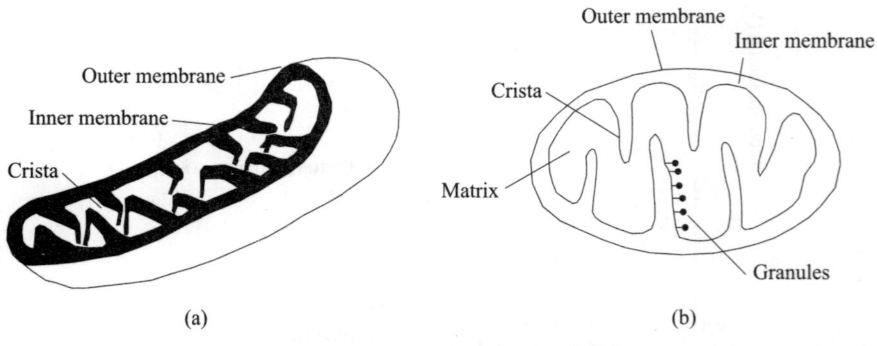

Fig. 14.11 Schematic (a) three-dimensional and (b) two-dimensional representations of a mitochondrion.

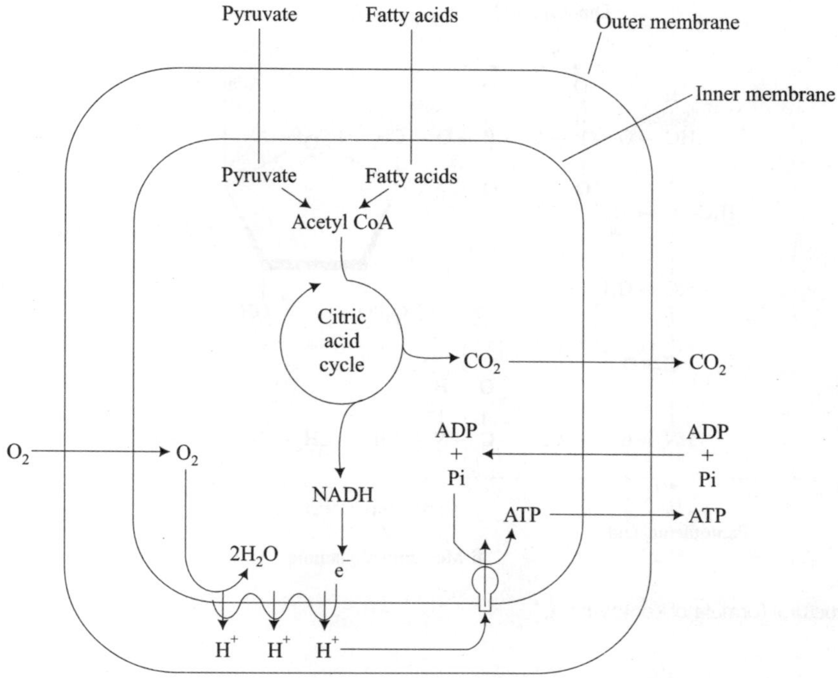

Fig. 14.12 Location of enzyme complexes of respiratory and oxidative phosphorylation reactions in the mitochondrial matrix and the inner membrane.

14.2.2 Electron-Transport Processes

The oxidation process in mitochondrion (in the matrix) begins when pyruvate $(H_3C-\underset{\underset{\displaystyle O}{\|}}{C}-COO^-)$ and fatty acids are converted into acetylcoenzyme A (acetyl CoA) (Fig. 14.13), which is the starting product for the citric acid cycle (Krebs cycle). In the citric acid cycle, the acetyl group of acetyl CoA is oxidised to produce end-products NADH and $FADH_2$ (Fig. 14.14), which are in turn the starting products for the electron-transport chain, redox reactions in the mitochondrial inner membrane, that transport electrons down the respiratory chain to reduce molecular oxygen, O_2, to water.

The oxidative and phosphorylative reactions in the mitochondrial inner membrane can be functionally compartmentalised into five enzyme complex groups. Four of these are respiratory chain complexes, I, to IV (complexes I and II are dehydrogenases, complex III is cytochrome **b-c1** complex and complex IV is cytochrome **c**-oxidase) and the Vth complex, ATP synthase, is coupled to the electron-transport chain. Each of the complexes of the electron-transport chain has increasing affinity for electrons (starting from complexes I to IV) and electrons are thus transported vectorially down the electron-transport chain to O_2, which has the greatest affinity for electrons. The Vth complex, ATP synthase, catalyses synthesis as well as hydrolysis of ATP (Fig. 14.15a).

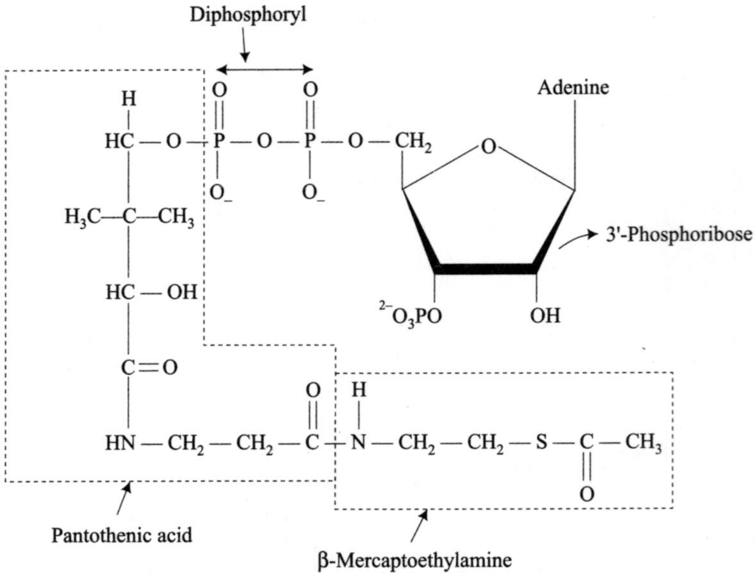

Fig. 14.13 Structural formula of conezyme A.

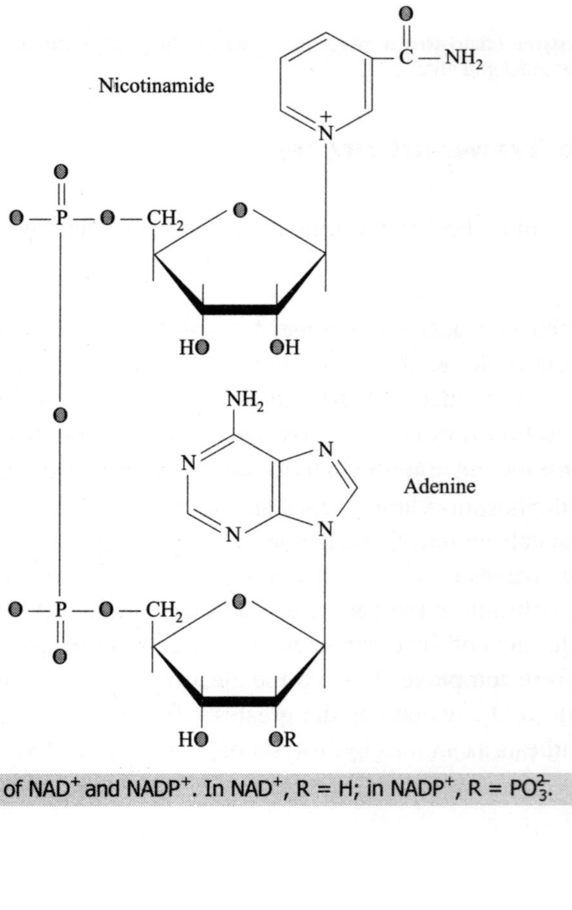

Fig. 14.14a Structural formulae of NAD^+ and $NADP^+$. In NAD^+, R = H; in $NADP^+$, R = PO_3^{2-}.

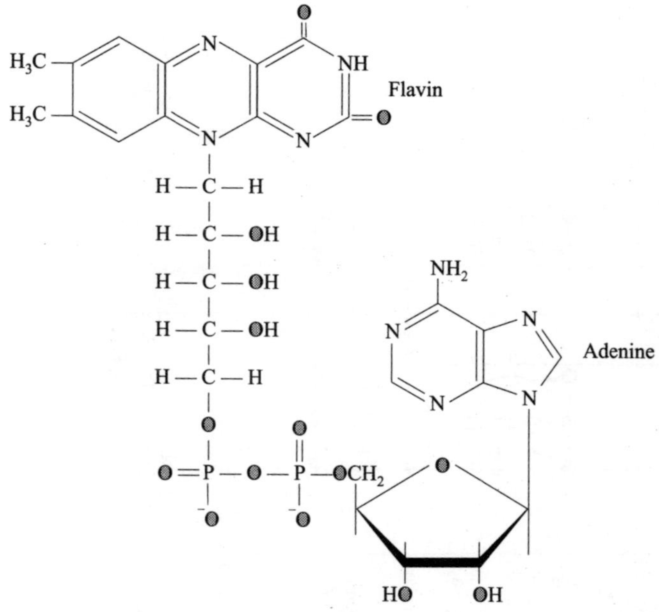

Fig. 14.14b Structural formula of FAD.

The Electron-Transport Chain (Respiratory Chain). The electron-transport chain (located in the inner membrane) comprises the enzymatic machinery for electron transport. Electron transport starts with NAD^+, reducing it to NADH (high negative redox-potential, $\Delta G = -200$ kJ/mol) and ends in oxidising cytochrome **c**-oxidase ($\Delta G = 0$), which in turn reduces O_2 to H_2O. The electron-transport chain is schematically represented in Fig. 14.15 (b).

There are three types of carrier proteins, differing in their prosthetic groups (in addition to ubiquinone, Q)—flavoproteins (flavin-nucleotides), cytochromes (heme-proteins) and metallo-proteins (iron-sulphur centres or copper-proteins).

(i) Complex I (NADH-Q Reductase): Complex I is a large molecular mass (>500 kilodaltons) flavoprotein with about 15 subunits. It transfers two electrons from NADH to coenzyme Q (Fig. 14.16). NADH is oxidised in the matrix face of the membrane by the FMN-containing component, NADH-dehydrogenase. There are about 4 to 7 Fe-S centres.

(ii) Complex II (Succinate dehydrogenase-Q Reductase): Complex II is an enzyme complex of flavoprotein (~70,000 daltons) and iron-protein (27,000 daltons) and transfers electrons from succinate, oxidising succinate to fumarate in the matrix during the operation of the Krebs cycle. The complex consists of 7 to 8 non-identical subunit proteins and three Fe-S centres. The role of ubiquinone (Q) subunit is a "**hydrogen pool**", i.e., a mobile ($2H^+$ and $2e^-$) carrier, transferring electrons from complexes I and II (from the flavoproteins) to complex III (cytochrome **b-c1** complex) (Fig. 14.17). Cytochrome **b** appears to be essential for the anchoring of the complex to the membrane.

(iii) Complex III (QH_2-Cytochrome Reductase): The complex III, QH_2-cytochrome reductase (**b-c1** enzyme complex) is a multisubunit enzyme complex of molecular mass ≈280,000 daltons (≈ eight subunits that include two molecules of cytochrome **b** and one molecule of cytochrome **c1** and an electro-positive Fe-S protein). Whereas cytochrome **b** is an integral protein, cytochrome **c1** is a

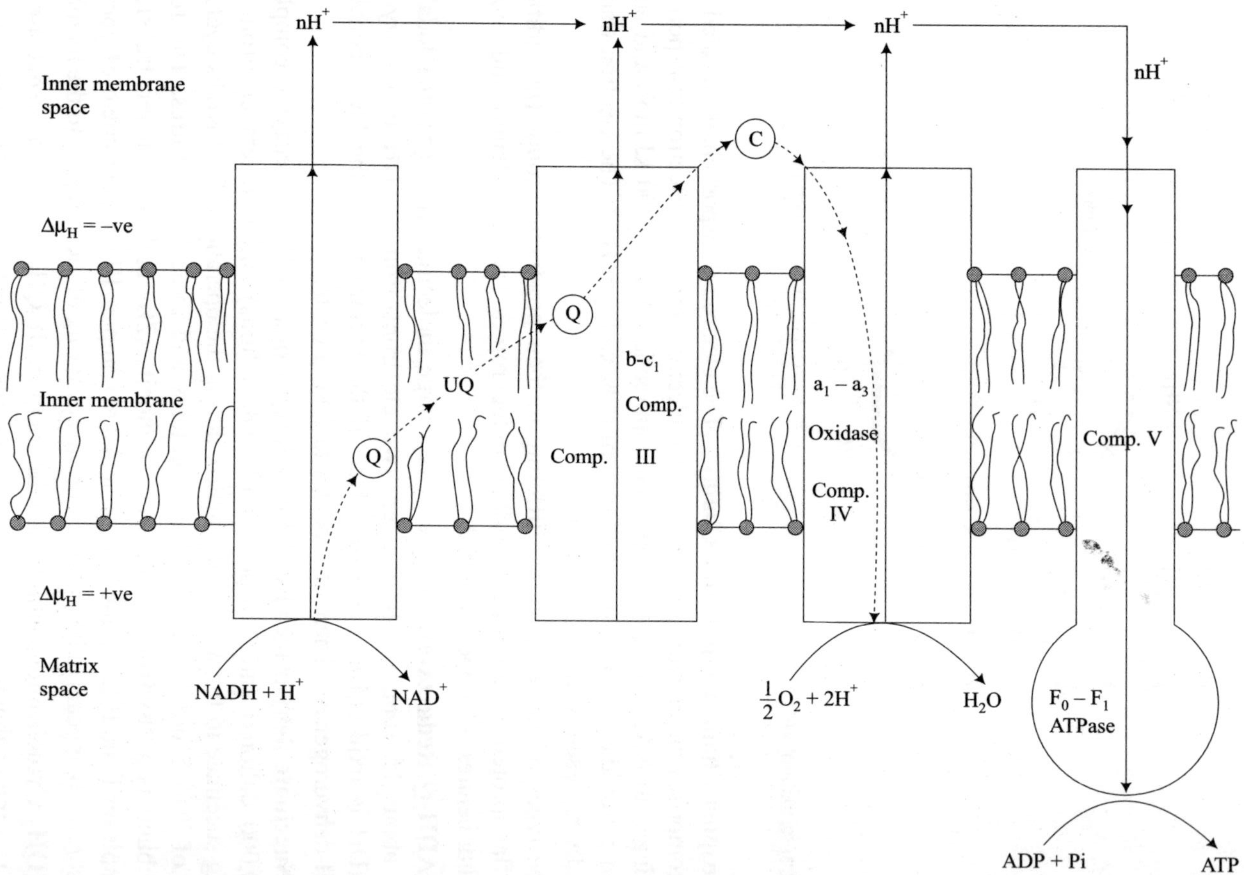

Fig. 14.15a Schematic representation of the organisation of the electron transport chain (complexes I to IV) ATP synthase (complex V) in mitochondria.
UQ = Ubiquinone; *C* = Cytochrome **c**.
— — → Electron transport by electron carriers *Q* & *C*.

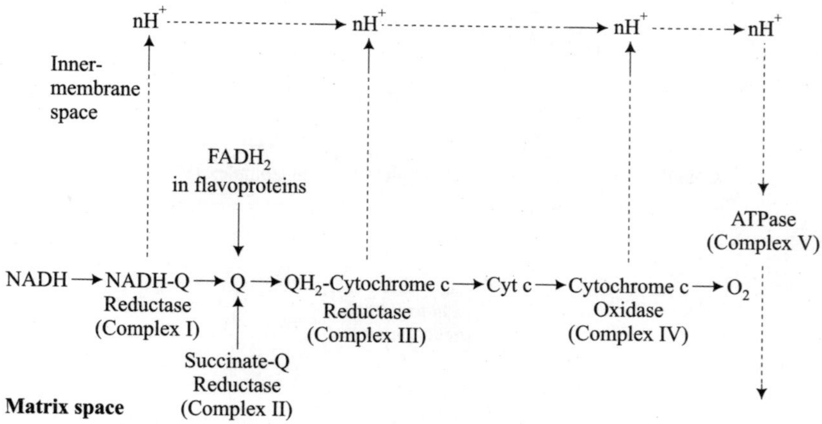

Fig. 14.15b Schematic representation of the electron-transport chain.

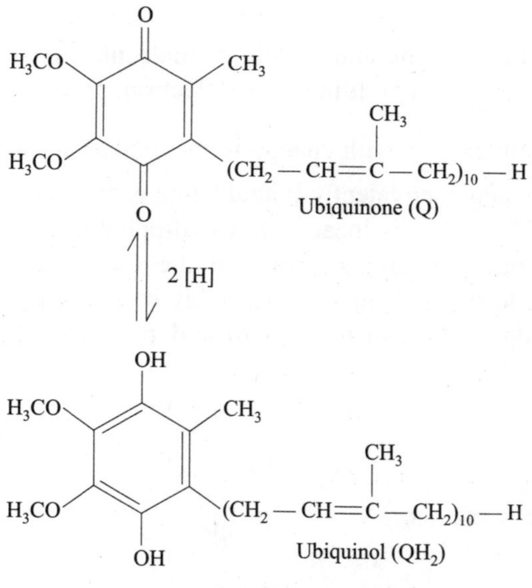

Fig. 14.16 Structural formulae of ubiquinone and ubiquinol—Q & QH_2. Q and c act as electron carriers.

peripheral protein. Complex III catalyses the reduction of cytochrome by QH_2 (translocation of electrons to cytochrome **c** *via* QH_2). Two molecules of cytochrome **c** are reduced while one molecule of QH_2 is reoxidised.

$$QH_2 + 2\text{Cyt } \mathbf{c}\ (Fe^{3+}) \longrightarrow Q + 2\text{Cyt } \mathbf{c}\ (Fe^{2+}) + 2H^+ \qquad ...(14.9)$$

The **b-c1** complex is similar in function to the **b_6-f** complex in photosynthetic reaction centres. They provide the only pathway for transferring electrons from quinol on the reducing side of the respiratory and photosynthetic electron-transport to c-type cytochrome or plastocyanin on the oxidising side.

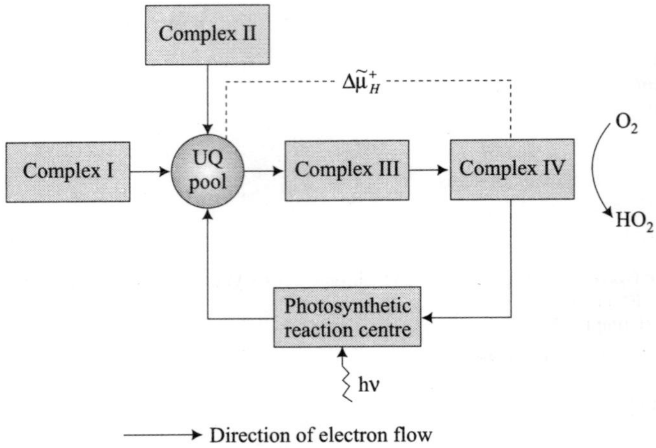

Fig. 14.17 The central role of "ubiquinone pool" in the electron-transport chain.

Cytochrome **c** is a peripheral membrane protein of molecular mass 18-15 kilodaltons, with an iron-porphyrin chromophore (Fig. 14.18). It transports electrons from cytochrome of the complex III to cytochrome **c** oxidase (complex IV), with change in the oxidation state of iron ($Fe^{2+} \rightleftharpoons Fe^{3+}$).

Cytochrome **c** contains a single, covalently-bound heme **c**, attached *via* two thioether linkages to a single polypeptide chain. The heme is located in a hydrophobic crevice surrounded by nonpolar amino acid residues. Structural features relevant to the mechanism of electron-transport are: (i) exposure of a heme edge to the solvent and (ii) highly-conservative positive charge distribution about this exposed heme edge. Electron transfer to and from cytochrome **c** occurs through the

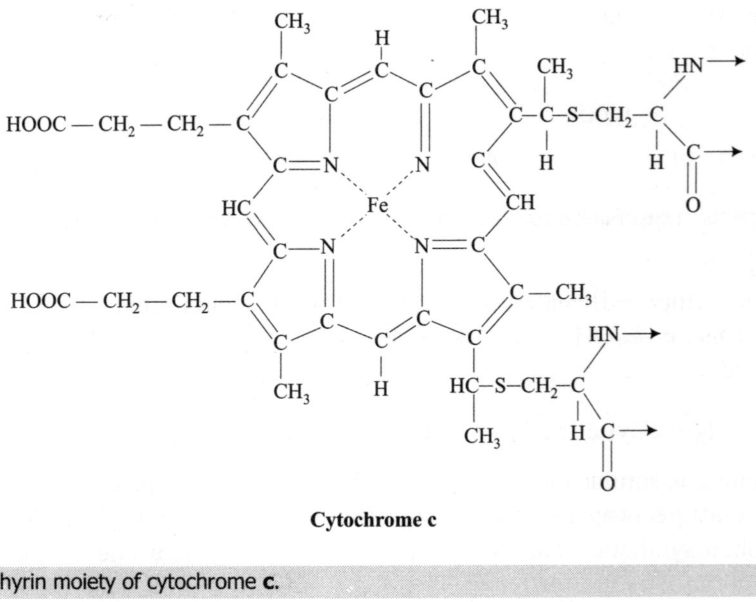

Cytochrome c

Fig. 14.18 Fe-Porphyrin moiety of cytochrome **c**.

exposed heme edge of Cyt **c**, with positively-charged **Lys** residues providing the main ionic interactions to the acceptor/donor redox protein(s). The electron acceptors of Cyt **c1**, Cyt **f** and Cyt **c** are in mitochrondria, Cyt **c2** in chromatophores and plastocyanin in chloroplasts. Cyt **c**, Cyt **c2** and plastocyanin are on the positive $\Delta\tilde{\mu}_{H^+}$ -side of the membrane, i.e., on the cytoplasmic, periplasmic and lumen sides of the mitochondrial, chromatophore and chloroplast membranes, respectively.

Ubiquinone has a central role to play (ubiquinone pool) in electron transport. The "quinone pool" is the branching point for electron transport (Fig. 14.17).

(iv) Complex IV (Cytochrome c Oxidase): Complex IV is the terminal complex of the respiratory chain, which catalyses the crucial step in the respiratory chain of electron-transport from Cyt **c** from the outside of the inner mitochondrial membrane to reduce molecular oxygen to water. It is a multi-subunit enzyme complex with cytochromes **a** and $\mathbf{a_3}$ constituents. The component which is susceptible to CO and CN^- coordination is designated cytochrome $\mathbf{a_3}$. The sixth ligand of cytochrome $\mathbf{a_3}$ is molecular oxygen which is reduced to H_2O. The copper-containing subunits are similar to plastocyanin (PC) and azurin. Four molecules of Cyt $\mathbf{a_3}$ catalytically reduce a molecule of O_2 to H_2O and in this process the electron donor is cytochrome **c**.

$$4\ \text{Cyt}\ \mathbf{a_3}\ (Fe^{2+}) + O_2 + 4H^+ \longrightarrow 4\ \text{Cyt}\ \mathbf{a_3}\ (Fe^{3+}) + 2H_2O \qquad ...(14.10)$$

The cytochrome oxidase not only catalyses the reduction of O_2 to H_2O, but it also acts as a redox-linked proton pump, with four protons translocated for every O_2 consumed. The enzyme complex is ~200,000 daltons (manomer) and spans the membrane as an asymmetric Y-shaped molecule. The stalk of the molecule extends upto ~50 Å towards the cytoplasmic side of the inner-mitochondrial membrane and the arms cross the lipid membrane and penetrate ~20 Å inside the mitochondrial matrix.

Subunit I of cytochrome oxidase has twelve membrane-spanning helices. Heme **a** and $\mathbf{a_3}$ and Cu_B bind to subunit I. Subunit II traverses the membrane twice and is the binding site of Cyt **c**. The Cu_A site is in subunit II. Subunit II shows homology with 'blue' copper proteins, such as azurin and plastocyanin. Subunit III has five helices spanning the membrane. Its role is, presumably, to assist proper folding of subunits I and II. Electron transfer to the heme **a** site is followed by a series of intra-molecular electron transfer steps enabling them to reach the binuclear Cu_B-heme $\mathbf{a_3}$ site.

The mechanism of proton translocation is presumed to be *via* hydrogen-bonded chains (HBC) and/or through transient trans-protein-water channels. Amino acids such as **Asp**, **Glu**, **His**, **Ser**, **Lys**, **Tyr** and **Trp** etc., provide the polar groups for HBC formation. The mechanism of the cytochrome oxidase proton pump may be similar to bacteriorhodopsin, driven by redox-linked processes.

As mentioned earlier, electron transport over a long range down the respiratory chain with speed and specificity is accomplished by the hierachal and proper organisation of enzyme complexes and redox intermediaries, so that electron transfer is achieved stepwise and thermodynamically unidirectional from the strong electron donor, NADH, to a strong electron acceptor, O_2 (that is, redox potentials change gradually from more negative to more positive).

Quinones and cytochromes act as electron carriers in the electron-transport chain. Fe-heme-containing redox-active proteins are cytochromes, undergoing one-electron oxidation-reduction change in the iron of their heme groups ($Fe^{2+} \rightleftharpoons Fe^{3+} + e^-$). The coordination of iron is six, with four of the ligands being the nitrogen atoms of the porphyrin ring. In the **b** and **c** cytochromes, the 5th and 6th ligands are N and S of amino acid side chains of the protein. In the $\mathbf{a_3}$ cytochromes, the 6th ligand is O_2. That is, many invariant amino acids of the protein/enzyme complexes, the oxidation state of the metal ions, sulphur ligands and disulphide bridges are intimately involved in the electron-transport process.

Even the functions of the enzyme, cytochrome **c** oxidase (complex IV), reduction of O_2 and translocation of H^+ are structurally, kinetically and thermodynamically coupled. In the case of O_2, the ground state has two extra electrons in the π_g orbital $[\pi_x^* (\uparrow), \pi_z^* (\uparrow)]$ imparting to oxygen inertness on the one hand and high oxidation potential on the other. That is, the chemistry of O_2 makes it thermodynamically reactive but kinetically stable. In addition, the H^+-translocation is coupled to electron transfer, if and only if O_2 is ligated at the proper heme-centre to cytochrome $\mathbf{a_3}$. A brief summary of the enzyme complexes involved in oxidative phosphorylation is given in Table 14.1.

Table 14.1 Enzyme complexes in electron-transport and oxidative phosphorylation.

Enzyme complex	*Features*	*Redox potential (mV)*	*Function*
Complex I:	Flavoprotein (> 500 kD).		Transfer of two electrons from NADH to Q (Q_{10})
NADH-Q Reductase	≈ 15 subunits.		
	Prosthetic groups— Fe-S proteins; FMN	(–370 to –20)	NADH is oxidised on the matrix side.
Complex II:	Flavoprotein (90 kD).		Transfer of electrons from succinate
Succinate-Q Reductase	Prosthetic groups–IFAD and several Fe-S proteins.	(–270 to +140)	to Q (Q_2)
Ubiquinones, Q	Lipid-soluble quinones. (Ubiquinone, Q_{10} & Q_2)		Mobile electron carriers from complexes I & II to complex III.
Complex III:			
QH_2–Cytochrome **c** Reductase	**b–c₁** enzyme complex (280 kD) ≈ 8 subunits.		Transfer of electrons to cytochrome **c**.
	Prosthetic groups—two Cyt **b**, one Cyt $\mathbf{c_1}$ and Fe-S protein.	(–90 to +50) (+220 to +290)	Reduction of Cyt **c** with the oxidation of QH_2.
Cytochrome **c**	Peripheral membrane protein (13 kD), with Fe-Porphyrin group.		Carrier of electrons from complex III to complex IV. One electron oxido-reduction *via* $Fe^{2+} \rightleftharpoons Fe^{3+}$.
Complex IV:			
Cytochrome **c** Oxidase	Terminal complex of the respiratory chain. Multi-subunit lipoprotein complex. (200 kD). Prosthetic groups— Cyt **a** and Cyt $\mathbf{a_3}$.	(+300 to +400)	Oxidises Cyt **c** and reduces O_2 to H_2O. Process is coupled to oxidative phosphorylation.
Complex V:			
F_0F_I-ATPase	H^+-translocating ATPase. F_I has five subunits—$\alpha_3\ \beta_3\ \gamma\ \delta\ \varepsilon$ (380 kD). F_0 has 3 to 5 subunits. Stalk ≈ 18 kD.		Synthesis of ATP by vectorial translocation of H^+, powered by the proton-motive force generated by the electron-transport chain.

kD = kilodalton

14.2.3 Oxidative Phosphorylation

Oxidative phosphorylation is the single-most important function of mitochondria. It is the process whereby the energy released from the oxidation reactions of the electron-transfer chain is coupled to the synthesis of ATP. The enzyme complex catalysing this reaction is called ATPsynthase or H^+-translocating ATPase. ATPsynthase is a multisubunit enzyme complex with three morphologically-distinct segments (F_0, F_1 and stalk). The sites of synthesis and hydrolysis of ATP are in he β-subunit of the F_1-part (soluble part of ATPase; the knobs protruding towards the matrix in the inner membrane of mitochondria). For the synthesis of ATP proper assembly of F_0, F_1 and the stalk is crucial, whereas for the hydrolysis of ATP only the F_1-part is involved (F_0 and stalk are not essential).

The electron-transport system of mitochondria is coupled with the phosphorylating system. The protons, H^+, originating from the electron-transport at three places (major enzyme complex sites— NADH-dehydrogenase complex (complex I), **b-c** complex (complex III) and cytochrome oxidase complex (complex IV) are translocated by the respiratory chain across the inner membrane, from the matrix side to the intermembrane space. This creates a pH difference and electro-potential gradient across the inner membrane (Fig. 14.19).

This asymmetric environment across the inner membrane creates a net electrochemical potential gradient. It exerts a proton-motive force (ΔG) that tends to move H^+ from the inter-membrane space into the matrix side, which is coupled to the synthesis of ATP from ADP and P_i in the matrix (Fig. 14.15a). For each molecule of NADH oxidised, six H^+ are translocated through the inner membrane, giving rise to three ATP molecules. The free energy (ΔG) of the process due to pH as weli as electro-potential differences is given by

$$\Delta G = 2RT\,(pH_{in} + pH_{out}) + z\,\mathcal{F}\,\Delta\psi \qquad ...(14.11)$$

For protons $$\Delta\mu_H{}^+ = \mathcal{F}\,\Delta\psi - 2RT\,\Delta\,(pH)$$

where z = charge on proton; $\mathcal{F}$ = Faraday constant.

Proton translocation is similar in mitochondria and chloroplasts. However, in mitochondria the contribution to proton-motive force is mainly from the membrane potential gradient, while in chloroplasts, the major contribution is due to the pH gradient (H^+ is pumped out from the stroma (pH = 8.0) into the thylakoid space (pH = 5.0). H^+-ATPase can also function as hydrolase, hydrolysing ATP to ADP and P_i by reversing the direction of H^+-translocation (pumping H^+ out from the matrix into the intermembrane space).

Mechanisms of Oxidative phosphorylation. Energy-transducing membranes have two distinct enzyme/protein assemblies: (i) primary energy source (absorption of light in photosynthesis or respiration, i.e., oxidation of substrates, like NADH) and (ii) ATPsynthase to catalyse the uphill synthesis of ATP from ADP and P_i by utilising available energy. Several hypotheses have been proposed to explain the coupling between respiration and phosphorylation— (i) direct coupling (chemiosmotic) mechanism by **Mitchell** and (ii) indirect (conformational change) mechanisms by **Boyer** and others. The chemiosmotic hypothesis proposed by Mitchell seems to give a better account of the experimental observations.

According to the chemiosmotic hypothesis the coupling is essentially electrochemical in nature.

(i) ATP synthesis is driven by a proton electro-chemical potential, $\Delta\tilde{\mu}_{H^+}$.

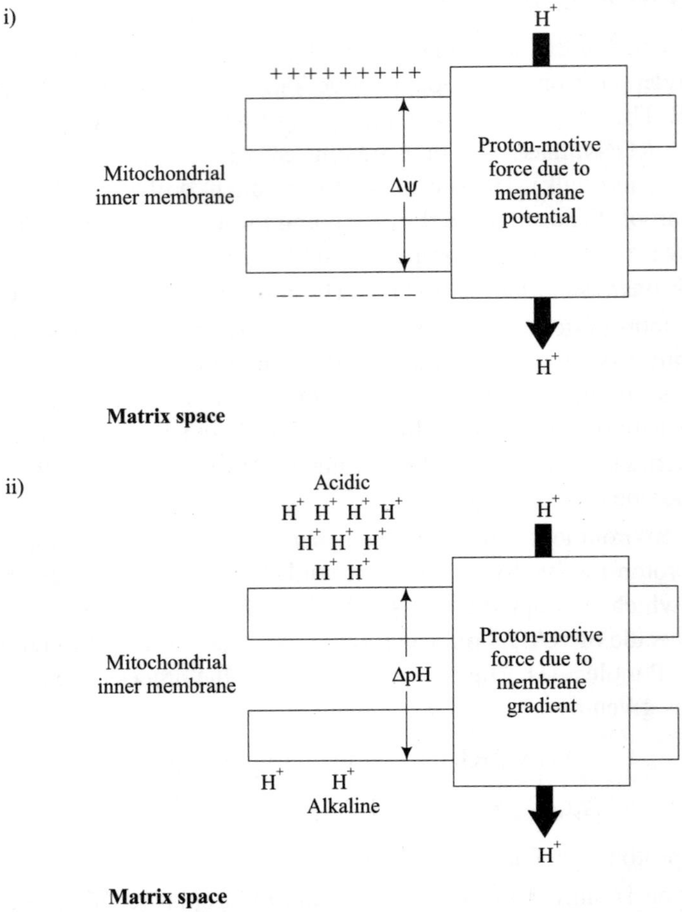

Fig. 14.19 Components of proton-motive force (ΔG).
(i) membrane potential gradient (Δψ)
(ii) proton concentration gradient (ΔpH)

(ii) The energy, $\Delta\tilde{\mu}_{H^+}$ is formed by vectorial translocation of H^+ and e^- carriers in the electron-transport chain.

(iii) The energy is stored in the inner membranes and it is

$$\Delta\tilde{\mu}_{H^+} = \mathcal{F} \,.\, \Delta\psi - 2RT\,\Delta\,(pH) \qquad ...(14.12)$$

(iv) H^+-translocation is coupled to ATP synthesis/hydrolysis, and is catalysed by the multi-subunit F_0F_1-ATPase complex.

The electron-transport chain components are organised in the inner-membrane in stepwise "redox-loops", such that during oxidation-reduction there is an asymmetric distribution of pH and electrical potential gradients. This combined pH and electrical potential gradient, called the proton-motive force, acts as the primary source of energy for the synthesis of ATP. ATP synthase translocates

H^+ (reducing the H^+-gradient) with simultaneous synthesis of ATP. That is, the electron-transport chain is coupled to ATP synthesis by a proton-chemical potential ($\Delta\mu_{H^+}$) across the energy-transducing membrane. In this hypothesis, $2H^+$, driven through the F_0 segment of ATPase by $\Delta\mu_{H^+}$, attack one of the O atoms of P_i, forming H_2O and a reactive species that reacts directly with ADP to form ATP.

$$ATP + P_i \rightleftharpoons ATP + H_2O \qquad ...(14.13)$$

Indirect models place emphasis on the conformational changes of F_1, which cause the synthesis of ATP by altering the relative affinities of F_1 to substrates and products. In these hypotheses, ATP release is the major energy-requiring step, not the synthesis step. Purely conformational models do not adequately explain the coupling of ATP synthesis to proton translocation. Actually, conformational models can be accounted for as one of the features of the chemiosmotic model.

Many of the reactions that occur in photosynthesis and oxidative phosphorylation are similar, but certain features are different—the main objective of photosynthesis is the synthesis of organic products and these end products are the starting products for oxidative phosphorylation to generate ATP. The photosynthetic electron-transport chain is responsible for the tapping of energy from sunlight to form NADH. In chloroplast, the protons move inwards (lumen side) in response to photosynthetic electron transport. Therefore, the chloroplast ATPase is the outer face of the membrane in contrast to mitochondrial ATPase (see Figs. 14.8 and 14.15a). The interior of chloroplast is more acidic than the outside due to inward movement of H^+ during photosynthesis. A comparison of photosynthetic and oxidative phosphorylation reactions is given in Table 14.2 and schematically represented in Fig. 14.20.

Table 14.2 Comparison of photosynthetic and oxidative phosphorylation reactions.

Photosynthesis	*Oxidative phosphorylation*
In chloroplasts	In mitochondria
In thylakoid membranes	In inner-mitochondrion membranes
Light-dependent	Not light-dependent
Liberates O_2	Uses O_2
Uses H_2O & CO_2	Liberates CO_2
Splits H_2O	Forms H_2O
$CO_2 + H_2O + \xrightarrow{h\nu}$ (Organic compounds) end product	Organic compounds + $O_2 \longrightarrow CO_2 + H_2O + H^+$ (end product)
NADH $\longrightarrow O_2$ Flow of electrons	$H_2O \longrightarrow NADPH_2$ Flow of electrons
Redox potential change: $-0.6 \longrightarrow +0.8$ V	$+0.8 \longrightarrow -0.6$ V
CF_0F_I-ATPase	F_0F_I-ATPase (similar function)
Some of the redox-protein complexes— **b_6-f** complex and plastocyanins	**b–c_1** complex and Cu-binding protein components in complex IV, similar to plastocyanin and azurin

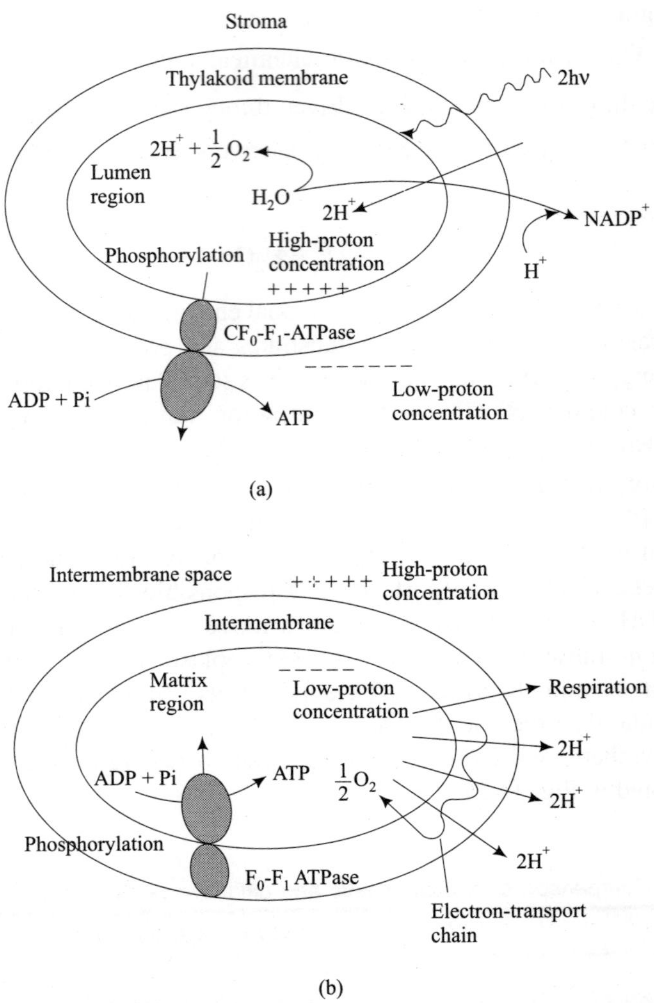

Fig. 14.20 Schematic representation of photosynthetic and oxidative phosphorylation reactions in thylakoid and mitochondrial inner membranes.
(a) thylakoid membrane (ΔpH is the domainant component).
(b) mitochondrial inner membrane (Δψ is the dominant component).

Suggested Reading

Books

Clayton, R.K. "*Photosynthesis: Physical Mechanisms and Chemical Patterns*". Cambridge University Press/Cambridge. (1980).

Crammer, W.A. & Knoff, D.B. "*Energy transduction in Biological Membranes: A Textbook of Bioenergeties*". SpringerVerlag/Berlin. (1990).

Danks, S.M., Evans, G.H. & Whittaker, P.A. "*Photosynthetic Systems*". Wiley/New York. (1983).

Eisenberg, D. & Crothers, D. "*Physical Chemistry with Applications to the Life Science*". Benjamin-Cummings Menlopark/CA (1979).

Ernster, L (Ed.). "*Bioenergetics*". Elsevier/New York (1985).

Finean, J.B., Coleman, R. & Mitchell, R.H. "*Membranes and Their Cellular Functions*" Blackwell Sci/Oxford. (1978).

Foyer, C.H. "*Photosynthesis*". Wiley/New York. (1984).

Hoober, J.K. "*Chloroplasts.*" Plenum Press/New York (1984).

Hoppe, W. et. al (Eds.). "*Biophysics*" Springerverlag/Berlin (1983).

Martinosi, A.N. (Ed.). "*Bioenergetics of Electron and Proton Transport*". 2nd Edn. Plenum Press/New York (1985).

Nicholls, D.G. & Ferguson, S.J. "*Bioenergetics*". Academic Press/New York (1992).

Stryer, L. "*Biochemistry*". 4th Edn. Freeman Press/New York (1995).

Tzogoloff, A. "*Mitochondria*". Plenum Press/New York. (1982).

Voet, D. & Voet, J.G. "*Biochemistry*". John Wiley/New York. (1990).

Zuby, G. (Ed.) "*Biochemistry*". 2nd Edn. (McMillan/New York. (1988).

Articles

Amzel, L.M., Mc Kinney, M., Narayanan, P. & Pederson, P.L. "*Structure of the Mitochondrial F1-ATPase at 9-Å resolution*". Proc. Natl. Acad. Sci., USA. 79; 5852 (1982).

Amzel, L.M. & Pederson, P.L. "*Proton ATPases: Structure and Mechanism*". Annu. Rev. Biochem. 52; 801 (1983).

Deisenhofer, J., Mitchell, H. & Huber, R. "*The Structural Basis of Photosynthetic Light Reactions in Bacteria*". TIBS. 10; 243 (1985).

Deisenhofer, J. & Chan. V. "*Structures of Bacterial Photosynthetic Reaction Centres*". Annu. Rev. Cell Bio., 7; 1 (1991).

Feher, G. et al. "*Structure and Function of Bacterial Photosynthetic Reaction Centres*". Nature 339; III (1989).

Gust, D. & Moore, T.A. "*Photosynthesis*". Science 244; 35 (1989).

Hinkle, P.C. & McCarthy, R.E. "*How Cells Make ATP*". Sci. Amer. 238 (3); 104 (1978).

Huber, R. "*Structural Basis of Light Energy and Electron Transfer in Biology*". Eur. Biochem. 187; 283 (1990).

Knoff, *D.B.*"*The photosystem I reaction centre*". TIBS 13; 157 (1988).

Krauss, N. et al. "*3-D Structure of System I of Photosynthesis*". Nature 361; 326 (1993).

Miller, K.R. "*The Photosynthetic Membrane*". Sci. Amer. 241 (4); 102 (1979).

Mitchell, P. "*Coupling of Phosphorylation to Electron and Hydrogen Transfer by Chemiosmotic Type Mechanisms*". Nature 191; 144 (1961).

Pederson, P.L. & Amzel, L.M. "*ATP Syntheses: ....*" J. Bio. Chem. 268; 9937 (1993).

Rees, D.C. et al. "*The Bacteria Photosynthetic Reaction Center as a Model for Membrane Proteins*". Annu. Rev. Biochem. 58; 607 (1989).

Rees D.C. et al. "*Hydrophobic Organisation of Membrane Proteins*". Science. 245; 510 (1989).

Rutherford, A.W. "*Photosystem II. The Water-splitting Enzyme*". TIBS 14; 227 (1989).

SECTION IV

Biological Systems

OVERVIEW

The external world (surroundings) is perceived by an organism from the stimuli (signals) it receives. These stimuli—mechanical, chemical, thermal etc., are ultimately transduced into electrical signals and communicated to the central nervous system (CNS). Generation of the impulse in the cell membrane is due to membrane characteristics—selective permeability to different ions and asymmetry in the concentration gradient across the cell membrane leading to membrane potential. While ion channels facilitate the passive diffusion of particular ion species, ion pumps (ATPases) generate and maintain the "uphill" ion concentration gradient to provide metabolic energy for the cellular transport processes.

Neurons are signal transduction apparatus. Nerve impulses are transmitted by (i) endocrine, (ii) paracrine and (iii) synaptic signalling processes. Synaptic signalling is the mode of transmission of signals by the nervous system from one cell to the other. Receptor cells are membrane-protein complexes. They can be classified according to their function—chemo-, mechano-, photo-receptors etc. The transduction of impulse by receptor cells is, in general, a two-step process. Sensory cells (mechano-, chemo-, photo- etc.) are specialised receptor cells, on account of their structural development to respond selectively to particular kind of stimulus.

Cell mobility can either be confined to movements within a cell (contraction) movement or the entire cell (locomotion). Contraction and locomotion in living organisms are carried out by contractile (actin-myosin) and motility (tubulin-dynein) complexes. Contraction (motion) is regulated by Ca^{2+} ions and accessory proteins. All sensory stimuli are transduced to the CNS as electrical impulses to the PNS, and vice versa. Biomedical sensors and transducers are employed to monitor fluid transport, blood pressure, heartbeats, biopotentials (e.g. ECG, EEG) and other body functions.

Exposure of organisms to high-energy radiations can lead to a complex variety of reactions, which are dependent on the energy and the dose of the radiation as well as the target. High-energy radiation interacts with matter via (i) photoelectric, (ii) compton and (iii) pair production processes, depending on the energy of the incident radiation. Biological effects of radiation are due to radiation-induced chemical and biological changes. These effects can manifest at molecular, cellular and organism levels. Some types of radiations are biologically more hazardous than others (such as α-rays and neutrons). Exposure to radiation due to natural sources cannot be completely eliminated. Radiation doses below the minimum level do not create health hazards. The main sources of industrial radiation are: X-ray equipment, nuclear installations and radioactive components used in laboratories and hospitals. Radioactive isotopes (radionuclides) are increasingly used in nuclear medicine for diagnostic as well as therapeutic purposes.

15 Neurobiophysics

Synopsis: Nerve cells (neurons) receive and transmit signals from sensory cells. A neuron consists of a cell body, long neuron fibre (axon) and dendrites. Insulated neurons are covered by a myelin sheath, with only regions of the axon in between myelin sheaths, called nodes of Ranvier, exposed to extracellular activities. Myelination gives neurons better cable properties, increased speed and efficiency in signal transmission and reduction in energy consumption.

Nerve cells, like all other cells, are surrounded and separated from each other by cell membranes. The transport of small molecular solutes/ions across cell membranes is controlled by the selective permeabilities of membranes to various ions and resultant concentration gradients. The same principles of molecular transport across cell membranes also apply to signal transmission in nerve cells. Molecular transport can be uniport, symport, antiport, electro-neutral or electrogenic etc. The cell in a resting state has an asymmetric distribution of electropotential—negative inside with respect to outside the cell membrane. This potential is called membrane potential, arising mainly from concentration gradients and free diffusion of particular ion(s) across the cell membranes due to selective membrane permeabilities. The concentration and potential gradients are interrelated and the situation at equilibrium is given by the **Nernst** and **Goldman** equations.

The membrane potential of a cell at rest, called the resting potential, is close to the equilibrium potential of K^+ (≈ – 70 mV) and this asymmetry of potential across the cell membrane is maintained by Na^+/K^+ transport system, which is electrogenic; for every $3Na^+$ transported out $2K^+$ are pumped in, the process is coupled to the energy-giving process, hydrolysing ATP. That is, cellular membrane potential for a resting cell depends largely on the selective membrane permeability to K^+ ions and the K^+ ion concentration gradient across the cell membrane.

The origin of nerve impulse is action potential—arising out of a rapid and transient local change in the resting potential of the membrane, from the polarised state (negative membrane potential) to the depolarised state (change in membrane potential towards positive value) due to steep increase in the membrane permeability towards Na^+ and thereby shifting the membrane potential towards the Na^+ equilibrium potential (≈ + 50 mV).

Nerve signals are transmitted from cell to cell via synapses and in chemical synaptic transmission, the chemical neurotransmitters diffuse across the synoptic cleft and bind to specific channel-linked receptor proteins in the post-synaptic membranes. A neurotransmitter can be either excitatory or inhibitory, depending on its mode of action. Acetylcholine is a well known chemical neurotransmitter.

Macromolecules, cell debris and other large-size matter are transported through the cell membranes by **exocytosis, endocytosis** and **transcytosis**.

The receptors for neurotransmitters (in the post-synaptic cell) can be either channel-linked (ligand-gated) or non-channel-linked proteins. Whereas, action of channel-linked processes is fast and specific, action through non-channel-linked receptors is slow, spatially diffuse and long-lasting. Receptors can be classified according to their sensory functions. **ACh**-receptor is an example of chemo-receptor. The hearing organ of the vertebrates functions on rnechano-reception—mechanical stimulus directly opens ion channels in the auditory system in the process of mechanical to electrical signal transduction to the brain. The organisation of the constituents in the photoreceptor cells imparts optical wave-guide properties and, in specialised retinal cells of organisms, greater sensitivity to polarised light due to oriented-arrangement of chromophore molecules. Photopigment molecules with different spectral absorption characteristics (as in cones) enable colour vision. Gustatory and olfactory receptors are chemo-receptors. Their functions are interrelated. Many of the substances are not only "tasted" but also "felt". Echo-location by sonoreception and receptors for geomagnetic and gravitational forces are utilised by several organisms for directional finding and site locations etc.

15.1 THE NERVOUS SYSTEM

The vertebrate nervous system consists of a central nervous system (CNS)—brain and spinal cord, linked to a number of peripheral nervous systems (PNS), which contain all the sensory nerve cells and local ganglia (nerve cell clusters). The nervous system provides communication between various parts of the body and thereby controls reactions to stimuli, information processes and behavioural pattern—regulation, preservation and adaption.

Performance of an organism depends on reception and correct interpretation of information from its internal and external environments.

15.1.1 Nerve Cells

Nerve cells, also called **neurons,** are cells specialised to receive, conduct and transmit the signals of the excited-state sensory cells. Heterogeneity and specificity are the characteristics of nerve cells. They show great diversity in shape. A neuron essentially consists of a cell body with processes—**soma**, **axon**, **dendrites** and **microtubules**. Soma is that part of the neuron that contains the nucleus and surrounding cytoplasm. The axon is a long single neuron fibre, usually covered with an insulating myelin sheath. The axon behaves like a poorly-insulated cable.

The main function of an axon is to conduct the nerve impulse. Dendrites are branched-regions of an axon, extending from the cell body, like an antenna, and providing extended surface area to receive signals from the axons of other nerve cells. That is, the dendrite is the receptor process and axon is the transmitter process of the neuron. Microtubules are longitudinal and rod-like and unbranched with a wall composed of spiral strings of globular proteins. In addition, there are non-nervous cells having some relationship to nerve cells, called neuroglia or glial cells.

Neuron Categories. Anatomically, nerve cells are either unipolar, bipolar or multipolar, depending on the number of processes emerging from the soma. Also, neurons with distinguishable axon and dendrites are called heteropolar. Generally, neurons in invertebrates are unipolar and in vertebrates they are multi- and hetero-polar.

Functionally, neurons can be categorised as motor, sensory or neuronal, depending on the *receptor* cells, which are developed to respond to particular kinds of stimuli. Motor meurons carry nerve signals for muscle contraction, sensory neurons carry information in specific stimuli (**mechano**, **photo**, **chemical** etc.). However, whatever may be the diversity of different-categories of neurons, the fundamental role of any neuronal activity is transmission/transduction of nerve impulses in the form of electric signals. This basic and important function of neurons is effected by the local changes in the electrical potential across the plasma membrane surrounding the cells. Though neurons can transmit normally for a short distance, an active signalling mechanism (action potential/ spike potential) is necessary for long-distance conduction.

Myelinated-Neurons. Propagation of nerve signals along an axon depends on its cable properties (capacitance and conductance). Conduction through bare axons, as has been shown in giant squid axons, is not efficient. Increasing the thickness of the neuron and insulating it imparts greater efficiency and speed of signal transmission. **Myelin** sheaths, wrapped around axons, provide neurons effective electric insulation, increased capacitance, reduction in transverse current leakage and reduction in energy consumption. Myelin has high percentage of lipid ($\approx$ 70 % of dry weight) — mostly phospholipids, glycolipids (**cerebrosides**) and cholesterol. Cerebroside is found in the brain only in myelin sheaths.

The myelin sheath is a lipoprotein-membrane that surrounds the axon of the nerve fibre. It is special membrane structure where lipids are more abundant than proteins. The myelin sheath in the peripheral nervous system is formed by **Schwann cells** (cell-surface membrane) and in the central nervous system by oligodendroglia (oligodendrocytes, astrocytes and microglia), respectively. Small regions of axon membrane are left uninsulated between myelin sheath segments, exposing only these regions of axons to the extracellular environment. These exposed regions are called **nodes of Ranvier** and practically all ion channels are concentrated at these nodes. The cellular organisation of myelinated axon imparts electrical passivity to the insulated regions of the axon membrane with excellent cable properties (passive electrical properties). That is, myelinated neurons will have much higher transverse resistance and much lower transverse capacitance than normal cell membranes. When a neural membrane is depolarised at a node of Ranvier, the local electrical circuit generated cannot flow through the adjacent membrane. Instead it excites the membrane at the next node. That is, the signal propagates along the axon, jumping from one node of Ranvier to the other and the conduction of signal is faster in myelinated neurons than in bare neurons. While passive flow of current is along the axon, active flow is from the nodes of Ranvier, *via* voltage-dependent ion channels. This mode of impulse conduction is termed as *saltatory* (*saltare* = to leap) conduction. A schematic representation of nerve cell is given in Fig. 15.1.

Nerve cells, like other cells, are compartmentalised by membranes surrounding them and all the processes of transport, signal transmission and reception etc., can be understood from the underlying principles of transport across cell membranes.

15.2 MOLECULAR TRANSPORT ACROSS CELL MEMBRANES

The lipid bilayer of the biological cell membranes acts as an electrical capacitance. Nerve impulse generation and transmission are carried out *via* neuronal apparatus. Asymmetric distribution of ions

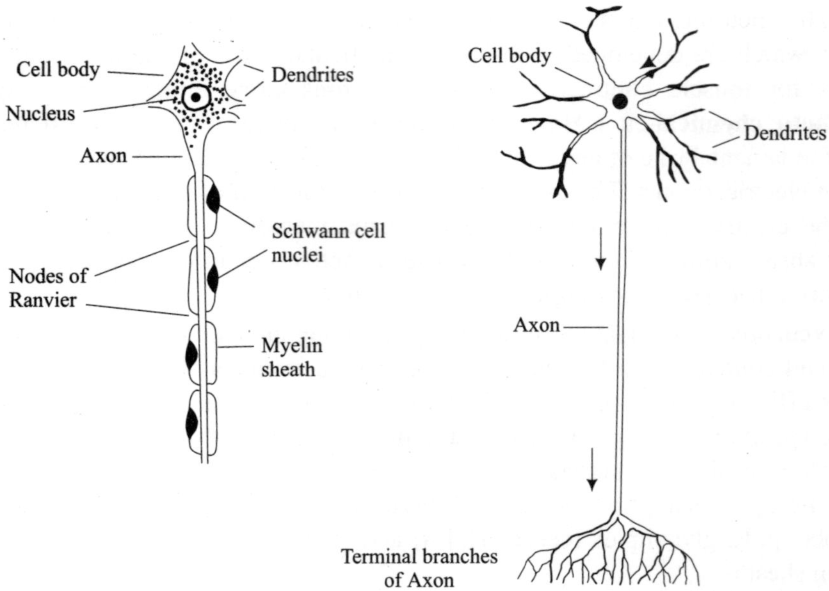

Fig. 15.1 Schematic representation of a nerve cell.

across the cell membrane gives rise to transmembrane potential; and ion channels (permeases), embedded in the lipid bilayer, provide selective conductance for membrane transport processes.

In any lipid-water system, the lipids tend to form bilayers with polar-head groups at the two aqueous interfaces. These lipid bilayers, due to their hydrophobic composition in the interior of the membrane, are easily permeable to small hydrophobic molecules such as benzene, ether and steroids etc., but are highly impermeable to charged ions such as Na^+, K^+, Ca^{++}, CI^- etc. Small neutral molecules such as water and urea can also diffuse through the bilayer, but larger neutral molecules such as glucose cannot permeate the lipid bilayer.

Biological membranes exhibit selective permeability characteristics towards the transport of small molecules. As all cells/organisms are separated from their surroundings by membrane barriers. the process of transport across cell membranes is of fundamental importance to life processes. The transport of small molecules across the cell membrane depends on the nature of the solute, its charge and its structural features etc., and transport is mediated by membrane-bound proteins (channel- and carrier-proteins). Mass transport phenomenon across biological cell membranes is based on the following assumptions–

1. Cells are compartmentalised with lipid bilayer matrices.
2. Lipid bilayers act as selective barriers in mass transport processes.
3. While lipid-soluble and few polar substances, like O_2, H_2O and alcohol, can pass through the lipid matrix unhindered, the matrix is selectively permeable to particular ions and polar molecules.
4. Transport of ions and polar molecules are mediated *via* membrane proteins, which act as transport channels.

5. The asymmetric distribution of components across the cell membranes (Table 15.1) creates electropotential gradient across the cell membrane, which is the basis for cellular transport processes.
6. Transport of macromolecules (bulk matter), and cell debris is by endo and exocytosis.

15.2.1 Mass Transport

Cellular transport is achieved by diffusion (simple and facilitated) and active transport. The factors influencing mass transport across the cell membranes are (i) diffusion, (ii) membrane permeability, (iii) membrane potential, and (iv) active transport against the membrane potential.

Diffusion and Osmosis: Random transport of matter (solute) from a higher concentration to lower concentration is called diffusion. Transport of solvent across a semi-permeable membrane is called osmosis. Features of mass transport by diffusion and osmosis are given in Table 15.2.

Facilitated Transport: Facilitated transport, mediated by specific carrier-proteins, occurs due to the existence of a concentration gradient of molecules/ions and is labelled according to the type of transport (Table 15.3 and Figs. 15.2 and 15.3). Facilitated transport is wide-spread in organisms—transport of sugars, amino acids etc. A **uniport** carrier transports a single solute across the membrane (e.g. glucose transport). In coupled-transport, the transport of two or more molecules in the same direction simultaneously is called **symport** or cotransport and if the simultaneous transport of two molecules is in the opposite directions across the membrane, it is called **antiport** (exchange-

Table 15.1 Distribution of chemical constituents in cellular fluids.

Constituent	*Extracellular Side*	*Intracellular (Cytosol) Side*
Cations		
Na^+	150 mM	15 mM
K^+	5	150
Ca^{2+}	5	20
Mg^{2+}	3	60
Anions		
Cl^-	120 mM	5 mM
HCO_3^-	30	10
HPO_3^-	2	100
pH	7.4	7.0
Potential	0	– 80 mV

Table 15.2 Feature of mass transport by diffusion and osmosis.

Diffusion	*Osmosis*
Solute transport down the concentration gradient.	Solvent transport to equalise the solute concentration across the membrane.
Membrane partition is not a prerequisite.	Semi-permeable membrane partition is a prerequisite.
Equal diffusion as a function of distance.	Unequal diffusion phenomenon.
Hydrostatic pressure is not developed.	Hydrostatic pressure is developed.

Table 15.3 Categories of solute transport across cell membranes

Type of transport	Transport mechanism	Representation(A)
(A)		in | out
(i) Uniport	Single molecule transport	→ Glucose
(ii) Coupled-transport		
(a) Symport	Parallel transport of several molecules simultaneously	→ H^+, → H^+ Lactose
(b) Antiport (exchange transport)	Transport of a molecule linked with the transport of another molecule in the opposite direction	→ Na^+; K^+ ←
(B)		
Electroneutral	No charge transfer across the cell membrane	→ Na^+; H^+ ←
Electrogenic	Net charge transfer across the membrane	→ $3Na^+$; $2K^+$ ←
Energy-coupled transport	Solutes transported against the gradient	nH^+ ← (O_2); nH^+ ← (ATP → ADP + Pi)

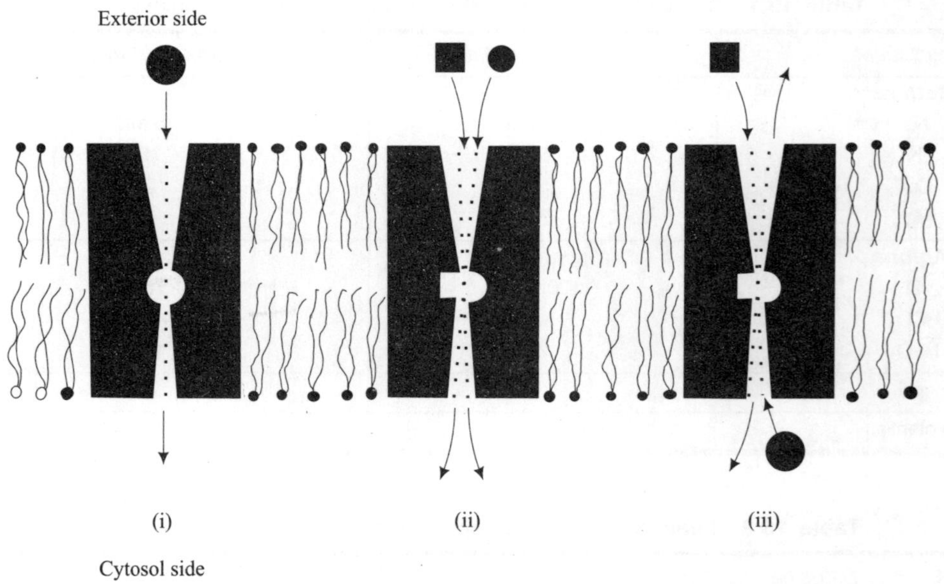

Fig. 15.2a Schematic diagrams of transport across cell mambrane by carrier-proteins. (i) Uniport; (ii) Symport; (iii) Antiport.

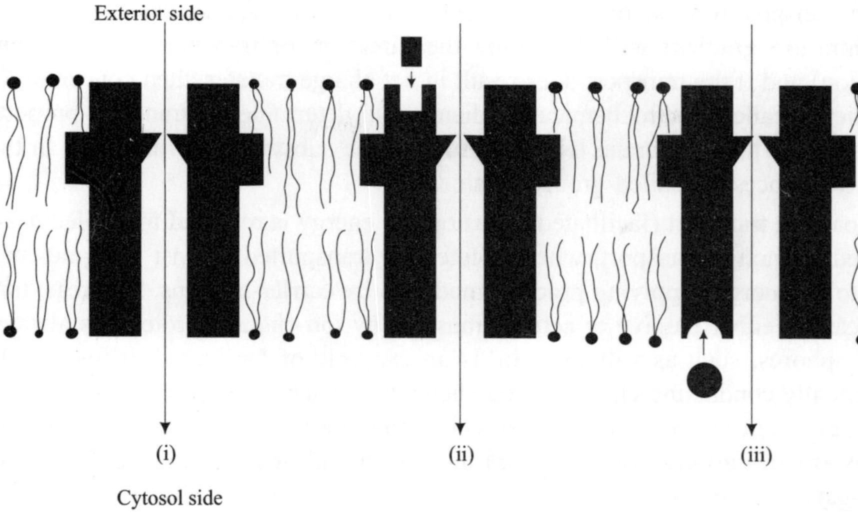

Fig. 15.2b Schematic diagram of transport across cell membrane by channel poteins.
(i) Voltage-gated channel: (opens momentarily in response to change in the membrane potential).
(ii) Lignd-gated channel: (opens in response to an extracellular signal).
(iii) Signal-gated channel: (opens in response to an intracellular signal).

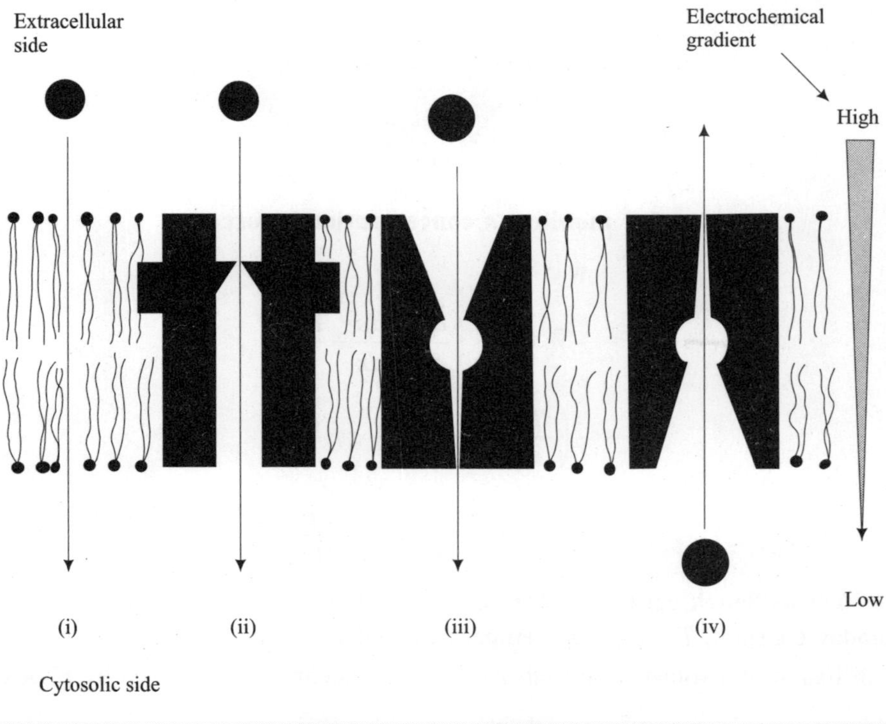

Fig. 15.2c Transport across cell membrane down an electrochemical gradient.
(i) Simple diffusion; (ii) Channel-mediated facilitated diffusion; (iii) Carrier-mediated facilitated diffusion; (iv) Active transport.

transport). If the transport process of solute molecules does not result in net charge transfer, then only the concentration gradient will determine the direction of transport (passive transport or facilitated diffusion) and if the transport does result in net charge transfer, then concentration as well as potential gradients, called electrochemical gradients, will determine the transport properties of the solute molecules across the membrane (**electrogenic**). If the substrates are modified in the process of transport, such a process is called group translocation.

Whereas in passive transport (facilitated diffusion) no energy is required for molecular transport, energy is required for active transport, where solutes are transported against the gradient, and it is always coupled to an energy-supplying process, mediated by carrier-proteins. Whereas, transport by carrier-proteins can be either passive or active, transport by ion-channel proteins is always passive. Transport of ionophores, such as valinomycin, is an example of facilitated diffusion with carrier molecules. Electrically conducting channels (ion channels), which form conducting pores, similar to gramacidin **A** pores, seem to occur in nearly all cell membranes and may be mainly responsible for the passive transport of cations (ion channels) and electrically-conducting channels with gating mechanism (ion-gated-channels).

15.2.2 Concepts of Membrane Transport

The forces responsible for molecular transport across cell membranes are concentration gradient, pressure/temperature gradient and electrical potential gradient. If the forces, *F,* and resistances, *R,* are known the magnitude and direction of the fluxes, *J*, can be computed. The converse is also valid, that is

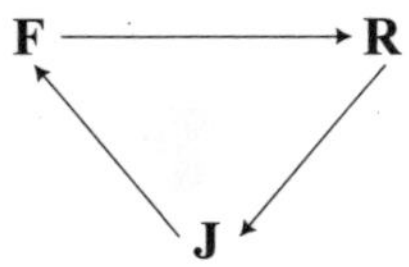

J = mobility × concentration × force

$$J = \text{w}\cdot\text{c}\cdot\text{F} \qquad \text{...(15.1)}$$

$$F = -\nabla.\text{V} = -\left(\frac{d\mathbf{c}}{dx} + \frac{dp}{dx} + \frac{d\psi}{dx}\right)$$

$$F = -\left(\frac{RT}{\mathbf{c}}\frac{dc}{dx} + \bar{v}\frac{dp}{dx} + z\mathcal{F}\frac{d\psi}{dx}\right)$$

$$J = -\,\omega\mathbf{c}RT\left(\frac{1}{\mathbf{c}}\frac{dc}{dx} + \frac{\bar{v}}{RT}\frac{dp}{dx} + \frac{z\mathcal{F}}{\mathcal{RT}}.\frac{d\psi}{dx}\right) \qquad \text{...(15.2)}$$

which is the **Nernst-Planck** equation, where **c** = concentration, R = gas constant, z = valency of the ion, $\mathcal{F}$ = Faraday constant, T = absolute temperature and $\bar{v}$ = specific volume.

Simple diffusion of a solute from higher to lower concentrations is given by **Fick's** equation

$$J = -D\left(\frac{dc}{dx}\right) \qquad \text{...(15.3a)}$$

$$D = -\frac{kT}{\zeta} \qquad \text{Einstein relation} \qquad ...(15.3b)$$

$$\frac{1}{A}\frac{d\mathbf{n}}{dt} = -D\left(\frac{dc}{dx}\right) - = -\text{ P.dc} \qquad ...(15.3c)$$

$$D = P \cdot dx$$

$$\left(J = \frac{1}{A}\frac{d\mathbf{n}}{dt} \text{ by definition}\right)$$

where ζ = frictional coefficient; **n** = number of solute molecules moving across the area A; P is the permeability coefficient, and D is the diffusion coefficient.

Diffusion can be correlated to the permeability coefficient of the solute; permeability is a directly measurable parameter, but its molecular significance is a complex. The permeability coefficient of a particular solute in a membrane is a function of solute concentration, partial resistance coefficient and the structural state of the membrane. Introduction of solute molecules can significantly alter the permeability properties of biological membranes to other small molecular solutes (e.g. Na^+, K^+, Ca^{++}, CI^- etc.). So biological membranes can be permeable, less permeable or even impermeable to some molecular species in the presence of other ions in chemical reactions.

Long-range coulombic forces cause non-randomness in ion distribution so that a particular ion is on the average surrounded by opposite ions. Thus, the position and velocity of any specific ion is intimately connected with the positions and velocities of all other ions, so that the coupling between the positions and velocity of an ion species with the positions and velocities of other ions surrounding that ion is quite strong within the membrane. The motion of an ion in a non-homogeneous diffusion barrier, such as that exists in a membrane, is influenced by two different fields—the external field and the field due to other ions present in the barrier (**Debye-Hückel sea**). In addition, if the time taken by a particle to cover the thickness of the membrane is faster than the relaxation time, then that particle experiences a friction due to polarisation. In such a situation the interaction between two ions will no longer be pure coulombic but will have a screening effect, expressed in terms of dielectric coefficient (polarisation produced in the ions of the medium by external field). The dielectric coefficient of a non-homogeneous medium is not a constant. The permeability of membranes to ions also markedly depends on the dielectric coefficient of the membrane. At least three parameters are required to characterise fully the behaviour of a membrane to a particular solute— (i) membrane permeability to that solute, (ii) interaction between solute and the membrane and (iii) solvent-membrane interactions. Concepts from thermodynamics have been invoked to address these parameters.

Membrane Potentials. Transport of charged molecules is governed not only by the concentration gradients, but also by transmembrane potentials, electrochemical gradients, giving rise to electrochemical potentials and selective membrane permeabilities. In living cells an electrical potential difference across the plasma membrane does perforce exist, arising out of unequal distribution of ions inside and outside the membrane. This potential difference between the inside and outside of the membrane is called membrane potential and several factors contribute to the existence of membrane potential—(i) **Donan** potential, (ii) selective membrane permeability to solute and (iii) active transport (chemical potential gradient) to maintain a concentration gradient.

The Donan potential does not contribute much to membrane potential and plays a very minor role in membrane transport. If the membrane is permeable to a single neutral (electrically) species, then those molecular species tend to diffuse out, creating a concentration gradient and at equilibrium state, the chemical potential different, $\Delta\mu$,

$$\Delta G = \Delta\mu = RT \ln \frac{[X]_i}{[X]_o} \text{ for a neutral molecule} \qquad ...(15.4a)$$

In the case of an ionic species, the ion diffusing down the concentration gradient is opposed by oppositely-charged ions, creating in that process a potential difference on either side of the membrane and the electrochemical potential, $\Delta\mu$ is

$$\Delta\mu = RT \ln \frac{[X]_i}{[X]_o} + z\,\mathcal{F}\,\psi \qquad ...(15.4b)$$

where $[X]_i$, $[X]_o$, ψ are concentrations of ion X inside and outside the membrane and membrane potential, respectively.

If $\Delta\mu < 0$ transport can be passive and for $\Delta\mu > 0$ transport must be active, coupled to an energy-providing process. The Nernst equation represents the diffusion potential at equilibrium and explains only passive diffusion. That is,

(concentration gradient)$_{\text{ion}}$ = (potential gradient)$_{\text{ion}}$

$$\psi = \text{RT} \ln \frac{[X]_o}{[X]_i} \qquad ...(15.5)$$

Diffusion potential contributes maximally to the maintenance of membrane potential.

In actuality, biological membranes are always permeable to several ion species in varying degrees and the concentration gradient is maintained by active ion transport, even if the plasma membrane is permeable to several ion species and the membrane potential is dependent on the relative permeabilities of the membrane to different ion species. A general equation relating membrane potential to the concentrations of permeating ions is represented by the Goldman equation, (when no current is flowing through the membrane)

$$\psi = \frac{RT}{z\mathcal{F}} \ln \frac{P_a[A^+]_o + P_b[B^+]_o \ldots + P_q[Q^-]_i + P_r[R^-]_i + \ldots}{P_a[A^+]_i + P_b[B^+]_i + \ldots + P_q[Q^-]_o + P_r[R^-]_o + \ldots} \qquad ...(15.6)$$

for $\quad z = 1; \quad \psi = 58 \log_{10} \quad (\quad " \quad\quad " \quad)$

where P_a, P_b etc., are permeabilities of species with concentrations [A], [B] etc.

The concentration gradient of ions is maintained by active transport of ions against the concentration gradient by ion pumps (e.g. Na^+/K^+ transport), but ion pumps do not contribute much to membrane potential maintenance. Na^+ ions play a small part in determining the membrane potential of nerve and muscle cells; the effect is greater when the external K^+ ion concentration $[K^+]_o$ is low.

The most important transport system in all living systems is that of Na^+/K^+ transport, which is also a fairly well understood process and highlights the basic underlying principles of molecular transport of small molecules across cell membranes.

15.3 PERMEASES

Ion channels and ion pumps are known collectively as *permeases*. The simplest of the permeases, the membrane channels (voltage-gated, ligand-gated etc.), provide openings through which specific ion species can diffuse across the cell membrane. These channels are selectively passive to particular type of ions to move down their concentration gradients. Such ion channels "open" and "close" in response to the level of membrane potential, rendering the membrane permeability voltage-sensitive. The biological features of voltage-sensitive ion channels can explain the ion transport phenomenon across the cells.

Gated-channel mass transport is due to induced conformational changes in the receptor molecule (protein complex) upon binding a transmitter molecule (e.g. acetylcholine, ACh). This conformational change allows the gate to open for mass transport of particular ion species.

Ion pumps are protein complexes (ATPases) that transport solutes (ions) across the cell membranes against the electrochemical potential gradient and maintain ion concentration gradients with the expenditure of metabolic energy. Some functional features of permeases (ion channels and ion pumps) are given in Table 15.4.

Table 15.4 Some functional features of permeases.

Ion channels	*Ion pumps*
Selective passage of ions down the concentration gradient.	Ions are pumped against the concentration gradient.
Depletion of concentration gradient by diffusion.	Generation and maintenance of concentration gradient across the cell membrane.
Passive transport down the concentration gradient.	Active ("uphill") transport against the concentration gradient.
Avalanche diffusion.	Slow and steady process.
No energy is expended.	Metabolic energy expended.

15.3.1 Aqueous-channel (Membrane-pore) Diffusion

Aqueous-channel (membrane-pore) diffusion is through intramolecular spaces within the structures of protein molecules line along the membranes. The membrane pores are ~0.8 nm in diameter, which facilitate the free passage of water and other molecules. The permeability of pores depends on (i) nature of the charge of ions compared to the charges of the pores, (ii) the hydration size compared to the pore diameter, and (iii) potential difference across the cell membrane.

Channel Proteins: Channel proteins, also known as ion channels, form, in their 'open' conformation, highly-selective water-filled pores across cell membranes, thereby allowing only specific ion species to cross by passive diffusion: This selectivity of 'opening' and 'closing' is achieved by a gated mechanism that is controlled by transient changes in the membrane potential. There are several types of ion channels—voltage-gated (electro), mechano-gated, ion- or ligand-gated (molecular), G-protein-gated etc., (Table 15.5).

Table 15.5 Types of gated ion channels.

Gated effecter	*Ions transported*	*Examples*
Voltage	Na^+, K^+, Ca^{2+} etc.	Neurons.
Mechanical	K^+	Auditory hair cells.
Ligand	Na^+, K^+	Acetylcholine receptor (AChR).
G-protein mediated	K^+	Muscarinic ACh receptor; signal transduction.
Chemical (odour; taste)	Na^+, K^+	Olfactory; gustatory reception.

15.3.2 Active Transport

Active transport is the transport of solute (ion) across the electrochemical potential gradient, mediated by specific enzyme complexes (ATPases) to provide metabolic energy for the "uphill" transport of solute. It maintains steady-state ionic imbalance between extra- and intracellular (cytosolic) media. The cell requires (even in the resting state) the continuous transport of solute against the electrochemical potential gradient.

The acetylcholine receptor is an example of a ligand-gated cation channel. K^+-leak, Na^+-leak and Ca^{++}-leak channels are examples of voltage-gated ion channels in the plasma membrane. K^+-leak and Na^+-leak channels give high permeability of the membrane to their respective ions allowing free passage of those ions in and out of the cell and thereby controlling the cell membrane potential. In the case of the Na^+ channel, the amino acid sequence data indicate that in the α-subunit there are four internally homologous repeat unit with positively-charged $\mathbf{S_4}$ regions which are implicated in the voltage-sensory process for channel 'opening' and the region between repeats III and IV is thought responsible for 'closing' (Fig. 15.3).

The operation of the voltage-gated Na^+ channel has three distinct conformational states, which is general to any voltage-gated ion-channel (Fig. 15.4). In the closed, but not inactivated state, the membrane is depolarised. The channel-blocking segment of the protein closes the channel thereby blocking the influx of Na^+ from outside. The cytosol side is negative and the channel-inactivating segment is in the cytosol. When the membrane is depolarised the outside becomes negative and the voltage-sensing α-helices move towards the outer side of the membrane thereby opening the channel (open state) and influx of Na^+ occurs. At this stage the membrane is depolarised and the reversal process starts and the channel-inactivating segment in the cytosol side moves into the blocking position and prevents the influx of Na^+ (refractory state). When the cytosol side becomes negative, it again (the membrane is polarised) riverts the channel back to the original closed but not-inactive state.

15.3.3 Secondary Active Transport

Na^+/K^+ transport and H^+ transport are examples of primary active transport. Secondary active transport is initiated by a passive influx of solute, coupled to a primary active transport mechanism. Secondary active transport is found in practically all epithelial cells—transport of sugars, amino acids etc. For example, the passive transport of glucose coupled to the primary active transport of Na^+ ions inside the cell are pumped out by the Na^+/K^+ pump. Similarly, transport of a lactose molecule is accomplished by coupling its transport with that of H^+. The H^+ gradient needed to drive the active transport is generated by the flow of electron through the respiratory chain.

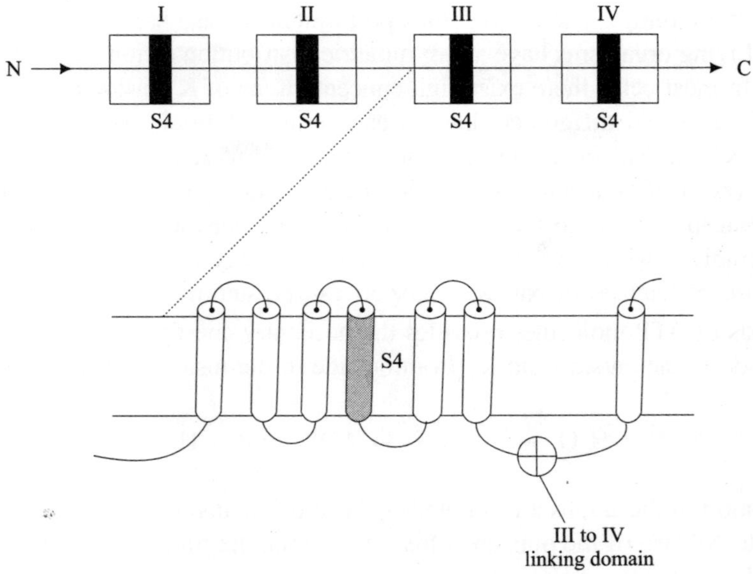

Fig. 15.3 Internal homologous repeat units in the α-subunit of the Na^+-ion-channel.
Each repeat unit has six-membrane spanning regions (multipass). S_4 regions have basic amino acids (+vely charged) involved in gating.
(After: Miller, C. Science. 252; 1092 (1991)).

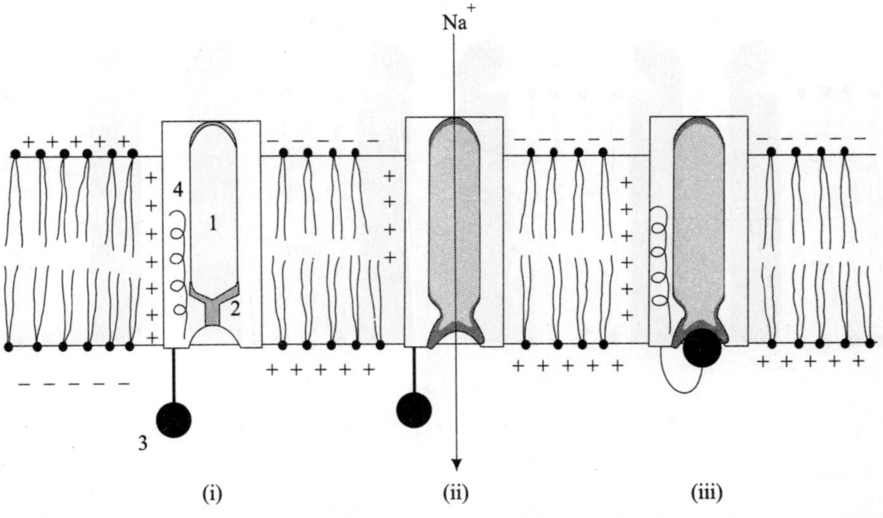

Fig. 15.4 Schematic representation of the conformations of a voltage-gated ion (Na^+ channel) channel.
(1) Na^+-channel; (2) channel-blocking segment; (3) channel-inactivating segment; (4) voltage-sensing α-helix.
(i) Closed but not inactivated (membrane polarised).
(ii) Open state and flux of ions (Na^+) (membrane depolarised).
(iii) Inactive (refractory) state (membrane depolarised).

Na^+/K^+ Transport System: The Na^+/K^+ transport system is also known as the Na^+/K^+ pump and Na^+/K^+ ATPase. Living organisms have an asymmetric distribution of ion concentrations across their cell membranes. In most cells, there exist high concentrations of K^+ inside and Na^+ outside the cell, respectively. This potential difference, known as resting potential, arises due to i) difference in concentrations of K^+ and Na^+ inside and outside the cell and ii) selective permeability of the plasma membrane to K^+ (ratio of concentrations of (intra/extracellular) is ≈ 20 for K^+ and ≈ 0.1 for Na^+). Thus, the active transport of K^+ to the inside and Na^+ to the outside of the cell by the Na^+/K^+ pump creates an ionic gradient, which is the source of potential energy to drive various transport processes (Fig. 15.5). Functional features of Na^+/K^+ transport can be summarised:

(i) Hydrolysis of ATP molecules provides the necessary energy for the active transport of Na^+ from inside to the outside and K^+ from outside to the inside of the cell.

$$ATP^{4-} + H_2O \xrightarrow{Mg^{2+},\, Na^+/K^+} ADP^{3-} + P_i + H^+ \quad ...(15.7)$$

(ii) The transport of these ions are coupled—Na^+ must be inside and K^+ must be outside the cell to activate ATPase (if the outside is free of K^+, then the transport of Na^+ is stopped and *vice versa).*

(iii) ATP is hydrolysed only if Na^+ and K^+ are transported.

(iv) ATP is a substrate for the ATPase inside the cell only.

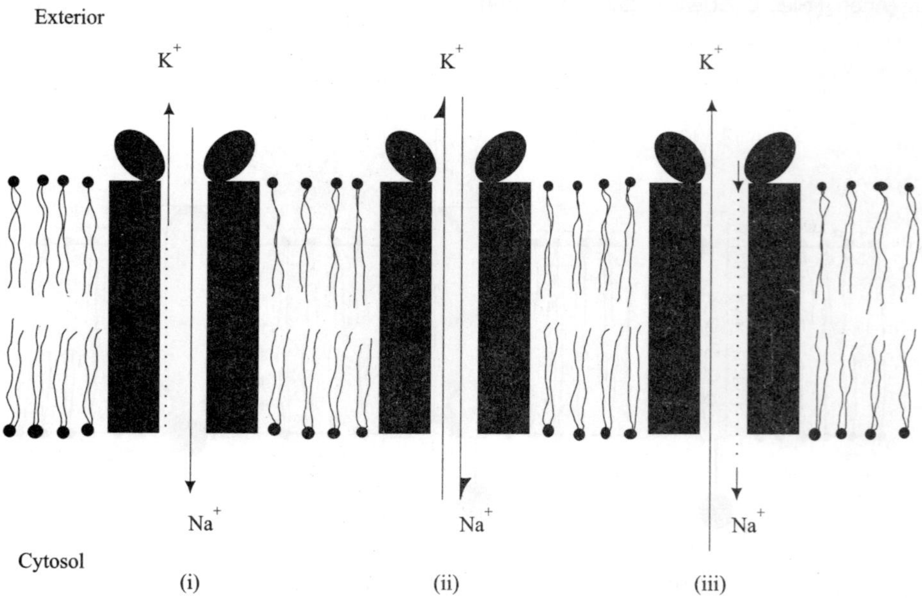

Fig. 15.5 Na^+/K^+ transport as a function of membrane potential.
(i) $\psi \approx -70$ mV. Net inward current ($\therefore$ excitation)
(ii) $\psi \approx 0$ mV. Net current = 0
(iii) $\psi \approx +30$ mV. Net outward current.

$$E + ATP \underset{}{\overset{Mg^{2+}\ Na^+}{\rightleftharpoons}} E\ .\ P + ADP$$

$$E \cdot P + H_2O \xrightarrow{K^+} E + P_i \qquad ...(15.8)$$

(v) Na^+/K^+ transport is **electrogenic antiport**—for every three Na^+ ions pumped out, two K^+ ions are pumped in.

(vi) The Na^+/K^+ pump acts as reversible pump also—it can be used to synthesise ATP **(it is an ATPase)**.

Na^+/K^+ Transport Process. Na^+/K^+ ATPase is a transmembrane plasma protein complex— α and β-subunits of 110, and 55 kilodaltons each, and the functional unit is a dimer, $(\alpha, \beta)_2$. ATP-binding sites are located in the α-subunits, towards the cytosol side (Fig. 15.6). The enzyme exits in two distinct conformational states, E1 and E2, with several intermediate states and with different ligand specificities (Fig. 15.7). Salient features of the transport mechanism are:

(i) In the dephosphorylated E1 state, only the sites for Na^+-binding are available. E1.3Na^+-bound enzyme is phosphorylated in the presence of Mg^{++} and hydrolysing ATP to form phosphorylated E1-P.3Na^+ complex. This intermediate state undergoes comformational change forming an E2-P.3Na^+ complex and facilitates the transport of 3Na^+ (three moles of Na^+ for every mole of ATP) from inside to outside of the cell membrane. The phosphorylation step requires Na^+, Mg^{++} and ATP but not K^+. E1 → E2 conformational change leads to outward transport of Na^+.

(ii) The E2-P complex is specific for K^+ only (only K^+-binding sites are available) and binds 2K^+ from outside of the membrane to form an E2–P.2K^+ complex. Binding of K^+ induces the dephosphorylation of the complex to form E2.2K^+, which upon conformational change translocates 2K^+ to the inside of the cell and finally reverts back to the original EI-conformation state, with the binding sites of Na^+ again available. That is, E2-E1 conformational change leads to inward (towards cytosol) transport of K^+.

Since the Na^+/K^+ pump drives out three moles of Na^+ for every two moles of K^+ pumped in (for every mole of ATP), it is an electrogenic process, creating a potential gradient with the inside of the cell membrane negative with respect to the outside (Fig. 15.8a). Anions, like Cl^- are kept outside the

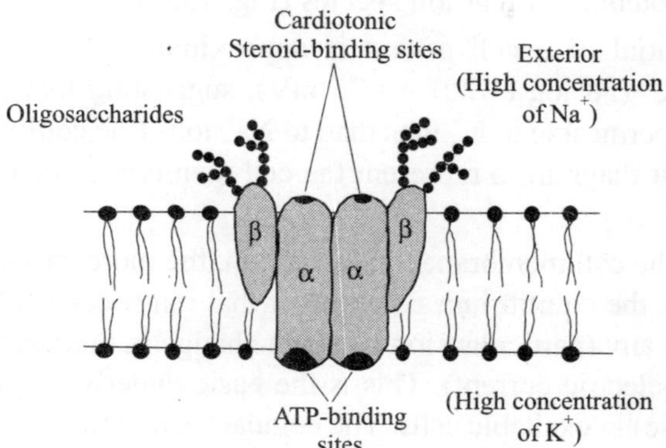

Fig. 15.6 Schematic representation of the subunit organisation $(\alpha_2\ \beta_2)$ of Na^+/K^+. ATPase.

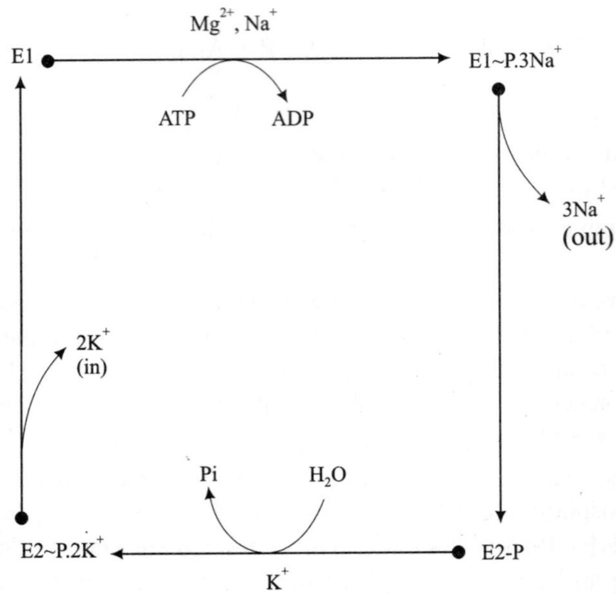

Fig. 15.7 Mechanistic view of Na^+/K^+ transport system function.
E1. conformational state: high affinity to Na^+(cytosol side).
E1 → conformational change: transport of Na^+ outward.
E2. conformational state: high affinity to K^+(outside).
E2 → E1 conformational change: inward transport of K^+.
E1.P and E2.P: Phosphorylated intermediates. Hydrolysis of intermediates powers the transport of ions.

cell due to this negative membrane potential. The equilibrium state, where there is no net movement of charge across the membrane is the resting membrane potential. For each ion species there is a distinct equilibrium potential, at which potential there is no net transport of that ion species across the membrane. When the membrane potential is not equal to the equilibrium potential of a particular ion species, ions of that species will tend to diffuse across the membrane only if the membrane is permeable to that ion species Transport of any particular ion species will shift the membrane potential towards the equilibrium potential of that ion species (Fig. 15.8b).

The membrane potential of a cell at rest is approximately –70 mV, which is close to the equilibrium potential of K^+ (58 log (1/20) = –75 mV), suggesting that in the resting state the cell membrane is more highly permeable to K^+ ions than to Na^+ ions (the equilibrium potential of $Na^+ \approx 58$ log 10 = 58 mV). A circuit diagram to represent the cell membrane potential, due to the presence of ions, is given in Fig. 15.9.

The more permeable the cell membrane to a given ion, the more strongly the membrane potential tends to be driven towards the equilibrium potential of that ion species. Consequently any change in membrane permeability to any (particular) ion causes a change in the membrane potential, enabling a net movement of charge (electric current). This is the basic underlying principle of electrical signal transmission (nerve impulse) in excitable cells. The cellular membrane potential depends largely on K^+ -leak channels (for passive diffusion of K^+) and the K^+ concentration gradient across the membrane barrier.

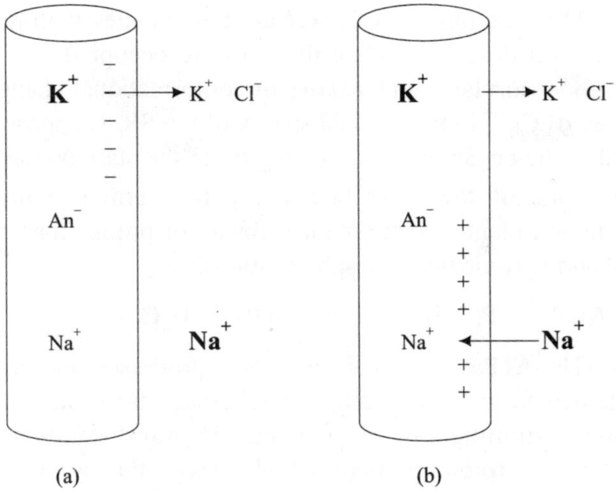

Fig. 15.8 Membrane states due to selective permeabilities to K^+ and Na^+ ions.
(a) In a cell permeable only to K^+, the diffusion of K^+ down its concentration gradient leaves the interior of the cell negatively charged,
(b) In a cell permeable only to Na^+, the diffusion of Na^+ down its concentration gradient leaves the interior of the cell positively charged.

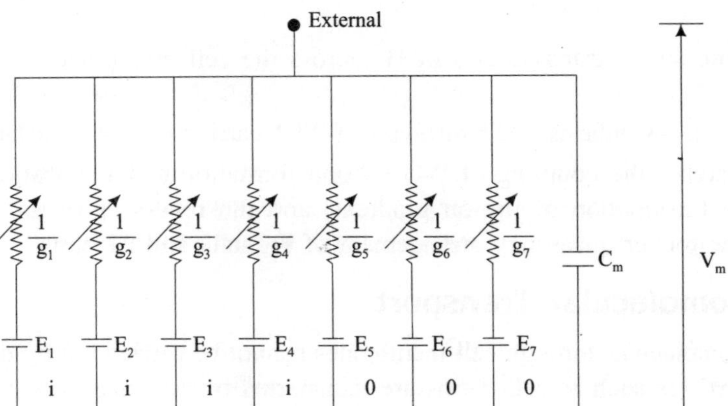

Fig. 15.9 Circuit diagram for membrane potential due to the presence of ions inside (i) and outside (o) of the membrane. g_1, g_2, ..., and E_1, E_2,... are the conductances and potentials of the ions. C_m and V_m are the membrane capacitance and potential.

Other Transport Systems:

Ca^{++}-Transport System. The next most important transport system is the Ca^{++} -transporting ATPase of the sarcoplasmic reticulum. This ATPase belongs to the family of P-type ion pumps and operates during mucle relaxation transporting Ca^{++} for the cell plasma into luminal spaces of the reticulum. The transport is driven by the energy obtained from the hydrolysis of ATP (2 moles of

Ca^{++} for every mole of ATP). The binding of two Ca^{++} ions to sites within the Ca^{++} ATPase is the essential first step in Ca^{++} translocation. When the sites are occupied by Ca^{++}, ATP hydrolysis is triggered, the enzyme is phosphorylated and a series of conformational changes in the enzyme result in the transport and release of Ca^{++} in the luminal spaces of the SR. It appears that high-affinity Ca^{++} -binding sites are located in the negatively-charged region of the stalk domain.

H^+-Transport System. One of the most fascinating biochemical pathways involving living systems—animal, bacteria and plant—is the main pathway of production of ATP that involves the phosphorylation of ADP and P_i (oxidative phosphorylation).

$$ADP^{3-} + P_i + H^+ \rightleftharpoons ATP^{4-} + H_2O \quad ...(15.9)$$

H^+ -translocating ATPase (H^+ -ATPase) is involved in these processes, and the energy required for the synthesis of ATP is obtained from the oxidation of substrates from the electron-transfer reactions (respiratory-chain reaction or from light quanta in plants. H^+ translocation across the cell membrane generates an electrochemical proton gradient which drives the synthesis of ATP (**Mitchell's** hypothesis). H^+-ATPase can also catalyse the reverse process namely the hydrolysis of ATP to ADP and P_i, with concommitant establishment of an electrochemical proton gradient across the cell membrane. That is, H^+-ATPase is a reversible pump—ATP synthase as well as ATP hydrolase (ATPase).

All H^+-ATPase are classified morphologically into three units—water-insoluble sector F1, water-soluble Fl part and stalk. The main functions of the enzyme can be conceptually organised into three parts. These are:

(i) Tanslocating step—translocation of H^+ across the cell membrane, carried out by F_o and stalk.

(ii) Catalytic step—synthesis and hydrolysis of ATP, carried out by the F1-part.

(iii) Coupling step—the coupling of P-O-P bond formation to translocation of H^+ across the membrane (dissipation of proton gradient) and the reverse process. The coupling step requires the proper quaternary arrangement of F_o, stalk and F1-parts.

15.3.4 Macromolecular Transport

Transport of macromolecules through cell membranes cannot be carried out by diffusion. These and other large-size particles such as cell debris are transported by (i) **exocytosis**. (ii) **endocytosis** and (iii) **transcytosis**.

(i) Exocytosis: In response to extracellular signals, secretory vesicles containing particular molecules (e.g. hormones) fuse with the plasma membrane and release the contents of the vesicle to the exterior. This process is called exocytosis. Exocytosis can either be constitutive or regulated; regulated-exocytosis is a localised response of plasma membrane/cytoplasm.

(ii) Endocytosis: The contents to be transported are encapsuled in a packet of the cell membrane and pinched off, so that it enters the interior of the cell; This is called endocytosis. Receptor-mediated endocytosis is one of the methods used to import specific extracellular macromolecules (e.g. import of cholesterol by taking up low density lipoproteins (**LDL**)). Endocytosis is of two types — **pinocytosis**, that involves ingestion of matter via small vesicles and **phagocytosis** which involves ingestion of matter *via* large endocytic vesicles (**phagosomes**).

(iii) Transcytosis: Macromolecules are also transported from one extracellular space to another across epithelial cell sheets by transcytosis (e.g. transfer of antibodies).

15.4 NERVE IMPULSE GENERATION AND TRANSMISSION

Nerve cells are surrounded by a surface membrane which acts as a barrier to the diffusion of ion species (and polar solutes). Most electrical signals in the nerve cells are produced by selective changes in the permeabilities of the cell membranes to solute molecules (Na^+, K^+, Ca^{++} and $C1^-$).

15.4.1 Action Potential (Nerve Impulse Generation)

Electrically-excitable cell membranes contain voltage-gated ion channels, which are responsible for the generation of action potentials—rapid, transient and self-amplifying electrical excitation of the membrane. The steady state potential of cell at "rest" is called the *resting potential.* In nerve and muscle cells, the resting potential is –80 mV, which is the value for K^+-equilibrium potential (Table 15.6).

The resting membrane potential of the cell (≈ –70 mV) is close to the K^+ equilibrium potential. As such nerve cells at rest are highly permeable to K^+ ions (relative permeability, $P_{K^+} = 1$; $P_{Na^+} = 0.02$; $P_{Cl^-} = 0.05$) and the cell membrane of a nerve cell exists in a polarised state (towards the negative membrane potential). A nerve impulse is a transient electric impulse affecting this state, whose effect is dependent on membrane permeability and membrane potential, and is produced by a rapid change in the permeability of the cell membrane to Na^+ ions that leads to a total depolarisation of the cell membrane (less negative membrane potential), allowing easy passage of Na^+ across the cell membrane. That is, a change in membrane potential is accompanied by a change in the relative permeability of the membranes to ions (K^+, Na^+ and Cl^-). This sequence of transient changes in the impulse and the accompanying change in the membrane voltage is called the *action potential (AP),* also called the *nerve impulse.* An action potential is triggered when the membrane is depolarised above a threshold value (Fig. 15.10).

Impulses below the threshold level (–30 mV) lead to local depolarisation and not to an action potential. Initially $g_{Na} < g_{K^+}$ and the resting potential is nearer to E_{K^+}. Membrane depolarisation increases Na^+ conductance (g_{Na^+}) rapidly ($g_{Na} > g_K$). A depolarisation above the threshold level drives the voltage-gated Na^+ channels in the membrane to open and allow the influx of Na^+ ions down the electrochemical potential gradient, which in turn steeply increases the permeability of the already depolarised membrane in favour of Na^+ influx and in this process amplifying further the depolarisation of the membrane and thereby driving the membrane potential towards the Na^+ equilibrium potential. In a situation where the membrane depolarisation crosses beyond the

Table 15.6 Ionic concentration across the cell membrane and equilibrium potentials.

Ion	*Cytosolic side (In) mM*	*Extracellular side (Out) mM*	*Ratio: (Out/In)*	*Potential (mV)*
Na^+	15	150	10	+ 60
K^+	150	5	0.03	– 80
Cl^-	7	100	~15	– 70
Ca^{2+}	0.5	10	20	—
Mg^{2+}	10	50	5	—

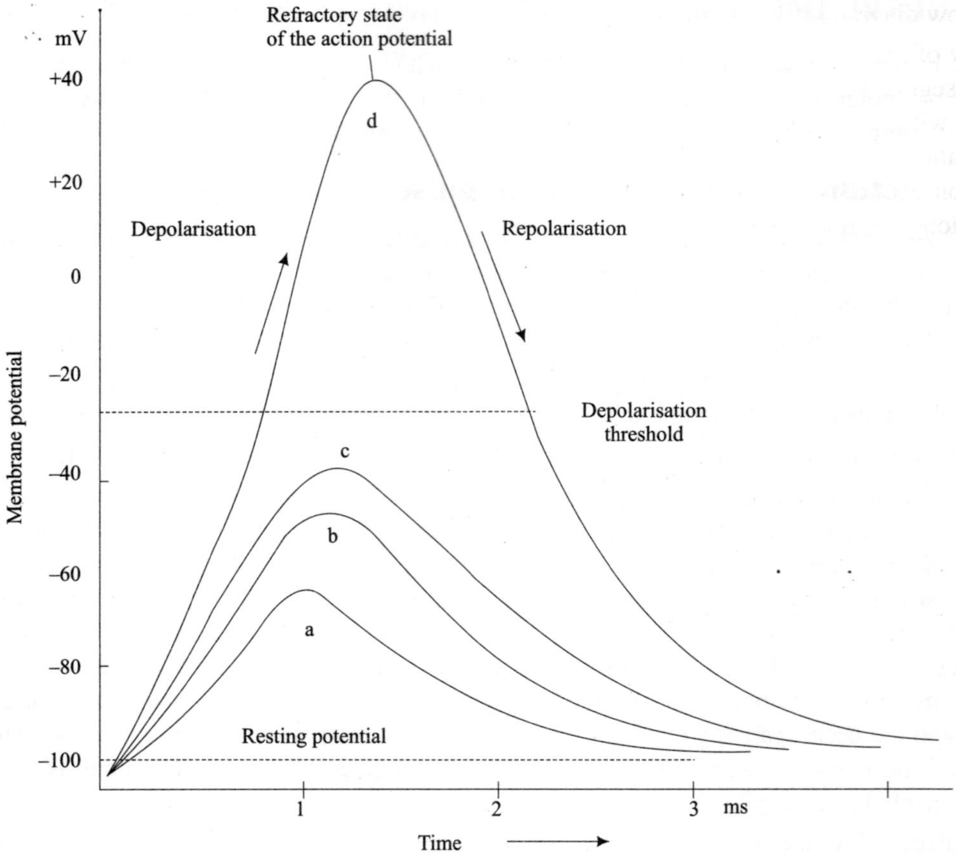

Fig. 15.10 Pattern of nerve impulse as a function of membrane potential. *a, b, c* impulses below the threshold level leading to local depolarisation; *d* impulse above the threshold level leading to all-or-none action potential.

'threshold' level, such that the ionic currents $I_{Na^+} > I_{K^+}$, the positive feedback action (self-amplification) drives the action potential to completion in an '**all-or-none**' manner. The nerve impulse conduction is *self-excitatory* and the impulse is propagated along the axon without decrement. That is, once the action potential is initiated n a neuron, it travels along the axon, providing same amount of change in membrane potential (V_m) at every point it propagates. As the membrane potential approaches the Na^+ equilibrium potential (≈ +40 to +50 mV) in the refractory phase, the conductance of Na^+ steeply decreases with concommitant increase in the conductance of K^+ ($g_K > g_{Na}$). The Na^+ channels spontaneously revert to the 'inactivated' conformation, to avoid permanent permeability to Na^+ diffusion and K+ channels 'open' and with the efflux of K^+, the membrane potential is restored to the polarised state (Fig. 15.11). The action potential above threshold voltage is called *excitatory*. Following every action potential is a brief period during which time a new stimulus cannot induce an excitation. This period is called the *refractory* period, which may last for several seconds.

A flow diagram of the action potential is given in Fig. 15.12.

Flow of ions through ion channels is controlled by gate molecules (segments). The extension of the gate-segment into an ion channel is monitored by a 'sensor' which responds to the electric field strength within the membrane. At a particular 'threshold' level, the gate 'opens', altering the conformation of the ion channel from an 'open' to 'closed' state, so that ion species that are specific to that ion channel can diffuse freely across the cell membrane. The 'opening' is reversed by the inactivation processes to the 'inactive' conformation (resting and closed) to avoid the possibility of

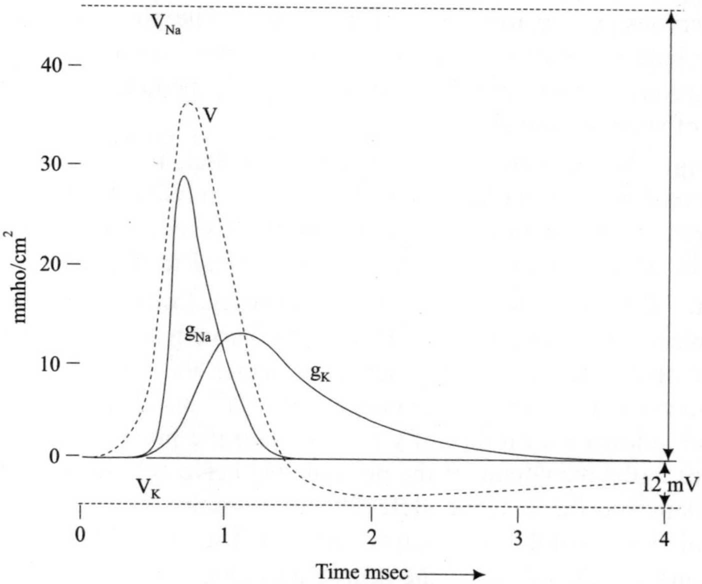

Fig. 15.11 The action potential (V) and Na^+ and K^+ conductances. (From Hodgkin, A.L. & Huxley, A.F. *J. Physiol.* 117; 500 (1952).

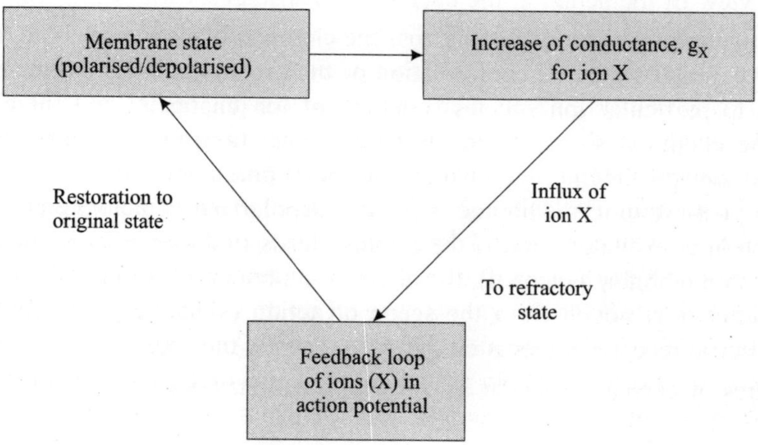

Fig. 15.12 A flow chart of the action potential.

permanent permeability to those ion species. A voltage-gated ion channel adopts, therefore, at least three distinct functional states— '**closed**', '**open**' and '**inactivated**' (open but resting) (Fig. 15. 4).

15.4.2 Signal Transmission

The function of the nervous system is to transport impulses spatially. There are three major ways of signal transmission in animal cells — (i) **endocrine** signalling, which acts throughout the body via hormones, (ii) **paracrine** signalling, which acts only on local cells and (iii) **synaptic** signalling, which acts on the adjacent post-synaptic cell. Synaptic signalling is the mode of transmission/transduction of signals by the nervous system from one cell to another. The physical stimuli, which stimulate neuronal action, include a variety of energy forms– pressure, chemical, light, heat etc., all the responses of the nervous system– motion, perception, pain, memory– are from the physical and chemical processes of neuronal signalling.

Synaptic Signalling. Nerve cells are interconnected but not in direct contact and the interconnecting neuronal junction for the neuronal signal transmission from one neuron (presynaptic) to an adjacent neuron (postsynaptic) is called a **synapse**. Nerve signals pass from cell to cell at synapse which can be either electrical (gap junctions) or chemical (Fig. 15.13a).

Neurotransmission. One of the ways of signal transmission from one neuron to another across a synapse is by chemical neurotransmission. Pre-synaptic nerve terminals contain synaptic vesicles within which are the neurotransmitters. In a chemical synapse, the depolarisation of the pre-synaptic cell membrane by a nerve impulse opens the voltage-gated Ca^{++} ion-channels in the plasma membrane of the axon terminal, allowing an influx of Ca^{++} ions into the pre-synaptic cell. This triggers the fusion of the vesicles to the membrane of the pre-synaptic nerve cell, releasing the neurotransmitter molecules they contain into the synaptic cleft (20 ≈ nm width). Release of transmitter molecules depends on the depolarisation of the pre-synaptic terminal. The influx of Ca^{++} ions due to membrane depolarisation is coupled to the release of the neurotransmitter.

Depolarisation of the pre-synaptic terminal ⟶ Influx of Ca^{++} ions into Pre-synaptic cell ⟶ Release of the transmitter into the Synaptic cleft.

The mechanistic view of the action at the chemical synapse is shown in Fig. 15.13(b).

The released neurotransmitter diffuses across the cleft and binds to a specific receptor protein of the post-synaptic cell and alters the conformation of the protein to make the membrane selectively more permeable to particular ion species (opening of ion channels) and thereby triggering the conversion of the chemical signal to an electrical signal (**action potential**). Neurotransmitters reaching the post-synaptic membrane change the membrane conductivity to ions (post-synaptic potential). If the post-synaptic membrane is locally depolarised (a positive change in membrane potential), the action is excitatory and if the impulse leads to hyper-polarisation of the membrane (negative change in membrane potential), the effect is inhibitory, preventing the nerve impulse. The chemical transmitter is removed from the scene of action (synaptic cleft) immediately after its binding to the specific receptor proteins on the post-synaptic membrane.

Salient features of events that occur at neuronal junctions (synapses) due to neurotransmission are:

(i) At the neuromuscular junction, the action potential triggers the influx of Ca^{++} ions into the pre-synaptic cell.

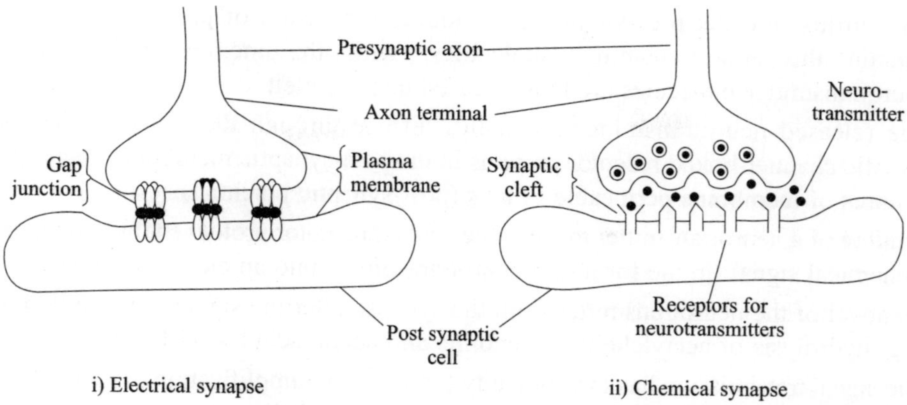

Fig 15.13a Synaptic junctions.

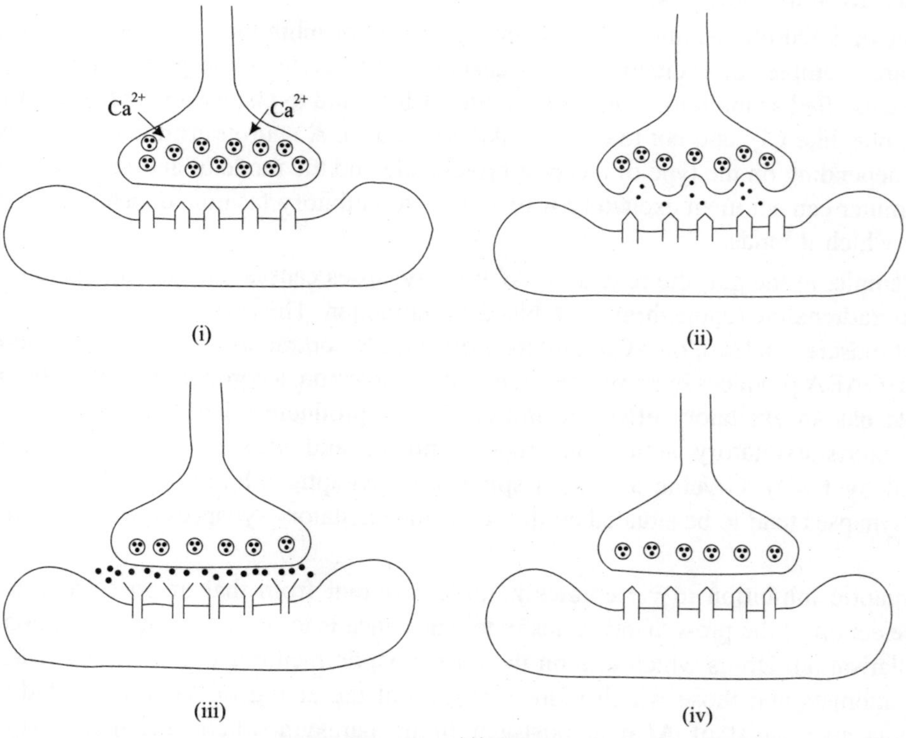

Fig. 15.13b Schematic representation of events in a chemical synapse.
(i) Action potential triggers entry of Ca^{2+} into pre-synaptic terminal.
(ii) Synaptic vesicles release neurotransmitter into the synaptic cleft.
(iii) Neurotransmitter binds to receptor in the post-synaptic membrane.
(iv) Transmitter is removed from the cleft and the initial state is restored.

(ii) Ca^{++} influx into the pre-synaptic cell induces the fusion of pre-synaptic vesicles, which contain the neurotransmitter molecules, with the membrane by exocytosis and neurotransmitter molecules are thus released into the cleft.

(iii) The released neurotransmitter molecules diffuse through the synaptic cleft and bind to specific channel-linked receptor proteins in the post-synaptic membrane, thereby altering the features of membrane permeable to ions (post-synaptic potential).

(iv) Binding of a neurotransmitter to a channel-linked receptor protein enables the transmission of a chemical signal (in the form of a neurotransmitter) into an electrical signal.

(v) Removal of the neurotransmitter from the synapse after the signal transmission/transduction (e.g. hydrolysis of acetylcholine to choline and acetic acid by **ACh**-esterase).

(vi) The signal transmission has directionality (precession), amplification (sensitivity), adaptation, integration, and discrimination and inhibition characteristics.

Chemical neurotransmitters. Some of the known chemical neurotransmitters are—acetylcholine (**ACh**). epinephrine, aspartate, glutamate, glycine and γ-aminobutyric acid (**GABA**). Their chemical formulae are given in Fig. 15.14.

The role of a neurotransmitter can be either excitatory or inhibitory. Acetylcholine, aspartate and glutamate are examples of excitatory neurotransmitters (generate action potentials) and glycine and GABA are classified as inhibitory neurotransmitters (they bind to channels which are permeable only to small anions, like Cl^-, and not to cations, like Na^+, Ca^{++}). A neurotransmitter can be excitatory or inhibitory, depending on the type of receptor protein site and the mode of action. That is, a particular neurotransmitter can act in an excitatory way and in an inhibitory fashion, depending on the receptor proteins to which it binds.

For example, in the gut, the release of **ACh** by synapses causes muscle contraction whereas the release of noradrenaline (epinephrine) inhibits the contraction. The reverse is the situation in the case of the heart muscle—release of **ACh** inhibits heart muscle contraction while noradrenalin stimulates contraction. GABA produces hyperpolarisation or depolarisation according to the membrane potential. *L*-glutamate has an excitatory effect on muscle fibres, producing depolarisation and contraction. Serotonin exerts excitatory action on smooth muscle and sensory nerves (these actions are antagonised by **LSD**). Glycine acts as a spinal post-synaptic inhibitory transmitter. In general, inhibitory synapses tend to be situated on the soma and excitatory synapses tend to be situated on the dendrites.

Pre-synaptic inhibition in vertebrates is quite different from that of post-synaptic inhibition, because the action of the pre-synaptic transmitter substance is to produce a depolarisation rather than a hyper-polarisation. Drugs which acts on the post-synaptic membrane in a similar manner to **ACh** are called agonists and those which reduce or prevent the acting of **ACh** are called antagonists. Muscarine is an agonist of **ACh** at post-ganglionic parasympathetic synapses. Atropine is an antagonist to **ACh** at synapses where its action is muscarinic and curare acts as an antagonist to **ACh** at the neuro-muscular junction. Many convulsant drugs act by depressing the post-synaptic inhibitory processes in the spinal cord. Examples are strychnine and tetanus toxin (glycine antagonists?). Botulinum toxin depresses spinal post-synaptic inhibition at cholinergic synapses.

Excitatory:

Acetylcholine (ACh)

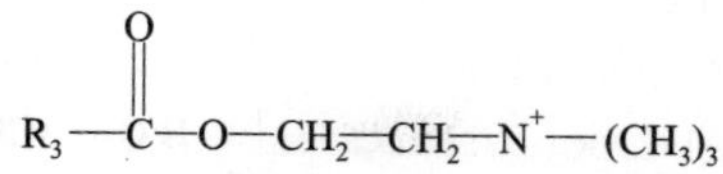

Aspartate

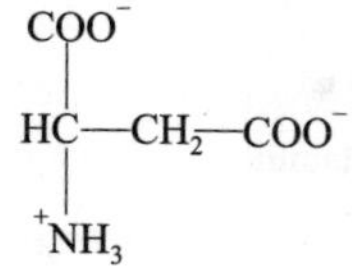

Glutamate

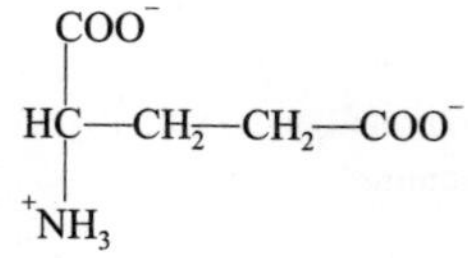

ATP

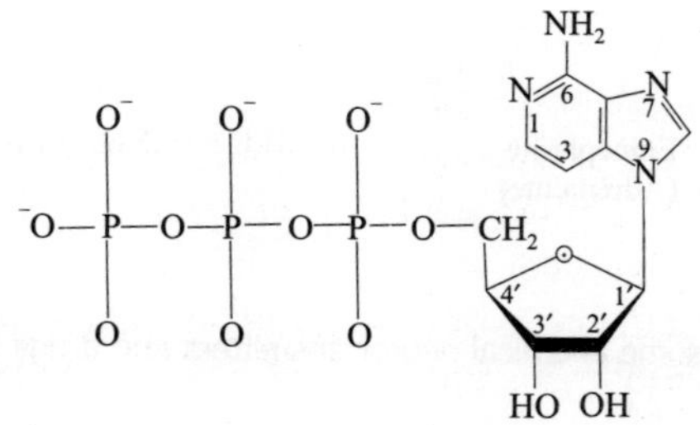

Norepinephrine (Noradrenaline)

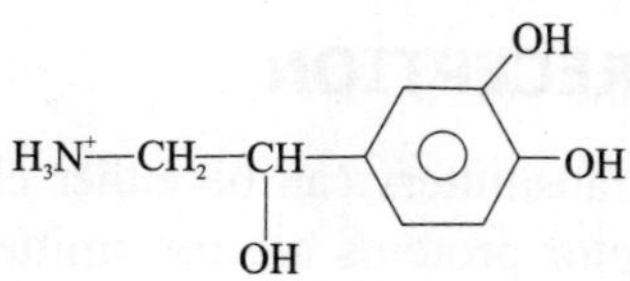

Serotonin (5-hydroxytryptamine)

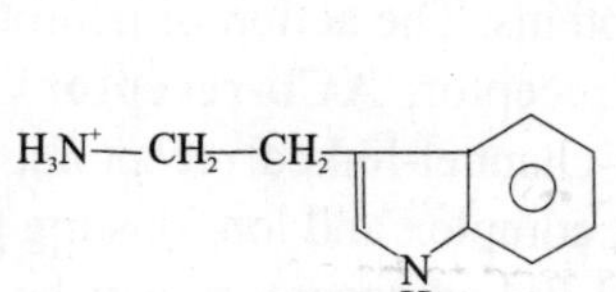

Histamine

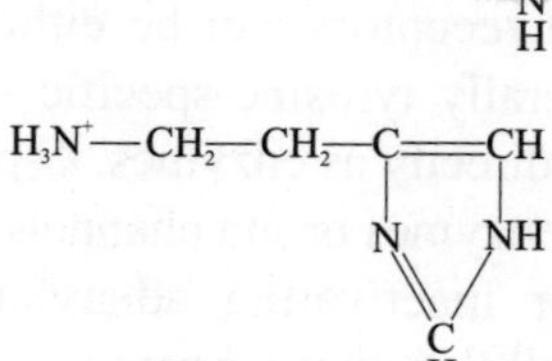

Inhibitory:

γ-Aminobutyrate (GABA) $H_3N^+—CH_2—CH_2—COO^-$

Glycine $H_3N^+—CH_2—COO^-$

(continued)

Other neurotransmitiers:

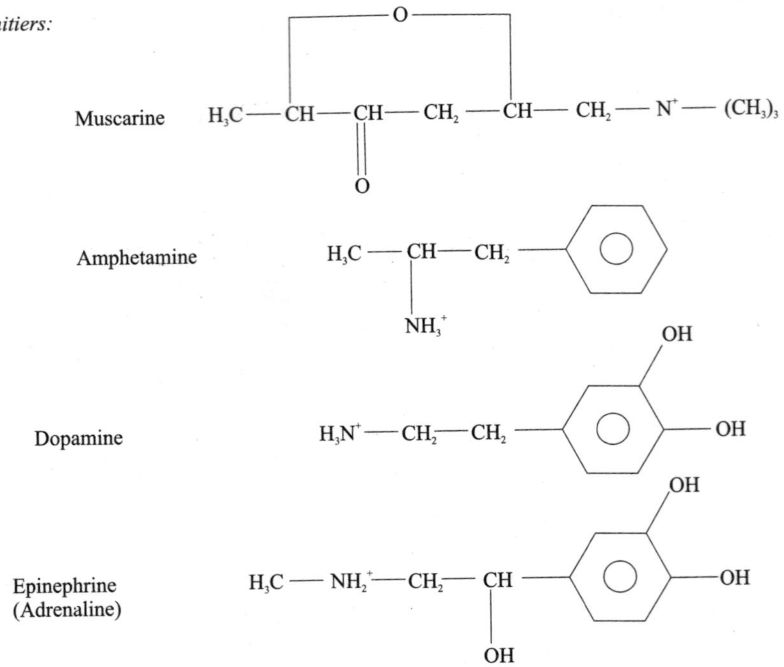

Fig. 15.14 Formulae of some chemical neurotransmitters and drugs (agonists and antagonists).

15.5 SIGNAL RECEPTION

Receptors for neurotransmitters can be either channel-linked (ligand-gated) or non-channel-linked. Channel-linked receptor proteins are transmitter-gated ion channels and are invariably multipass, transmembrane proteins. The action of neurotransmitters involving channel-linked receptors (such as the acetylcholine receptor, **ACh-receptor**) is fast, direct and specific, acting on a single post-synaptic cell. Non-channel-linked receptors, on the other hand, are involved in slow, spatially and temporally diffuse, complex and long-lasting processes and a single neurotransmitter can act on many cells. Non-channel-linked receptors can be either (i) catalytic or (ii) *G*-protein-linked. Catalytic receptor proteins are generally tyrosine-specific single-pass transmembrane proteins. They act, on activation by their ligands directly as enzymes. *G*-protein-linked receptors exert and indirect action on plasma-membrane-bound enzymes or ion channels, mediated through GTP-binding regulatory protein (*G*-protein), activating or inactivating adenylate cyclase and in that process controlling the concentrations of intracellular messengers — Ca^{++} and/or cyclic AMP (Fig. 15.15). These intracellular messengers act to alter the functions of other target proteins in the cell. The generator potential of any sensory receptor is a graded potential change, and not an 'all-or-nothing' potential.

Receptors can also be classified broadly according to their functions in sensory cells— chemo-, electro-, mechano- photo- and sono-receptors etc., which are specialised neural elements to detect and translate special forms of external impulses. Sensory cells (organs) are specialised transducers, (like all neurons), converting stimuli of different kinds (muscle stretch, sound vibrations and light etc.) into electrical signals. The outer segments of sensory cells, where stimuli are received, are

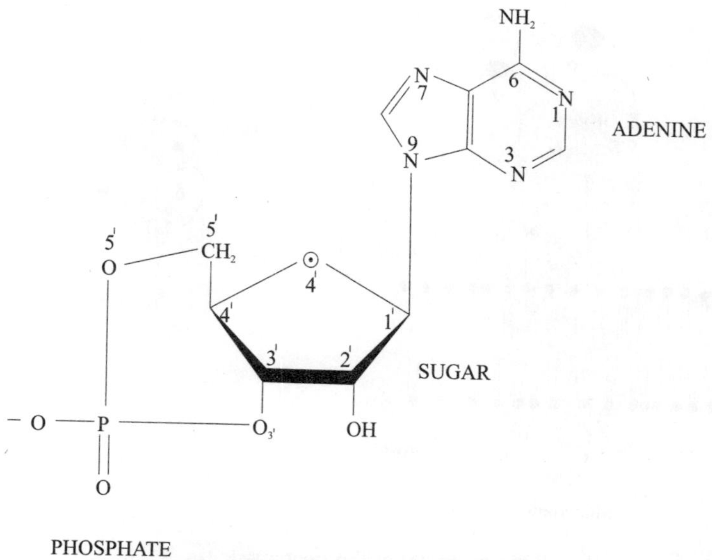

Fig. 15.15 Chemical formula of cAMP.

differently structured, according to stimulus characteristics. For example, in mechano-receptors of auditory cells, the tip of the outer segment contains a special group of microtubules with an integral membrane unit; in photoreceptors, the outer segment has an enlarged membrane surface to accommodate a sufficient number of photopigment molecules.

The transmission of impulses by receptor cells is in general a two-step process. First. the stimuli bring about conformational (structural) changes in the receptor membrane-protein complexes. In the second step, as a consequence of structural changes in the membrane-protein complexes there arises a change in the permeability of the receptor membrane, leading to ion fluxes and receptor potential. The conversion of the receptor potential to action potential (both electrical signals) is an analogue-analogue conversion process. The response of many receptors consists of two parts—a tonic response, which is proportional to the intensity of stimulus, and a phasic response, which is produced by changes in stimulus intensity.

15.5.1 Chemo-receptors

Chemo-receptors are involved in the processes of chemical signalling in the form of free molecules, effecting the signal transduction, by binding to specific receptor proteins and getting inactivated after transduction. The processes involve allosteric conformational changes in the receptor protein molecules and the amplification processes are coupled to the transduction processes. Examples of chemo-receptors are: **ACh**-receptor, **cAMP**-receptor and olfactory-receptors.

Acetylcholine-Receptor (ACh-receptor). The **ACh**-receptor is a ligand-gated cation channel. It is an allosteric protein composed of four different transmembrane subunits assembled with the stoichiometry, $\alpha_2\ \beta\ r\ \delta$. A major portion of the protein mass is extracellular (Fig. 15.16).

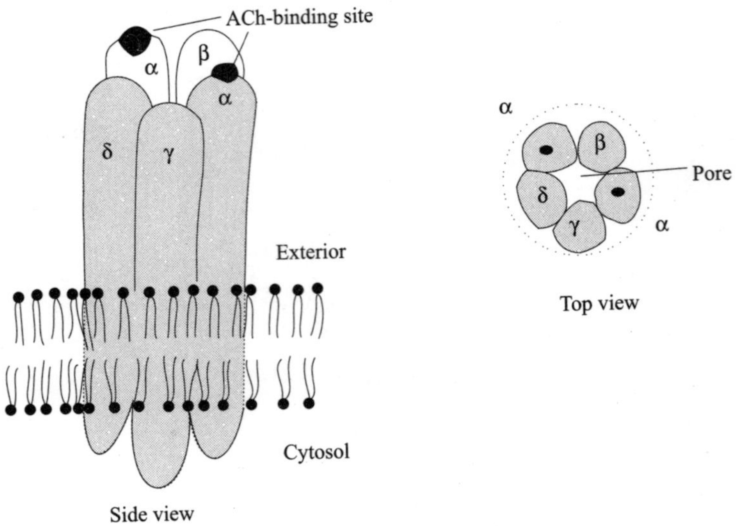

Fig. 15.16 Schematic view of the subunit assembly of the pentameric (α_2 β γ δ). Acetylcholine-receptor protein complex. Each α subunit contains an acetylcholine-binding site. (Ref: Unwin, N., *Nature* 373; 37 (1995).

The **ACh**-receptor has three functionally different conformational states — (i) resting, (ii) active (open to ions) and (iii) desensitised (closed and inactive) (Fig. 15.17).). The mechanistic features are:

(i) In the absence of acetylcholine (**ACh**, the receptors are in the 'resting' (closed) state.

(ii) **ACh**, released in the synaptic cleft, binds to the active conformation of the receptor (two **ACh** molecules to α-subunits), altering the **ACh**-receptor from 'closed' to 'open' state. The 'open' state of the **ACh**-receptor allows a large influx of Na^{++} or Ca^{++} ions (but not anions, like Cl^{-}), causing depolarisation of the post-synaptic membrane.

(iii) With prolonged exposure to agonists, the receptor shifts at a lower rate to a 'desensitised' state (closed and inactive).

(iv) Antagonists (blockers of neuromuscular transmission) tend to stabilise 'closed' or 'inactivate' states, but not the 'open' state.

Acetylcholine is hydrolysed by acetylcholinesterase to choline and acetate.

Neurotransmitters, such as **ACh** and glutamate, 'open' ligand-gated ion channels of the receptor, that are permeable only to small cations (Na^{+}, Ca^{++}) and thereby act as excitatory neurotransmitters. GABA and glycine, on the other hand, open up ligand-gated ion channels that are permeable only to anions (Cl^{-}) and thereby produce inhibitory post-synaptic action. However, a single neurotransmitter can bind to different receptors and, therefore, can exhibit different effects in different target cells. For example, **ACh** binds to skeletal and heart muscle cells, in a channel-linked and non-channel-linked way, respectively. In channel-linked receptors, the binding of **ACh** stimulates skeletal muscle contraction (known as nicotinic actions), and in non-channel-linked receptors, the binding of **ACh** inhibits the contraction of heart muscles (called muscarinic action). Nicotinic agonists resemble **ACh**, with the carbonyl side similar and the methyl side altered; and in muscarinic agonists the carbonyl side is altered while the methyl side is similar.

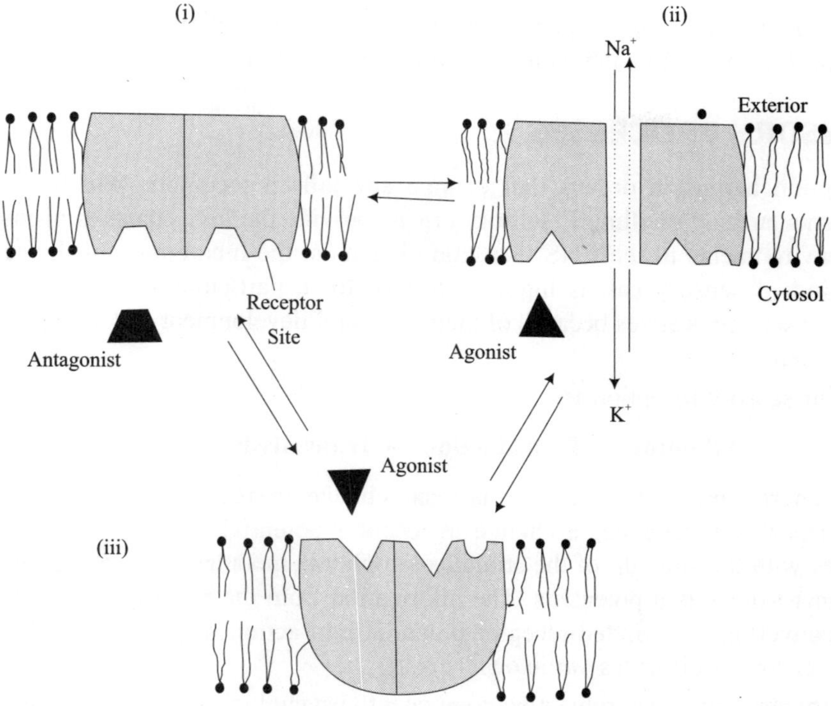

Fig. 15.17 Three conformational states of the ACh-receptor.
(i) Resisting state (channel is closed, but not inactivated).
(ii) Active state (channel is open).
(iii) Desensitised state (channel is closed and inactivated).
(Ref: Changeux, J.P. et al. *Seience* 225; 1335 (1984)).

15.5.2 Metabotropic Reception

Metabotropic (metabolic-related) receptor proteins (e.g. G-proteins, rhodopsin) differ fundamentally from the ionotropic receptors in that they affect the ion channels indirectly *via* the activation of G-proteins). When a receptor protein has been activated by a specific signal, it activates G-proteins, and they in turn activate some other membrane-bound proteins (e.g. adenylate cyclase, phopholipase c, phosphodiesterase). Activation of one of these enzymes leads to increase or decrease in the availability of specific ion or molecule (e.g. second messenger), which in turn controls some critical event in the cellular process.

15.5.3 Modulation and Potentiation

The functional effect of neuromodulation is its long-term effect on post-synaptic neuron. Neuromodulation allows neurons to adapt to changing circumstances. A neuromodulator is usually applied to a neurotransmitter that acts *via* second messenger system.

Facilitation is a transient increase in synaptic strength that occurs when two or more action potentials arrive at the pre-synaptic terminal in close succession, resulting in more neurotransmitter release with each succeeding AP. Potentiation is a form of facilitation in which the increase in the amplitude of post-synaptic potential continues after the stimuli that induced have ceased.

15.6 SENSORY PROCESSES

Sensory reception begins in organs that contain specialised receptors, which are developed to respond to particular kind of stimuli. Sensory organs provide the *only* channels of information from the external environments to the CNS. Sensation begins at the membranes of the sensory receptor cells. Each kind of sensory cell is highly selective for a particular stimulus (Table 15.7). The selectivity of sensory cells arises because of their structural development and arrangement to respond to particular stimuli.

The path of sensory reception is–

Stimulus → Transduction → Transmission

Stimulus energy produces a conformational change in receptor protein, and a change in membrane permeability produces a change in receptor potential. The magnitude of the receptor potential varies with the strength of the stimulus (amplitude modulated). That is, receptor potentials are quite distinct from action potentials. The information from the receptor potential is conveyed to the CNS by converting the graded-receptor potential into action potential, in order to convey the information over longer distances *via* axons.

The sensory perception has subjective component– psychological. The magnitude of perception, ψ, which is subjective, to physical energy, E, of the stimulus, is given by

$$\psi \approx K (E - E_0)^n \qquad \text{...(15.10)}$$

$$\begin{aligned} n &= 0.33 && \text{for visual perception} \\ &= 1.66 && \text{for thermal perception} \\ &= 3.5 && \text{for electrical perception} \end{aligned}$$

(K = constant; E and E_0 = magnitudes of stimulus and threshold stimulus, respectively).

15.6.1 Mechano-Receptors

Mechano-receptors are found in muscles/connective tissues, skin and in hearing organs. Free nerve endings may also function as mechano-receptors (regions of transduction). The mechano-sensitive

Table 15.7 Receptor cells and function of sensory neurons.

Receptors	*Stimulus*	*Sensation*
Chemo	Chemical	Chemical signaling; olfaction and gestation.
Mechano	Mechanical	Touch; auditory; echo location.
Nociceptors	Mechanical/chemical	Pain reception.
Photo	Light	Visual perception.
Electro	Electrical	In skin; and marine organisms.
Thermo	Heat	In skin; and pit organs of snakes.
Magneto	Magnetic field	In birds and insects.

receptor molecule and ion channel are closely coupled and the basal membrane, attached to the cilia, helps in the filtering of stimuli and enhancement of receptor sensitivity. The stimulus that activates the mechano-receptive membrane is the distortion of the surface membrane.

The Hearing Organ. The human ear is a sensory organ for hearing (in the frequency range 20-20,000 Hz) and coordination of body movements and balance. It acts on the basis of mechano-reception-detection of auditory vibrations in the environment by sensory cells of the ear (hair cells). The ear has external, middle and internal compartments to carry out those functions (Fig. 15.18a).

The external part of the ear channels the sound energy (vibrations) to the eardrum (tympanic membrane) through the auditory canal. The middle ear contains three small bones (the ossicles—malleus, incus and stapes Fig. 15.18b). The eardrum, attached to the malleus, acts as a pressure receiver and the vibrations of the ear drum are transmitted by ossicles to the flat plate of the stages that forms the entrance to the inner ear.

Sound vibrations are then transmitted to the **cochlea** of the inner ear. The sound energy is amplified about twenty-fold in the middle ear by the participation of the tympanic membrane (diaphragm) and ossicles. This amplification arises due to the difference in the surface areas of the malleus that is attached to the ear drum (~70 mm^2) and the stapes (~3 mm^2 and due to the lever action of the ossicles. The spiral-shaped cochlea is filled with fluids and energy transduction by the hair cells (receptor cells) is carried out in the cochlea of the inner ear.

The organ of **Corti** (resting on the basilar membrane of the cochlea) includes sensory cells (**hair cells**), inserted into the tectorial membrane, and auditory nerve endings. (Basilar and tectorial membranes and 'hair cells' form the organ of Corti) (Fig. 15.18c). The function of the organ of Corti is to transform mechanical oscillations in the inner ear to electrical signals. Movement of the stapes induces travelling waves in the basilar membrane *via* the cochlear fluid with a frequency equal to the sound frequency.

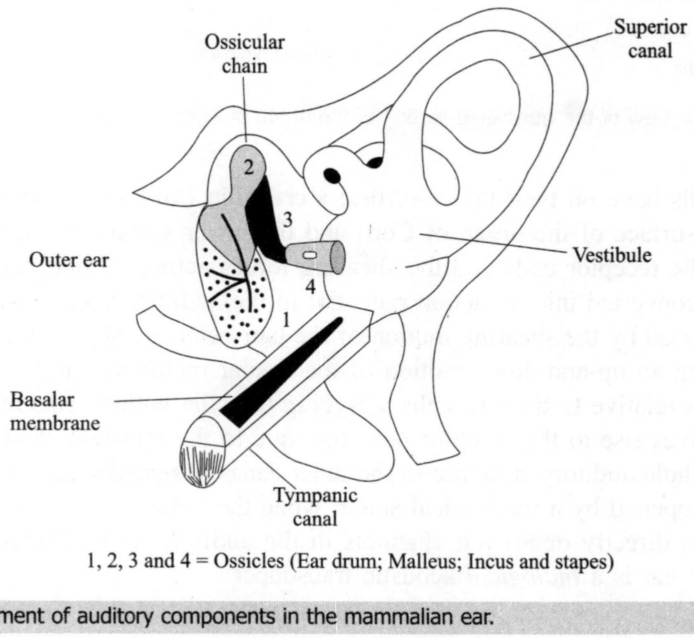

1, 2, 3 and 4 = Ossicles (Ear drum; Malleus; Incus and stapes)

Fig. 15.18a Arrangement of auditory components in the mammalian ear.

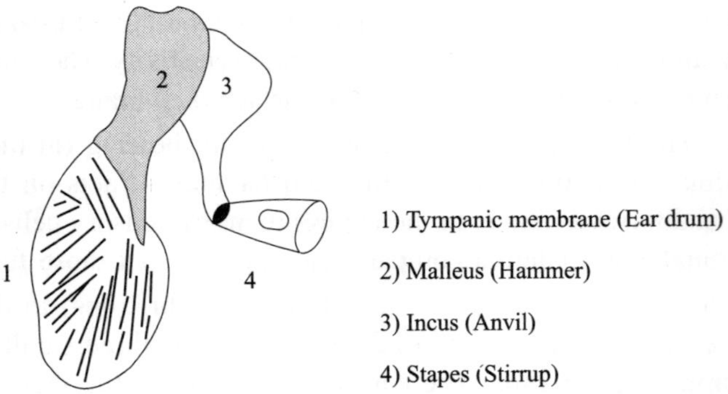

Fig. 15.18b Arrangement of ossicles (ossicular chain).

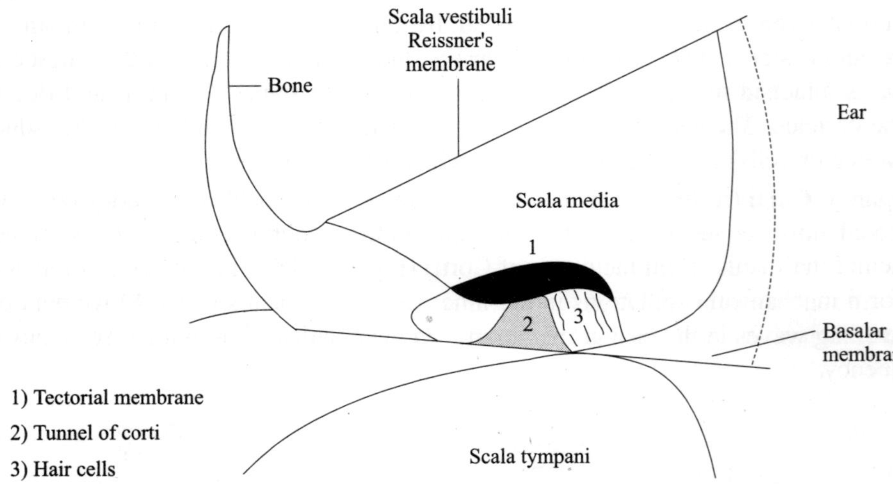

Fig. 15.18c Schematic view of the transverse section of the organ of Corti.

The sensory cells have on their upper surface **stereocilia** (tufts of microvilli) that project out-between the upper surface of the organ of Corti and the lower surface of the tectorial membrane. The hair cells are the receptor cells and the shearing forces acting on hair cells generate cochlear potential, which is converted into an action potential in the auditory nerve ending at the hair cells. Stereocilia are deflected by the shearing motion of the hair cells and supporting tectorial membrane (which occurs during an up-and-down motion of the basilar membrane and its overlying organ of Corti), which moves relative to the hair cells, dragging the cilia within. This mechanical deflection (tilt) of hair cells gives rise to the receptor potential, due to the transient change in the membrane permeability. The whole auditory structure in the inner ear is designed so that ion channels near the tips of the cilia are opened by a mechanical action when the bundle of stereocilia is tilted. That is, mechanical stimulus directly opens ion channels in the auditory system (mechanically-gated ion channels). Thus, the ear is a *biological* acoustic transducer.

In addition, the mechanoreceptors (gravity receptors), in the otolith organs of the vestibular system in the ear, transduce inertial and acceleration and non-uniform (force) effects. In this way, the vestibular system acts as an inertial guidance device and as an accelerometer.

In mammals, mechano-reception includes bathoreceptors that deal with the sense of touch and pressure, and nociceptors and thermo-receptors that deal with the perception of pain. Stimuli applied to the skin and nerve endings affect the ionic permeability of the receptor membrane. Changes in the membrane permeability produce membrane depolarisation (receptor potential) that triggers the action potentials.

Thermo-reception as source of information about the environ has been primarily utilised by some snakes that have heat-sensitive receptors (pit organs).

15.6.2 Photoreceptors

Photosensory cells transduce visual stimuli to electrical signals. Upon absorption of photons by a photo-receptor molecule, a cascade of molecular processes is initiated that alter the membrane potential of the sensory cell (**receptor potential**) and the electrical signal is transmitted *via* synaptic junctions to adjacent nerve cells. A key intermediate in this cascade processes is a GTP-binding protein ($G_{t\alpha\beta\gamma}$), called **transducin**.

Photosensory Cells. Photoreceptors are highly differentiated cells (**rods** and **cones**) that have a lipoprotein complex that is specialised for photo-reception (rods contain **rhodopsin** and the cones, **iodopsin**). A photo-receptor cell in the vertebrate eye consists of (i) an outer segment that contains a stack of membrane discs in rods and infolding of the cell membrane in cones, which function as antennae for receiving photons to initiate a nerve impulse, with a membrane surface containing photopigment molecules, rhodopsin, arranged in stacks parallel to the surface membrane, (ii) an inner segment system containing metabolic and nuclear sources and (iii) a synaptic region at the base of the cell (for communication) which makes synaptic contact with the nerve cells of the retina (Fig. 15.19). Rod cells mediate the formation of black-and-white images in dim light, while rod cells mediate colour vision in bright light. The special sensitivity rod (**scotopic**) vision differs from that of cone (**photopic**) vision. Scotopic vision is most sensitive to blue-green light and insensitive to red light. On the other hand, photopic vision covers the whole visible spectrum with highest sensitivity in the yellow region, this shift in maximum sensitivity to the longer wavelength regions, which accompanies the change from sceptic to photoic vision, is known as the **Purkinje** shift.

Rhodopsin Complex. The outer segment of the rod cell is a narrow tube filled with a stack of membrane discs, containing rhodopsin protein molecules involved in photochemical reactions. Rhodopsin is a transmembrane chromoprotein complex, composed or a glycoprotein, **opsin**, and a chromophore, **retinal** (Fig. 15.20). Opsin is a polypeptide comprising 348 amino acid residues. It is a heptahelical transmembrane protein and in bacterial rhodopsin the retinal molecule is bound covalently through a protonated **Schiff's** base to the ε–amino group of the lysine residue of opsin in the hydrophobic moiety (to one of the α– helices) (Fig. 15.21).

Absorption of light leads to the photoisomer of *cis*-**retinal** to the *trans* isomer. The *trans* isomer complex (**Rh***) dissociates spontaneously resulting in **all-*trans*-retinal** and **opsin**. The general scheme of photoreaction is:

$$\text{11-}cis\text{-}\mathbf{retinal} + H_2N\text{-}(CH_2)_4\text{-}\mathbf{Opsin} \xleftrightarrow{\text{Rhodopsin, Rh}}$$

(lysine side chain)

$$\downarrow h\nu \ (H^+)$$

$$R - \overset{15}{\underset{H}{C}} = \overset{H_+}{N} - (CH_2)_4 - \mathbf{Opsin} + H_2O$$

cis-**retinal** ↓ Isomerisation

$$\mathit{Trans}\text{-}\mathbf{retinal} - C = N - (CH_2)_4 - \text{Opsin}$$

↓ spontaneous dissociation

All-*trans*-retinal + Opsin.

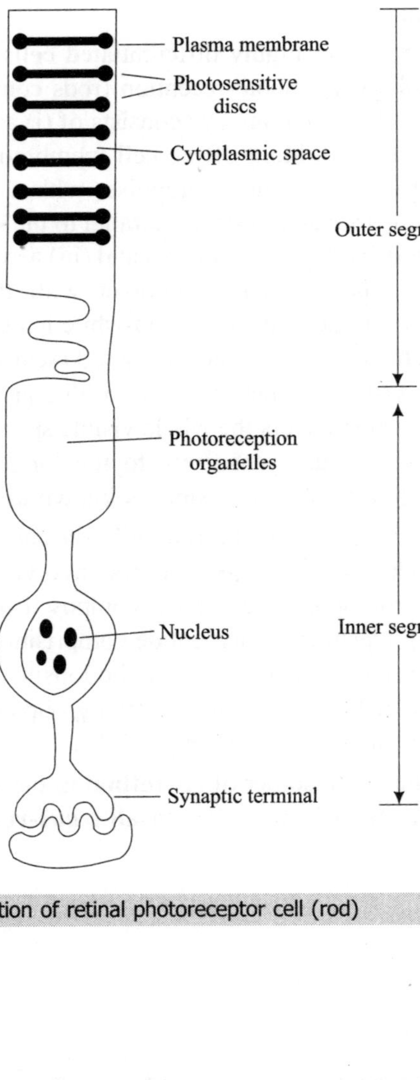

Fig. 15.19 Schematic representation of retinal photoreceptor cell (rod)

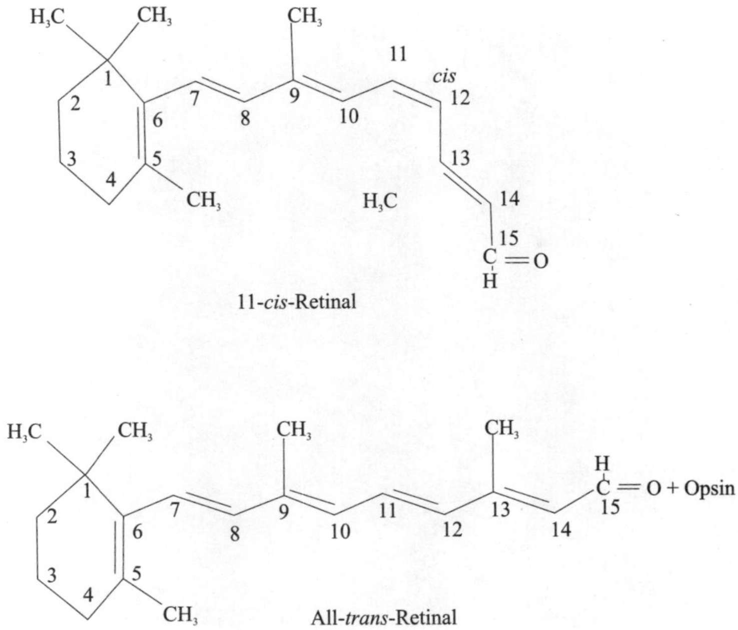

Fig. 15.20 Chemical formulae of *cis*- and *trans*-**retinal** molecules.

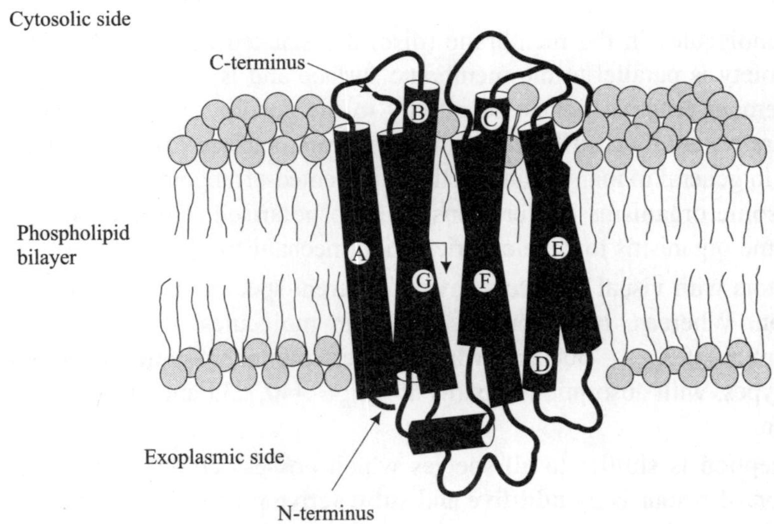

Fig. 15.21a Schematic diagram of bacterial rhodopsin with the seven *trans* membrane α-helices (A to G). **Retinal** molecule (⅂) is covalently attached to lysine residue of G-helix. (After: Henderson, R. et al., J. *Mol. Biol.* 21.3; 899 (1990)).

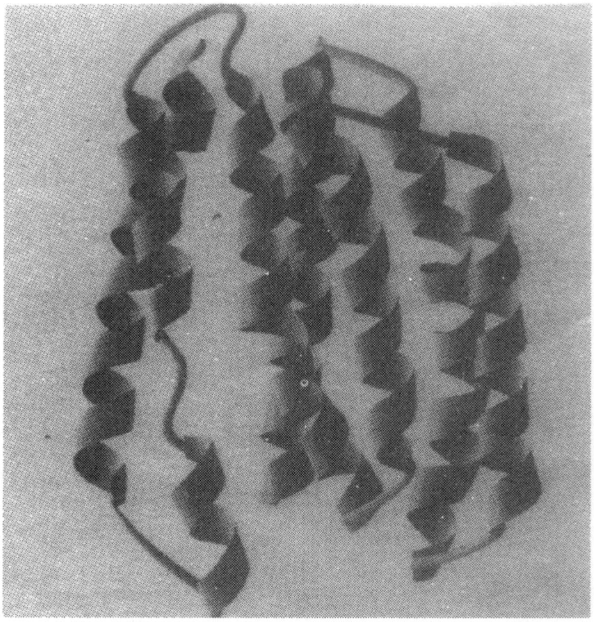

Fig. 15.21b Ribbon diagram of seven-helix bundle in bacteriorhodopsin*.
(Source: Chou, K.C. et al. (1 BAD; 7/92)
*The ribbon diagram has been generated at the Bioinformatics Centre, Indian Institute of Science, Bangalore, from the BNL Protein Data Bank coordinates of the indicated source.

Rhodopsin molecules in the membrane (disc) are stacked in such a way that the axis of the chromophore moiety is parallel to the membrane surface and is perpendicular to the incident light. Such an arrangement renders highest sensitivity to the absorption of unpolarised light when the chromophores are randomly distributed in the membrane surface (azimuthal random distribution). Such is the case in general in many retinal cells. An oriented-arrangement of chromophores in special retinal cells in some organisms confers sensitivity to polarised light and this property is used to advantage by some organisms to develop orientation mechanisms.

Photo-receptors with visual pigments having different spectral absorptions are the pre-condition for colour vision. Whereas, **rods** contain rhodopsin molecules of a particular type, with light absorption maximum at $\lambda_{max} = 500$ nm, which renders monochromatic vision, cones contain three visual pigment types, with absorption maxima at $\lambda_{max} \approx 440$, 535 and 600 nm (Fig. 15.22) enabling colour perception.

Colour perception is similar to all species which possess colour vision apparatus. In human beings, perception of colour is by **additive** and **subtractive** processes. i) In the additive process the perception of colour is characterised by mixing of **primary** colours, **red**, **green** and **blue**. Mixing together of all three primary colours produces white colour. Combination of any two primary colour gives the corresponding secondary colour (e.g. red + blue = **magenta**; red + green = **yellow**; and blue + green = **blue-green**). ii) In the case of subtractive process (mixing of coloured pigments, such as paints and dyes), pigments absorb *(subtract)* certain colours, and reflect back other colours;

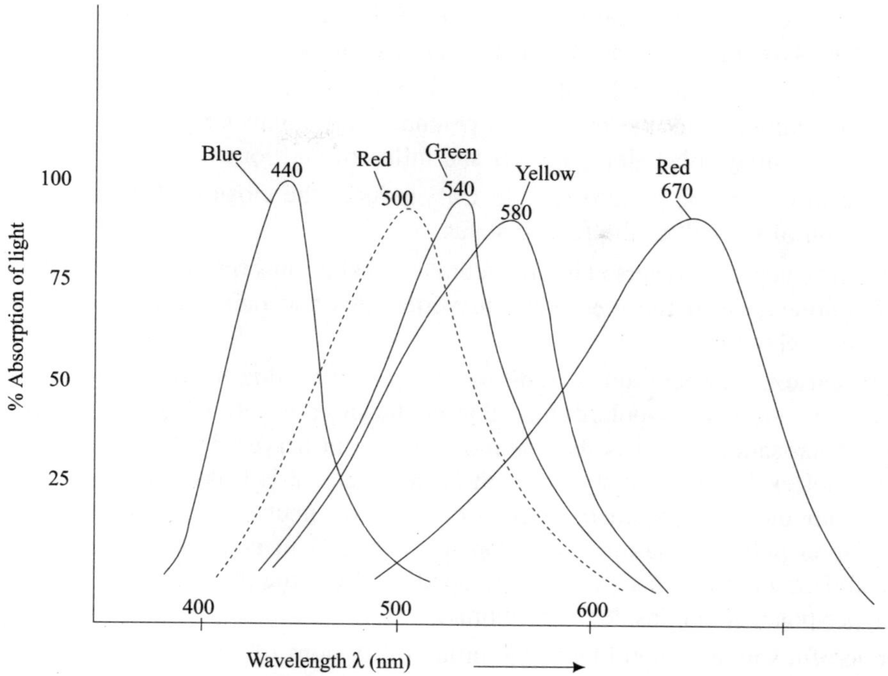

Fig. 15.22 High absorption spectrum of rods and cones.

and the eye perceives the *colour* from mixing of these reflective colours. In this *subtractive* process the **primary** colours are: **magenta, yellow** and **blue-green**. Each pigment absorbs (subtracts one of the *primary* colours of light, and reflects back the other two. E.g. yellow pigment absorbs blue light and reflects back red light and green light and these two colours mixed by the eye to perceive it as yellow (red + green = yellow); blue-green pigment absorbs red light and reflects back green light and blue light (blue + green = blue-green); and magenta pigment absorbs green light and reflects back blue light and red light (blue + red = magenta). That is, in human beings, the perception of colour is by **additive** as well as **subtractive** processes. Mixing of **primary** colours with **secondary** colours and mixing of **secondary** colours with secondary colours (by *additive* and *subtractive* processes) give rise to various colour combinations and hues. Perception of colour is quantitative as well as subjective. Not all animals perceive an object in the same colours and it differs from species to species and may differ among individuals of the same species.

Signal Transduction. Signal transmission begins with the absorption of a photon by 1-*cis*-**retinal**, triggering the enzymatic activity of **opsin** and subsequent cascade of molecular reactions. A general scheme of photoreactions is —

(i) Absorption of a photon of light causing 11-***cis*-retinal** to **all-*trans*-retinal** form and conformational changes in the **opsin** moiety ($\pi^* \leftarrow \pi$ and $\pi^* \leftarrow n$ transitions).

(ii) Cleavage of **all-*trans*-retinal** from opsin. That is, photo-isomerisation to all-trans-retinal converts rhodopsin, **Rh**, to an activated intermediate form, **Rh***. The H^+-translocation is driven by the photochemical ***cis-trans*** isomerisation of retinal and the deprotonation of the

Schiff's base nitrogen of the lysine residue. Specific arrangements of amino acids, such as **Asp, Glu, His, Thr, Tyr, Ser** are also considered important in translocation.

(iii) The photoreceptor potential in the rod cell is maintained by the closure of Na^+-ion channels. Photoabsorption alters the membrane potential and conductance of Na^+ ions, resulting in transient opening of Na^+-ion channels and influx of Na^+ ions.

(iv) Change in membrane permeability to Ca^{2+} induces the closure of photo-ion channels and restoration of the cell to the resting states.

The above mechanism is reversed in invertebrates. That is, absorption of light quanta leads to the reduction of conductance of the membrane in vertebrates and increase in the conduction of the membrane in invertebrates.

Signal amplification. Generation of rod-receptor potential, due to the excitation of rods upon photon absorption causes hyper-polarisation, that is, increased negativity of the membrane potential (rather than depolarisation which is the common process in nerve impulse signalling). In the dark when **rods** are not excited the membrane of the outer segment is leaky to Na^+-ions, and Na^+-ions, pumped out from the inner segment, leak back, thereby neutralising the possibility of hyper-polarisation. Upon photo-excitation of rhodopsin molecules in **rods** (**Rh** $\rightarrow$ **Rh***), the electrical balance is disturbed and more Na^+-ions are pumped out from the inner segment, creating increased negativity (hyper-polarisation) inside the membrane.

Absorption of a single photon blocks the influx of millions of Na^+ ions by closing hundreds of Na^+ channels in the plasma membrane. This signal amplification is achieved by intracellular messengers involving enzyme cascade processes, mediated by cyclic GMP. The cleavage of cGMP by phosphodiesterase leads to spontaneous closure of Na^+-ion channels, allowing hyperpolarisation of the cell (negative polarisation of the transmembrane voltage).

Heterotrimeric guanine nucleotide-binding protein, **transducin** (a *G*-protein with three subunits, $\alpha\,\beta\,\gamma$) functions as the molecular switch. Nucleotide-binding and hydrolysis perform the transduction of the external signal (light) into intracellular effects. Rhodopsin, **Rh,** is activated to **Rh*** by photon absorption. Binding of **Rh*** to transducin releases the GTP-bount α-subunit of transducin $(G_{t\alpha})$ from the $\beta\,\gamma$ part $(G_{t\,\beta\,\gamma})$ and the $G_{t\,\alpha}$. The GTP complex in turn binds to and activates a cyclic GMP-phosphodiesterase (PDE) complex. The binding of the inhibitory part of phosphodiesterase (γ -subunit) to α-transducin complex ($G_{t\alpha}\cdot$GTP) amplifies the catalytic activity of phosphodiesterase (activated from the enzyme without the inhibitory γ-subunit) manifold. The transduction of signals depends on the ability of the α-subunit of G-protein $(G_{t\,\alpha})$ to shuttle between 'active' (GTP-bound) and 'resting' (GDP-bound) conformations.

The mechanistic view of the role of transducin in the photoexcitation processes is: (i) activation of rhodopsin, **Rh to Rh*** by a photon, (ii) binding of activated **Rh*** to GDP -$T_{\alpha\beta\gamma}$ (iii) binding of GTP to T_α subunit, concommitantly releasing $T_{\beta\,\gamma}$ part of transducin, (iv) binding of GTP- T_α to PDE-γ subunit, thereby exposing the cleavage site of cyclic GMP in PDE-$\alpha\,\beta$ part. The GTPase activity of the α-subunit of transducin hydrolyses the GTP to GDP. The GDP-bound α-subunit ($G_{t\alpha}\cdot$GDP) has greater affinity for the $\beta\gamma$ complex of the proteins and thereby brings the components back to heterotrimeric G protein $(G_{t\,\alpha\,\beta\,\gamma})$ again.

The intracellular cGMP concentration regulates the phosphorylation of gating-proteins which control the 'open', 'closed' and 'inactivated' states of Na^+-ion channel in rod cells. The synthesis of cyclic GMP from GTP is controlled by a 23 kilodalton Ca^{2+}-binding protein, called **recoverin**. X-ray structure determination of α-transducin has been able to provide 'freeze frame' features of several reaction intermediate states to suggest and clarify possible molecular mechanisms in the visual signal transduction process.

Photoreceptor Characteristics. The retinal photo-receptor apparatus consists of rods and cones that contain stacks of photoreceptive membrane discs. The discs contain photo-sensitive rhodopsin molecules and phospholipids (in the liquid-phases). The discs are optically anisotropic — they exhibit dichroic as well as birefringence properties.

Absorption of a photon by the photopigment molecule, rhodopsin, is influenced by the structure of the receptor. Rhodopsin molecules are arranged in the layered photoreceptor region (outer segment) which has a higher refractive index than the surrounding medium. Such a region exhibits optical wave-guide properties (total internal reflection) for all rays incident at an angle, ϕ, satisfying the condition

$$\sin \phi > \sin \phi_c = \frac{\mathbf{n}_2}{\mathbf{n}_1} \quad \text{for } \mathbf{n}_1 > \mathbf{n}_2 \qquad ...(15.11)$$

where ϕ_c is the critical angle and $\mathbf{n}_1$ and $\mathbf{n}_2$ are the refractive indices of the system and the surrounding medium, respectively.

15.6.3 Chemoreceptors

Taste (gustatory) and smell (olfactory) receptors are chemoreceptors. The gustatory receptors detect ingested saliva-soluble chemicals, and olfactory system detects volatile molecules. They are interconnected. Many substances are not only 'tasted' but also 'felt'.

Organs of taste (the taste buds) in the tongue react to stimuli and each class of taste stimuli activates a distinctive cellular pathway in the receptors that respond to it.

Site of olfactory transduction is the olfactory cilia, located in the upper region of the nasal passage. Smell has been described as 'taste at a distance'.

15.6.4 Other Type of Receptors

There are many other types of receptors found in insects, birds and animals—sono-receptors in bats used for echo-location, electro-receptors in fishes and thermo-receptors in some snakes/etc. and electromagnetic and gravitational receptors which are special to migratory birds and animals for navigational purposes. Discussion of these topics is beyond the scope of this book.

Suggested Readings

Books

Aidely, D.J. "*The Physiology of Excitable Cells*". 3rd Edn. Cambridge University Press/Cambridge, UK. (1989).

Barlow, H.B. & Mollon, J.D. (Eds). "*The Senses*". Cambridge University Press/Cambridge, UK. (1982).

Bullock, T.H., Orkand, R. & Grinell, A. "*Introduction to the Nervous System*", Freeman/New York. (1979).

Ceve, G. & Marsh, D. "*Phospholipid Bilayers: Physical Principles and Models*". John Wiley/New York (1987).

Cone, R.A. & Douling. J.E. (Eds) "*Membrane Transduction Mechanism*". Raven Press/New York (1979).

Delcomyl, F. "*Foundations of Neurobiology*". Freeman/New York. (1998).

Finean, J.B, Coleman, R. Mitchell, R.H. "*Membranes and Their Cellular Functions*". 2nd Edn. Blackwell Sci. Pb/Oxford. (1978).

Hille, B. "*Ionic Channels on Excitable Membranes*". 2nd Edn. Sinauer/New York (1992). Hoppe, W. et al. "*Biophysics*". Springerverlag/Berlin, (1983).

Kandel, E.R. & Schwartz, J.H. "*Principles of Neural Science*". 2nd Edn. Elsevier/Amsterdam (1985).

Kandel, E.R. et al. "*Principles of Neural Science*". Elsevier/New York. (1991).

Katz, B. "*Nerve, Muscle and Synapse*", McGraw Hill/New York (1966),

Miller, W.H. (Ed.) "*Molecular Mechanisms of Photoreceptor Transduction*". Academic Press/New York. (1981).

Purves, D. et al., "*Neuroscience*". Sinauer Associates/Sunderland; MA. (1997).

Robertis, E.D.P. & Robertis, E.M.F. "*Cell and Molecular Biology*" 7th Edn. Saunder College/Philadelphia. (1980).

Shephard. G. "*Neurobiology*". 2nd Edn. Oxford University Press/Oxford. (1988).

Stein, W.D. (Ed.) "*Transport and Diffusion Across Cell Membranes*". Academic Press/New York (1986).

Stein, W.D. (Ed.) "*lon Pumps: Structure, Function and Regulation*" (1988).

Articles

Amzel, L.M., Narayanan, P., Pedersen, P.L. & Sygusch, J. "*The Three-dimensional Structure of F1-ATPase*". Annal. New York Acad. Sci. 402; 21(1982).

Berg, H.C. & Purcell, E.M. "*Physics of Chemoreception*". Biophys. J. 20; 193 (1977)

Berridege, M. "*The Molecular Basis of Communication within the Cell*". Sci. Amer. 253(4); 142 (1985).

Borg, E. & Counter, S.A. "*The Middle Ear Muscles*". Sci. Amer. 261 (2); 62. (1989).

Brisson, A. & Unwin, P.N.T. "*Quaternary Structure of the Acetylcholine Receptor*". Nature 315; 474. (1985).

Brown, M.A. & Birnbaumer, L. "*Ionic Channels and Their Regulation by G-Proteins*". Annu. Rev. Physiol. 52; 197 (1990).

Catterall, W.A. "*Structure and Function of Voltage-gated ion Channels*". Science. 242; 50 (1988).

Changiux, J.P., Thiery, A.D. & Chemoulli, P. "*Acetylcholine Receptor; An Allosteric Protein*" Science 225; 1335 (1984)

Doyle, D.A. "*The Structure of the Potassium Channel: Molecular Basis of K^+ Conduction and Selectivity*". Science. 280; 69 (1998).

Eisenberg, D. "*Three-dimensional Structure of Membrane and Surface Proteins*". Annu. Rev. Biochem. 53: 595 (1984).

Goody, R.S. "*How G-proteins Turn Off*". Nature. 372; 220 (1994).

Henderson, R. et al. "*Model for the Structure of Bacteriorhodopsin based on High-resolution Electron Cryo-microscopy*". J. Mol. Biol. 213; 899 (1990).

Hildago, C. "*Lipid-protein Interactions and the Function of the Ca^{++}-ATPase of the Sarcoplasmic Reticulum*". CRC Crit. Rev. Biochem. 21; 319 (1987).

Horisberger, J.D. et al. "*Structure-function Relationship of Na^+/K-ATPase*". Annu. Rev. Physiol. 53; 565 (1991).

Hudspeth, A.J. "*The Cellular Basis of Hearing: The Biophysics of the Hair Cells*". Science. 230; 745 (1985).

Jenks, W. "*How Does a Calcium Pump Pump Calcium?*". J. Biol. Chem. 264; 18855 (1989).

Jennings, M.I. "*Topography of Membrane Proteins*". Annu. Rev. Biochem. 58; 999 (1989).

Lambright, D.G. et al. "*Structural Determinants for α-subunit of a Heterotrimeric G-protein*". Nature. 369; 624 (1994).

Lester, H.A. "*The Response to Acetylcholine*". Sci. Amer. 236 (2); 106. (1977)

Linder, M.E. & Gilman, A.G. "*G-proteins*". Sci. Amer. 247 (4); 38. (1992).

Noel, J.P. et al. "*2.2 Å Structure of Transducin-α Coupled with GTP γ S*". Nature. 366; 654 (1993).

Ovchinnikov, Yu. A. "*Rhodopsin and Bacteriorhodopsin: Structure-function Relationships*" FEBS. Lett. 148; 179 (1982).

Pedersen, P.L. & Amzel, L.M. "*ATPsynthases:.....*" J. Biol. Chem. 268; 9937 (1993).

Rapport, T. "*Protein Transport Across the Endoplasmic Reticulum Membrane*". TIBS. 15; 355 (1990).

Sathev, W.A. et al. "*Structural Basis of Ion Channel Permeability and Selectivity*". Curr. Opin. Neurobiol. 4 (3); 313 (1994).

Snyder, S.H. "*The Molecular Basis of Communication between Cells*". Sci. Amer. 253 (4); 132.

Spencer, R.H. et al. "*Feeling the Pressure: Structural Insights into a Gated Mechano-sensitive Channel*". Curr. Opin. Struct. Biol. 9 (4); 448. (1999).

Stryer, L. "*The Molecules of Visual Excitation*" Sci.Amer. 257 (1); 42. (1987)

Unwin, N. & Henderson, R. "*The Structure of Proteins in Biological Membranes*". Sci. Amer. 250 (2); 78 (1984).

Unwin, N. "*Acetylcholine Receptor Channel Imaged in the Open State*". Nature 373; 37 (1995).

16 Biomechanics

Synopsis: Cell motility can be either movement within the cell (contraction) or movement of the entire cell (cell locomotion). Contraction and locomotion in living organisms are carried out by three types of protein complexes—(i) **actin-myosin**, (ii) **tubulin-dynein** and (iii) **intermediate filaments**. Skeletal and cardiac muscles, which exhibit striated features, are involved in fast contraction process, whereas in smooth muscles the contraction process is slow. The functional unit of a muscle myofibril, the **sarcomere**, exhibits alternating dark and light bands, due to the unique assembly of myosin (thick) and actin (thin) filaments. During muscle contraction, thick filaments slide past the thin filaments unidirectionally. Muscle contraction is regulated by Ca^{2+} ion concentration, and accessory protein complexes of actin-tropomyosin and troponin (TnT, TnI and TnC).

Cardiac muscles have some characteristics similar to the striated muscles. Cardiac muscles exhibit automatic and functional syncytium and possess inherent rhythmicity. Cardiac muscles do not exhibit tetanus.

In smooth muscles, the actin and myosin filaments lack "striated" organisation. They exhibit tonus and plasticity.

A crucial feature of cell locomotion is directionality. Locomotion of organelles is realised by the action of the tubulin-dynein protein complex in microtubules. The bending force in the microtubular apparatus, **axoneme**, is due to the relative sliding motion of microtubule doublets. Cell motility in bacteria is accomplished by flagella which rotate like propellors and thus move the cell.

The dynamics of aquatic life are determined by hydrostatics, viscosity and other environmental parameters. At low **Reynolds** numbers, viscous forces are dominant and the locomotion of organelles is by wiggling and twisting movements. Motion in fluids of high Reynolds numbers is turbulent and aquatic creatures in regions of high Reynolds numbers are conditioned to the effects of turbulence, viscous and pressure drags. Gliding and flying are the processes of locomotion by flying-creatures. For aeroflight, lift and drag components have to be optimised. Shapes of the wings and feathers of birds etc., are in accordance with the environmental requirements (gliding and flying motions).

16.1 CELL CONTRACTILITY

The cytoskeleton is a complex network of proteins involved in biochemical processes—(i) contraction, which is the ability of the cell to generate force (for relative motion) and ii) motility, i.e., locomotion. There are three types of protein complexes and accessory proteins involved in contraction and locomotion. These are (i) **actin-myosin** complex and accessory proteins that are involved in muscle contraction, (ii) **tubulin-dynein** complex and accessory proteins that are involved in ciliary and flagellar motion and (iii) **intermediate filaments**.

There are three types of vertebrate muscles—skeletal, cardiac and smooth muscles. Myofibrils of skeletal and cardiac muscles show alternating dark, birefringent (aniosotropic) and clear, isotropic zones. Striated muscles are involved in fast contraction. In smooth muscles (in the intestines, stomach wall etc.), contraction is slow and the myofibrils do not have striations. The process of contraction and motility are intimately coupled to the transduction of chemical energy into mechanical work. The chemical energy due to the break down of glycogen in muscle contraction leads to the formation of pyruvic acid as an intermediate which is broken down into CO_2 and H_2O in aerobic conditions and into lactic acid in anaerobic conditions. While skeletal muscles do not have electrical synapses (gap junctions) and are under control of somatic nervous system, the cardiac and smooth muscle systems have gap junctions and are controlled by the autonomic nervous system.

The function of muscle is to contract (or relax) and the events leading to muscle contraction are: (i) conduction of impulse along an axon to its termination on the muscle's motor-end plate, (ii) liberation of neurotransmitter from the axon terminal and (iii) phase transmission of the muscle fibre, called twitch. These events are referred to as neuromuscular transmission.

16.1.1 Sketetal Muscle

Vertebrates employ muscle-contraction mechanisms for biochemical transport and work. A muscle is a bundle of longitudinally-arranged muscle fibres, attached to tendons at its ends. A muscle fibre is about ≈0.05 nm thick and ≈500 mm long, comprising parallelly-arranged cylindrical filaments (of about 1 μm diameter), called myofibrils. Myofibrils have alternate dark and light bands that give rise to the striated appearance of skeletal muscles observed under optical microscopes Fig 16.1.

In a striated myofibril, the structure and functional repeating unit is called a **sarcomere** (≈2.3 μm), sandwiched by two adjacent **Z**-lines. The dark, anisotropic band in the central part of the sarcomere, is called the **A**-band. Dark **A**-bands and light **I**-bands alternate regularly, punctuated by **Z**-lines. The central region of the **A**-band, which is less dense is called the **H**-zone and in the middle of the **H**-zone, is the **M**-line, which divides the **A**-band (Fig. 1.6 .2)

16.1.2 Proteins of the Contractile Apparatus

The structural architecture of a striated muscle fibre is due to an assembly of two kinds of protein filaments— (i) thick filaments, which are composed entirely of myosin and (ii) thin filaments, which are composed of actin filaments and accessory protein components, tropomyosin and troponin (TnT, TnI and TnC). At the **AI** region, where the thick and thin filaments overlap, each myosin is surrounded by six actin filaments and each actin filament by three myosin filaments so that the myosin–actin ration is 2 : 1. This ratio varies in other muscles. Observed under polarising microscope, *A*-bands exhibit birefringence (anisotropic environment), because it is an overlap region of both thick

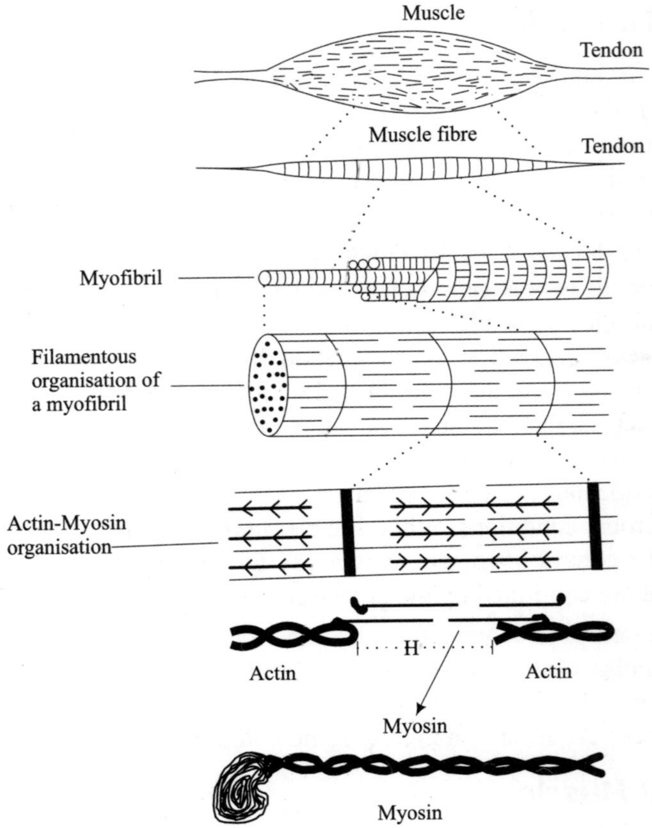

Fig. 16.1 A schematic representation of muscle organisation from cellular to molecular levels.

and thin filaments. Both the *H*-zone and *I*-bands are isotropic, as they contain only thick and thin filaments, respectively. The *Z*-line is a dense amorphous material to which thin filaments (**F-actin**) are attached. *M*-protein is located in the *M*-line. Myosin fibres meet tail to tail at the *M*-line, heads pointing away from the *Z*-end. Some of the contractile proteins of skeletal muscle and their functional roles are listed in Table 16.1. The associations of thick and thin filaments in the **AI** regions are responsible for striated muscle contraction, due to the formation of acto-myosin cross-bridges.

Myosin. Thick filaments of the muscle fibre consist of myosin molecules. A myosin molecule is composed of two identical heavy chains (200 kilodalton) and four light chains ($\approx$20 kilodalton each). Heavy chains consist of a long α-helical coiled coil, rod-like filament ($\approx 140 \times 2$ nm). Two globular myosin heads ($\approx 11 \times 4$ nm), called heavy meromyosin subfragments (HMMS1), are joined to the helical rod. The globular HMMS1 have ATP-hydrolysing capacity. The hinge-region, HMMS2 fragment, allows the head (S1 fragment) to bend away from HMMS2, which is an important feature in generating force (torque) in muscle contraction. A pair of light chains are bound to each of the two heavy-chain heads (HMMS1 fragments) by nonbonded interactions. The myosin molecule is schematically represented in Figs. 16.3 a & b.

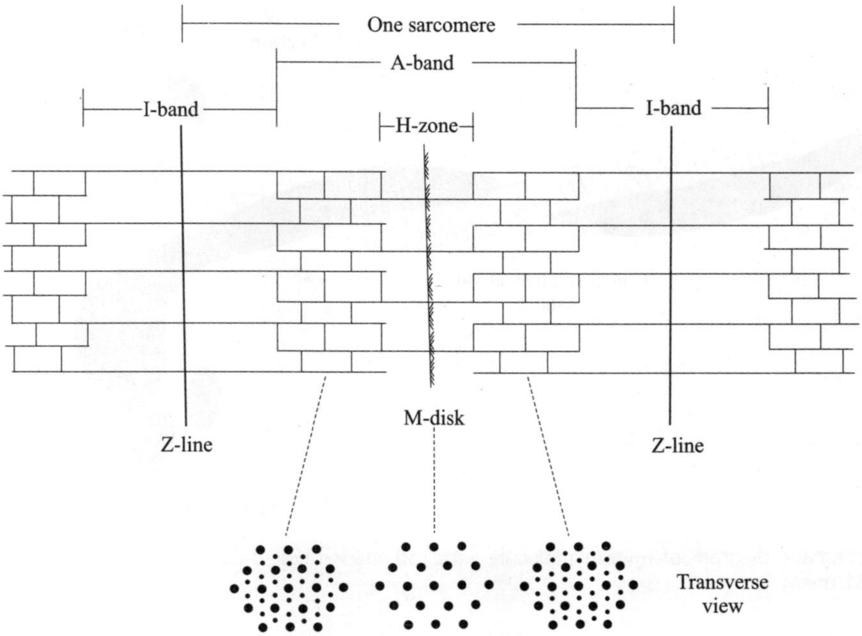

Fig. 16.2 Schematic representation of a sarcomere (a functional unit between two Z-lines) with the arrangement of actin–myosin filaments. Arrangement of actin and myosin filaments leads to regions — **I**-band containing only thin filaments (actin); **H**-zone containing only thick filaments (myosin) and **A**-band that contains thick and thin filaments.

Table 16.1 Contractile muscle proteins and their functions.

Protein	*Structural features*	*Functions*
Myosin (Thick filament)	An enzyme with two heavy chains (2 × 220 kD) and two light chains (2 × 40 kD).	Interacts with thin filaments (F-actins) to generate the mechanical force.
F-actin (Thin filament)	A polymer of seven G-actin monomers (7 × 40 kD).	Component of thin filaments of the myofibril. Interacts with thick filaments to generate force.
Tropomyosin	Coiled coil, rod-like protein (2 × 35 kD).	Binds to the actin filament grooves, regulating actin-myosin binding.
Troponin	Heterotrimeric protein complex; TnT = 38 kD, TnI = 25 kD, TnC = 18 kD.	Three-subunit protein complex, placed at regular intervals along the actin filament. TnC regulates the binding of troponin complex to the tropomyosin.
Other:		
α-Actinin	200 kD	One of the components of the Z-band.
β-Actinin	60 kD	—
c-Protein	150 kD	Structural protein; binds to the myosin chain, functioning as a clamp.
M-Protein	100 kD	Structural protein.
Gelsolin	90 kD	Actin-capping protein.
Ca^{2+}-ATPase	100 kD	Regulates Ca^{2+} ion concentration.
Calciquestrin	40 kD	Ca^{2+}-regulatory protein in SR membrane.

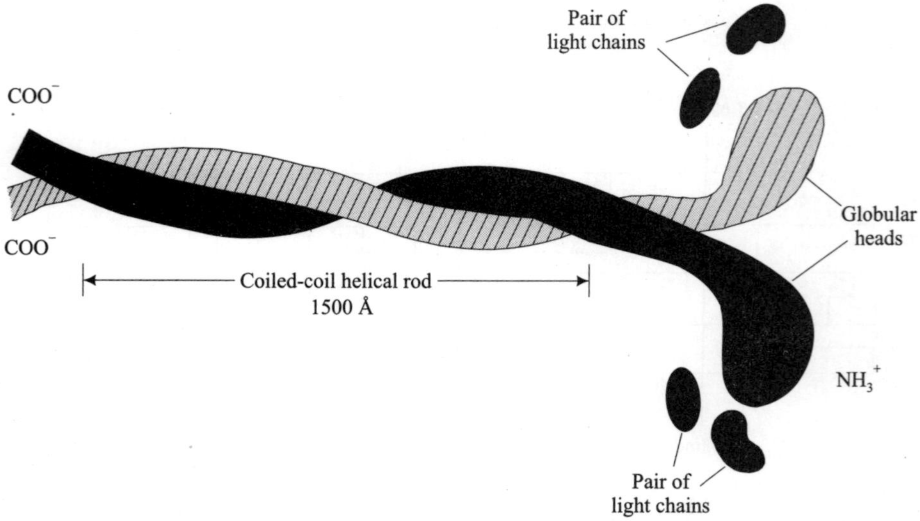

Fig. 16.3a Schematic diagram of myosin molecule with rod-shaped coiled-coil tail (≈1500 Å) and globular heads (**S1** units)

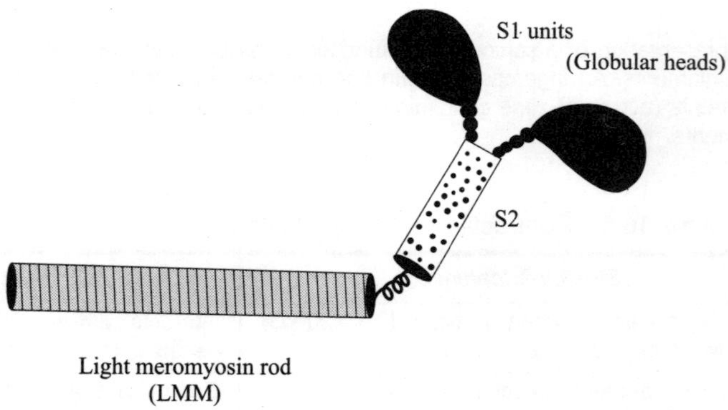

Fig. 16.3b Hinge regions in myosin molecule allowing conformational flexibility to globular heads (**S1**) during muscle contraction process.

Actin. Actin is the major component of thin filaments. The actin monomer, called **G**-actin, is a globular α-helical polyeptide of ≈42 kilodaltons. Polymeric actin strands, comprising seven **G**-actin monomers, called **F**-actin, wound around each other forming a rod-like double-stranded coiled coil (2 × 7 monomeric *units*), with a repeating unit of ≈ 72 nm. Each actin monomer (**G**-actin) is in contact with four neighbours. *F*-actin has high affinity for myosin and for the myosin-ADP-P_i complex, but has low affinity for the myosin-ATP complex. The actin assembly further contains other regulatory proteins for structural stability and function.

Tropomyosin. Tropomyosin is a component of the thin filament of the muscle fibre. It is a long (≈42 nm), rod-like, double-stranded α-helix, attached along the length of the double-helical **F**-actin filament in the grooves (stretching over seven **G**-actin monomers). Similar to keratin, there is a repeating unit of seven amino acid sequence, $(\mathbf{a}-\mathbf{b}-\mathbf{c}-\mathbf{d}-\mathbf{e}-\mathbf{f}-\mathbf{g})_n$, with **a** and **d** residues being nonpolar and **e** and **g** residues being charged. The molecule is divided into 14 regions of ≈2.8 nm. Within each of these regions there are segments of polar and nonpolar residues. These segments provide binding sites for the other protein components of the muscle — actin and troponin. The contractile function of actomyosin complex (actin-myosin complex) is coordinated by the tropomyosin-troponin complex and Ca^{2+}-ions.

Troponin Complex (TnT, Tn1 and TnC). Troponin is a heterotrimeric protein complex of three subunits located in thin filaments at intervals of ≈38.5 nm. TnT is a 37 kilodalton polypeptide which binds strongly to tropomyosin and is a Ca^{2+} -independent. TnI is a 24 kilodalton polypeptide that binds to actin and inhibits actin-myosin interactions (cross-bridge formation) and TnC is an 18 kilodalton protein that makes the troponin complex sensitive to Ca^{2+} ions. A molecule of TnC can bind four molecules of Ca^{2+} and it is closely related to calmodulin. Conformational change of TnC upon binding of Ca^{2+}-ions is central to the regulatory process of muscle contraction. It is a dumbbell-shaped molecule, consisting of *N*- and C-terminal domains joined by a nine-turn α-helical link (Fig. 16.4a). This linker-helix is exposed to solvent and presumably plays a major role in "fine-tuning" of Ca^{2+} in the function of TnC. Organisation of proteins in thin filaments (actin, tropomyosin and troponin complex) is schematically represented in Fig. 16.4b.

Other Accessory Proteins. α-actin is a rod-shaped molecule (2 × 100 kilodalton) found in **Z**-lines. It binds to actin filaments, forming actin bundles. **Filamin** (2 × 270 kilodalton) cross-links actin filaments forming a gel-like three-dimensional network. The protein, **spectrin** (2 × 260 kilodalton)

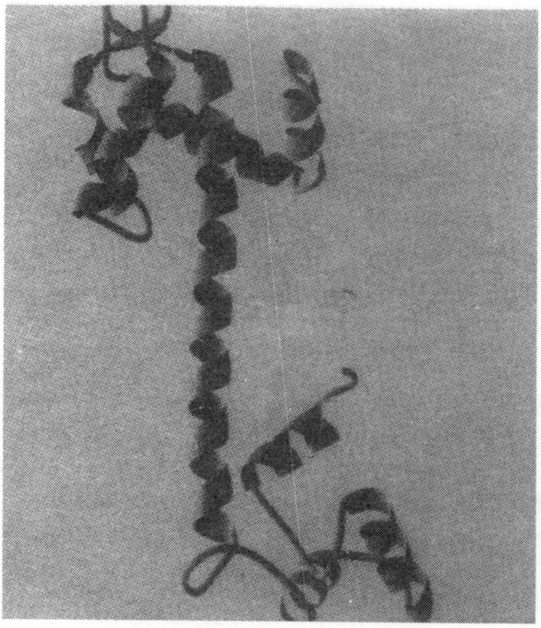

Fig. 16.4a Ribbon diagram of troponin, TnC (the Ca^{2+}-binding subunit of troponin complex) of chicken*
(Source: Sundaralingam, M. J. 1 Top; 8/93)
*The ribbon diagram has been generated at the Bioinformatics Centre, Indian Institute of Science, Bangalore, from the BNL Protein Data Bank coordinates.

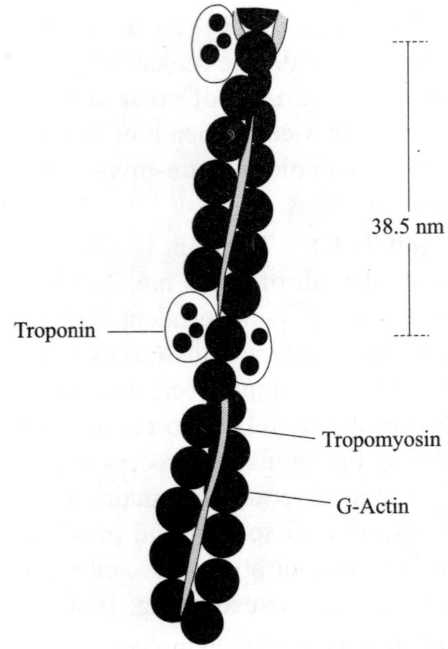

Fig. 16.4b Schematic view of the organisation of **F**-actin and double helical tropomyosin complex. Troponin complex (TnT, TnI & TnC) at ~ every 7th **G**-actin residues controls the contraction process.

anchors the sides of actin filaments to the plasma membrane. **C-protein** is present in the **A**-band and **M**-protein in the **M**-line region. These proteins appear to have structural roles.

16.1.3 Bioenergetics of Muscle Contraction

The molecular basis of muscle contraction can be explained on the basis of the **"sliding filament"** model. During muscle contraction the thick filaments slide past the thin filaments unidirectionally towards the plus-end of the actin filament, resulting in the contraction of the sarcomere. In muscle contraction, the lengths of thick and thin filaments are unaltered; only the length of the sarcomere is altered. The length of dark bands is unaltered, while the length of the light bands is altered. (Fig. 16.5).

Thick filaments (heads) can form cross-bridges which make contract with adjacent thin filaments during the sliding motion. The cross-bridges protrude from the thick filament myosin at every 14.5 nm interval. During contraction, myosin heads hydrolyse ATP, while binding to actin. A rapid release of ADP and P_i from the myosin heads produces a series of allosteric changes in the conformation of the myosin molecule, similar to the "power-stroke" of a piston in an automobile (Fig. 16.6).

Myosin (an ATPase) transduces chemical energy (ATP) into mechanical motion bycyclic attachment and detachment of the myosin head from the actin filament and the movement of the myosin head. The sequence of events in muscle contraction is (Fig. 16.6) :

(i) In the absence of bound ATP, myosin binds tightly to actin; the head at ~45° to the axis of the actin filament (**"rigor"** state).

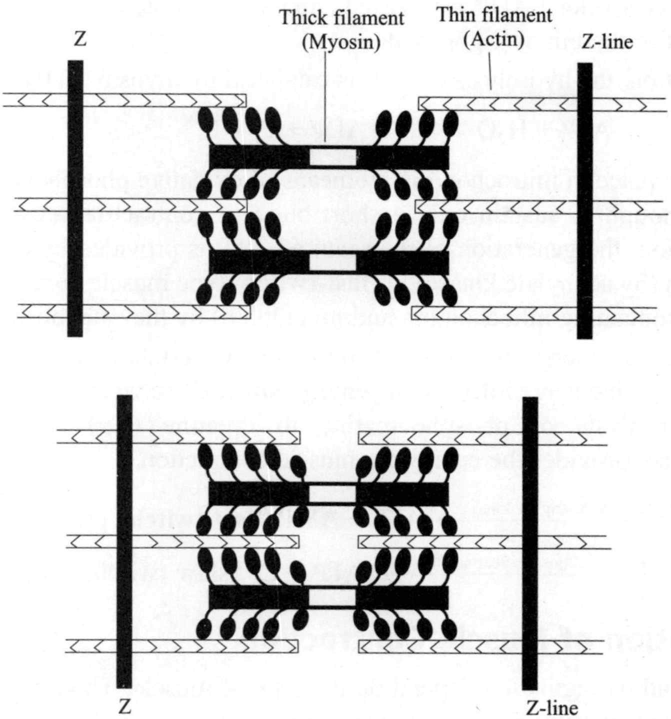

Fig. 16.5a Schematic view of the sliding interaction of actin and myosin filament in muscle contraction.

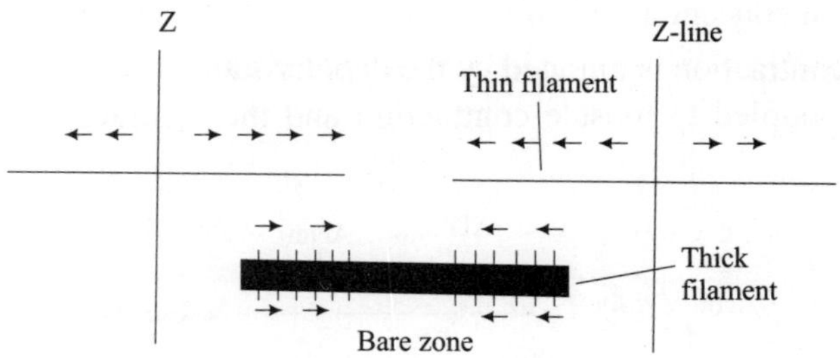

Fig. 16.5b Polarity change of thick and thin filaments halfway between Z-lines.

(ii) Binding of ATP to myosin head (Step 1 in Fig. 16.6a) effects the dissociation of the myosin head from the adjacent actin filament.

(iii) The energy for muscle contraction comes from the hydrolysis of the bound-ATP by the myosin head, S1 (Step 2). The hydrolysis of ATP leads to a conformational change in the tightly bound ADP + P_i + myosin head (myosin head is at 90° conformation with respect to the axis of the actin filament).

(iv) The myosin head binds to the actin filament (at a new position) at 90° conformation (step 3) and releases P_i first and then undergoes 45°-orientational conformation change at the end of which ('**power stroke**') ADP is released (step 4). After step 4 the myosin-actin complex is brought back to the initial 'rigor' state.

In musle contraction, the hydrolysis of ATP is catalysed by myosin-ATPase.

$$ATP + H_2O \rightleftharpoons ADP + P_i + H^+ \quad ...(16.1a)$$

ATP production takes place in mitrochondria by means of oxidative phosphorylation. The amount of ATP in muscle is enough to sustain only a short burst of contractile activity. During anaerobic (**ischaemic**) contraction, the generation of the required ATP is provided by anaerobic glycolysis of glucose and glycogen (by adenylate kinase) in '**fast-twitch**' type muscle contraction (fast-acting type ('red' muscle fibres containing mitochondria and myoglobin) by the catabolism of glucose and fats.

The major source of energy in muscle contraction in vertebrates is phosphocreatine (**PCr**) (Fig. 16.7). Phosphocreatine is regarded as an "**energy shuttle**" between the sites of ATP production and utilisation. Hydrolysis of phosphocreatine to creatine (**Cr**) and ATP (catalysed by phosphocreatine kinase) provides the energy for muscle contraction.

$$2ADP \xrightarrow{\text{glucose \& Adenylate kinase}} ATP + AMP \text{ “}\textbf{fast twitch}\text{” process} \quad ...(16.1b)$$

$$PCr + ADP + H^+ \xrightarrow{\text{Phosphocreatine kinase}} ATP + Cr \text{ “}\textbf{slow twitch}\text{” process}$$

16.1.4 Regulation of Muscle Contraction

Muscle contraction and its regulation depend on the type of muscle. Though the molecular basis of impulse generation, transmission and muscle contraction is the same, the biophysical characteristics and the response towards impulse conduction differ in different types of muscle cells.

Skeletal (Striated) Muscle Contraction: The contractile and relaxed states of 'striated' muscles are controlled by Ca^{2+} ion concentrations in cytosol.

In all muscle cells, contraction is initiated by the depolarisation of the cell membrane. That is, the membrane potential is coupled to muscle contraction and the contractile tension is related to the

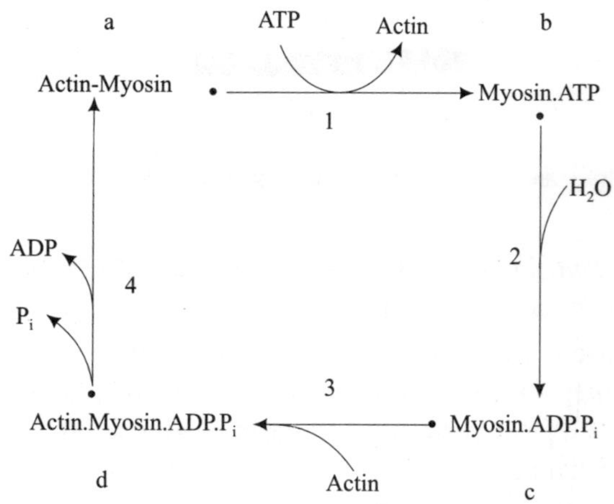

Fig. 16.6a The reaction mechanism in muscle contraction powered by actomyosin-catalysed ATP hydrolysis.

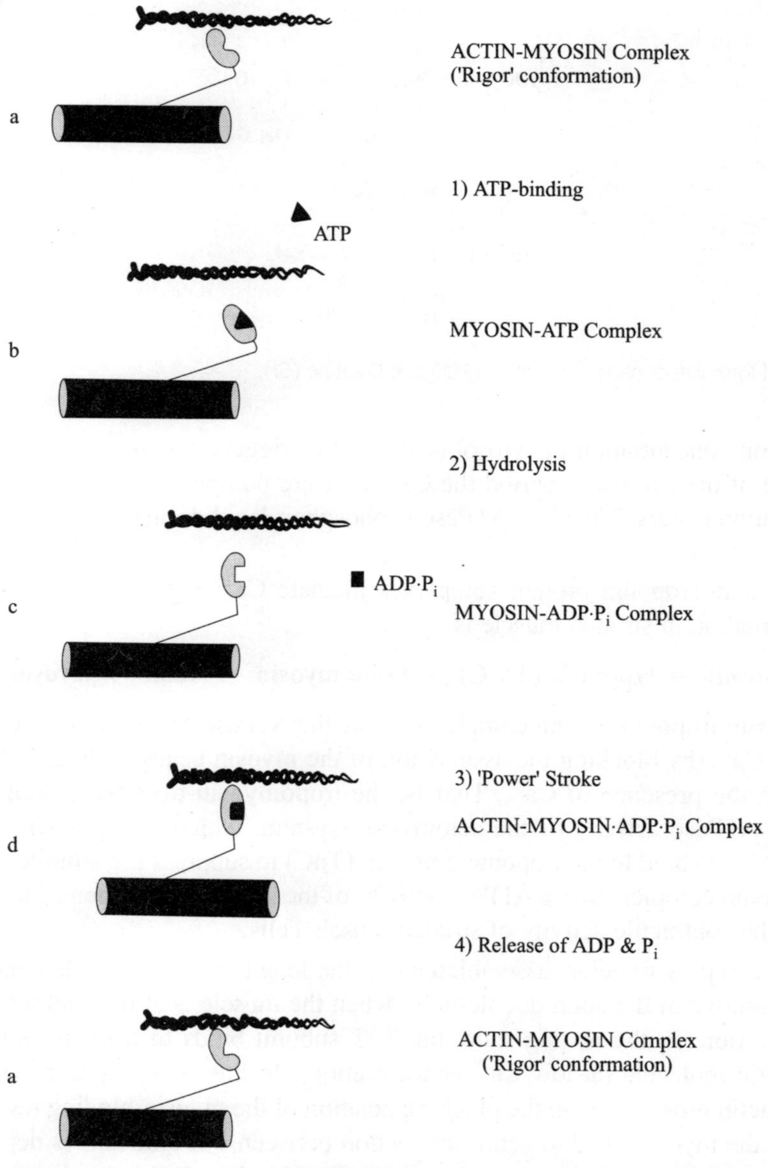

Fig. 16.6b Conformational changes of Actin-Myosin complex in muscle contraction with changes in the tilt of the myosin head.

degree of depolarisation of the cell membrane. No muscle contraction occurs if the membrane potential is below ~−50 mV.

In the resting muscle, ATP and Mg^{2+} are present in required concentrations (≈3 and 10 mM), but the Ca^{2+}-ion concentration is low in the cytoplasmic side of the sarcoplasm (≈ 10^{-5} mM). A nerve signal triggers an action potential in the nerve cells, inducing the opening of the Ca^{2+}-channels in the sarcoplasmic reticulum (**SR**) and the influx of Ca^{2+}-ions from the SR into the cytosol. The sudden

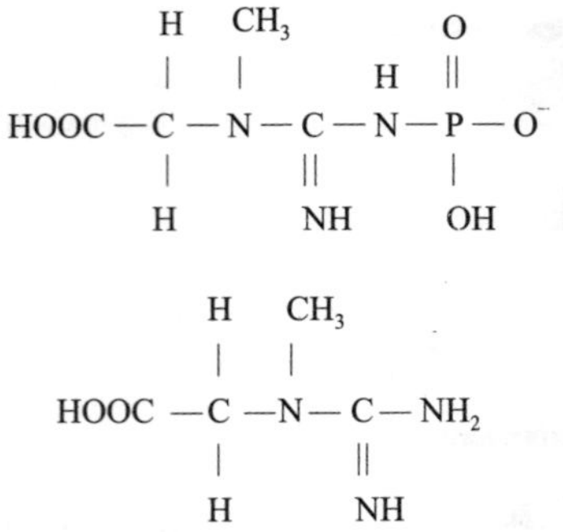

Fig. 16.7 Chemical formulae of phosphocreatine (PCr) and Creatine (Cr).

increase in Ca^{2+}-ion concentration in cytosol ($\approx 10^{-3}$ mM) triggers the contraction process of muscle fibrils. At the end of the excitation period the Ca^{2+}-ions are pumped back into the lumen of the **SR** and muscle relaxation occurs. The Ca^{2+}-ATPase is phosphorylated during Ca^{2+}-stimulated hydrolysis of ATP.

Tropomyosin and Troponin protein complexes mediate Ca^{2+}-regulation in muscle contraction. The flow of information in striated muscle is

$$\mathbf{Ca^{2+}\ (in\ cytosol) \rightarrow Troponin\ (Tn\ C) \rightarrow Tropomyosin \rightarrow Actin \rightarrow Myosin} \quad ...(16.2)$$

The tropomyosin-troponin protein complex inhibits the ATPase activity of the actomyosin system in the absence of Ca^{2+} (by blocking the association of the myosin heads with actin filaments), but it does not do so in the presence of Ca^{2+}. That is, the tropomyosin-troponin complex regulates the sensitivity of the ATPase activity of the actomyosin system, which is dependent on the Ca^{2+}-ion concentration. Ca^{2+} ions bind to the troponin complex (TnC) to suppress the inhibitory function of the tropomyosin-troponin complex on the ATPase activity of the actomyosin system. This effect is crucial in the control of the contractile activity of striated muscle cells.

The troponin complex proteins, assembled along the length of the actin filament, are positioned differently in the groove of the actin double-helix when the muscle is at rest and when the muscle is undergoing contraction. In the resting state, the TnT subunit binds to tropomyosin and TnI binds strongly to the actin molecule (at low Ca^{2+} concentration). In this state, the tropomyosin lies at the outer edge of the actin groove, i.e., in the blocking position of the myosin-binding region and prevents the attachment of the myosin head to actin. Interaction between TnI and TnC is dependent upon the free Ca^{2+} ion concentration. On saturation of Ca^{2+}, the TnC binds to TnI, releasing the binding of TnI to actin and thereby allowing the tropomyosin molecule to move, from the "blocking" position, deeper into the groove of the actin double-helix and exposing the myosin-binding region on the actin filament. In this "on" position the cross-bridges of the myosin head bind to the actin; ATP is split and the muscle contracts. The binding of Ca^{2+} to TnC causes tighter interaction between TnI and actin. Reduction of Ca^{2+} concentration results in Ca^{2+}-free TnC, which can no longer bind to TnI and this "loosening" of the complex makes the TnI subunit rebind strongly to actin. That is, tropomyosin physically blocks the actin-myosin binding region in the relaxed state, and exposing the binding-region to make contraction possible (Fig. 16.8). Tropomyosin amplifies the **TnI-blocking** effect, extending it over seven **G**-actin monomers.

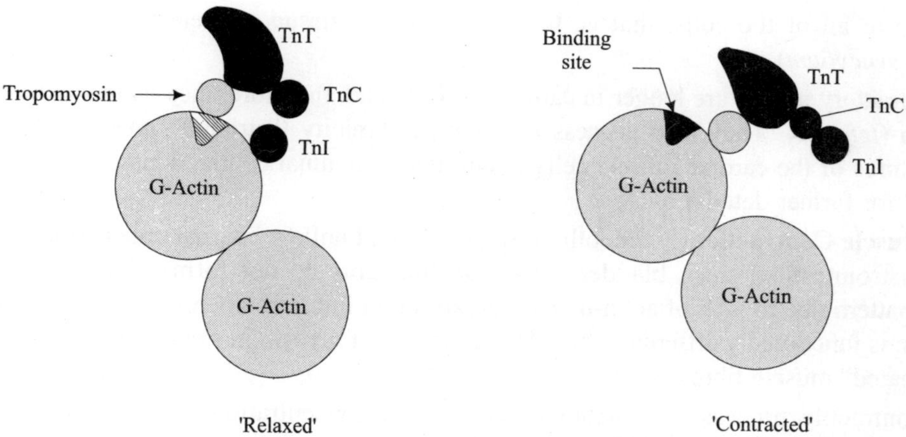

Fig. 16.8 Schematic diagram of the positions of actin, tropomyosin and troponin complex in 'relaxed' and 'contracted' states.

That is, withholding or making available of Ca^{2+} ions controls muscle contraction. A summary of the events in "striated" muscle contraction and relaxation is given:

1. The depolarisation of the surface membrane of the muscle leads to an action potential (AP).
2. AP is transmitted deep into the muscle fibre along the T-tubules.
3. Depolarisation of the T-tubule membrane leads to conformational changes in the voltage-sensitive receptors in the T-tubule membrane. This process in turn induces conformational changes in the receptor proteins, located in the SR membrane, leading to the opening of the Ca^{2+}-ion channels in the SR membrane.
4. Ca^{2+} ions that enter the myoplasm bind to the troponin C subunit of the troponin protein complex, inducing a conformational change in the troponin complex.
5. The conformational change in the troponin complex removes the steric hindrance of the tropomyosin (positional shift), thereby allowing the myosin to bind to the binding-sites on the actin filament.
6. Binding of the myosin "cross-bridges" to actin (forming of actomyosin complex) leads to cyclic rotation of myosin "heads" against the actin filament, producing torque ("power stroke").
7. Binding of ATP to the myosin "head" at the ATP-binding sites causes the detachment of actomyosin complex (separation of actin and myosin).
8. Hydrolysis of ATP is the source of energy for the conformational changes in the myosin "head" and movement of "cross-bridges" along the actin filament.

Cardiac Muscle Contraction: Cardiac muscles have many characteristics common with the 'striated" muscle fibres, but differ considerably in some important features. Cardiac muscle cells are connected in series with each other, separated by *intercalated* membrane discs, and thereby functioning as a single unit during the conduction of an AP from one fibre to another. The cardiac muscle cells are so tightly bound that when one of these cells becomes electrically excited, the whole

AP spreads to all of the cells; that is, the whole cardiac tissue acts electrically as a single cell (*functional syncytium*).

The refractory phases are longer in cardiac muscles and, therefore, they *cannot* exhibit sustained contraction (*tetanus*). They also possess inherent rhythmicity ("striated" muscles do not). These special features of the cardiac muscle cells ensure that individual contractions *cannot* summate (see Chapter 17 for further details).

Smooth Muscle Contraction: Smooth muscles surround hollow internal organs (lung walls, blood vessels, gastrointestinal tract, bladder etc.). The filaments do not form myofibrils. There is no "striated" pattern due to lack of actin-myosin organisation into sarcomeres. They also lack troponin. The myosin is functionally different– the ATPase activity of myosin in smooth muscle cells is 10% of that in "striated" muscle fibres.

The contractile process is controlled by Ca^{2+} ion concentration, mediated by Ca^{2+}-binding protein, calmodulin. In these control processes, the Ca^{2+} ion, *via* calmodulin, acts as a messenger that transmits extracellular signals across the interior of the cell.

The contractile process of smooth muscles differs markedly from that of skeletal muscles. Smooth muscles do not have all-or-nothing APs. The contraction process is slower and may persist for several seconds. They exhibit tonus (involuntary resistance to passive stretch). They exhibit plasticity. Some of the features of vertebrate muscles are given in Table 16.2.

Table 16.2 Features of vertebrate muscles.

Feature	*Skeletal*	*Cardiac*	*Smooth*
"Striated" pattern	Yes	Yes	No
Troponin	Yes	Yes	No
Tropomyosin-troponin complex	Yes	Yes	No
T-tubules	Yes	Yes	No
Sarcoplasmic reticulum (SR)	Yes	Yes	No
Calmodulin	No	No	Yes
Innervation	Somatic	Autonomic	Autonomic
Action	Fast	Fast	Very slow
Contractile process	Very short	Long	Longer
Refractory period	Short	Long	Longer
Tetanus	Possible	Not possible	Possible
Tonus	Depends on innervation	Depends on innervation	intrinsic

16.2 CYTOSKELETAL MOTILITY

The cytoskeletal fabric is formed by microtubules and microfilaments (formed by actin). Microtubules and microfilaments are some of the main components involved in cell motility in many organelles such as cilia and flagella. They are also involved in mitosis and cell cleavage.

16.2.1 Microtubule Apparatus

One of the major components in the motility of cilia, flagella etc., is the axial microtubular structure, called the **axoneme**.

Tubulin. Microtubules are present in eukaryotic cytoplasm. They are hollow, cylindrical tubes of ~25 nm diameter. The wall of microtubules is made up of 5 nm diameter filaments, called **tubulins**. Tubulin is a globular, heterodimer protein, with α- and β-subunits (2 × 50 kilodaltons). Tubulin molecules assemble to form protofilaments with an α-subunit followed by a β-subunit. There are thirteen parallel protofilaments in a full tubulin, arranged in a tubular form (Fig. 16.9).

Axoneme. The core of a cilium, that is the axoneme, contains a bundle of parallel microtubules arranged in a '9 + 2' pattern — a central pair of microtubules (A-type with 13 subunits) surrounded by nine outer doublets, consisting of **A**-type (13 sub-units) and **B**-type (11 subunits) microtubules fused together (Fig. 16.10). The bending force in the axoneme is produced by the relative sliding of microtubules and the energy source is the hydrolysis of ATP (Fig. 16.11).

The microtubules of the axoneme are linked to structures that generate force and harness it to translate to motility. These proteins are **dynein, kinesin, nexin, tectin** and **radial spokes** (Fig. 16.10). Some of the proteins involved in the structure and function of the axoneme are listed in Table 16.3. Dynein shows ATPase activity and it generates force using microtubules as a substrate. **Kinesin** is different from dynein. It generates force for moving organelles along axonal microtubules. Its function might be to serve as a motor for organelle transport and for microtubules-based movements during mitosis. **Nexin** connects tubule **A** of one unit to the tubule of a neighbouring doublet. Radial spokes connect the **A** tubules of the doublets to the central structure.

Dynein. Just as actin and myosin are involved in muscle contraction in eukaryotes, tubulin and dynein are involved in motility and locomotion in cilia and flagella. Dynein is a high-molecular mass

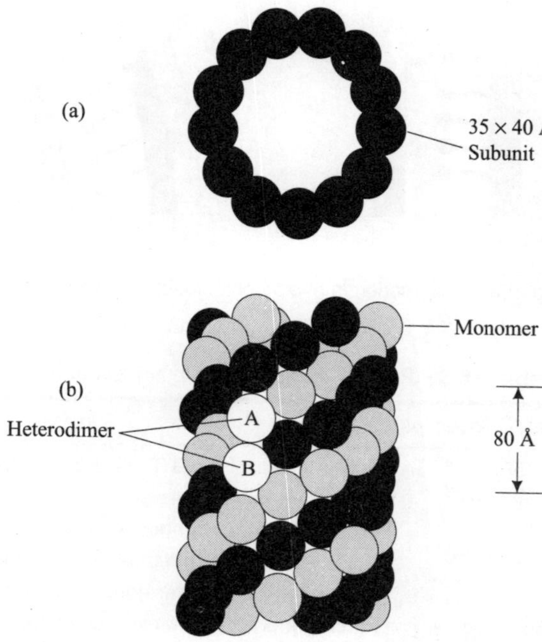

Fig. 16.9 Diagram of the molecular structure of a microtubule showing the β subunits (a) in cross-section and (b) the monomers and heterodimers in longitudinal view.
The spacing of 8 nm may correspond to the heterodimer.

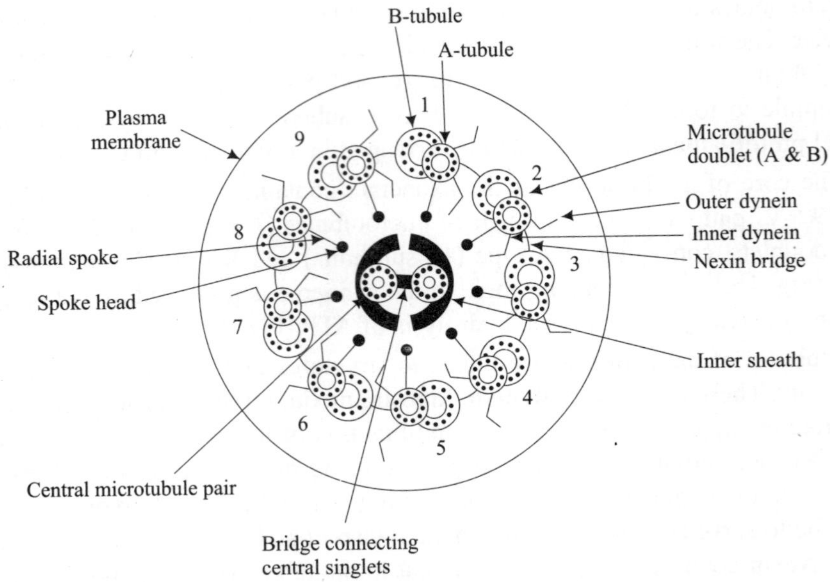

Fig. 16.10 Schematic diagram of the structure of an axoneme.

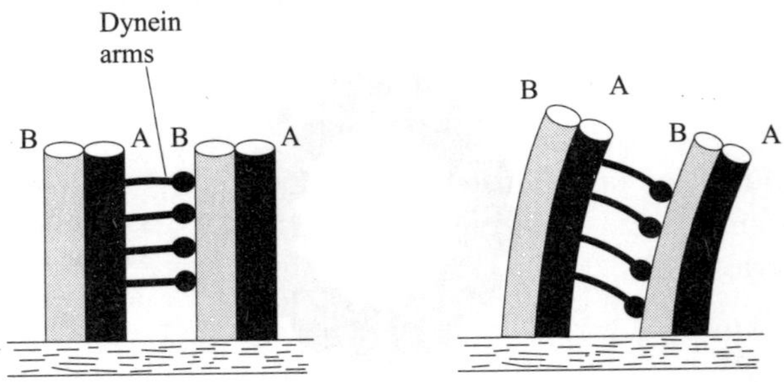

Fig. 16.11 Conversion of sliding to bending motion in microtubule doublets (axoneme).

Table 16.3 Protein components in an axoneme.

Protein	*Molecular mass (kilodaltons)*	*Function*
Tubulin	(2 × 50)	Acidic protein. Has GTPase activity. Component of microtubules.
Dynein	≈ 40	Arms form microtubule doublets. Interact with adjacent doublets.
Tectin		Protein filament arranged longitudinally fusing **A** & **B** microtubules.
Nexin	≈ 150	Holds adjacent microtubule doublets together like a rubber band.
Radial Spokes		Extend from nine outer doublets inward towards the central pair.
Kinesin	2 × 110 + 65/70 doublet	Force-generating protein complex for moving organelles along microtubules.

ATPase, with a globular base (attached to A-tubulin) and globular heads (two to three) linked by thin filaments. Each head has ATPase activity (similar to the S1 head of myosin).

The mechanism of sliding motion of microtubules is presumed to be similar to the process that occurs in the sliding motion of myosin in muscle contraction — a regular cycle of conformational changes in dynein heads, driven by ATP binding and hydrolysis, moves the dynein heads along a microtubule unidirectionally, from the plus-end towards the minus-end of the microtubule.

Microtubule Assembly and Control. Microtubules assemble in the acidic pH range (~6.6), in the presence of Mg^{2+} and GTP, and assembly is temperature-dependent. Ca^{2+} acts as an inhibitor. The role of GTP is to induce proper conformation of the tubulin for polymerisation. Microtubule-associated proteins have a regulatory function in the tubulin polymerisation process. Microtubules generate movement by active sliding and dynein is crucially involved in the process of movement. Dynein forms a fixed bridge between the **A**-tubule and the **B**-tubule of the adjacent doublet, the bridge is cleaved upon the hydrolysis of ATP, leading to change in orientation and causing the sliding motion.

16.2.2 Intermediate and Microfilaments

Intermediate filaments. Intermediate filaments (**IFs**) of ~10 nm diameter, intermediate between thin and thick filaments of muscle cells, are protein fibres in cytoplasm. There are four types of intermediate filaments — Type I similar to keratins (**k – m – e – f type**), Type II are found in epithelia and Type III in muscle and glial cells, Type III and Type IV are found in neural cells and nuclear lamina. Intermediate filaments may be physically associated with microtubules and they may play an important role in motility in mammalian cells.

Microfilament Assembly and Function. Microfilament of eukaryotic cells, present in cytoplasm, are long thread-like structures (~5 nm diameter), the major constituent being actin. Microfilaments of non-muscle cells are responsible for local motility—amoeboid movement, alterations in cell shape, cell division, and phagocytosis etc.

Microfilaments can generate movement either (i) by a sliding mechanism, which is the case in muscle contraction, or (ii) by assembly and disassembly of microfilament bundles. Microfilaments of non-muscle cells can be polymerised or depolymerised according to the needs of the cell, which is the second mechanism for cell motility. ATP accelerates actin polymerisation by activating **G**-actin to add preferentially to a particular end of a growing actin filament. Microfilament assembly and disassembly processes are controlled by actin-binding proteins (e.g. **profilin**, a 16,000 dalton protein, binds to **G**-actin and prevents polymerisation).

16.2.3 Cellular Motion

Cell locomotion depends on two principal types of movement — flagellar/ciliary and amoeboid. Flagella and cilia (of eukaryotic cells) are cylindrical organelles which undulate resulting in the movement of cells. The flagellar/ciliary movement is generated by the action of the dynein arms of microtubular doublets in the presence of ATP (ATP-powered **'walking'** of dynein arms along an adjacent subfibre).

Flagellar and Ciliary Motion. Bacterial flagella are long and thin hair-like structures (~20 × 0.1 μm), protruding from the bacterial cell. A flagellum is hollow tube-like and consists of three segments — the filament, hook and basal body. The flagellum wall comprises the filaments made up of eleven longitudinal rows of flagellin protein (~ 50,000 daltons) subunits. The basal body consists of ring-like components and the '**hook**' may function as a 'universal joint' for transmission of the movement. The synthesis of flagellar organelles is controlled by environmental and chemical factors.

The basal body forms the rotatory part of the flagellar motion and the hook and the filament are the mechanical converters of rotary motion of the basal body to linear thrust (propeller motion). Flagellar movement is realised by the rotation of the basal body, like a propeller, and the control mechanism of the direction of rotation is the basis for the directionality of the movement of the cell. Change of direction is achieved by the reversal of flagellar rotation. During the reversal of rotation there is change in the conformation of the filament, from helical to random-coil, accompanied by a change in the direction of the movement of the cell. In the absence of stimuli (chemotaxis, thermotaxis etc.) bacterial motility results in tumbling (random walk).

At the molecular level, the axoneme of the flagellum is responsible for the movement, due to interactions between the dynein arms and microtubule doublets. The sliding movement of the flagellum is transformed into bending movement by the microtubule-associated proteins—nexin links and radial spokes.

The movement of cilia (ciliary undulation) is like the undulating, wavy patterns of grass as the wind blows. Ciliary movement involves a 'power stroke' of cilia and reversal of the ciliar body. Locomotion in flagellar/cilial cells is controlled by stimuli—chemical, electrical, mechanical, photochemical etc. Further discussion on cell locomotion due to various stimuli—chemi, - thermo- and mechanotaxis etc., is beyond the scope of this book.

16.3 DYNAMICS OF AQUA- AND AERO-MOTIONS

Underwater creatures have to adapt to the static and dynamic properties of the aquatic environment for their existence and locomotion—hydrostatic pressures and viscous and other forces.

16.3.1 The Aquatic Environment

Ideal fluids obey **Berrnoulli**'s law.

$$P + \frac{1}{2}\rho v^2 + \rho g h = \text{constant} \qquad \text{Berrnoulli's law} \qquad ...(16.3a)$$

For normal (**Newtonian**) fluids, the stress is proportional to strain. The elastic modulus for solids, η (rigidity modulus) is

$$\eta = \frac{\text{Stress}}{\text{Strain}} = \frac{F / A}{dx / dz} \qquad ...(16.3b)$$

Biological fluids, like cytoplasm, are non-Newtonian (flow is dependent on pressure). Fluids do not produce constant shearing deformation, instead they produce a constant velocity gradient and the viscosity, η, is

$$\eta = \frac{\text{Shearing stress}}{\text{Velocity gradient}} = \frac{F / A}{dv / dz} \qquad ...(16.3c)$$

Aquatic creatures have to generate propulsive thrust (by swimming) to balance drag and also for massive creatures dynamic buoyancy should be maintained.

Reynolds Number, Re. There are two forces that come into play in the flow of fluids (under an external force). These are (i) inertial force (*I*) and (ii) viscous force (*V*); the ratio of inertial to viscous force is the Reynolds number, **Re**.

$$\text{Re} = \frac{\text{Inertial force}}{\text{Viscous force}} = \frac{I}{V}$$

$$\text{Re} = \frac{\rho v l}{\eta} \quad ...(16.4)$$

where ρ = density, v = velocity, l = length and η = coefficient of viscosity. In laminar flow, the viscous force is proportional to velocity, as given by **Hagen-Poiseuille** law

$$(P_1 - P_2)\, r^2\, h = 8\,\pi\,\eta\,\text{lv} \qquad \text{Hagen-Poiseuille Eqn.} \quad ...(16.5a)$$

In turbulent flow, the viscous force is $\propto V^2$ ...(16.5b)

The velocity above which the laminar flow becomes turbulent is the critical velocity given by the **Reynolds** equation

$$v_c = \text{Re}\,\frac{\eta}{\rho r} \qquad \text{Reynolds Equation} \quad ...(16.5c)$$

$$\text{Re} \ll 1 \qquad \text{Laminar flow}$$

$$\text{Re} \gg 1 \qquad \text{Turbulent flow} \quad ...(16.6)$$

(i) **Low Re:** At low Re velocity of flow is laminar. Under such conditions transport of solids in a viscous medium is given by **Stokes** equation

$$F_d = 6\,\pi\,v\,\eta\,r$$

where F_d is drag force.

The flow of liquid through a cylindrical tube of radius r is given by **Poiseuille's** Equation

$$\text{Flux} = J = \text{volume of flow/second} = \text{A} \cdot \text{v}$$

$$J = \text{A} \cdot \text{v} = \frac{\pi r^4}{8\eta}\,\frac{\Delta P}{l} \quad ...(16.7)$$

($\Delta P/l$ is the pressure gradient).

Microscopic particles, such as bacteria, algae and spermatoza, swim at very load Reynolds numbers ($\text{Re} < 10^{-3}$) where the viscous forces are dominant and, therefore, the ability of these organisms to glide is lost. Wriggling and whipping movements are employed to produce forward motion (flagellar movements). The drag, *D,* is proportional to velocity, *v,* and the velocity is dependent upon the size, shape and density of the particle; a body should have low drag coefficients.

(ii) **High-Re:** High Re (Re > 100) corresponds to low viscosity; flow is not uniformly laminar, and the drag for F_d is dependent upon velocity and the size of the solid

$$F_d \propto v^2 \rho l \qquad ...(16.8)$$

Swimming of fish, flight of birds and aeroplanes are examples.

In the regions of high Re, the effects due to viscous and pressure drags have to be taken into account. Rigid bodies moving in a fluid of high Reynolds numbers (Re $\geq 10^6$) will experience turbulent flow. The influence of pressure drag can be minimised by streamlining the shapes—elongated ellipsoidal body form with widest point toward the back (e.g. tuna fish) and creation of maximal laminar boundary layer combined with 'little' vortex turbulence (by tails). By optimising the interplay of drag and propulsion effects, swimming by aquatic creature with minimum exertion is possible (e.g. eels, snakes, flagella etc.). Concepts on the movement of particles in Newtonian and non-Newtonian liquids, under laminar and turbulent flow conditions have relevance in the study or the movement of erythrocytes suspended in blood, flow of body fluids in organisms. Swimming motion also helps in preventing the formation of a boundary layer.

16.3.2 Aeroflight

Gliding and flying are the processes of locomotion by birds and other flying objects. For aeroflight, combination of lift and drag components have to be optimised. A wing (of birds or aeroplanes) at an angle α (angle of attack) to the direction of the flow of airstream experiences drag and lift forces and for an object to fly in air the lift component should be naturally greater than the drag component. The design of the wing should be such that the coefficient of lift is high and the coefficient of drag is minimal. Maximum lift occurs at the angle of attack $\approx 10°$. Above this angle, the drag on the wind increases, eddy currents ensure and stalling occurs. In addition, induced drag also comes into play at low speeds. Induced drag is least for long and narrow wings. So birds that soar high have long and narrow wings to reduce the effect of induced drag. Birds that fly by flapping should have short and wide wings. Short wings lead to greater maneuverability and feathers at the wing tips (many smaller wing tips) further reduce the induced drag.

16.3.3 Biostatistics of Locomotion

Locomotion features of creatures depend not only on their physical structures (size, shape and mass) but also on their adaptation to the static and dynamical properties of the environment. Shape with respect to size (slenderness factor, λ) determines the structural stability of the terrestrial species. Whereas the mass, M, increases at the cube root of the linear dimension, d, the surface area increases only as the square. The slenderness factor root, λ, is related to the height, h, of the object.

$$\lambda = h/d \qquad ...(16.9)$$

As the mass of a vertical column increases with height, faster than its resistance, ë, the structure collapses of its own weight beyond a limiting height, if its girth is held constant. This feature is born out from the biological species (ë for bamboo = 130; sequoia tree = 15). That is, slenderness decreases rapidly with increasing size in terrestrial creatures (e.g. giant dinosaurs, mastodons, elephants).

Changes in body mass, size and shape have direct effect on the energetics of locomotion, and the metabolic rates of the organisms. There exists an inverse relationship between the body size and metabolic rate. The metabolic rate of unit mass of a tissue decreases with increasing body mass. That is, locomotion is metabolically less expensive for large size creatures. Therefore, large creatures expend less metabolic energy to move a given distance.

In addition, under oblique force, a structure is subject to compression, shear and bending stress. Biological structures like bones are resistant to compression and shear, but break under bending stress. In skeletal structures, stretching muscles, connective tissue and tendons compensate the bending stresses.

Suggested Readings

Books

Bagshaw, C.R. "*Muscle Contraction*". Chapman-Hall/New York. (1993).

Capuccinelli, P. "*Motility of Living Cells*. Chapman-Hall/London. (1980).

Dustin, P. "*Microtubules*." 2nd Edn. Springerverlag/Berlin (1984).

Harrison, R. & Lunt, G.G. *Biological Membranes: Their Structure and Function.* 2nd Edn. Blackie & Sons/ Glasgow. (1980).

Huxley, A.F. "*Reflections on Muscle*" Princeton University Press/Englewood, New Jersey (1980).

Hyams, J.S. & Lloyd, C. W. "*Microtubules*". Wiley-Liss/New York (1994).

Junge, D. "*Nerve and Muscle Excitation*". 3rd Edn. Sinauer-Sunderland/MA. (1992).

Lackie, J.M. "*Cell Movement" and Behavior*" Allan & Unwin/London (1986).

Robertis, E.D.P & Robertis, E.M.F. "*Cell and Molecular Biology*", 7th Edn. Saunders College/Philadelphia. (1980).

Sheterline, P. & Sparrow, J.C. "*Protein Profile: Actin*" Vol., I. Academic Press/New York (1994).

Squire, J.M. & Viber, P.J. (Eds). "*Fibrous Protein Structure*", Academic Press/New York (1989).

Traub, P. "*Intermediate Filaments*", Springerverlag/Berlin (1985).

Articles

Amos, L.A. & Backer, T.S. "*The Three-dimensional Structure of Tubulin Protofilaments*". Nature 299; 607. (1979).

Avila, J, "*Microtubule Functions*". Life Sci. 50; 337 (1992).

Bretscher, M.S, "*How Animal Cells Move*". Sci. Amer. 257(6): 72 (1987)

Cohen, C. & Parm, D.A.D. "*α-helical Coiled-coils—A Wide Spread Motif in Proteins*". TIBS. 11; 245 (1986).

Cooke, R. "*The Mechanism of Muscle Contraction*". CRC Rev. Biochem. 21; 53. (1986).

Cooke, R. "*Force Generation in Muscles*". Curr. Opin. Cell Biol. 2; 62. (1990).

Eisenberg, E. & Itill, T.L, "*Muscle Contraction and Free Energy Transduction in Biological Systems*". Science, 227; 999 (1985),

Geiger, B. "*lntermediate Filaments*", Nature, 329; 392 (1987).

Heizmann, C.W. & Hunziker, W. "*Intracellular Calcium-binding Proteins*". Trends Biochem. Sci. 16; 98. (1991).

Herzberg, O. et al. "*A Model for the Ca^{2+}-induced Conformational Transition of Troponin C: A Trigger for Muscle Contraction*". J. Biol. Chem. 261; 2638 (1986).

Holmes, K.C. et al. "*Atomic Model of Muscle Contraction*". Nature. 347; 44. (1990).

Huxley, H.E. "*The Mechanism of Muscle Contraction*". Science, 164; B56 (1969). Kikkawa, M. et al. "*Three-dimensional Structure of the Kinesin Head-microtubule Complex*". Nature. 376; 274 (1995).

Kretsinger, R.H. "*Structure and Evolution of Calcium-modulated Proteins*", Crit. Rev. Biochem. 8; 119 (1980).

Mannherz, H.G. & Goody, R.S. "*Proteins of Contractile Systems*". Annu. Rev. Biochem. 45; 427 (1976).

Matsudaira, P. "*Modular Organisation of Actin Cross-linking Proteins*", TI BS, 16; 87 (1991).

Murry, J.M. & Weber, A. "*The Cooperative Action of Muscle Proteins*", Sci. Amer. 230 (2); 59 (1974).

Namba, K. Yamashita, I & von der Viszt, F. "*Structure of the Core and Control Channel of Bacterial Flagella*". Nature 342; 649 (1989).

Payne, R. & Rudnick, S.E. "*Regulation of Vertebrate Striated Muscle Contraction*". TIBS. 14; 357 (1989).

Pollard, T. D. & Cooper, J,A, "*Actin and Actin-binding Proteins*", Annu, Rev, Biochem. 55; 987 (1986),

Porter, M. "*Dynein Structure and Function*". Ann. Rev. Cell Biol. 5; 19 (1989).

Raymont, I. "*Structure of the Actin-myosin Complex and Implication for Mucile Contraction*". Science. 261; 58 (1993).

Rozycki, M.D. et al. "*Structural Aspects of Actin-binding Proteins*". Curr. Opin. Cell Biol. 6; 87 (1994).

Spencer, M. "*Sliding Microtubules*" Nature, 273; 595 (1978).

Spudich, J.A. "*How Molecular Motors Work?*". Nature. 372; 515 (1994).

Stewart M. "*Intermediate Filament Structure and Assembly*" Curr. Opin. Cell Biol. 5; 3 (1993).

17 Bioelectronics

Synopsis: All forms of sensory stimuli (mechano-, chemo-, photo- etc.) are transduced into electrical signals and transmitted to CNS, and *vice versa* from CNS to PNS. There are varieties of sensors and transducers, based on mechanical, electrical, thermal, optical, chemical and forms of energy for measuring physiological variables/parameters and biopotentials, in order to monitor organ and body functions and for diagnostics detection of abnormalities in their functions. These form a part of biomedical engineering.

Some of these sensors/transducers are displacement sensors, strain gauges, piezoelectric, thermal (e.g. thermistors), optical (e.g. photodiodes, scintillation counters) transducers. Electrochemical transducers include ion-selective electrodes (pH meter) and biosensors.

Biopotentials are ionic voltages produced by body organs as a result of electrochemical activities of various parts of the body (skin, heart, brain etc.), which are monitored by various recordograms (e.g. ECG, EEG).

17.1 SENSORS AND TRANSDUCERS

Data acquisition about a system is the basic step towards understanding its structure and function. In all biomedical measuring devices the detecting processes involve the use of sensors and transducers and output signals are recorded/displayed and analysed.

There is no interconversion of energy from one form to the other in sensor devices (mechanical to mechanical, electrical to electrical etc.), whereas transducers are devices that convert one form of energy (signal) to another form. In biomedical instruments, the transducers are employed to detect input stimuli (signal transduction), whose amplitudes and changes are proportional to the input signals. Signal processing includes amplification, modification and output display. Some of the functions of the transducers and their applications are given in Tables 17.1 and 17.2.

17.1.1 Mechano-transducers

Some of the mechano-transducers are based on resistance, capacitance, induction and piezoelectric effects.

Table 17.1 Some transducers and their functions.

Input energy	*Transduction*	*Effects*
Mechanical	Electrical	Magnetic/electrical induction.
Pressure/Vibration	Electrical	Piezoelectric
Thermal	Electrical	Thermoelectric & Seebeck effects.
Light	Electrical	Photovoltaic; photoelectric.
Chemical	Electrical	Voltaic/pH; ion-sensors; biosensors.
Sound	Electrical	Microphone.
Electrical	Mechanical (pressure)	Piezoelectric.
Electrical	Thermal	Peltier effect.
Electrical	Light	Light-emitting diode (LED).

Table 17.2 Biomedical applications of some sensors and transducers.

Sensor/ transducer	*Measurement*
Strain-gauges	Blood, bladder and other pressures.
Displacement/Impedance	Volume change (Plythesmography).
Flow meters	Fluid flow and velocity.
Thermocouples/Thermistors	Temperature/Thermography.
Ion-specific Electrodes	pH; pO_2; pCO_2 etc.
Surface and Needle Electrodes	Biopotentials (E.g. ECG; EEG).
Microphone	Phonocardiography.
Spirometer	Respiratory function.

Strain-gauges: Resistance to flow is universal. For a homogeneous conductor, the resistance to flow, R, is

$$R = \rho . L/A \qquad ...(17.1a)$$

The basic principle of resistance strain-gauges is based on the premise that the changes in the electrical resistance, ΔR, of a conductor is proportional to its linear dimensional change, ΔL, given by the gauge factor, G.

$$G = (\Delta R/R) \cdot (L/\Delta L) \qquad ...(17.1b)$$

$$\Delta R/R = \underbrace{(\Delta L/L) - (\Delta A/A)}_{\text{Dimensional}} + \underbrace{(\Delta\rho/\rho)}_{\text{Piezoresistive}} \qquad ...(17.1c)$$

(L = linear dimension; A = cross-sectional area; ρ = resistivity).

Resistance strain-gauges are metallic filaments or foils, either piezoresistive or semiconductor devices. When strain is applied to piezoelectric materials (e.g. quartz) they give rise to a voltage, the effect as the piezoelectric effect. The reverse effect—the application of electric field to piezomaterials—causes mechanical deformations (dimensional changes). The piezoresistance is change in electrical resistance as a function applied stress. Piezoelectric transducers are used mainly in sonic devices (in microphones and sonography).

Resistance strain-gauges are extensively used in cardiovascular and pulmonary measurements. Semiconductor strain-gauges have many advantages, but they are temperature-sensitive and nonlinear in their response.

Capacitive Transducers: Capacitive transducer is a pressure sensor, which measures the variation of capacitance with stimuli, by measuring the deflection of a dielectric diaphragm positioned between two capacitive plates.

$$dC/C = -dx/x \quad ...(17.2)$$

(C = capacitance; x = separation of capacitor plates).

Change in capacitance is a measure of pressure change. This type of transducer is most commonly used as a microphone pick-up of respiratory and cardiac sounds.

Induction Transducers: If an electrical conductor is moved in a magnetic field, altering its magnetic flux, a voltage is induced. The induction process is a transduction of mechanical to electrical, and *vice versa.* Electrical motors, generators and loudspeakers are based on Faraday's electromagnetic induction principle. Important biomedical appliances are microphones, fluid flow meters and pulse transducers.

17.1.2 Thermal Transducers

Thermal transducers absorb radiation and transfer into heat, causing a rise in temperature. Resistance thermometers are used to detect heat radiation, based on the property that a thermal conductor (Pt-wire) varies approximately linearly with temperature. These transducers are of low sensitivity. More sensitive devices are thermistors, thermocouples and thermopiles and infrared sensors.

Thermistors are temperature-sensitive semiconductors. These devices, with high precision, small size and low cost, are extensively used in biomedical thermometry. Thermocouples and thermopiles are devices transducing thermal to electrical energy, based on the Seebeck effect that two dissimilar metals kept at different temperatures generate a voltage proportional to the temperature difference. They have several advantages—fast response, ease of fabrication and stability. They can be miniaturised to be inserted into catheters and hypodermic needles.

Infrared (IR) sensors are employed in thermography to map cellular and metabolic activities, based on the Stefan-Boltzmann relationship between the surface temperature and radiation power ($S \alpha T^4$). The input for an IR sensor is radiant energy from the human body and the output is an electrical signal. One of the applications of radiation thermography is the measurement of core body temperature of human body by measuring the magnitude of IR radiation emitted from the tympanic membrane (eardrum) and surrounding canal (*ear thermometry*). Ear thermometry offers several clinical advantages over glass-bulb and devices. It is a non-contact thermometer and the readings provide actual temperature of the subject, and not of the sensor.

17.1.3 Optical Transducers

Devices that convert light energy into electrical signal (photoconductive), thermal energy to light energy (thermoluminescence) and electrical energy to light (LEDs) and fibre optical are classified under optical transducers.

Light impinging on a photosensitive material (photocell) releases electrons and the amplitude of current is proportional to the intensity of the incident light. Photoemissive transducers are photodiodes that include vacuum photodiodes; photomultiplier tubes (PMTs), in which light impinging on a photoemissive surface releases photoelectrons proportional to the intensity of the incident light.

Fibre Optics: Fibre optics deals with the transmission of visible light through thin transparent materials (glass, plastics) and transduction of optical signal to electrical signal. Optical sensors (*optrodes*) are analogous to electrical sensors but they are free from electromagnetic interference and are explosion-proof. Their capacity for handling higher density information content than possible in electrical counterparts finds applications in diverse fields, especially in information technology and telecommunications (*optoelectronics*).

Fibre optic transducers have brought tremendous changes in biomedical fields. Instrumental miniaturisation has found applications in miniature-microscopy (endoscopies), and other diagnostics and surgical devices. Fibre optics temperature sensor is the one that is ideally suited for temperature measurement in the environment of electromagnetic fields. Therefore, it finds applications in cancer therapy and patient rewarming.

17.1.4 Ionisation Transducers

Ionisation radiation is used in radiation measurements, particle counting in radiology and nuclear medicine. Film badges, ionisation (proportional and Geiger-Müller) and scintillation counters are used to carry out radiation detection and mapping (see Chapter 18: Radiation Biophysics).

17.1.5 Electrochemical Sensors

Ion-selective electrodes such as pH meters and gas-sensing (Clark and Sevringhouse) electrodes are used to measure ion and gas concentration in solutions. They are used in pH/ blood gas analysers, to monitor respiratory functions.

Biosensors: A biosensor devise comprises a biocatalyst (enzyme, cell or tissue) and transducers that convert the chemical signal to electrical one. Biosensors (enzyme electrodes) incorporate the specificity of enzyme-substrate (antibody-antigen) reactions and electrochemical transduction to provide highly selective biochemical transduction.

Biocatalyst + Transducer → Biosensor

17.2 BIOPOTENTIALS

Biopotentials are ionic voltages produced as a result of electrochemical activities of body organs. These ionic voltages are converted into electrical signals through transducers, which are recorded by various detecting devices. Biopotential currents (physiological currents) are transported by ions, while mobile electrons transport current in electric/electronic devices. Therefore, stimulation of nerves *via* surface electrodes requires an electrode-electrolyte interface. Electrolyte gels that contain anions and cations in suspension are used to accomplish this conversion.

Important biopotential measurements are related to cardiovascular and brain and other neuronal activities. Bioelectric potentials associated with other body organs (skin, retinal etc.) can also be obtained by various recordograms.

17.2.1 Cardiovascular Processes

Cardiovascular measurements include recording and monitoring of heartbeat rhythms, blood pressure and blood flow and velocity of flow.

Functional features of the heart are related to its structural organisation, and to the nature of the muscle fibres, which differ in many features from those of the skeletal (striated) muscle fibres.

Heart as a Pump: The mammalian heart is a pulsatile, viscoelastic and synchronous, but functionally isolated two-stage, pump with four chambers, composed of two upper chambers (atria) and two lower chambers (ventricles) (Fig. 17.1). The heart serves as a pump because of its ability to contract under electrical stimulus.

The atria serve as the reservoirs of blood, whereas the ventricles provide the main force that propels the blood through the lungs and through the peripheral circulatory system. Systemic circulation has high pressure gradient between the arteries and veins and pulmonary circulation has low pressure gradient. Hence left part of the heart can be considered as pressure pump and the right part as the volume pump. Rhythmic contractions of the heart muscles result in ejection of blood into the circulatory system in a pulsatile and synchronous manner.

Mechano-electrical Properties of the Heart: The cardiac muscle (myocardium) exhibits many properties intermediate between striated (skeletal) and smooth muscles, structurally and functionally. There are three classes of cardiac muscle fibres: (i) small, weakly contractile— slow-conductive and

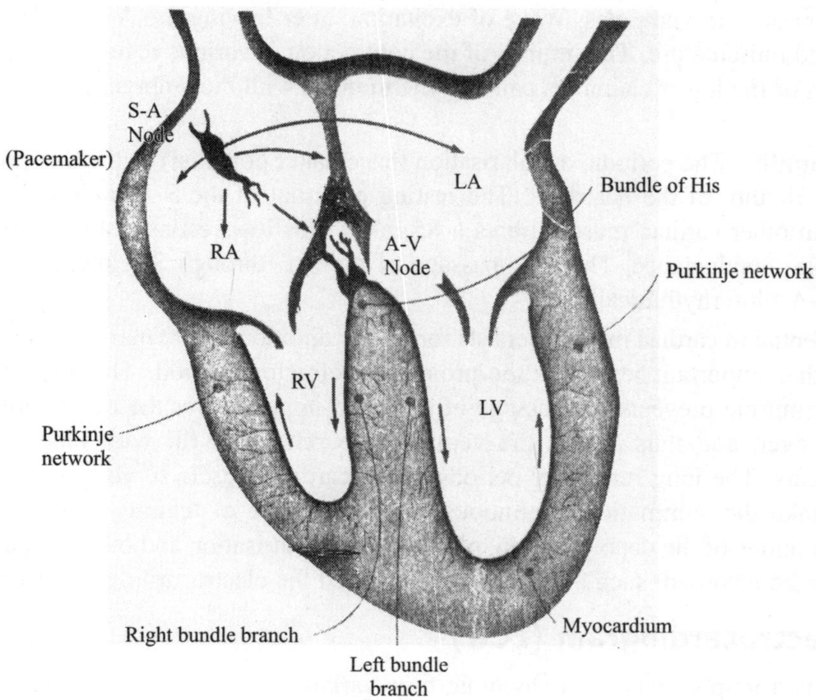

Fig. 17.1 Schematic representation of heart muscle apparatus RA = Right atrium; LA = Left atrium; RV = Right ventricle; LV = Left ventricle; S-A = Sinoatrial; A-V = A trioventricular

autorhythmic cells, found in sinoatrial (S-A) and atrioventricular (A-V) nodes; (ii) weakly contractile but fast-conductive cells, found in the largest myocardium cells, and (iii) strongly contractile cells, found in the intermediate-sized myocardial cells.

The important characteristics of the heart muscles include electrical excitability, rhythmicity, contractility and distensibility. To produce efficient pumping of the blood, the myocardial fibres must contract more or less simultaneously. Intercalated discs interconnect functionally the units of the heart muscle such that they function (electrically) as a cohesive, single mass of the tissue that quickly conducts electrical impulses— automatic and functional syncytium (behaviour as a single cell).

Cardiac muscle is also rhythmic in action. The fibrous skeleton of the heart separates various parts of the heart. In addition to the myocardium, the heart also consists of sinoatrial (S-A) node, a small mass of nodal tissue on the wall of the right atrium, atrioventricular (A-V) node, located between the right atrium and the ventricle, the bundle of His, a small bundle of nodal-tissue fibres that extend from the A-V node to the wall of the two ventricles, and ventricular Purkinje fibres, which extend in all parts of the musculature. S-A node is the internal pacemaker, a device that supplies an electrical impulse to the heart muscle to stimulate contraction of the mammalian heart. The A-V node and the Purkinje tissue constitute a bridge for electrical conduction of the impulse from the atria to the ventricles. The S-A node initiates the impulse responsible for contraction of the heart. Following each spontaneous depolarisation at the S-A node, the activation process is transmitted radially. Next, the A-V node is excited, which acts like a delay line (biological pulse delay) to retard the advancement of action potential to ventricle. The A-V node is the only means by which electrical impulse can be transmitted from atria to ventricles. Wave of excitation, after leaving the A-V node, passes through the ventricle and musculature. The timings of the contraction at various regions of the heart are such that contraction of the lower chambers can be synchronised with the emptying of blood in the upper chambers.

Spacing of Stimuli: The periodic depolarisation (pacemaker potential) of the S-A node membranes determines the rhythm of the heartbeat. The resting potential of the S-A node is ~ –55 mV, quite below to that in other cardiac muscle fibres (–85 mV). This low resting potential of the S-A node enables high Na^+ conductance. The easy passage of the Na^+ through S-A membranes causes self-excitation of S-A fibre rhythmically.

Action potential in cardiac muscle persists for longer duration (~200 ms) than in skeletal muscles (~1 ms). Another important feature is the prolonged refractory period. This prolonged refractory period in heart muscle prevents the passage of a second impulse over the heart until the preceding contraction is over, and thus allows the ventricles to relax and fill with blood before another contraction occurs. The long refractory periods of the cardiac muscle serve to preserve the cardiac rhythm and make the summation continuous and occurrence of tetanus (sustained contraction) impossible. Durations of the depolarisation, plateau and repolarisation and other biopotential features of the heart can be recorded; such a recordogram is called the electrocardiogram (ECG).

17.2.2 Electrocardiogram (ECG)

ECG or EKG is a graphic record of rhythmic time-variant voltages produced by the heart muscle (myocardium) during the cardiac cycle–– the voltage changes in the heart (in mV) as function of time (in seconds) from the time an electric impulse, initiated in the S-A node (pacemaker), travels through atria, A-V node, bundle of His and Purkinje fibres and finally through the ventricular muscles.

An ECG record of single heartbeat consists of three main components— P-wave, QRS cycle and T-wave (Fig. 17.2).

The essential features of the electrical activity of a normal ECG can be described under:

1. The P-wave (~ 0.25 mV) represents the electrical impulse (depolarisation of the atria) from the S-A node, spreading through the atria of the right heart (duration ~0.1 s).
2. The P-R interval (0.18 s) is due to the conduction delay in the A-V node. It represents the time taken for the impulse to reach the ventricles and initial ventricular depolarisation. It is the time between atrial systole and the start of the ventricular systole (~ 0.12 – 0.2 s).
3. The QRS complex represents the depolarisation of the ventricle (R-wave ~ 1.6 mV) and repolarisation of the atria (0.1 s).
4. The S-T segment represents the time period (~0.3 s) between depolarisation and repolarisation (recovery) of the ventricle muscle.
5. The T-wave (0.1 – 0.5 mV) represents the recovery phase (contraction) and repolarisation of ventricles (~0.4 s).

Through the ECG analysis, conditions associated with the heart and various heart diseases can be diagnosed. Both time and amplitude are to be taken into account in the interpretation of the ECG profiles. If the P-R interval is >0.2 s, it could be due to blockage of A-V node (obstruction of the spread of excitation from atria to ventricles). The larger duration of QRS cycle, >0.1 s, is indicative of bundle block. In coronary insufficiency (deficient blood circulation in the coronary arteries), the S-T wave is shortened and the T-wave is sometimes inverted. T-wave is one of the most important indicators of cardiovascular abnormalities, because changes in T-wave can result from even slightest change in the period of depolarisation of only one position of the heart.

While ECG provides useful information in the case of arrhythmia and myocardial infarction, it cannot shed light on disorders involving heart valves. These can be diagnosed from angiography and echocardiography.

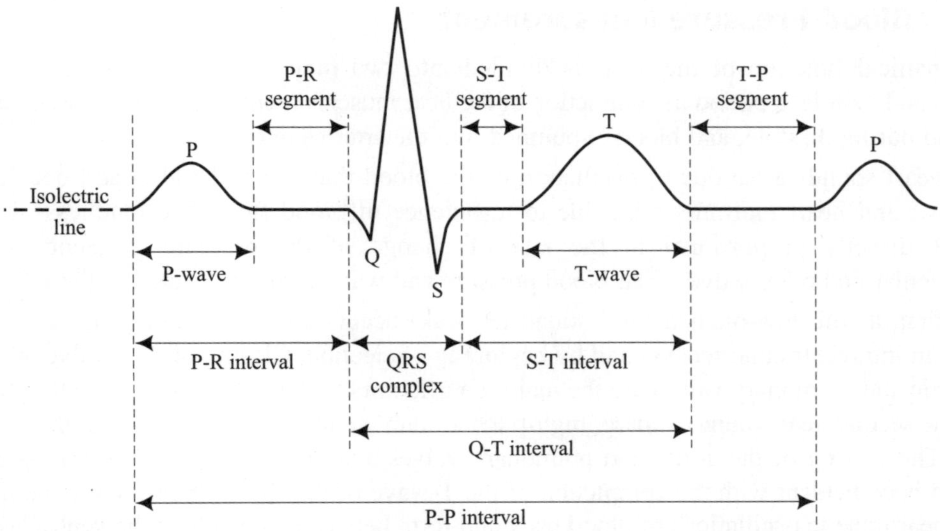

Fig. 17.2 Electrocardiogram (ECG) of heartbeat cycles.

17.2.3 Measurement of Heartbeats and Sounds

Heartbeats and sounds are due to rhythmic contractility of heart muscles and blood flow. Since the heart is a pulsative pump, blood enters the arteries intermittently, causing pressure pulses in the arterial system.

Heartbeat measurements can reveal cardiac disorders (abnormal heart murmurs, congenital defects and hypertension etc. Each heartbeat in humans originates in S-A node with periodicity ~ 70 beats/min. accelerate heartbeat, called trachycardia (atrial flutter), occurs when a portion of the heart becomes "irritated" and discharges independently at a rate that exceeds that of the S-A node. This situation leads to ventricular fibrillation. Abnormal cardiac sounds (gallop rhythm) are associated with serious heart ailments. Murmurs (due to narrowing of the heart vessels, called stenosis) are due to (i) constricted opening, which increases the velocity of the blood flow, and (ii) the roughened wall surface, which reduces the critical velocity at which turbulence is produced. Murmurs due to regurgitation occur if the valves do not close properly and on time, due to which turbulence collision occurs between the blood flowing back into and entering the chamber in the normal direction.

When the bundle of His is partially interrupted, incomplete heart block (A-V block) occurs. Though all atrial impulses reach the ventricles, the P-R interval is abnormally prolonged. This abnormality is called the first-degree heart block. In the second-degree heart block, not all atrial impulses are conducted to ventricles. In the third-degree heart block, the bundle of His is completely interrupted and the cells in the A-V node are dead. Therefore, activity cannot be transmitted from atria to ventricles, and atria and ventricles independently (no correlation of heartbeat), leading to complete heart attack.

Rhythmic disturbances can also arise at the ischemic regions due to deficient blood supply. Occlusion of blood supply leads to lowered activity of the metabolically-dependent Na^+:K^+ pump. These electrophysiological changes bring about pronounced changes in the ischemic regions, indicated in the altered profiles of the QRS-complex, S-T segment and the T-wave.

17.2.4 Blood Pressure Measurement

The mechanical function of the heart is divided into two major phases— diastole, a period of relaxation and systole, a period of contraction of the hear muscles (ventricles). The heart chambers fill with blood during diastole, and blood is pumped into the arteries during the systole.

The heart sounds arise due to oscillation of the blood due to acceleration and deceleration of blood flow; and heart murmurs arise due to turbulence of blood flow. The loudness of the heart sounds is directly proportional to the rate of change of the pressure gradients across the atrioventricular and other valves. The blood pressure and wave form are related to the ECG profile.

The first, a soft, low-pitched 'lub' sound (R-peak) occurs at the beginning of the systole with rapid rise in intraventricular tension and by beginning of ejection. Closing of A-V valve and opening of the aortic and pulmonary valves are the major contributors to the production of the first 'lub' heart sound. The second heart sound, a sharp, high-pitched 'dub' sound (T-wave) begins at the onset of the diastole. The closing of the aortic and pulmonary valves and opening of the A-V valves cause the sound and is coincident with the completion of the T-wave of the ECG. Occasionally the third heart sound is heard due to oscillations of blood back and forth between the walls of the ventricles initiated by inrushing blood from atria. The fourth heart sound (atrial heart sound) when atria contract and due to inrush of blood into the ventricles.

Auscultatory Method of Measurement: Blood pressure (heart sounds) measurements can be recorded by auscultatory and phonocardiography methods. Monitoring of Korotkoff sounds for blood pressure measurement is called auscultatory method, and relies on monitoring lub/dub sounds (Korotkoff sounds) using a stethoscope and a sphygmomanometer. The apparatus consists of an inflatable pressure cuff and a manometer to measure the pressure in the cuff. The cuff is wrapped around the patient's upper arm and the stethoscope is placed over an artery downstream (distal) to the cuff. To obtain the blood pressure measurements, first the cuff is inflated above the systolic pressure. At this pressure no sounds are heard through the stethoscope as the blood circulation is occluded in the lower part of the arm. Then, the pressure in the cuff is gradually reduced. As the cuff pressure falls just below the systolic pressure, small amount of blood once again spurts past downstream of the cuff. The pressure of the cuff (indicated in manometer) when the 1st Korotkoff sound is measured through the stethoscope is the systolic blood pressure (~120 mg Hg). As the cuff pressure is decreased gradually, the Korotkoff sounds vanish when the pressure in the cuff equals the minimum pressure in the artery, and the pressure at which the sound is no longer been heard is the diastolic pressure (~80 mg Hg). Auscultatory method can only provide systolic and diastotic pressure valves. Other heart sounds and continuous recording of the pressure variations can be carried out by phonocardiography.

Phonocardiography Method: Phonocardiography is a bioelectric graphical recording of heart sounds. Heart sounds occur in the audio range (20–200 Hz). The piezoelectric transducer is an acoustical microphone held against the patient's heart to record the heart sounds. An oscilloscope can be used for visual data display and strip charts for hardcopy.

17.2.5 Electroencephalogram (EEG)

A graphic record of the bioelectric potentials of the brain activity is called the electroencephalography (EEG). The EEG is a non-cyclical (unlike ECG) and is a profile of the complex wave patterns of brain activity. EEG electrodes, placed on the scalp, pick up the biopotentials and are displayed/recorded. The amplitude phase and frequency of EEG signals depend on the electrode placement, and 10–20 electrode placement scheme is widely employed. The EEG amplitudes range from 1–100 μV at frequencies 0.5–100 Hz. The EEG pattern is broadly classified into five frequency bands or rhythms (Table 17.3) and the frequency of the EEG seems to be affected by the mental activity of the person.

Most normal humans show α-waves (from the posterior brain) when they are relaxed with eyes closed (state of consciousness). Alert persons show higher frequency β-wave pattern, perhaps indicative of mental activity. γ-wave pattern is the result of sensory stimulation.

Table 17.3 Frequency bands of EEG signals.

Wave type	*Frequency range (Hz)*
δ	0.5 – 4
θ	4 – 8
α	8 – 13
β	13 – 22
γ	> 22

Sleep is defined behaviorally as a state of reduced consciousness and electrophysiologically by specific brainwave patterns— δ- and θ-waves. θ-wave pattern indicates emotional stress/activity. EEG sleep patterns show dramatic changes to the five stages of sleep— drowsiness, light sleep, moderately deep sleep and rapid eye movement (REM sleep) that usually follows deep sleep. EEG changes are also apparent in people with sleep disorders such as insomnia, hypersomnia, narcolepsy and nightmares. EEG sleep patterns δ- and θ-wave records are useful in assessing maturation in children (these waveforms are non-existent in children) and they show increasingly stronger signals as they mature.

EEG patterns are extremely used in the diagnosis of brain pathology— (i) local abnormalities (lesions and tumours), (ii) sleep patterns, (iii) sensory activity, (iv) brain responses to secondary stimuli, and (v) convulsions and mental disorders. One of the important clinical applications of EEG is the diagnosis of different types of epilepsy— grandmal, petitmal and psychomotor seizure (Fig. 17.3). EEG wave patterns show remarkable changes just prior to an epileptic seizure.

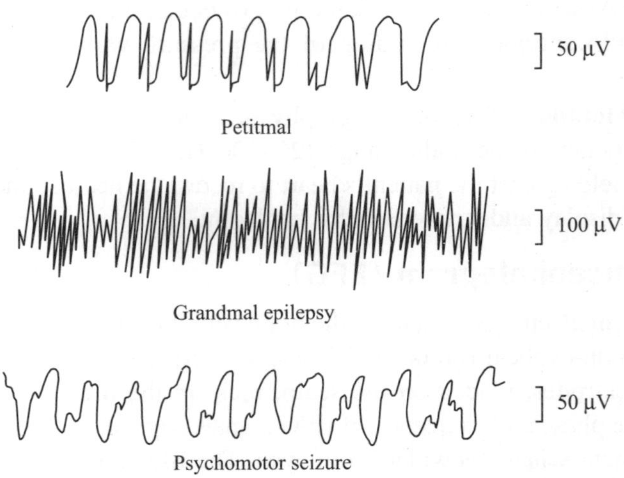

Fig. 17.3 EEG profiles of epileptic cases.

17.2.6 Electrical Source Imaging (ESI)

Electrical source imaging (ESI) is a technique for reconstructing brain or cardiac electrical activity from electrical potentials measured away from the brain or heart. The concept of ESI is to improve on electroencephalography (EEG) or electrocardiography (ECG) by determining the locations of sources of current in the body from measurements of voltages. ESI could improve diagnoses and guide therapy related to epilepsy and heart conduction abnormalities through its capability for monitoring the effects of drugs or for locating an electrical abnormality that is to be removed.

Differences in potential within the brain, heart, and other tissues reflect the segregation of electrical charges at certain locations within these three-dimensional conductors as nerves are excited, causing cell membrane potentials to change. While frequency analysis methods of EEG can indicate probable Alzheimer's disease by the abnormal distribution of spatial frequency bands, the true distribution of neuronal activity, knowledge of which could lead to more refined diagnoses, is masked or blurred by the conducting tissue layers between the central cortex and the electrodes.

Similarly, standard electrocardiography techniques such as electrocardiography (ECG) and vectorcardiography (VCG) are very limited in their ability to provide information on regional electrical activity and to localize bioelectrical events in the heart. In fact, VCG lumps all cardiac wave fronts into a single dipole located at the "center" of the heart and known as the "heart vector." Traditional ECG and VCG patterns of electrical activity cannot give the information required for characterizing the electrical activity of the heart. Non-invasive electrocardiography provides simultaneous recordings of electrical potential from 100 to 250 torso sites in order to map the body surface potential. These body surface potential maps (BSPMs) reflect the regional time course of electrical activity of the heart, information that is important for clinical treatment. The computer-controlled data acquisition system permits simultaneous recording from all 180 sites every millisecond throughout the cardiac cycle. Body surface potential distribution is a very low-resolution projection of cardiac electrical activity, and details of regional electrical activity in the heart cannot be determined merely from visual inspection of the BSPMs.

17.2.7 Other Bioelectric Potentials

Magnetoencephography (MEG)/Magnetocardiography (MCG): Ion currents arising in the neurons of the brain and the heart produce magnetic fields outside the body that can be measured by arrays of hypersensitive magnetic detectors, called superconducting quantum interference device (SQUID) detectors placed near the chest or head. The recording of these magnetic fields is known as magnetocardiography (MCG) or magnetoencephalography (MEG), which come under magnetic source imaging (MSI). An advantage of MSI over electrical source imaging (ESI) is that all body tissues are magnetically transparent and the magnetic fields propagate to the surface without distortion. MSI can be used in the surgical treatment of intractable epilepsy to locate the epileptigenic focus and functional areas of the brain that must be conserved during surgery. A major strength of MSI is that it can resolve events separated by milliseconds; whereas other imaging methods such as functional magnetic resonance imaging (fMRI), magnetic resonance spectroscopy (MRS), positron emission tomography (PET), and single photon emission computed tomography (SPECT) have time resolutions of seconds to many minutes.

Galvanic Skin Resistance (GSR): GSR is a record of the skin resistance profile, useful for studying the neurological responses of the skin.

Electrooculogram (EOG): EOG is a record of the corneal-retinal potential as affected by the position and movement of the eye. There is a linear relationship between the angle of gaze (eye movement within ±30°) and EOG output. For the horizontal gaze, surface electrodes are placed symmetrically near the left and right eyes, on the nose and temple. For vertical eye movement, electrodes may be placed above and below the eyes. The EOG is employed for recording reading ability, visual fatigue, and movement of eyes during sleep and dreaming. EOG can also help in detecting certain major diseases of retina.

Electroretinogram (ERG): ERG is a record of the bioelectric potentials, overall response of the eye to light stimulus, obtained from the retina of the eye due to visual stimulus. To record ERG, one electrode is mounted on a contact lens that fits over the cornea and the other electrode is attached above the ear or forehead. ERG has potential importance in assessing functional retinal behaviour.

Electromyography (EMG): EMG is a record of the bioelectric pattern associated with muscular activity. It is a clinical tool for studying muscular fatigue/disorders.

Electrogastrogram (EGG): EGG is a record of the intestinal response to peristaltic movements of the gastrointestinal tract.

Suggested Reading

Books

Barry, L. et al (Eds). *"Frontiers in Cardiovascular Imaging"*. Raven Press/New York. (1993).

Cameron, J.N. *"Principles of Physiological Measurement"*. Academic Press/New York. (1986).

Carr, J.J. & Brown, J.M. *"Introduction to Biomedical Technology"*. Prentice-Hall/Engelwood, NJ. (1998).

Cobbold, E.S.C. *"Transducers for Biomedical Measurements"*. Wiley/New York. (1978).

Demarre, D.A. & Michaels, D. *"Bioelectronic Measurements"*. Prentice-Hall/Engelwood, NJ. (1983).

Koryta, J. & Stulik, K. *"Ion-selective Electrodes"*. 2nd Edn. Cambridge University Press/Cambridge, UK. (1983).

Pilkington, T.C. & Plonsey, R. (Eds). *"Engineering Contributions to Biophysical Electrocardiography"*. IEEE, New York (1982).

Plonsey, R. & Barr, R.C. *"Bioelectricity"*. Plenum Press/New York. (1988).

Turner, A.P.P. et al. (Eds). *"Biosensors: Fundamentals and Applications"*. Oxford University Press/Oxford, UK. (1987)

Webster, J.G. (Ed). *"Medical Instrumentation: Application and Design"*. 3rd Edn. Wiley/New York.

18 Radiation Biophysics

Synopsis: Radiation biophysics is ordinarily concerned with the interaction of ionising radiations with matter and the effects on biological systems. Radiation in lower energy ranges interacts with matter by excitation processes but high energy radiation can ionise atoms and the liberated 'primary' electrons can further ionise or excite neighbouring atoms. α- and β-rays interact directly with matter, while high energy photons and neutrons interact with matter indirectly by photoelectric, Compton and pair formation processes; any particular process may predominate dependinç on the energy of the incident radiation.

Measurement of radiation is called dosimetry. The biological effects of radiation depend to a large extent on the energy and the penetration power of the radiation and the nature of the target. α-rays are more harmful to skin tissues than hard X-rays. The effects are ordinarily due to radiation-induced chemical/biological changes. As water is the most common substance in biological systems, the chemical species produced during the radiolysis of water play a major role in the biological effects of radiation damage. Presence of oxygen enhances the harmful effects. There is no clear-cut threshold level radiation. A dose below 0.25 Gy is considered the limiting dose. At ≈1 Gy dose, the immune system, bone marrow cells etc. get damaged and doses above 5–10 Gy are lethal. Radioactive isotopes are becoming increasingly important in radiotherapy and are routinely used in nuclear medicine.

18.1 IONISING RADIATION

All radiations of the electromagnetic spectrum have similar characteristics, but differ widely in their energies—from low energy radio and microwaves to infrared, visible and ultraviolet light to high energy *X*-rays and γ-rays. All types of radiation, irrespective of their origin, can affect living organisms. The interaction of radiation in the radio, microwave, infrared, visible and near-ultraviolet regions with matter is the concern of respective spectroscopies. Radiation biophysics is, in general, concerned with the interaction of ionising radiations (above 500 eV range), such as *X*-rays, γ-rays, β-(electrons and positrons) and α-rays, protons and neutrons, with matter and the ensuing biological effects due to photochemical, ionisation, and radiation-induced molecular alterations, and the resulting damage to the cell and organism. Radiation protection and therapy are also part of radiation biophysics.

18.1.1 Excitation/Ionisation

Interaction of radiation with matter can lead to either (i) excitation or ii) ionisation, depending on the energy of the incident radiation. If the energy of the radiation is below the threshold ionisation potential, it can lead to excitation from one state to another ($A \rightarrow A^*$ state), as incident energy is not sufficient to knock out the electron(s) from the attractive force of the nucleus of the atom. Excitation energies (less than 500 eV) are in the ultraviolet region; many of the photochemical reactions, mutation effects on nucleic acids (e.g. thymine dimerisation) and photo-carcinogenesis by UV radiation are examples of biological effects of radiation by molecular excitation processes.

High energy radiations such as X-rays and γ-rays (high-energy photons) have, on the other hand, sufficient and even excess kinetic energy above the ionisation threshold potential of atoms so that they can completely remove electrons from atoms, leading to ionisation.

$$A \rightarrow A^+ + e^- \qquad ...(18.1a)$$

Ionising radiations can be either pure radiation (X-rays and γ-rays) or particulate (α-rays, protons and neutrons) and the ionisation processes can be either direct or indirect. α and β-rays are directly-ionising radiations while X-rays and γ-rays are indirectly-ionising radiations, liberating ionising particles (e.g. electrons). On impact. (*Note: X-rays and γ-rays are synonymously used*).

18.1.2 Radiation Sources

Sources of ionising radiation are natural as well as man-made. Cosmic rays, radiations emitted by radioactive decay, such as α-particles, protons and neutrons and photon emission are natural sources. Some of these radiations and other (primarily X-rays) can be produced by man-made gadgets. The interaction of any radiation with matter depends on energy, penetration power and other characteristics specific to the radiation and to the target.

α-rays. α-rays are high-energy helium nuclei ($^4He^{++}$) released in α-decay, that occurs in heavy nuclei (generally for Z > 83) with excess of protons and neutrons, such as thorium, uranium and radium.

$$^{238}_{92}u \longrightarrow {}^{234}_{90}Th + {}^{4}_{2}He$$

$$^{236}_{88}Ra \longrightarrow {}^{232}_{86}Rn + {}^{4}_{2}He \qquad ...(18.1b)$$

The penetrating power of α-rays is very low (they can be stopped by a thin sheet of paper). However, they have very high ionisation power and linear energy transfer **(LET)** (which is defined as the energy deposited per unit length of track = dE/dl).

The α-particles can cause considerable radiation damage to skin tissues in organism, since all their energy is deposited in a very small tissue volume and is therefore radiotoxic.

β-rays. β-rays are high-energy electrons (β^- = electrons and β^+ = positrons), produced by the transformation of neutrons into protons with concomitant emission of electrons ($n \rightarrow P^+ + e^-$). Transformation of protons into neutrons induces the emission of positrons ($P^+ \rightarrow n + \beta^+$).

Neutron ⟶ Proton + electron + antineutrino.

Proton ⟶ Neutron + positron + neutrino.

$$^{14}_{6}C \longrightarrow ^{14}_{7}N + \beta^- + \nu \text{ (neutrino)}$$

$$^{65}_{30}Zn \longrightarrow ^{65}_{29}Cu + \beta^+ + \nu \qquad ...(18.1c)$$

High-energy β-rays interact directly with matter by excitation, ionisation and **Bremsstrahlung** processes. The penetration of β-rays, which is about ≈1 cm in biological tissues, is higher than that of α-rays. In general, β-rays have a limited range in tissues, so does not constitute an external hazard. They are a major hazard when ingested and feature as therapy agents (^{32}P; ^{131}I).

Protons and Neutrons. Protons are hydrogen nuclei ($^{l}H^{+}$). Neutrons transfer their kinetic energy through elastic, inelastic and particle-capture processes. In the inelastic collision process with an atomic nucleus, part of the kinetic energy is utilised to excite the nucleus and the excited nucleus emits radiation (γ – ray photon). Neutrons do not produce ionisation directly, but indirectly by ejecting protons of the atomic nuclei of the target material. In effect, the biological effects of neutrons are due to the protons they produce ($n \rightarrow P + e^- + \bar{\nu}$).

X-rays and γ-rays. Both X-rays and γ-rays are high-energy photons, irrespective of their mode of production, and their effects on biological systems are quite similar. While X-rays are produced due to atomic transitions of core electrons on bombardment of any heavy-element target by high-speed electrons, γ-rays are produced in nuclear reactions, whenever there is a transition of a nucleus to the ground state or by annihilation of an electron-positron pair (β^- β^+). Generally, machine-generated high-energy photons are referred to as X-rays.

When high-energy electrons are decelerated by a target (Cu, Mo, Ni, Fe are some of the anticathodes generally employed) kinetic energy is converted to produce a continuous X-ray spectrum (Bremsstrahlung), as well as a discrete line spectrum (characteristic of the anti-cathode target used) due to transitions between discrete energy levels of core electrons (*K*, *L*, *M*, *N*, ... shells) (Fig. 18.1). The *K*-series involves electron transitions from higher levels to the *K*-shell; the *L*-series from higher levels to *L*-shell. X-rays are also produced by radioactive decay processes. Electron capture (**k**-capture) by a nucleus leads to the formation an X-ray line spectrum characteristic of the newly formed element.

$$^{55}_{26}Fe \xrightarrow{\text{k-capture}} ^{55}_{25}M_n + K_\alpha \text{ line (2.1 Å)}$$

$$^{109}_{48}Cd \xrightarrow{\text{k-capture}} ^{109}_{47}Ag + K_\alpha \text{ line} \qquad ...(18.1d)$$

18.2 INTERACTION OF RADIATION WITH MATTER

There are three processes by which high energy radiation interacts with matter— (i) photoelectric, (ii) Compton and (iii) electron-positron pair production processes. The predominant process of interaction depends on the energy of the incident radiation (Fig. 18.2).

Photoelectric Effect. At lower energies (<30 KeV) the photoelectric effect comes into operation. In this process all the energy of the incident radiation is utilised in ejecting a bound electron resulting in ionisation. Photoelectrons are absorbed locally, causing ionisation and excitation. The photoelectric effect is predominant at low energies (<0.1 MeV) and for elements of high atomic numbers.

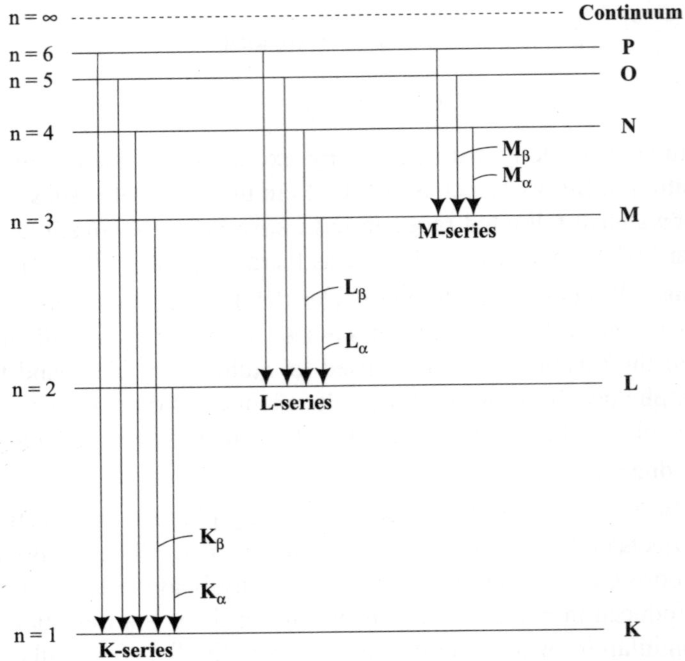

Fig. 18.1 Schematic representation of a characteristic X-ray line spectrum.

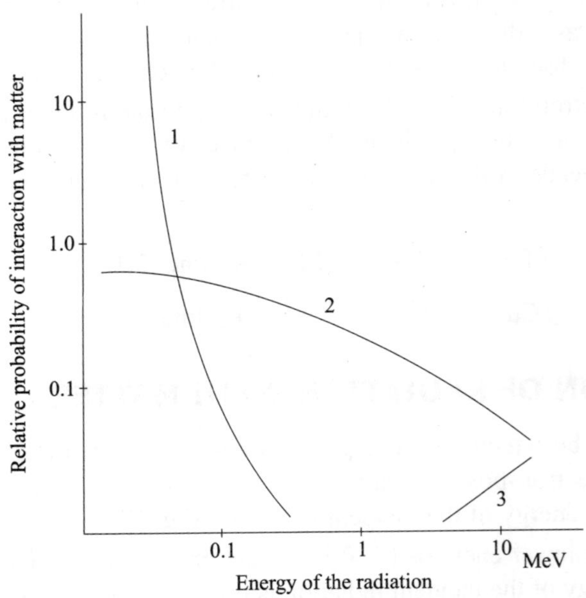

Fig. 18.2 γ-ray interaction with matter as a function of energy.
1 = Photoelectric process
2 = Compton process
3 = Pair-production process.

$$(\mu/\rho) \propto (Z/h\nu)^3 \qquad ...(18.2)$$

(μ/ρ) = mass attenuation; Z = atomic number; h = Planck's constant; ν = frequency).

Photoelectric effect plays an important role in the imaging process and radiation dosimetry.

Compton Effect. At medium energies (<10 MeV) the **Compton** effect is dominant. Elastic scattering of X-rays (photons) by matter is the basis of X-ray crystallography. The Compton effect, on the other hand, is inelastic scattering, wherein a photon transfers part of its energy to weakly-bound electrons ('recoil' electrons) and the scattered photon has reduced energy. The energy of the 'recoil' electrons, absorbed locally will be the cause for further ionisation and excitation processes. γ-rays (photons) of intermediate energies (20–100 KeV) can interact with matter both by photoelectric and Compton processes and in the 0.1–3 MeV energy range they interact almost exclusively by the Compton process. The Compton effect is prominent at moderately high energies and for materials of low atomic number, and therefore of relevance in its effects on biological tissues. It is the prominent process of interaction of radiation with matter in biological systems in the enregy range, 0.1 to 10 MeV.

Pair Production (e^-e^+). Radiations (photons) with energies >1.02 MeV passing close to a nucleus of an atom can create *de novo* matter, that is, an electron-positron pair ($e^- e^+$). Pair formation is greater in heavy elements ($\alpha\ Z^2$) than in lighter elements. The processes are summarised in Table 18.1.

18.2.1 Energy Transfer Processes

The radiation passing through matter is attenuated due to absorption and scattering. Attenuation due to absorption follows the **Beer-Lambert** law

$$I = I_0 \exp(-\mu\, \mathbf{x}) \qquad ...(18.3a)$$

where I and I_0 are transmitted and incident intensities of the radiation, **x** is the thickness of the material and μ is the linear attenuation (absorption) coefficient. The linear attenuation coefficient of the material is related to its atomic number and, therefore, the mass attenuation coefficient μ_m, is dependent on the density, ρ, given by

Table 18.1 Effect of ionising radiation on matter

Type of radiation (particle)	*Type of interaction*	*Resultants*
Photons (X-rays & γ-rays)	(i) Photoelectric effect (Z-dependent) $E_{ph} \geq h\nu - E_{be}$ (ii) Compton effect (Z-independent) (iii) (e^- e^+)-pair production (Z^2 -dependent) E_{tr} = ($h\nu$ – 1.02) MeV	Electrons. $h\nu$ < 0.1 MeV Electrons. $h\nu \approx$ 0.1 to MeV γ-rays $h\nu$ > 1.02 MeV
Neutrons	(i) Elastic scattering (ii) Inelastic scattering (iii) Particle (positron) capture (iv) Nuclear process	Protons γ-rays γ-rays Positive ions
Charged particles (Electrons/positrons/ α-particles)	Inelastic scattering	Ionisation and excitation

$$\mu_m = \frac{\mu}{\rho} \quad ...(18.3b)$$

The total attenuation is the sum of three components due to the three processes— photoelectric (μ_p), Compton (μ_c) and pair formation (μ_π). The contribution of u_p is significant below 50 KeV and is Z-dependent, $\mu_c \approx 0.1$ to 50 MeV and is Z-independent and $\mu_\pi > 1.1$ MeV and is Z^2-dependent (Z = atomic number).

$$\mu_{tot} = \mu_p + \mu_c + \mu_\pi \quad ...(18.3c)$$

The primary electrons liberated from the scattering processes of high energy radiations have sufficient energies to undergo further multiple scattering with the surrounding molecules of the medium, producing secondary electrons (called δ-rays). These secondary electrons (δ-rays) are the mediators in energy transfer to surrounding molecules, leading to ionisation and excitation. At high energies, above 10 MeV, the Bremsstrahlung process (emission of photon radiation due to deceleration of high-velocity charged particles by the medium) is also important. Because of high **LET**, the biological effectiveness of neutrons is very significant in energy transfer processes and radiation damage in biological systems.

18.3 MEASUREMENT OF RADIATION (DOSIMETRY)

Dosimetry deals with the various methods of radiation measurement, both physical and chemical.

Absorbed Dose (D). Absorbed dose, **D**, is the energy absorbed by unit mass of the material given by

$$\mathbf{D} = \frac{\text{Energy imparted}}{\text{mass}} = \frac{\widetilde{A}}{m} \sum_i \delta_i . \phi_i \quad ...(18.4a)$$

where $\widetilde{A}$ = accumulated dose; m = mass of the target; δ_i= equilibrium dose constant for the ith radiation and ϕ_i = fraction of the radiation (i^{th} type) absorbed by the target. For non-penetrating radiations (electrons, positrons and α-rays) the $\phi \approx 1$ and for penetrating radiations (X-rays and γ-rays) the value of ϕ depends on, among other parameters, the energy of the radiation. The unit of absorbed dose is Gray, $\text{Gy} = 1\ \text{Jkg}^{-1}$. (1 Gy = 100 rad).

Dose Equivalent (H). Radiation-induced biological effects depend not only on the amount of absorbed dose, **D**, but also on other parameters, like the type of radiation and the nature of the target etc., Dose equivalent, **H**, is the absorbed dose, weighted according to the damage potential of the incident radiation. For example, α-particles cause more biological damage to the skin tissues (due to its low penetration power) than γ-rays.

$$H = \sum_{i=1}^{n} D_i\, Q_i\, N \quad ...(18.4b)$$

where Q_i is the quality factor for the ith type of radiation and N is the product of other factors (dose rate etc.).

$N = 1$; and $Q \approx 1$ for photons, electrons and positrons
≈ 10 for protons and neutrons
≈ 20 for α-particles.

The unit of dose equivalent, H, is the Sievert, Sv = 1 J.kg^{-1}. (1 Sv = 100 rem).

Effective Dose Equivalent (H_{ef}). Not all parts of organisms are equally sensitive to a given radiation. So, different parts of the body should be given appropriate weightage. Therefore the effective dose equivalent, H_{ef}, is

$$H_{ef} = \sum_T W_T H_T \qquad ...(18.4c)$$

where H_T is the average dose equivalent in a tissue and the W_T is the weighting factor.

Radiation dose is measured by various methods— by physical dosimetry that includes ionisation chambers, **Geiger-Müller** counters, luminescence and photographic methods and by chemical dosimetry (**Fricke's** dosimeter).

Physical Dosimetry. High energy radiation can be measured by gas-filled, scintillation and semiconductor detectors. The measurement of the number of ionisations is the basic principle on which physical dosimetry is based. Dosimetry of photon radiation is conveniently carried out by methods based on ionisation of air (ionisation chambers, proportional and Geiger-Müller counters); a closed chamber provided with electrodes, and subjected to steady ionising radiation which will produce a fixed number of ions per second. Ionisation chambers are the basis of many radiation protection instruments. Photon detectors are good for measuring lower energy range radiations and electron detectors are useful in the higher energy radiation regions, where secondary electrons (δ-rays) play a prominent role in causing biological effects.

In all these detectors, the amplitude of the pulse is non-linearly related to the applied voltage and detecting devices perform with maximum efficiency in their appropriate voltage regions (Fig. 18.3).

The ionisation chamber is one of the commonly used radiation detectors. In any ionisation chamber device there is a closed chamber containing air or inert gas and the electric field is maintained between the two electrodes. There are three distinct regions of the operation of the counter— (i) ionisation region, (ii) proportional counter region and (iii) Geiger region. In the ionisation region (< 300 V), ionisation of gas in the chamber (ionisation current), caused by high energy radiation, is proportional to the incident radiation. In the proportional counter region (300 < V < 800), the applied voltage is above the ionisation level but below the Geiger voltage. In this region, the number of electrons reaching the anode increases rapidly with the applied voltage, due to formation of secondary ion-pairs (amplification $\approx 10^5$). But, the pulse amplitude is still proportional to the initial ionisation and the total number of secondary electrons is proportional to the number of primary ion-pairs produced by the original ionising radiation. Detectors operating under these conditions are called proportional counters. They are ideal for detecting high counting rates (>10^5 counts/sec) and are essential for the detection of particle of high ionising power, such as α-particles. In the Geiger region, the applied voltage is above the proportional counter voltage (>1000 V).

In a Geiger-Müller counter (G-M tube), a high potential, above 1000 V, is applied to the central anode wire which is surrounded by a cylindrical cathode. The tube is filled with a gas mixture (argon + ethane) and sealed. A thin mica or glass window acts as an entry area for incident radiation. An

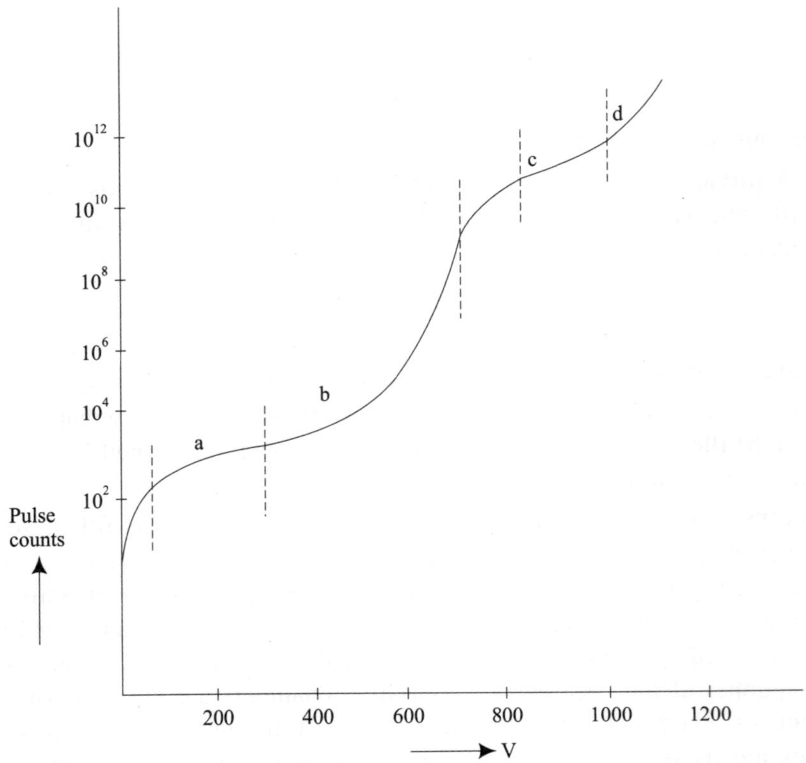

Fig. 18.3 Relationship of pulse amplitude to the applied voltage in a gas-filled detector. *a* = Ionisation chamber region; *b* = proportional counter region; *c* = Geiger-Müller counter region; *d* = continuous discharge.

incident ionising particle initiates ion-pair production in the G-M tube. These ion-pairs collide with neighbouring gas molecules and ionise then. Such cascade collisions of ions produce an avalanche of ionisation, and the electrons upon reaching the anode produce photons which in turn produce ionisation throughout the tube. The amplification reaches $\sim 10^{10}$ counts/sec. Geiger counters are simple, rugged and easy to operate but sluggish. They are used for detection and monitoring of high-energy radiation and β-particles.

A pulse-height analyser can be incorporated into detectors, such as proportional counters, to make them operational in the restricted range of X-ray or γ-ray energies. A pulse-height analyser eliminates all the pulses of energies other than the preset pulse amplitudes. By this means energy-discrimination can be achieved.

Semiconductor detectors (SiLi and GeLi) are also in use in physical dosimetry.

Scintillation Counters. The fluorescent effect, that is, the conversion of high energy radiation into a visible spectrum by chemicals, is measured by scintillating counters (photomultiplier tubes). Almost all nuclear radiation detectors (e.g. gamma camera) employed in nuclear medicine utilise scintillation technique. Silver-activated zinc sulphide is a good scintillator for α-particles and for counting γ-rays, sodium iodide, doped with thallium iodide (0.2%), is generally employed. Semiconductor detectors are well-suited for γ-ray detection (surface-barrier silicon and lithium-drifted germanium detectors).

The liquid scintillation technique is extensively employed in medicine, pharmacology, environmental science and biology. This technique is based on the fact that certain organic molecules in solution can be induced to fluoresce upon excitation by UV and high energy radiation. Aromatic-type molecules usually meet the energy and π-electron requirements. As these compounds are generally in a liquid state at room temperature, the liquid scintillation method has wide applications.

In liquid scintillation, the scintillation solution consists of a solvent or solvents, such as benzene, toluene, *p*-xylene and *p*-dioxane etc., and a solute or solutes, such as naphthalene, anthracene, *p*-terphenyl, pheanthrane, carbazole etc. The solvent acts as a medium for absorbing the energy of ionising radiations, resulting in the formation of excited solvent molecules, which transfer energy to the solute molecules, which act as a source of photons. Most solvent molecules are not by themselves good scintillators. So addition of suitable solutes that are efficient scintillators is essential.

Liquid scintillation counters employ a coincidence method of counting. That is, two photomultiplier tubes are used to measure the same count to reduce background noise. Photocathodes are cesium-antimony (Cs_3 Sb) or multialkali materials (Cs Na_2 K Sb; K_2Cs Sb).

In general, β-emitters (^{3}H, ^{14}C, ^{32}P, ^{35}S etc.) are counted in liquid scintillators, but γ-rays (^{125}I, ^{135}I, ^{60}Co, ^{99}Tc etc.) can also be counted in a liquid scintillator by counting the electrons produced by γ-rays upon interaction with matter by photoelectric (<20 KeV) and/or Compton (0.02 → 3 MeV) processes. The pair-production process by γ-rays is not very relevant to biological systems. Counting efficiency of γ-rays is in general very low as scintillation liquids have low atomic number species (generally H(1), C(6), N(7) and O(8) and the efficiency decreases with increase in γ-ray energies (and hence in general solid state scintillation is preferred). However, the counting efficiency of γ-ray detectors can be improved either by adding electron-dense compounds, such as ^{125}1, ^{131}1, ^{60}Co etc., to the scintillator solution or by using large-volume scintillators (whole body counters). Large-volume liquid scintillation counting is employed in the measurement of radioactivity in animals and humans and for monitoring the uptake of metabolites, hormones and drugs, radioactive products (contamination and environmental pollution) in food stuffs; milk, water etc.

High-energy charged particles (such as electrons) travelling with velocity, **v**, through a medium of refractive index, **n,** radiate part of their energy if their velocity, **v**, is above the threshold.

$$\mathbf{v} > \frac{C}{\mathbf{n}} \qquad (C = \text{velocity of light}) \qquad ...(18.5a)$$

This radiation is called **Cerenkov** radiation. This effect can also be employed for the measurement of radiation dose (by the liquid scintillation method).

Chemiluminescence and Bioluminescence. Chemiluminescence and bioluminescence result from internal reactions in a solution leading to products in excited states.

$$A + B \longrightarrow C^* + D$$

$$C^* \longrightarrow C + h\nu \qquad ... (18.5b)$$

In the normal course of liquid scintillation counting, chemiluminescence is undesirable as it acts as an impediment to accurate evaluation of only those counts that are due to the radioactive label of the compound, So, care must be taken to minimise the causes for chemiluminescence in scintillation counting, However, liquid scintillation counters can be employed to measure the concentration of reactants that create luminescence by chemical reactions. That is, chemiluminescence can also be studied by the liquid scintillation technique, This technique can be employed in the bioluminescence assay of ATP, FMN, NADP and other enzymatic substrates.

The degree of darkening of a photographic emulsion due to radiation is the basis of photochemical measuring devices (photographic plates, film and badges) which are routinely employed to detect and monitor ionising radiations.

Chemical Dosimetry. Oxidation of ferrous to ferric ions ($Fe^{2+} \rightarrow Fe^{3+}$) by the radiolysis products of water is one convenient method of chemical dosimetry. This is the basis for the **Fricke** dosimeter. The **Fricke** dosimeter has a dilute solution of a mixture of ferrous ammoniumsulphate (10^{-3} M) and NaCl (10^{-3} M) and H_2SO_4 (0.4 M). The reaction is carried out at pH = 1.5, The amount of ferric ions produced by irradiation is measured directly from the absorbance of the solution at 304 nm. The oxidation reaction can be represented as

$$2H_2O + O_2 + 4Fe^{2+} \xrightarrow[\text{radiation}]{\text{ionising}} 4Fe^{3+} + 4OH^- \quad ...(18.6)$$

18.4 RADIOACTIVE ISOTOPES (RADIONUCLIDES)

The another common source of ionising radiation is radioactive isotopes. Radioactive isotopes (**radionuclides**) are of considerable importance in the presentday world, in the diagnostics of radiation effects on biological systems as well as for therapeutic purposes in nuclear medicine. The source of radiation is not external but *within* the patient.

18.4.1 Radioactivity

Certain atomic nuclei (radionuclides) which are unstable achieve stability by radioactive decay, giving rise, in that process, to other types of isotopes and atoms. The β-decay processes transform neutrons to protons and protons to neutrons with the emission of electrons and positrons, respectively. The α-decay process emits α-particles and transition of the excited nuclide to ground state causes the emission of a γ-photon. γ-photons are also produced by (e^-e^+) fair annihilation.

The rate of radiaoactive decay is statistical (Poisson distribution) and is characteristic of that nuclide and is independent of physical and chemical environments. The rate of decay is given by

$$N = N_0 \exp p\,(-\lambda\, t) \quad ...(18.7a)$$

where N_0 is number of radioactive atoms at the initial time, $t = 0$ and N is the number at time t, and λ is the decay constant. The period during which the radioactive decay of a radionuclide is half of its initial value is called **half-life** (τ), which is related to the decay constant.

$$\lambda = \frac{0.693}{\tau} \quad ...(18.7b)$$

The amount of radioactivity A, ($A = \lambda N$), is measured in terms of disintegration rate. The unit is called the **Becquerrel, Bq**, which is one disintegration per second.

$$1 \text{ Bq} = 1 \text{ disintegration/second}$$

$$1 \text{ Bq} = \frac{1}{3.7} \times 10^{-10} \text{ curie.}$$

18.4.2 Production of Radionuclides

Radionuclides are isotopes (elements with same atomic number, but different atomic mass numbers) that reach stability by emitting radiation or particles. A radionuclide is characterised by its half-life, type of decay—transition—and the type of particle emission and its energy. The most important method of producing artificial radionuclides is by neutron bombardment of a required material in an atomic reactor. The transformation is denoted by

$$A\ (\delta, X)\ B \quad \text{...(18.8)}$$

where A is the initial nucleus, B is the final nucleus, δ is the bombarding particle and X is the emitted particle (radiation).

Choice of a radionuclide for any biological study depends on its half-life, the radiation it emits and the target effectiveness (high target/non-target ratio). Isotopes with short half-life and which are γ-emitters are ideally suited as γ-rays are absorbed to only a small extent by biological systems. Isotopes which are very relevant in biology, such as ^{3}H , ^{14}C, ^{32}P, are all pure β-emitters. The technitium isotope ($^{99}Tc^{m}$) satisfies many of the requirements of nuclear medicine. Some common radioisotopes employed in radiomedicine and tracer techniques are given in Table 18.2.

18.4.3 Applications of Radioactive Tracers

Radioactive-labelling has found extensive applications in biological and clinical fields in many assay methods (such as radioimmunoassays, autoradiography etc.) and in monitoring chemical reactions. Radioactive monitoring is simple and quantitative.

Radioimmunoassay (RIA). One of the important applications of radioactive tracers in biology and medicine is the Radioimmunoassay (**RIA**) method. This method can be employed to assay substance present in small quantities (10^{-9} gm /ml) with a high degree of specificity and can be used in the *in vitro* diagnostic monitoring of abormalities of body functions, enzyme catalysis and other biochemical reactions.

RIA is based on the high degree of affinity of an antibody for specific antigens; radioactive-labelling is employed to monitor its immunological specificity. The general principle of RIA is the quantitative measurement of the ratio of labelled antigen (**Ag***) molecules to an unknown quantity of assay antigen (**Ag**) competing for the binding sites of a known quantity of an antibody (**Ab**).

In radioimmunoassay a known quantity of labelled antigen (**Ag***) is added to a known quantity of antibody (**Ab**) (in addition to the unlabelled antigen, **Ag**).

When there is an excess of antigen molecule (**Ag** and **Ag***) for a known quantity of antibody, **Ab**, there arises a competition of **Ag** with **Ag*** for the same sites of the antibody, **Ab**. At equilibrium

Table 18.2 Radioactivity data on some common radioisotopes.

Element Symbol	*Atomic Number Z*	*Isotope*	*Type of Emission*	*Half-life (τ)*
H	1	^{3}H	β^-	12.3 Yrs.
C	6	^{11}C	β^+	20 Mts.
		^{14}C	β^-	5800 Yrs.
N	7	^{13}N	β^+	10 Mts.
O	8	^{15}O	β^+	2 Mts.
Na	11	^{24}Na	β^+; γ	15 Hrs.
P	15	^{32}P	β^-	14 Days
S	16	^{35}S,	β^-	87 Days
K	19	^{42}K	β^-; γ	12 Hrs.
Ca	20	^{45}Ca	β^-	163 Days
Fe	26	^{52}Fe	β^+; γ	8 Hrs
Co	27	60Co	β^-; γ	5.2 Yrs
Tc	43	$^{99}Tc^m$	γ	6 Hrs
		^{99}Tc	γ	2.1×10^5 Yrs
In	49	$^{113}In^m$	γ	1.6 Hrs
I	53	^{131}I	β^-; γ	8 days
		^{125}I	**k**; γ	60 days
Au		^{198}Au	β^-; γ	2.7 Days

k = electron capture

$$\mathbf{Ag + Ab \rightleftharpoons Ag \cdot Ab}$$

$$\text{Free} + \text{Bound} \qquad ...(18.9a)$$

$$\mathbf{Ag^* \rightleftharpoons Ag^* \cdot Ab}$$

In this reaction, increase of **Ag** leads to decrease in **Ag*.Ab** complex (therefore, decrease in its radioactive counts). The assay is based on the quantitative evaluation of the total labelled antigen, **Ag***, bound to the antibody (**Ag*. Ab** complex) from the radioactive counts of the free and bound-ligands (Fig. 18.4a), from which the amount of unlabelled antigen (**Ag**), bound to the antibody (**Ab**), can be calculated.

At equilibrium, the ratio of the antigen in the bound (B) and free (F) phase is

$$\frac{B}{F} = \frac{[\mathbf{Ag.Ab}]}{[\mathbf{Ag}]} \qquad ...(18.9b)$$

A dose-response plot of F/B versus the antigen concentration is a hyperbolic curve and the reciprocal of the receptor concentration (1/q) can be calculated from the slope of the asymptotic line (Fig. 18.4b).

In RIA, the selection of the radiolabel is determined by the structural and sensitivity requirements of the immuneresponse system. ^{131}I or ^{125}I (in oxidised state) is the generally employed radiolabel (attached to the tyrosine residues of the protein). For the procedure to be successful, radiolabelled antigen (**Ag***) of high specific activity is most desirable.

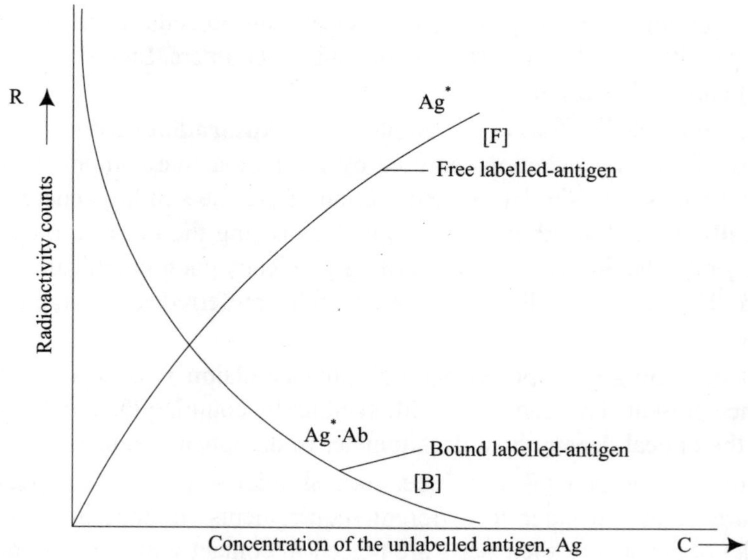

(a)

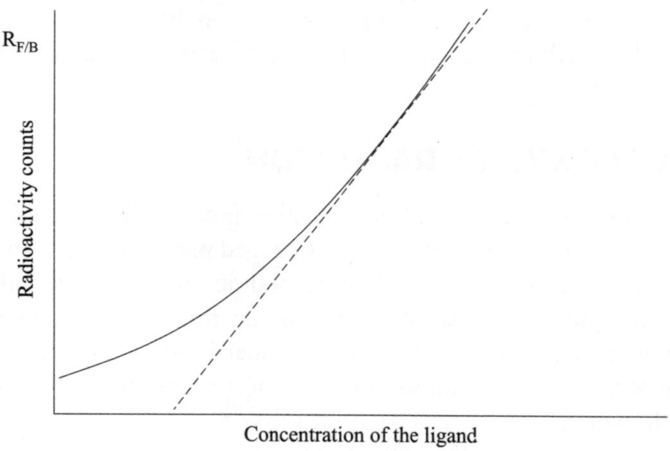

(b)

Fig. 18.4 (a) Plot of the concentration of the unlabelled antigen, Ag, versus radioactive counts, *R*.
(b) A dose-response plot of antigen concentration versus the radioactive counts (ratio of the free/bound phase).

Once the equilibrium state is reached, the components (free and bound ligand particles) are separated ('partitioning'). This can be carried out by several methods— chromatophoresis, gel-filtration (by **sephadex** and **sepharose** adsorbants), ion-exchange (with antibody-coated resins), salt-precipitation, immunoprecipitation (a double-antibody precipitation method where a second antibody is used to precipitate the primary antigen-antibody complex).

The choice of counting device depends on the separating procedure and on the type of isotope used for labelling. γ-emitters (^{132}I and ^{125}I) are counted by γ-counters. Liquid scintillators can also be used if the labelled anligen has β-emitters (^{3}H or ^{14}C).

Autoradiography. Radioactive tracers are employed in **Autoradiography**. Autoradiography is a record of the traces of ionising radiations emitted by the radioactive compound on a photographic emulsion. This technique is very similar to the detection of the paths of high energy particles (cosmic rays) photographically. In biological studies, (α particles (being the decay products of high atomic mass emitters) and γ-rays (having very poor recording efficiency photographically) are not employed. β-emitters, such as ^{3}H, ^{32}P and ^{35}S isotopes, are generally preferred in autoradiographic studies in biological sciences.

The degree of darkening of exposed photographic emulsion is a measure of the amount of radioactive substance present. This can be quantified either by counting the number of grains exposed or by determining the optical density by a densitometer or an optical scanner.

Various variations of the general technique, such as micro-, macro- and electron microscope-autoradiography, are in use to cater to different requirements. In macro- as well as in micro-autoradiography, the organisms or their parts are placed in contact with a photographic emulsion to produce and autoradiograph. This technique is useful to monitor metabolic products, drugs, hormones etc., in organisms. Electron microscope-autoradiography is employed to achieve higher resolution than that is possible with light microscopy. This technique enables one to visualise the detailed morphological features of intracellular organelles— transport, secretion, sites of synthesis of proteins, nucleic acids, carbohydrates etc.

18.5 BIOLOGICAL EFFECTS OF RADIATION

Biological effects of radiation result primarily from radiation-induced chemical/biochemical changes. Radiation can cause genetic mutation and cell damage. Charged particles such as α-rays and electrons damage living cells directly, whereas X-rays and γ-rays and neutrons damage indirectly (Fig. 18.5). X-rays and γ-rays have energies in excess of the ionisation threshold and, therefore, the ejected electrons are highly energetic and they in turn produce further ionisation due to secondary electrons (δ-rays) in neighbouring atoms. That is, ionisation is due to primary as well as secondary electron interaction processes with matter.

Absorbed radiation is converted into a variety of chemical and biochemical reactions leading to biological damage in the living tissues, cells and organisms. The relative biological effectiveness (**RBE**) of various radiations depends on their linear energy transfer (**LET**) capabilities.

$$\text{RBE} = \frac{D_x}{D_R} = \frac{\text{Biological effectiveness of the radiation}}{\text{Biological effectiveness of reference radiation}} \quad ...(18.10)$$

The reference radiation is generally ^{60}Co γ-rays. RBE depends on the dose, dose rate, oxygen presence, post-irradiation recovery/damage and other parameters.

18.5.1 Dose–Response (Dose–Survival) Relationships

The radiation harm to a living cell depends on the level of complexity— whether a single cell, group of cells or the whole organism, and frequency of cell division (Table 18.3).

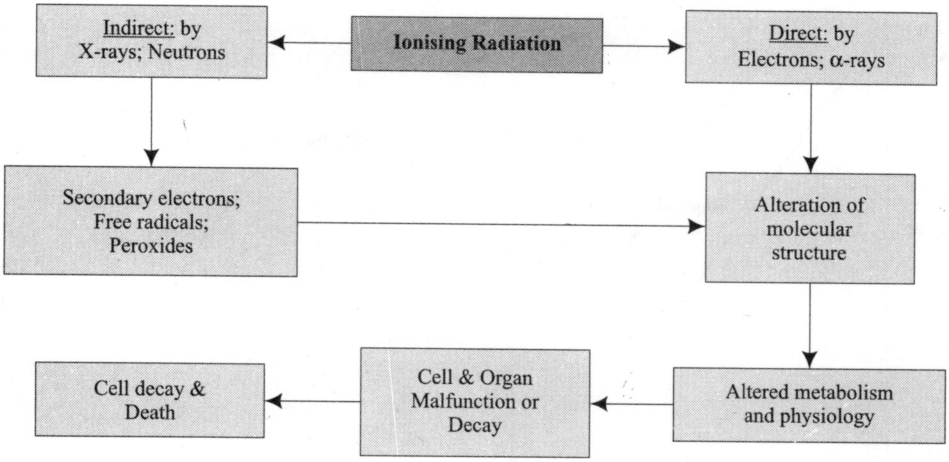

Fig. 18.5 Interaction of Ionising radiation with matter.

Table 18.3 Effects of radiation on biological materials.

Level of complexity	*Radiation effects/damage*
Molecular	Macromolecules (DNA; RNA; enzymes).
Sub-cellular	Cell membranes; nucleus; chromosomes; mitochondria.
Cellular	Impaired cell division; mutation; cell death.
Tissue/organ	Bone marrow; intestinal tract; nervous system; carcinogenesis.
Individual	Short life; death.
Population	Mutations; genetic disorders.

Biological radiation data are analysed by dose response relationship graphs. The end result of the dose–response is the result of many intermediary reactions. For chemical reactions where the radiation process is indirect (X-rays, γ-rays and neutrons) the relationship between dose and the observed chemical change depends upon the reaction product. If the product does not interact with free-radicals, then the number of molecules damaged is directly proportional to the dose applied. The overall effect of cellular damage is measured at the point where radiation dose causes a 50% cell loss (lethal dose for 50% death, LD_{50}-point). Haploid cells (bacterial) can survive high radiation doses (~200 Sv).

The radiation sensitivity of a cell depends on whether it is haploid (unpaired chromosomes— bacterial) or diploid (paired chromosomes— mammalian). While haploid cells (bacterial; viral) show a simple single-event exponential dose-response/survival curve (Fig. 18.6a), the dose–response relationship to radiation is more complex in the case of mammalian (diploid) cells (Fig. 18.6b).

Target Theory. The dose–response relationship is explained by the target theory, which is based on the assumption that within a cell a target has to be hit directly to induce cell damage. The radiation damage can be either due to single-hit dose or multi-hit dose response. The dose-response relationship for a single-hit dose is semi-logarithmic and the surviving fraction, **S**, is

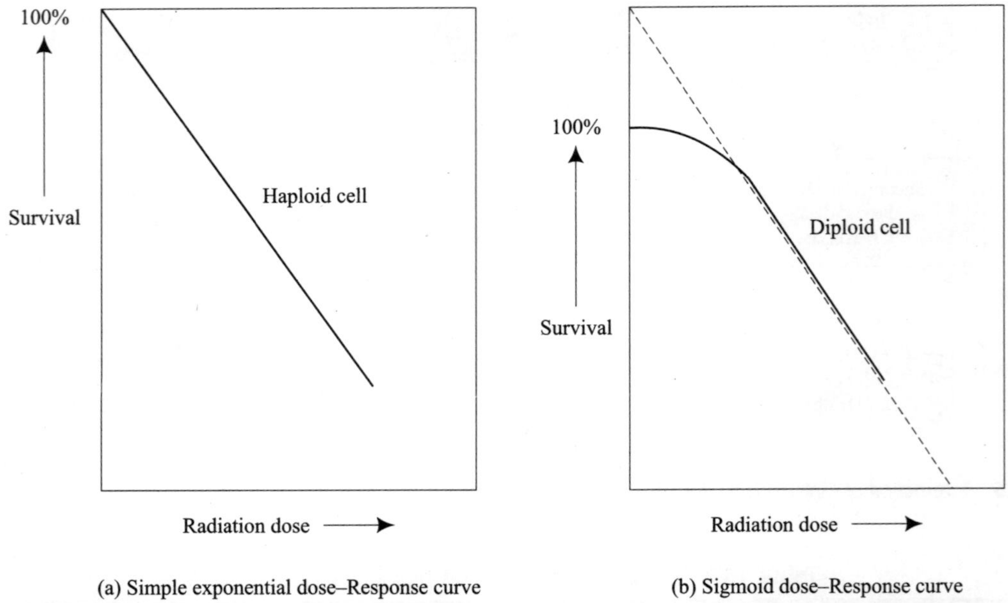

Fig. 18.6 Dose–Survival relationship in (a) Haploid and (b) Diploid cells.

$$\mathbf{S} = \frac{N_s}{N_o} = e^{-\alpha D} \qquad \text{...(18.10a)}$$

where N_o is number of targets exposed and N_s is number surviving after irradiation and α is the inactivation coefficient, which depends on the volume of the target, dose-rate etc.

The inactivation coefficient, α, of the single-hit dose-response relationship represents irreversible radiation damage and does not take into account other factors such as below-lethal damage and cellular repair processes. The genetic material, DNA, is taken as the principal target for radiation damage. Damage to the nucleic acid bases is rectified by several repair processes in the cell and therefore may not automatically lead to damage of the genetic material. Breakage of the sugar-phosphate backbone (single-chain scission) of DNA is more deleterious (sublethal), but repair processes do exist for such reactions in the cell. Double-stranded breaks are generally lethal. Therefore, damage-repair processes are due to a complex interplay of (i) sublethal damage and repair, (ii) potentially-lethal damage and its repair or further damage and (iii) irreversible damage.

Dose-response curves which are **sigmoid** in shape (Fig. 18.6b) are due to multi-hit processes (combined with molecular repair processes. An end-effect can be due to a combination of single-hit (point mutations) and multi-hit (double-hit chromosome reorganisation). In multi-hit processes, the damage is dose-rate dependent at higher doses. The shoulder of the sigmoid curve is taken as an indication of repair of damage.

A linear quadratic model has been proposed to account for repair processes in radiation dose–survival plots. The surviving function, *S*, is given by

$$S = \exp\,[-p(\alpha D + \beta\, D^2)]. \qquad ...(18.11)$$

where p is the biological effectiveness factor and β is the coefficient representing multi-hit damage.

The radiation sensitivity of a cell or tissue affects the shoulder profile of the curve and is related to its mitotic activity. Rapidly dividing populations within an organ (bone marrow) are most sensitive and tissues with little mitotic activity (brain; muscle) are more resistant to radiation.

Deviations from target theories can be attributed to differences in sensitivity to radiation in different targets. Target theories are inadequate to explain cellular damage, because post-irradiation conditions determine the magnitude of lesions and sub-lethal effects. However, the linear-quadratic equations which attempt to address radiation effects are in actuality similar to virial equations, which are empirical attempts to address the molecular interactions of real gases.

18.5.2 Radiolysis of Water

In biological systems, water is abundant and ubiquitous. Its high affinity for electrons makes the radiolysis of water a primary event in the initiation of radiation damage in living cells. Absorption of radiation by water leads to ionised species which further react with neighbouring water molecules or other polar molecules, to form a variety of new reactive species and free radicals. Free-radicals are not the primary products of radiolysis of water, but secondary products from subsequent decomposition of ions and excited species produced initially. They are $\mathbf{H_2O^*}$ and dissociation products $\mathbf{H^*}$, $\mathbf{OH^*}$, $\mathbf{H_2O^+}$ and $\mathbf{e^-}$. The radiation mechanism in aqueous media depends on the production of $\mathbf{OH^*}$, $\mathbf{H^*}$ radicals and hydrogen peroxide. The $\mathbf{OH^*}$ radical is the predominantly harmful radical (see eqn. 18.12).

$$H_2O \xrightarrow[\text{ionisation}]{h\nu} H_2O^+ + e^-$$

$$H^+ + OH^- + e^-$$

$$OH^+ + H + e^-$$

$$H_2O^+ + e^- \longrightarrow H_2O^* \text{ (excited state by the thermalised electrons)}$$

$$H_2O^+ \longrightarrow H^+ + OH^*$$

$$H^* + OH^+$$

$$H_2O^* \longrightarrow H^* + OH^* \text{ (radical species)}$$

$$H_2O + e^- \longrightarrow H^* + OH^-$$

$$H_2O^+ + H_2O \longrightarrow H_3O^+ + OH^- \qquad ...(18.12a)$$

Hydrolysis of free-radicals, $\mathbf{OH^*}$, which are very reactive can form hydrogen peroxide, which is a powerful oxidant)

$$OH^* + OH^* \longrightarrow H_2O_2 \qquad ...(I8.12b)$$

Free-radicals and H_2O_2 can react with genetic material, proteins and other biological molecules and alter their functions, leading to mutations, chemical lesions and biological complications like illness, cancer and ultimately death.

The Role of Oxygen. Oxygen is a highly reactive molecule. In the presence of oxygen the reactions produced by ionising radiation are different and are more harmful biologically. In the absence of oxygen, radicals $\mathbf{R}^*$ and $\mathbf{X}^*$ can react with one another or dimerise (or polymerise), $\mathbf{R_n}$.

$$R^* + X^* \longrightarrow RX$$

or

$$R^* + R^* + R^* \ldots \longrightarrow R_n \qquad \text{...(18.13a)}$$

Presence of oxygen blocks restoration processes and enhances radiation damage. In the presence of oxygen, the formation of the peroxy-radical, $\mathbf{RO_2^*}$ is the predominant reaction

$$R^* + O_2 \longrightarrow RO_2^* \text{ (damaged enhancement radical).} \qquad \text{...(18.13b)}$$

Peroxy radicals cannot dimerise or polymerise but give rise to hydroperoxide.

$$RO_2^* + H \longrightarrow RO_2H \qquad \text{...(18.13c)}$$

Irradiated water containing dissolved oxygen leads to the formation of relatively large amounts of perhydroxy radicals ($\mathbf{HO_2^*}$). Therefore, all biological systems are more radiosensitive in the presence of oxygen.

$$H_2O + O_2 \xrightarrow{h\nu} HO_2^* + OH_2^*$$

Each perhydroxy radical can oxidise three molecules.

$$3RH + HO_2^* \longrightarrow 3R^* + 2H_2O \qquad \text{...(18.13d)}$$

Therefore, a dose of radiation is more destructive to a biological system in the presence of oxygen than in its absence.

18.5.3 Effects of Radiation on Living Systems

Radiation-induced biological effects can manifest at the (i) molecular, (ii) cellular and (iii) organism levels.

Effects at the Molecular Level. Ionising radiations can break chemical bonds and radically alter the chemical structures of biological molecules. The principal reaction of amino acids in aqueous media is deamination with the release of water. Free-radicals, produced on radiolysis of water, can react with macromolecules (proteins, enzymes, membranes and genetic material) and damage them. One of the primary biological/biochemical lesions is the alteration of intercellular relationships between enzymes and substrates.

Presence of oxygen enhances the damage. Almost all proteins are denatured by ionisation which results in the loss of enzyme activity. The dose-response curve indicates that one reaction by a radical is sufficient to produce enzyme inactivation. Nucleic acids are the most radiosensitive molecules in the cell. Nucleic acid bases are readily reduced by electrons and react rapidly with hydroxyl radicals to produce addition products.

Effects at the Cellular Level. Radiation damage to the genetic material can be hereditary or somatic. Hereditary effects of radiation are mainly due to radiation-induced mutations in the ova and sperm cells. Ionising radiations increase the normal process of decay. Radiation can directly damage chromosome, considered to be the root-cause of cell damage.

A mutated cell may transmit flowed genetic message *via* its DNA to other cell generations. Stochastic effects cause potential harm even at low or all dose levels. If enough cells/tissues in an organ are damaged or prevented from normal functions, there will be loss of organ functions which is deterministic. A deterministic response is observed when the dose reaches a certain threshold.

Radiation may cause cell division not to occur properly or not at the proper time. In complex organisms certain cells never divide (central nervous system), some divide occasionally (kidney, liver etc.) and other cells divide continuously (bone marrow cells). Therefore, acute damage is most likely to occur in those cells which are undergoing rapid cell division (e.g., lymphoid cells). Consequently, infants are more susceptible to radiation damage than adults.

Irradiation may not only increase the frequency of deleterious somatic mutations, but may also be one of the causes of ageing as a result of slow accumulation of somatic mutations, altered biological functions and break-down of repair activities.

Radiation damage of cell membranes with abnormal changes in their permeabilities leads to cell death. Lesions in cell membranes lead to spatial disorganisation of enzymes, nucleus and cytoplasm, proteolytic and nucleic acid-attacking enzymes can then destroy the macromolecules (by phagocytosis). This enzyme-release hypothesis can explain many cellular phenomena of radiation damage. Higher doses of radiation lead to cell liquefaction and death. Shortening of life-span, cancer induction and damage to the embryo are the long-term effects of radiation damage.

High doses to the gastro-intestinal tract lead to damage of submucosal, vascular and connective tissues. In skin, the effects are dermal necrosis, loss of hair and dysfunctions of sweat glands. The spinal chord is highly radiosensitive. Radiation damage leads to demyelination and loss of glial cells. Opaqueness and cataract in the eye lens are the result of radiation (>2 Gy).

Radiation Carcinogenesis. Radiation carcinogenesis is a stochastic process (no dose threshold). The development of cancer appears to be dependent on many parameters. In general terms, carcinogenesis takes place in two steps—initiation (irreversible) and promotion. Ionising radiation acts as initiator as well as promoter. Estimation of the risk can be expressed in terms of a linear-quadratic function.

$$I_d = I_n + \alpha_1 D + \alpha_2 D^2 \qquad ...(18.14)$$

where I_d is the observed incidence of cancer for dose, D; I_n is the normal incidence of cancer without the radiation dose and α_1 and α_2 are linear and quadratic risk coefficients, respectively.

Somatic effects (carcinogenesis) occur mainly as a result of acute (short-time span) doses from atomic weapons or therapy.

Effects at the Organism Level. All forms of ionising radiation are harmful above certain minimum dose levels. There is no clear-cut criterion for threshold dose levels. Exposure to ionising radiations cannot be completely avoided as human beings are exposed to these from natural as well as man-made sources of radiation. Compared to other risks (smoking, chemical and biological pollutants), the biological effects of low-level ionising radiations are slight.

The most important, ubiquitous and harmful natural radiation sources are the decay products of radon, ^{222}Rn . ^{222}Rn is the gaseous daughter product of the decay of radium, ^{226}Ra ($^{226}_{88}Ra \longrightarrow ^{222}_{86}Rn + \alpha$), and ^{222}Rn once released into the atmosphere due to uranium, building, construction etc. pervades and being dense settles in enclosed spaces of buildings. Accumulation of radon also depends on the type of construction, the type of ventilation and the geographical location. In the temperate and arctic regions, buildings are well insulated thermally and with minimal ventilation; the populace in these regions is more prone to exposure to radon gas.

Randon radiation damage is due to the decay products of radon and α-rays (not due to radon gas directly). On inhalation of radon gas, the decay products are absorbed as aerosol particles by the bronchial tissues, leading to lung tissue damage and lung cancer.

Other naturally occurring nuclides of importance to humans are ^{42}K, ^{90}Sr, and ^{131}I being gaseous can spread over a large area and is absorbed by living organisms through the food chain. It causes minimal damage due to its very short half-life. Of all machine-produced sources, medical X-rays, γ-rays in nuclear medicine and nuclear reactors constitute by far the greatest health hazards.

Dose rates (accumulated doses) are of crucial importance in understanding the biological effects of radiation. Accumulated low doses can start off-chain reactions leading to genetic damage and cancer. As a rule of thumb, a dose of 0.25 Gy is considered the limiting dose. Doses in the range below 1 Gy affect sensitive organs and tissues, such as reproductive organs and bone marrow tissues, and cause sterility and lower production of lymphocytes. Radiation doses of ≈ 1 Gy damage eyes leading to blindness, cause deficiencies in the immune system and lead to anaemia due to non-formation of new erythrocytes by the bone marrow. The central nervous system and some tissues such as cartilage are fairly radioresistant to small or medium doses of ionising radiations. Doses in the range of 3–5 Gy lead to death within months due to bone marrow damage, haemorrhages, anaemia, infection and malnutrition. Doses of 10–40 Gy lead to death within weeks due to severe damage to the gastro-intestinal and other organs. Very high doses, >100 Gy, cause death within hours or days, due to damage to the central nervous system.

18.6 RADIATION PROTECTION AND THERAPY

Protection from, or neutralisation of, ordinary effects of radiation is not possible in biological systems, but protection from secondary effects at low and moderate doses of radiation is feasible.

Hormesis. Hormesis is positive stimulation by low doses of any poltentially harmful agent. The hormesis hypothesis states that at sub-harmful doses (below zero equivalent point, **ZEP,** which is the maximum quantity of an agent that can be administered with no discernible effect), ionising radiation is beneficial to living organisms instead of being hazardous; and low levels of ionising radiation may in fact be essential to well being of life by influencing increased life span, disease resistance and resistance to larger doses of radiation. The effect of any radiation at harmful doses is diametrically different from those of hormesis doses. The basis for this assumption is radiation stimulation. Examples are photo-reactivation of radiation-induced sporulation in bacteria, accelerated growth and reproduction in moulds and drought resistance in seeds etc.

18.6.1 Radiation Protection

As radiolysis of water produces free-radicals, addition of substances that compete for these free-radicals (free-radical scavengers) could minimise the random effect. Some of the molecules rendering chemical protection against radiation are cysteine and molecules containing free sulphydryl (-SH) groups and chelating agents such as EDTA.

Charge delocalisation reduces the number of radicals produced at the target, by diffusing them to wider areas. Therefore, aromatic agents are better protectors in charge-transfer mechanism. Macromolecules such as proteins can be protected by competitive removal of radicals or repair through a hydrogen-transfer reaction.

$$PH + OH^* \longrightarrow P^* + H_2O$$

$$P^* + XH \longrightarrow PH + X^* \qquad ...(18.15)$$

where P is a polymer, P^* is the polymer radical and XH is the protective agent. Colloidal sulphur is a protecting agent for enzymes.

Protecting of sensitive organs and sites from irradiation is another method of protection. As longer-wavelength X-rays are absorbed by the skin surface leading to skin damage, filtering of such radiations can be achieved by shields (Al, Cu etc.).

18.6.2 Radiation Therapy

In spite of the hazardous nature of ionising radiations, living organisms have to live with them and they are very much a part of nuclear medicine for diagnostic as well therapeutic treatments. X-rays are routinely used in Roentgenography and CAT-scan radiography. Radioactive isotopes (radionuclides) are employed as tracers in many biological processes. Labelled-radionuclides can easily be monitored due to their radioactivity and high sensitivity. Various functions of organs and their abnormalities can be monitored by such methods. Some of these are: diagnosis of the function of the thyroid gland by radioiodine; the function of kidneys by renography; and the uptake, distribution and elimination process of metabolites and drugs by pharmacokinetic methods.

Labelled proteins are used in clinical investigations—in the metabolic process of plasma proteins and the gastro-intestinal tract. By such investigations any abnormalities due to the effects of radiation or chemicals can be evaluated.

Rapidly dividing cells with undifferentiated structure (tumours) are more sensitive to radiation than strongly differentiated cells undergoing slow division. Use of ionising radiation in cancer therapy is based on the sensitivity of different types of cells to these radiations. Cancer cells react like ordinary cells in their response to oxygen. Whether the presence of oxygen should enhance the destruction of tumours is very difficult to evaluate due to various complex reactions. The role of reoxygenation in radiotherapy of human tumours is not properly explained.

Present day radiation therapy relies on the administration of radiation dose not in a single shot, but in fractions separated over a time period. This procedure is to allow the recovery of normal cells from the effects of radiation. The cell repair and repopulation factors are taken into account in models to explain the effects of radiation.

The effects of irradiation by the dose-fractionation procedure can also be expressed in terms of a linear-quadratic model

$$\log \mathbf{S} = -\,\mathbf{n}\,(\alpha D + \beta D^2) \qquad \text{...(18.16)}$$

where **S** is the survival fraction of the species and **n** is the number of fractions of the total dose. The ratio of single-hit and multi-hit coefficients, α/β, is a useful measure of dose specification in terms of the flexure dose (the minimum dose per fraction below which protection of normal tissues is not realised). The linear-quadratic dose-fractionation model is supposed to give a better understanding of the clinical radiotherapy so that effective tumour control can be realised with minimal damage to normal cells.

Suggested Reading

Alpen, E.L. *"Radiation Biophysics"*. Prentice Hall/Engelwood, New Jersey. (1990).

Bacq. Z.H. & Alexander, P. S. *"Fundamentals of radiobiology"*. Pergamon Press/New York. (1966).

Greening, J.R. "*Fundamentals of Radiation Dosimetry*". Adam Hilger/Bristol, UK. (1981).

Hughes, G. "*Radiation Chemistry*" Clarendon Press/Oxford (1973).

Parker, R.P., Smith, P.N.S. & Tayler, D.M. *"Basic Science of Nuclear Medicine"*. Churchill Livingston/London. (1978).

Ronto, G. & Tarjan, I. (Eds.) "*An Introduction to Biophysics with Medical Orientation*". Akademiai Kiado/Budapest (1991).

Sprawls, P. "*Physical Principles in Nuclear Medicine*". Med. Phys. Pubs/Madison, WI. (1993).

_________ "*Radiation Doses, Effects and Risks*" United Nations Environmental Programme. NEW UNEP Publication (1985).

A Appendix

APPENDIX A: FORMULAS AND FORMALISMS

A1 ALGEBRA AND MATRICES

A1.1 Linear Algebra

A general polynomial equation is of the form

$$f(x) = a_0 + a_1 x + ... + a_n x^n \quad ...(A1.1a)$$

where a_0, a_1 ... a_n are coefficients and x is the unknown variable.

(i) A linear equation:

$$ax + b = 0 (a \neq 0) \quad ...(A1.1b)$$

has solution $x = -b/a$.

(ii) A **quadratic equation:**

$$ax^2 + bx + c = 0 \quad ...(A1.1c)$$

has solution x_1 and x_2

$$x_{1,2} = -\frac{b}{2a} \pm \sqrt{\left(\frac{b^2 - 4ac}{2a}\right)} \quad ...(A1.1d)$$

x_1 and x_2 are real and different if $b^2 - 4ac > 0$

real and equal if $b^2 - 4ac = 0$

complex conjugate if $b^2 - 4ac < 0$

A1.2 Matrices

Matrices are arrays of numbers with distinct rules of addition and multiplication. A matrix A_{mn} has m rows and n columns.

$$A_{mn} = \begin{vmatrix} a_{11} & a_{12} & \cdots & a_{1n} \\ a_{21} & a_{22} & \cdots & a_{2n} \\ a_{31} & a_{32} & \cdots & a_{3n} \\ . & & & \\ . & & & \\ . & & & \\ a_{m1} & a_{m2} & \cdots & a_{mn} \end{vmatrix} \qquad \text{...(A1.2a)}$$

Whereas a scalar is a zero rank tensor and a vector a 1st rank tensor, a matrix is a 2nd rank (scalar) tensor.

$$A + B = B + A \qquad \text{Additive rule}$$

$$\left.\begin{aligned} AB = C = C_{ij} &= \sum_{K=1}^{n} a_{ik}b_{kj} \\ AB(C) &= A(BC) \end{aligned}\right\} \text{Multiplication rule} \qquad \text{...(A1.2b)}$$

$$AB \neq BA \text{ (usually noncommutative).}$$

Rows, of the matrix, represent vectors in space. For example,

$$\begin{aligned} a_{1i} + b_{1j} + c_{1k} &= V_1 \\ a_{2i} + b_{2j} + c_{2k} &= V_2 \\ a_{3i} + b_{3j} + c_{3k} &= V_3 \end{aligned} \qquad \text{...(A1.2c)}$$

where i, j, k are unit vectors and V_1, V_2 and V_3 are vectors represented by

$$\begin{vmatrix} a_1 & b_1 & c_1 \\ a_2 & b_2 & c_2 \\ a_3 & b_3 & c_3 \end{vmatrix} \begin{vmatrix} i \\ j \\ k \end{vmatrix} \qquad \text{...(A1.2d)}$$

Linear equations

$$\begin{aligned} a_{11}x_1 + a_2x_2 + \ldots a_{1n}x_n &= \alpha_1 \\ a_{21}x_2 + a_{22}x_2 + \ldots a_{2n}x_n &= \alpha_2 \\ a_{31}x_1 + a_{32}x_2 + \ldots a_{3n}x_n &= \alpha_3 \end{aligned} \qquad \text{...(A1.2e)}$$

$$\begin{matrix} \cdots & \cdots & \cdots \\ \cdots & \cdots & \cdots \end{matrix}$$

$$a_{n1}\,x_1 + a_{n2}\,x_2 + \ \ldots\ a_{nn}\,x_n = \alpha_n$$

is represented in the form (matrix form),

$$\begin{vmatrix} a_{11} & a_{12} & a_{13} & \cdots \\ a_{21} & a_{22} & a_{23} & \cdots \\ a_{31} & a_{32} & a_{33} & \cdots \\ \cdot & \cdot & \cdot & \cdots \\ \cdot & \cdot & \cdot & \cdots \\ \cdot & \cdot & \cdot & \cdots \\ a_{n1} & a_{n2} & a_{n3} & \cdots \end{vmatrix} \begin{vmatrix} x_1 \\ x_2 \\ x_3 \\ \cdot \\ \cdot \\ \cdot \\ x_n \end{vmatrix} = \begin{vmatrix} \alpha_1 \\ \alpha_2 \\ \alpha_3 \\ \cdot \\ \cdot \\ \cdot \\ \alpha_n \end{vmatrix} \qquad \text{...(A1.2f)}$$

That is, $$AX = \alpha \qquad \text{...(A1.2g)}$$

If $\alpha = 0$, the equations are homogeneous. Solution of non-homogeneous linear equations can be solved by the echelon method of matrix multiplication.

For any *(nxn)* square matrix, the 'spur' or 'trace' (*tr* (*A*)) is the sum of the main diagonal elements

$$tr(A) = \sum_{i=1}^{n} a_{ii} \qquad \text{...(A1.2h)}$$

A symmetric matrix is that in which $a_{ij} = a_{ji}$ for all pairs of (*ij*).

$$\begin{pmatrix} a_{11} & a_{12} & a_{13} \\ a_{12} & a_{22} & a_{23} \\ a_{13} & a_{32} & a_{33} \end{pmatrix} \quad \text{Symmetric matrix} \qquad \text{...(A1.2i)}$$

Antisymmetric matrix or skew-symmetric matrix is one for which

$$a_{ij} = -\,a_{ji} \qquad \text{...(A1.2j)}$$

Symmetric and skew-symmetric matrices must be square matrices, and diagonal of a skew-symmetric matrix must be zero.

A transpose of a matrix A^t is achieved by interchanging rows and columns.

That is,

If $$A = [a_{ij}],\ \text{then } A^t = [a_{ji}] \qquad \text{...(A1.2k)}$$

For a symmetric matrix $a_{ij} = a_{ji}$.

$$\therefore \qquad A = A^t \qquad \text{...(A1.2}l\text{)}$$

For a skew-symmetric matrix $a_{ij} = -a_{ji}$.

$$\therefore \quad A^t = -A \qquad \text{...(A1.2m)}$$

$$(A + B)^t = A^t + B^t$$

$$(AB)^t = B^t A^t$$

$$(A - A^t)^t = -(A - A^t)$$

If A and B are symmetric matrices, then the product $[AB]$ is symmetric if and only if

$$[AB] = [BA] \qquad \text{...(A1.2n)}$$

The inverse of a matrix, A^{-1}, is one that satisfies the condition

$$AA^{-1} = A^{-1}A = 1 \qquad \text{...(A1.2o)}$$

A diagonal matrix is a square matrix with non-zero elements only on the main diagonal elements

$$\begin{pmatrix} 2 & 0 & 0 \\ 0 & 7 & 0 \\ 0 & 0 & 4 \end{pmatrix} \quad \text{A diagonal matrix} \qquad \text{(A1.2p)}$$

A unit matrix (identity matrix) is a square matrix with 1 on the main diagonal elements $(a_{ii} = 1)$ and 0 on all other cross-elements $(a_{ij} = 0)$.

$$\text{Unit matrix } I_n = \begin{pmatrix} 1 & 0 & 0 & 0 \\ 0 & 1 & 0 & 0 \\ 0 & 0 & 1 & 0 \\ 0 & 0 & 0 & 1 \end{pmatrix} \qquad \text{...(A1.2q)}$$

Eigenvector: A vector X is an eigenvector of matrix A, if

$$AX = \lambda X \qquad \text{...(A1.2r)}$$

$$(A - \lambda) X = 0$$

which represents a set of homogeneous linear equations, where λ is a scalar, called eigenvalue.

$$\begin{vmatrix} a_{11} - \lambda & a_{22} - \lambda & \dots & a_{1n} - \lambda \\ a_{21} - \lambda & a_{22} - \lambda & \dots & a_{2n} - \lambda \\ \dots & \dots & \dots & \dots \\ \dots & \dots & \dots & \dots \\ \dots & \dots & \dots & \dots \\ a_{n1} - \lambda & a_{n2} - \lambda & \dots & a_{nn} - \lambda \end{vmatrix} \begin{vmatrix} x_1 \\ x_2 \\ . \\ . \\ . \\ x_n \end{vmatrix} = 0 \qquad \text{...(A1.2s)}$$

A1.3 Vector Algebra

A scalar is a zero-rank tensor and has a quantitative value without any directional conditions— e.g. heat, speed etc. A vector, on the other hand, which is a 1st-rank tensor, has a quantitative value as well as spatial orientation— e.g. velocity, momentum, acceleration etc. Scalars are represented by normal letters, a, s etc., while vectors are represented either by bold letters, **v**, **a**, **p** etc., or by $\vec{v}, \vec{a}, \vec{p}$ etc.

In an orthogonal coordinate system (**Cartesian system**), a vector $\vec{a}$ can be represented by its components a_x, a_y and a_z together with the corresponding unit vectors *i. j, k* (= 1, directional cosines) as (Fig. A1)

$$\vec{a} = ia_x + ja_y + ka_z \tag{A1.3a}$$

$$\vec{a} = \begin{pmatrix} a_x \\ a_y \\ a_z \end{pmatrix}\begin{pmatrix} i \\ j \\ k \end{pmatrix} \quad \text{...(A1.3b)}$$

A sum of a set of vectors is a vector

$$\vec{a} + \vec{b} = \vec{c}$$

with components

$$c_x = a_x + b_x;\ c_y = a_x + b_y \text{ and } c_z = a_z + b_z \quad \text{...(A1.3c)}$$

Product of a vector with a scalar, λ, is a vector

$$\vec{a} + \lambda\vec{b} \quad \text{...(A1.3d)}$$

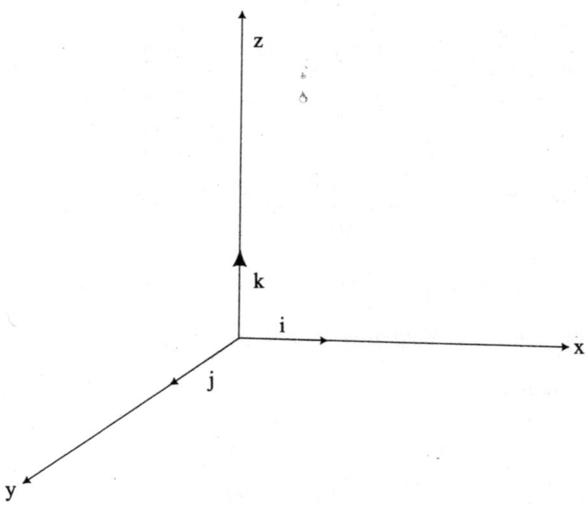

Fig. A1 Orthogonal (Cartesian) coordinate system.

The scalar product (dot product) of two vectors, $\vec{a}$ and $\vec{b}$ is a scalar. It is the product of the projection of the vector $\vec{a}$ on the vector $\vec{b}$.

$$\vec{a}\,.\,\vec{b} = ab \cos \phi \quad \text{...(A1.3e)}$$

where ϕ is the angle between the vectors.

$$ab = a_x b_x + a_y b_y + a_z b_Z \quad \text{...(A1.3f)}$$

For orthogonal vectors, the scalar product is zero.

$$\vec{a}\,.\,\vec{b} = ab \cos 90^\circ = 0 \quad \text{...(A1.3g)}$$

The scalar product can be thought of as a one-dimensional matrix.

The tensor product (cross-product) of two vectors, $\vec{a}$ and $\vec{b}$ is a vector $\vec{c}$ perpendicular to the plane of *ab*. Vectors $\vec{a}$, $\vec{b}$, $\vec{c}$ form a right-handed system.

$$\vec{c} = \vec{a} \times \vec{b} = -\vec{b} \times \vec{a} \quad \text{...(A1.3h)}$$

$$|c| = |a \times b| = ab \sin \phi$$

where ϕ is the angle between the vectors $\vec{a}$ and $\vec{b}$.

The tensor product is a square array of numbers — a square matrix with components

$$\begin{aligned} c_x &= (a \times b)_x = a_y b_z - a_z b_y \\ c_y &= (a \times b)_y = a_z b_x - a_x b_z \\ C_z &= (a \times b)_z = a_x b_y - a_y b_x \end{aligned} \quad \text{...(A1.3i)}$$

The gradient of a scalar quantity, $\psi (x, y, z)$ is a vector.

$$\nabla \psi = i \frac{\partial}{\partial x} \psi + j \frac{\partial}{\partial y} \psi + k \frac{\partial}{\partial z} \psi \quad \text{...(A1.3j)}$$

where operator, ∇ is

$$\nabla = i \frac{\partial}{\partial x} + j \frac{\partial}{\partial y} + k \frac{\partial}{\partial z} \quad \text{...(A1.3k)}$$

$$\left.\begin{aligned} \nabla (\phi + \psi) &= \nabla \phi + \nabla \psi \\ \nabla (\phi + \psi) &= \phi \nabla \psi + \psi \nabla \phi \end{aligned}\right\} \quad \text{...(A1.3l)}$$

Divergence of a vector is a scalar.

$$\left.\begin{aligned} d_{iv}\, \vec{a} &= \frac{\partial}{\partial x} a_x + \frac{\partial}{\partial y} a_y + \frac{\partial}{\partial z} a_z \\ d_{iv}\, \vec{a} &= \nabla a \end{aligned}\right\} \quad \text{...(A1.3m)}$$

A2 TRIGONOMETRY/GEOMETRY

For a triangle (orthogonal), with base, *b,* height, *h,* and hypotenuse, *t,* the trigonometric functions (Fig. A2) are:

$$\sin\theta = h/t; \qquad \cos\theta = b/t$$

$$\tan\theta = h/b = \sin\theta/\cos\theta = 1/\cot\theta$$

$$\cot\theta = b/h = \cos\theta/\sin\theta = 1/\tan\theta \qquad \text{...(A2.1a)}$$

$$\sin^2\theta + \cos^2\theta = 1;$$

$$1/\sin^2\theta = 1 + \cot^2\theta; \; 1/\cos^2\theta = 1 + \tan^2\theta$$

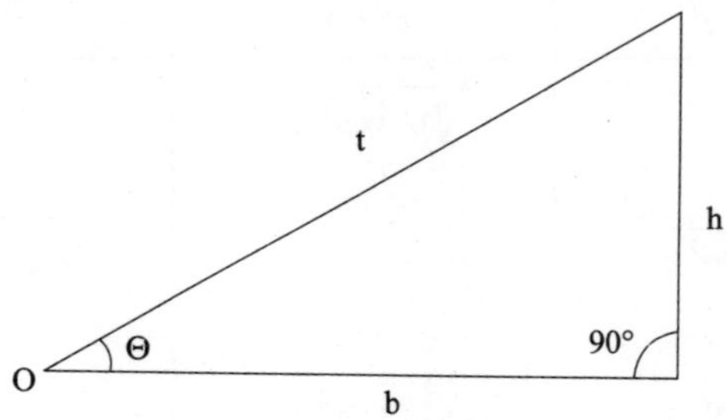

Fig. A2 Orthogonal triangle for circular trigonometric functions.

$$\tan(-\theta) = -\tan\theta; \; \cot(-\theta) = -\cot\theta$$

$$\sin(\alpha \pm \beta) = \sin\alpha\cos\beta \pm \cos\alpha\sin\beta \qquad \text{...(A.2.1b)}$$

$$\cos(\alpha \pm \beta) = \cos\alpha\cos\beta \mp \sin\alpha\sin\beta$$

$$\tan(\alpha \pm \beta) = \frac{\tan\alpha \pm \tan\beta}{1 \mp \tan\alpha\tan\beta}$$

$$\cot(\alpha \pm \beta) = \frac{\cot\alpha\cot\beta \mp 1}{\cot\beta \pm \cot\alpha}$$

$$\sin\alpha + \sin\beta = 2\sin((\alpha+\beta)/2)\cos((\alpha-\beta)/2)$$

$$\sin\alpha - \sin\beta = 2\cos((\alpha+\beta)/2)\sin((\alpha-\beta)/2) \qquad \text{...(A2.2)}$$

$$\cos\alpha + \cos\beta = 2\cos((\alpha+\beta)/2)\cos((\alpha-\beta)/2)$$

$$\cos\alpha - \cos\beta = -2\sin((\alpha+\beta)/2)\sin((\alpha-\beta)/2)$$

These trigonometric functions are circular functions because they parametrise the circle $x^2 + y^2 = 1$. Their values in the four Cartesian quadrants are as in Table A1.

Table A1 Circular trigonometric functions in four quadrants.

Angle in deg.	*sin θ*	*cos θ*	*tan θ*	*cot θ*
$\pm\theta$	$\pm\sin\theta$	$+\cos\theta$	$\pm\tan\theta$	$\pm\cot\theta$
$(90\pm\theta)$	$+\cos\theta$	$\mp\sin\theta$	$\mp\cot\theta$	$\mp\tan\theta$
$(180\pm\theta)$	$\mp\sin\theta$	$-\cos\theta$	$\pm\tan\theta$	$\pm\cot\theta$
$(270\pm\theta)$	$-\cos\theta$	$\pm\sin\theta$	$\mp\cot\theta$	$\mp\tan\theta$

Interconversion of circular trigonometric functions are as in Table A2.

Table A2 Interconversion of circular trigonometric functions.

Function	*sin θ*	*cos θ*	*tan θ*	*cot θ*
sin θ	–	$\sqrt{1-\cos^2\theta}$	$\frac{\tan\theta}{\sqrt{1+\tan^2\theta}}$	$\frac{1}{\sqrt{1+\cot^2\theta}}$
cos θ	$\sqrt{1-\sin^2\theta}$	–	$\frac{1}{\sqrt{1-\tan^2\theta}}$	$\frac{\cot\theta}{\sqrt{1+\cot^2\theta}}$
tan θ	$\frac{\sin\theta}{\sqrt{1-\sin^2\theta}}$	$\frac{\sqrt{1-\cos^2\theta}}{\cos\theta}$	–	$\frac{1}{\cot\theta}$
cot θ	$\frac{\sqrt{1-\sin^2\theta}}{\sin\theta}$	$\frac{\cos\theta}{\sqrt{1-\cos^2\theta}}$	$\frac{1}{\tan\theta}$	–

Trigonometric functions of some special angles are given in Table A3.

Table A3 Trigonometric functions of some special angles.

Angle *in deg*	*in radians*	*sin θ*	*cos θ*	*tan θ*	*cot θ*
0 or 360	0 or 2 π	0	1	0	± ∞
30	π/6	$\frac{1}{2}$	$\frac{1}{2}\sqrt{3}$	$\frac{1}{3}\sqrt{3}$	$\sqrt{3}$
45	π/4	$\frac{1}{2}\sqrt{2}$	$\frac{1}{2}\sqrt{2}$	1	1
60	π/3	$\frac{1}{2}\sqrt{3}$	$\frac{1}{2}$	$\sqrt{3}$	$\frac{1}{3}\sqrt{3}$
90	π/2	1	0	∞	0
180	π	0	– 1	0	± ∞
270	3π/2	–1	0	± ∞	0

In any triangle of sides *a, b, c* and angles α ($\hat{bc}$), β ($\hat{ca}$) and γ ($\hat{ab}$) the following relationships (Fig. A3) apply:

(i) Law of sines:

$$\frac{a}{b} = \frac{\sin\alpha}{\sin\beta}; \frac{b}{c} = \frac{\sin\beta}{\sin\gamma}; \frac{c}{a} = \frac{\sin\gamma}{\sin\alpha} \quad \text{...(A2.3a)}$$

(ii) Law of cosines:

$$\begin{aligned} a^2 &= b^2 + c^2 - 2bc\cos\alpha \\ b^2 &= c^2 + a^2 - 2ca\cos\beta \\ c^2 &= a^2 + b^2 - 2ab\cos\gamma \end{aligned} \quad \text{...(A2.3b)}$$

(iii) Sides of the triangle:

$$\begin{aligned} a &= b\cos\gamma + c\cos\beta \\ b &= c\cos\alpha + a\cos\gamma \\ c &= a\cos\beta + b\cos\alpha \end{aligned} \quad \text{...(A2.3c)}$$

(iv) Area of the triangle, Δ:

$$\Delta = \frac{1}{2}ab\sin\gamma = \frac{1}{2}bc\sin\alpha = \frac{1}{2}ca\sin\beta.$$

$$\Delta = \sqrt{s(s-a)(s-b)(s-c)} \quad \text{...(A2.3d)}$$

where $s = (a + b + c)/2$

(v) Radius R of the circle encompassing a triangle:

$$R = \frac{a}{2\sin\alpha} = \frac{b}{2\sin\beta} = \frac{c}{2\sin\gamma} \quad \text{...(A2.3e)}$$

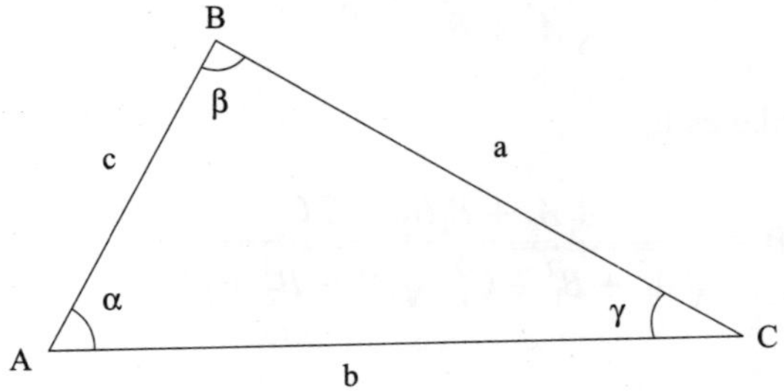

Fig. A3 Triangle with sides *a, b* and *c* and angles α, β and γ.

A2.2 GEOMETRY

A2.2.1 Straight Line

A straight line is an intersection of two planes.

The general form of a straight line (in Cartesian coordinates) is

$$Ax + By + Cz + D = 0$$

$$A(x - x_0) + B(y - y_0) + C(z - z_0) = 0 \qquad \text{...(A2.4a)}$$

Equation of a straight line with slope, *m*, is

$$y = mx + c \qquad \text{...(A2.4b)}$$

Distance, **d**, between two points (x_1, y_1, z_1) and (x_2, y_2, z_2) is

$$\mathbf{d} = \sqrt{(x_1 - x_2)^2 + (y_1 - y_2)^2 + (z_1 - z_2)^2} \qquad \text{...(A2.4c)}$$

A line through two points (x_1, y_1, z_1) and (x_2, y_2, z_2) is

$$\frac{x - x_1}{x_2 - x_1} = \frac{y - y_1}{y_2 - y_1} = \frac{z - z_1}{z_2 - z_1} \qquad \text{...(A2.4d)}$$

A2.2.2 Plane

A plane through three points (x_1, y_1, z_1), (x_2, y_2, z_2) and (x_3, y_3, z_3) is

$$\begin{vmatrix} x & y & z \\ x_1 & y_1 & z_1 \\ x_2 & y_2 & z_2 \\ x_3 & y_3 & z_3 \end{vmatrix} = 0 \qquad \text{...(A2.5a)}$$

Distance, d_i, of a point, P_i (x_i, y_i, z_i) from the plane $Ax + By + Cz + D = 0$ is

$$d_i = \frac{\mathrm{A}x_i + By_i \_ Cz_i + D}{\sqrt{A^2 + B^2 + C^2}} \qquad \text{...(A2.5b)}$$

Angle, θ, between two planes is

$$\cos\theta = \frac{A_1A_2 + B_1B_2 + C_1C_2}{\sqrt{A_1^2 + B_1^2 + C_1^2}\,.\sqrt{A_2^2 + B_2^2 + C_2^2}} \qquad \text{...(A2.5c)}$$

A2.2.3 Conics

Circle, ellipse, parabola and hyperbola are the conic sections (conics), formed by the intersections of a plane and a right circular cone. While circle and ellipse are closed curves, parabola and hyperbola are open curves.

i) Circle:

A circle is a special case of conics where the intersection of the plane to the axis of the cone is orthogonal (90°).

General equation of a circle:

$$x^2 + y^2 + 2Ax + 2By + C = 0 \quad \text{...(A2.6a)}$$

$$(x - x_1)^2 + (y - y_1)^2 = R^2$$

where (x_1, y_1) are the coordinates of the origin and R is the radius of the circle.

$$\left.\begin{aligned} \text{Circumference } S &= 2\pi R \\ \text{Area} &= \pi R^2 \end{aligned}\right\} \text{ of a circle} \quad \text{...(A2.6b}$$

$$\left.\begin{aligned} \text{Area } A &= 4\pi \mathrm{R}^2 \\ \text{Volume } V &= \frac{4}{3}\pi \mathrm{R}^3 \end{aligned}\right\} \text{ of a sphere} \quad \text{...(A2.6c)}$$

General conics: ellipse, parabola and hyperbola.

A conic section is the locus of a moving point, P, such that its distance from a fixed point (**focus**, S) is in constant ratio to its distance from a fixed straight line (**directrix**) (Fig. A4), and the constant ration is called the **eccentricity, e.**

If **e < 1;**	the conic is an ellipse
If **e = 1;**	the conic is a parabola
If **e > 1;**	the conic is hyperbola.

ii) Ellipse:

The distances from the focus and the directrix are < 1 (e < 1) in an ellipse.

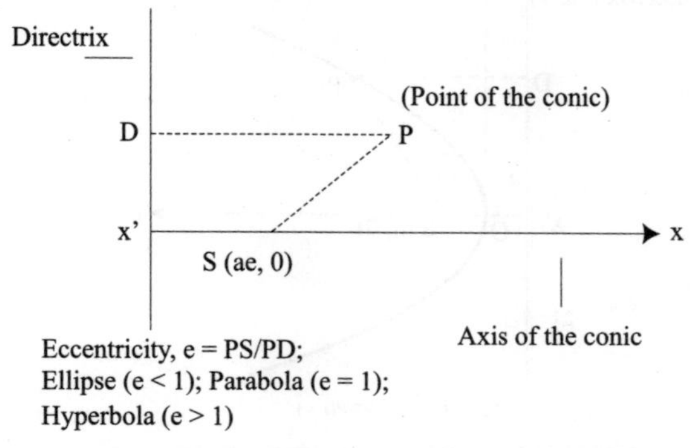

Fig. A4 Geometrical feature of conic sections.

Equations of ellipse:

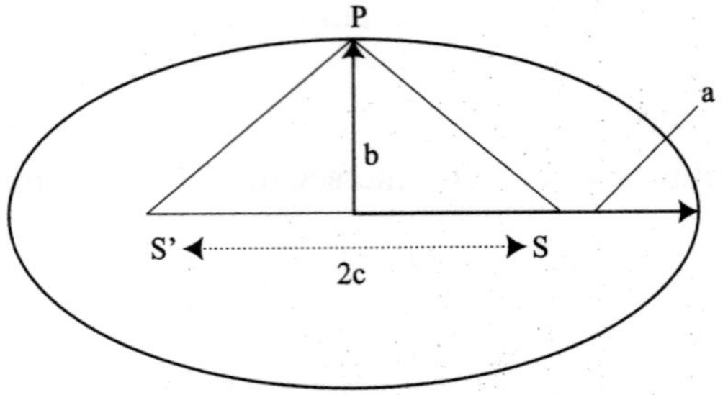

2a & 2b = Major and minor axes

Fig. A5 Ellipse.

$$\frac{x^2}{a^2} + \frac{y^2}{b^2} = 1 \quad \text{...(A2.7a)}$$

$$c^2 = a^2 - b^2$$

$$\text{Area } F = \pi\, ab$$

$$\text{Volume} = \frac{4}{3}\pi\, abc \quad \text{...(A2.7b)}$$

where $2a$ and $2b$ are the major and minor axes (Fig. A5).

iii) Parabola

Distances from the focus and the directrix are equal in a parabola (Fig. A6). Equation of a parabola is:

$$y^2 = 4ax \quad \text{...(A2.8)}$$

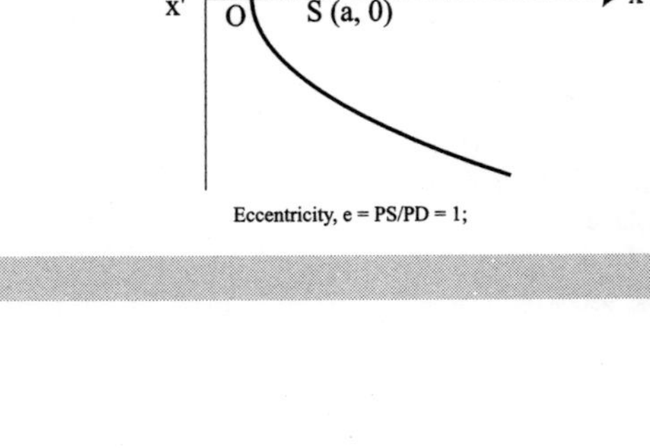

Fig. A6 Parabola.

iv) Hyperbola

A hyperbola is the locus of a point, P, where the eccentricity, e (distance from the focus, S/distance from the directrix) is > 1 (Fig. A7).

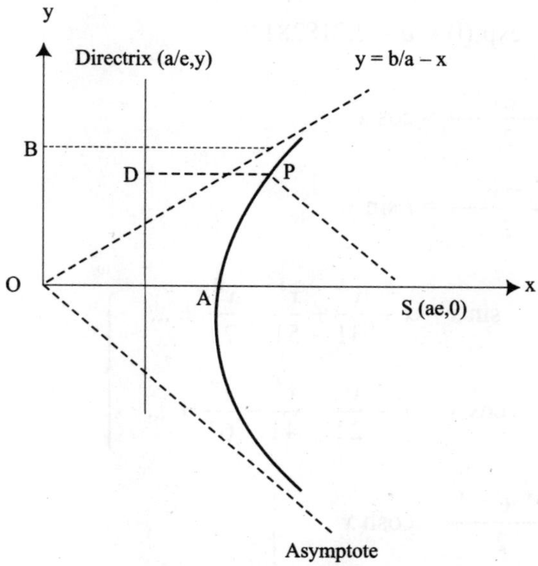

Eccentricity, e = PS/PD > 1; OA = a; OB = b

Fig. A7 Hyperbola.

$$\frac{x^2}{a^2} - \frac{y^2}{b^2} = 1 \qquad \text{...(A2.9)}$$

$$\text{Foci } c^2 = a^2 + b^2$$

A3 SPECIAL FUNCTIONS

A3.1 Exponential Functions/series

exp *(x)* is a transcendental function –

$$\exp(x) = e^x = 1 + \frac{x}{1!} + \frac{x^2}{2!} + \frac{x^3}{3!} + \ldots$$

$$= \lim_{n \to \infty}\left(1 + \frac{x}{n}\right)^n \qquad \text{...(A3. 1a)}$$

$$\exp(x) \cdot \exp(y) = \exp(x + y) \qquad \text{...(A3. 1b)}$$

$$\left.\begin{aligned} \exp(\pi i) &= -1; \\ \exp(2\pi i) &= i; \\ \exp(l) &= e = 2.718281 \,.. \end{aligned}\right\} \qquad \text{...(A3.1c)}$$

$$\begin{aligned} \frac{e^{ix} + e^{-ix}}{2} &= \cos x \\ \frac{e^{ix} - e^{-ix}}{2} &= i \sin x \end{aligned} \qquad \text{...(A3.1d)}$$

$$\left.\begin{aligned} \sin x &= x - \frac{x^3}{3!} + \frac{x^5}{5!} - \frac{x^7}{7!} + \ldots \\ \cos x &= 1 - \frac{x^2}{2!} + \frac{x^4}{4!} - \frac{x^6}{6!} + \ldots \end{aligned}\right\} \qquad \text{...(A3.1e)}$$

$$\left.\begin{aligned} \frac{e^{x} + e^{-x}}{2} &= \cosh x \\ \frac{e^{x} - e^{-x}}{2} &= \sinh x \end{aligned}\right\} \qquad \text{...(A3.1f)}$$

where **sinh** and **cosh** are hyperbolic functions, because they parametrise one branch of the hyperbola $x^2 - y^2 = 1$.

A3.2 Logarithms

The logarithmic function, $f(x) = \ln(x)$ is defined as the inverse function of the exponential function. That is, it is the exponent **y** to which the base, b, must be raised so that

$$\begin{aligned} x &= b^y \\ y &= \log_b x \end{aligned} \qquad \text{...(A3.2a)}$$

y is called logarithm to the base b

$$x = \ln e^x = e^{\ln x}$$

$$\ln x = \lim_{n \to 0} \left[\frac{x^n - 1}{n} \right] \qquad \text{...(A3.2b)}$$

$$\begin{aligned} \ln_a (x \cdot y) &= \ln_a x + \ln_a y \\ \log_a (x/y) &= \log_a x - \log_a y \\ \log_a x^p &= p \log_a x \end{aligned} \qquad \text{...(A3.2c)}$$

A4 DIFFERENTIATION AND INTEGRATION

A4.1 Differentiation

A function and its derivative are denoted by $f(x)$ and $f'(x)$ and rules of differentiation are given in Table A4.

Table A4 Rules of differentiation.

Function f(x)	*Differentiation* $f'(x) = \frac{df}{dx}$
$C.f(x)$	$C.f'(x)$
$f(x) \pm g(x)$	$f'(x) \pm g'(x)$
$f(x)\,g(x)$	$f(x)\,g'(x) + g(x)f'(x)$
$f(x)/g(x)$	$\frac{\{f'(x)\,g(x) - g'(x)\,f(x)\}}{g(x)^2}$

Some common functions and their derivatives are given in Table A5.

Table A5 Some functions and their derivatives.

Function f(x)	*Derivative* $\frac{df(x)}{dx}$
Const C	0
x^n	$n\ x^{n-1}$
a^x	$a^x \ln a$
e^{cx}	ce^x
$\ln x$	$1/x$
$\sin x$	$\cos x$
$\cos x$	$-\sin x$
$\tan x$	$1/\cos^2 x$
$\cot x$	$1/\sin^2 x$

A4.2 Integration

Rules of integration are as in Table A6.

Table A6 Rules of integration.

Function (x)	*Integration,* $\int f(x)$
$C.f(x)$	$C\int f(x)dx$
$[f(x) \pm g(x)]$	$\int f(x)\,dx \pm \int g(x)\,dx$
$[f(x)\,.\,g'(x)]$	$f(x)\,g(x) - \int f'(x)\,g(x)\,dx$

Some functions and their integrals are given in Table A7.

Tab1e A7 Integrals of some functions.

Function f(x)	*Integral f(x)*
$ax^n (n \neq -1)$	$\frac{a}{n+1} x^{n+1} + c$ (c = constant.)
$1/x$	$\ln \lvert x \rvert + c$
e^x	$e^x + c$
$e^{\lambda x}$	$\frac{1}{\lambda} e^{\lambda x} + c$
a^x	$\frac{a^x}{\ln a} + c$
$\ln x$	$x \ln x - x + c$
$\sin x$	$-\cos x + c$
$\cos x$	$\sin x + c$
$\tan x$	$-\ln \cos x + c$
$\cot x$	$\ln \sin x + c$
$1/\cos^2 x$	$\tan x + c$
$1/\sin^2 x$	$-\cot x + c$

B Appendix

APPENDIX B: STATISTICS

B1 BASIC CONCEPTS

Some of the common terms used in statistics are:

Population, *P*: A group of terms under study.

Sample, *S*: It is a subset of population, *P*.

$$S \subset P$$

Variable, *x*: It is an element of a given set of numbers.

Parameter, *p*: A measure that describes some aspect of population.

Static, *s*: An estimate of population parameter.

B1.1 Range, *R*

It is the difference between the highest and the lowest values of data

$$R = (\text{Highest}) - (\text{Lowest}) \quad \text{...(B1.1)}$$

Mean, median and mode are values which give an indication of population. Measurement of dispersion gives the deviation from the average, which is called standard deviation. A measure of dispersion is called parameter.

B1.2 Median, $\tilde{x}$

The median is the middle value of an ordered array in a set of data. 50% of the population will be higher and the other 50% will be lower than the median value.

B1.3 Mean, μ

It is the arithmetic average of a variable, x (expected value or a measure of central tendency)

$$\mu = \frac{\sum_{i=1}^{N} x_i}{N} \quad \text{...(B1.2a)}$$

Sample mean $$\bar{x} = \frac{\Sigma f_i X_i}{n} \quad \text{...(B1.2b)}$$

where n = number of measurements and f_i = frequency of X_i

Weighted average (Fig. B1)

$$\bar{f}(x)_{[a,\, b]} = \frac{\sum_{i}^{N} K_i \Delta x_i}{\sum_{i}^{N} \Delta x_i} \quad \text{...(B1.3a)}$$

for $$\Delta X_i = b - a$$

$$\bar{f}(x)_{[a,\, b]} = \frac{1}{(b-a)} \int_a^b f(x)\, dx \quad \text{...(B1.3b)}$$

As $\int_a^b f(x)\,dx$ represents the area under the curve, equal to $(b - a)\,Y_m$.

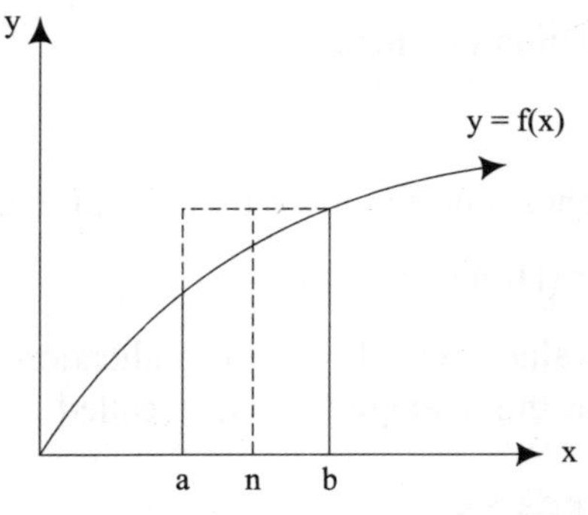

Fig. B1 Weighted average plot.

B1.4 Variance, σ^2

Mean-square deviation is called variance.

$$\left.\begin{aligned} \sigma^2 &= \frac{\sum_{i=1}^{N}(x_i-\mu)^2}{N} \text{ for pupulation} \\ s^2 &= \frac{\Sigma(x_i-\bar{x})^2}{(n-1)} \text{ for sample} \end{aligned}\right\} \quad \text{...(B1.4)}$$

B1.5 Standard Deviation, σ

It is the square root of variation. In computationally simple form

$$\left.\begin{aligned} \sigma &= \sqrt{\frac{\Sigma x^2-(\Sigma x)^2/N}{N}} \text{ for population} \\ &\sqrt{\frac{\Sigma x^2-(\Sigma x)^2/n}{n-1}} \text{ for sample} \end{aligned}\right\} \quad \text{...(B1.5)}$$

where $\Sigma x^2 = \sum_{i=1}^{N} x_i^2;\ (\Sigma X)^2 = \left(\sum_{i=1}^{N} x_i\right)^2$

The most useful measure of variability is standard deviation, σ.

B1.6 The Z-Test

The Z-test represents the number of standard deviations the value of the random variable X deviates from the mean, μ, of the distribution. It is given by

$$Z = \frac{(X-\mu)}{\sigma} \quad \text{...(B1.6a)}$$

X = observed value (variable); μ = arithmetic mean and σ = standard deviation.

As $\sigma\bar{x} = \sigma/\sqrt{N}$

$$Z = \frac{(\bar{x}-\mu)}{\sigma_x} = \frac{(\bar{x}-\mu)}{\left(\dfrac{\sigma}{\sqrt{N}}\right)} \quad \text{...(B1.6b)}$$

B2 GOODNESS OF FIT TESTS

Comparison of a distribution (observed) with some other distribution (expected) is called a "goodness of fit" test.

B2.1 Null Hypothesis, H_o

It is represented by

$$(\mu 1 - \mu 2) = 0 \qquad \text{...(B2.1)}$$

B2.2 χ^2-goodness of Fit

It indicates how far a sample distribution deviates from a theoretical distribution. For this, (i) calculate the expected number (value) from the null hypothesis, H_o, and (ii) subtract the expected number (value) from the observed number (value) and the χ^2-goodness of fit is defined as

$$\chi^2 = \sum_{i=1}^{n} \left\{ \frac{(\text{observed value} - \text{expected value})^2}{\text{expected value}} \right\}$$

$$\chi^2 = \sum_{i=1}^{n} \left\{ \frac{(O_j - E_j)^2}{E_j} \right\} \qquad \text{...(B2.2)}$$

where O_j and E_j are the observed and the expected frequencies of the jth cell. That is, in a goodness of fit test, the sample distribution is compared with the theoretical distribution.

B2.3 Measure of Diversity, H

$$H = -\sum_{i=1}^{k} P_i \log P_i \qquad \textbf{Shannon}\text{ Equation} \qquad \text{...(B2.3a)}$$

where k = number of categories; P_i = proportion of the observations found in category i.

If N is the sample size, then the frequency, f_i, of observation in category j is

$$P_i = f_j / N \qquad \text{...(B2.3b)}$$

$$\therefore \qquad H = \frac{(N \log N - \Sigma f_j \log f_j)}{N} \qquad \text{...(B2.3c)}$$

As $\qquad H_{max} = \log k$

Relative diversity, $\qquad J = \dfrac{H}{H_{\max}} \qquad$ ...(B2.3d)

B2.4 Regression

A relation, R, is a linear relation if

$$R = \{(x, y); y = a + bx\} \qquad \text{...(B2.4a)}$$

The best regression line is the line in which the sum of the squares of the displacement values is minimum. This method is called the "Least Squares"

$$(Y_j - \bar{Y}_j)^2 \equiv \text{error sum squares} \quad \text{...(B2.4b)}$$

$$\sum_{j=1}^{n} (Y_j - \bar{Y}_j)^2 \text{ is minimum (least squares fit).}$$

For
$$y = a + bx$$

$$\left.\begin{aligned} b &= \frac{n(\Sigma xy) - (\Sigma x)(\Sigma y)}{n(\Sigma x^2) - (\Sigma x)^2} \\ a &= \frac{\Sigma y - b(\Sigma x)}{n} \end{aligned}\right\} \quad \text{...(B2.4c)}$$

Standard error of the estimate (**esd**) is

$$S_{xy} = \sqrt{\left\{\frac{\Sigma (y - y')^2}{(n-2)}\right\}} \quad \text{...(B2.4d)}$$

B3 PROBABILITIES

Probability, $P(E)$, of an event, E, in the sample space, S, is

$$P(E) = \frac{N(E)}{N(S)} \quad \text{...(B3.la)}$$

where $N(E)$ is the number of points in E and
$N(S)$ is the number of points in S.

$$\text{If } E = 0; \quad P(E) = 0 \quad \text{...(B3.lb)}$$
$$\text{If } E = S; \quad P(E) = 1$$

For $E \neq 0$ and $E \neq S$, the probability of an event, $P(E)$, should be between 0 and 1.

B3.1 Single Probability (P ($E1$ or $E2$))

For any two events, $E1$ or $E2$, the probability is given by

$$P(E1 \text{ or } E2) = \frac{N(E1 \cup E2)}{N(S)} \quad \text{....(B3.lc)}$$

where $N(E1 \cup E2)$ is the number of points in the union of $E1$ and $E2$ and $N(S)$ is the number of points in the same space.

B3.2 Joint Probability, $P(E1$ and $E2)$

$P(E1 \,\&\, E2)$ is the probability that both events $E1$ and $E2$ occur in the sample space, S, and is given by

$$P(E1 \ \& \ E2) = \frac{N(E1 \cap E2)}{N(S)} \qquad ...(B3.2a)$$

where $N(E1 \cap E2)$ is the number of points in the intersection of $E1$ and $E2$. For two mutually exclusive events, $E1$ and $E2$

$$P(E1 \ \& \ E2) = 0 \qquad ...(B3.2b)$$

If the probability of occurrence of any event, $E1$, is independent of the occurrence of another event, $E2$, then

$$P(E1/E2) = P(E1) \qquad ...(B3.2c)$$

The joint probability of independent events, $P(E1 \ \& \ E2)$ is

$$P(E1 \ \& \ E2) = [P(E1)]\ [P(E2)] \qquad ...(B3.2d)$$

B3.3 Conditional Probability, *P*(*E*2/*E*1)

$P(E2/E1)$ is the conditional probability of two events, wherein $E2$ will occur after the event, $E1$, has already taken place.

$$P(E1/E2) = \frac{N(E1 \cap E2)}{N(E1)} \qquad ...(B3.3a)$$

and $N(E1) \neq 0$

The addition rule in probability involves: This and That. Multiplication rule of probability of events:

$$P(E1 \ \& \ E2) = [P(E1)]\ [P(E2/E1)]$$

or ...(B3.3b)

$$[P(E2)]\ [P(E2/E1)]$$

B4 DISTRIBUTIONS

B4.1 The Binomial Distribution

The **binomial** distribution is a probability function dealing with two mutually independent populations (e.g. male and female populations etc.) and the outcome for each trial is: either/or (yes/no)— e.g. living/dead, dominant/recessive etc.

If p is one category (say success), then $q = (1 - p)$ is the other category (say failure). The binomial probability distribution is given by

$$p(x) = \sum_{x=0}^{n} \frac{n!}{x!(n-x)!} \qquad ...(B4.1a)$$

where n = size of each sample, x = number of successes, $(n - x)$ = number of failures, p = probability of success and q = probability of failure. For large values, the binomial series expansion is

$$(1 + x)^n = 1 + nx + \frac{n(n-1)\,x^2}{2!} + \frac{n(n-1)(n-2)\,x^3}{3!} + \ldots \qquad \text{...(B4.lb)}$$

For large values, the normal approximation can be introduced

$$\mu = n \cdot p$$

$$\sigma_x = \sqrt{\sum_{i=1}^{n} (x_j - \mu)^2\, p(x_j)} \qquad \text{...(B4.1c)}$$

where μ and σ are average number of successes and standard deviation, respectively.

As p is the probability of success and $q = (1 - p)$, the probability of failure for a binomial experiment of n trials, the standard deviation, σ is

$$\sigma = \sqrt{n.p.q} \qquad \text{...(B4.1d)}$$

Normal random variable, X, can be evaluated in terms of μ and σ by Z-test

$$Z = \frac{(X - \mu)}{\sigma} \qquad \text{...(B4.le)}$$

B4.2 Normal (Gaussian) Distribution

The binomial formula is useful for evaluating small samples; as the sample size, n, increases, the binomial distribution approaches the normal (**gaussian**) distribution.

Normal distribution has the following features (Fig. B2):

(i) It is a continuous, symmetrical, 'bell-shaped' curve.

(ii) The arithmetic mean, median and mode are identical.

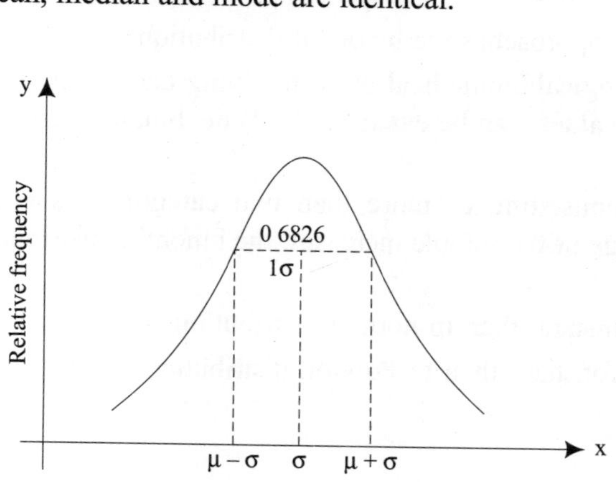

Fig. B2 Normal (gaussian) distribution curve.

(iii) The mean, μ, and the standard deviation, σ, determine the centre of the curve and the variation (symmetrical measure of the spread of the curve) around the mean.

The normal distribution function, $F(x)$, is given by

$$Y = F(x) = \frac{1}{\sigma\sqrt{2\,\pi}} \exp\left(-\frac{1}{2}Z^2\right)$$

where the variate $$Z = \left(\frac{x-\mu}{\sigma}\right) \quad \text{...(B4.2a)}$$

representing any observation, x, from the centre, μ, as a multiple of σ.

For a standard normal curve, $\mu = 0$ and $\sigma = 1$

$$\therefore \quad Y = F(x) = \frac{1}{\sigma\sqrt{2\,\pi}} \exp\left(-\frac{x^2}{2}\right) \quad \text{...(B4.2b)}$$

The normal distribution is applicable in many random events, such as Brownian motion.

B4.3 Poisson Distribution

The **Poisson** distribution describes the sampling distribution of isolated random and rare events in space or time— e.g. radioactive disintegration counts, bacterial population in liquid media, red-blood cell counts (per chamber), flow of gas molecules from a chamber through a hole, nerve fibre excitation etc.

The Poisson probability is given by

$$p(x) = \frac{e^{-\mu}\,\mu^x}{x\,!} \quad \text{...(B4.3)}$$

where $x = 0, 1, 2, \ldots$ and μ is population mean and the variance $\mu = \sigma^2$.

The Poisson distribution approaches the binomial distribution when **n** is large and **p** is small.

Experimental data (biological/biomedical etc.) involving categories, namely a set of observed values and that of expected values, can be evaluated by i) the binomial distribution test, (ii) Z-test or iii) χ^2-goodness of fit test.

A population of data consisting of more than two categories is evaluated by multinomial distribution test(s). As the size of the sample increases, the binomial distribution leads to two limiting cases:

(i) If **n** → ∞ at **p** = constant, then to normal distribution.

(ii) If **n** → ∞ at **np** = constant, then to Poisson distribution.s

C Appendix

APPENDIX C: TABLES & CONVERSIONS

The macroscopic and microscopic quantities in System Internationale (SI) Units are represented in multiple of $10^{\pm 3}$ with prefixes as given in Table C1.

Table C1 Macroscopic and microscopic quantities.

Macroscopic			*Microscopic*		
Multiplier	*Name*	*Symbol*	*Multiplier*	*Name*	*Symbol*
10^3	kilo	k	10^{-3}	milli	m
10^6	mega	M	10^{-6}	micro	µ
10^9	giga	G	10^{-9}	nano	n
10^{12}	tera	T	10^{-12}	pico	p
10^{15}	peta	P	10^{-15}	femto	f
10^{18}	exa	E	10^{-18}	atto	a

Some fundamental and derived constants in System International (SI) are presented in Table C2.

Table C2 Fundamental and derived constants in SI units.

Parameter	*Unit*	*Symbol*	*Value*	*Dimensionality*
Length	Metre	m	–	–
Mass	Kilogram	kg	–	–
Time	Second	s	–	–
Temperature	Kelvin	K	– 0°C = 273.16 K	–
Frequency	Hertz	Hz	–	s^{-1}

Contd.

Parameter	*Unit*	*Symbol*	*Value*	*Dimensionality*
Atomic mass	amu or u	μ	1.6606×10^{-27} kg	–
Velocity of Light	–	c	299.792×10^{6} $m.s^{-1}$ (in vacuum)	–
Amount of substance	Mole	mol	–	–
(Concentration)	Molarity	M	–	$mol.\ lit^{-1}$
	Molality	$\overline{M}$	–	$mol.kg^{-1}$
Avogadro Number	–	N_A	0.60225×10^{24} mol^{-1}	–
Force	Newton	N	–	$kg.m.s^{-2}$
Energy (work)	Joule	J	–	N.m
			10^{7} ergs	
			0.239 cal	
			6.242×10^{18} eV	
Pressure	Pascal	Pa	–	$N.m^{-2}$
			10×10^{-6} bar	
			7.5×10^{-3} torr	
			10 dynes. cm^{-2}	
			9.869×10^{-6} atm	
Power	Watt	W	–	$J.s^{-1}$
Planck's constant	–	h	$0\text{-}6626 \times 10^{-33}$	J.s
Gravitational constant	–	G	66.72×10^{-12}	$N.m^{2}.kg^{-2}$
Boltzmann constant	–	K_B	13.807×10^{-24}	$J.K^{-1}$
Gas constant	–	R	8.314	$J.mol^{-1}.K^{-1}$
Electric current intensity	Ampere	A	–	–
Electric charge	Coulomb	C	–	A.s
Elementary charge	–	e	0.1602×10^{-18}	C
Dipole moment	Debye	D	3.336×10^{-30}	C.m
Electric potential	Volt	V	–	$W.A^{-1}$
Electric resistance	Ohm	Ω	–	$V.A^{-1}$
Electric conductance	Siemens	S	–	$A.V^{-1}$
Magnetic flux	Weber	Wb	–	V.s
Electric inductance	Henry	H	–	$V.s.A^{-1}$
Electric capacitance	Farad	f	96.485×10^{3}	$C.mol^{-1}$
Magnetic flux density	Tesla	T	–	$V.s.m^{-2}$
Bohr magneton	–	μ_B	9.274×10^{-24}	$J.T^{-1}$
Electron **g**-factor	–	g_e	2.002532	
Luminosity	Candella	Cd	–	–
Radioactivity	Becquerel	Bq	0.027×10^{-9} curie	s^{-1}
Absorbed dose of radiation	Gray	Gy	100 rad	$J.kg^{-1}$
Dose-equivalent	Sievert	Sv	100 rem	$J.kg^{-1}$

D Appendix

APPENDIX D: IMPORTANT ELEMENTS IN BIOLOGY

Atomic numbers and masses of biologically relevant elements are given in Table D.

Table D Atomic numbers and masses of elements relevant to biology.

Element	*Symbol*	*Atomic number, **Z***	*Atomic mass, **M***
Aluminium	Al	13	26.98
Antimony	Sb	51	121.75
Argon	Ar	18	39.95
Arsenic	As	33	74.92
Barium	Ba	56	137.35
Beryllium	Be	4	9.02
Bismuth	Bi	83	209.0
Boron	B	35	10.82
Bromine	Br	35	79.92
Cadmium	Cd	48	112.40
Calcium	Ca	20	40.08
Carbon	C	6	12.01
Cesium	Cs	55	132.91
Chlorine	Cl	17	35.45
Chromium	Cr	24	52.0
Cobalt	Co	27	53.94
Copper	Cu	29	63.55
Fluorine	F	9	19.00
Gadolinium	Gd	64	157.0
Gallium	Ga	31	69.72
Germanium	Ge	32	72.60
Gold	Au	79	197.0
Hafnium	Hf	72	178.5

Contd.

Element	*Symbol*	*Atomic number,* ***Z***	*Atomic mass,* ***M***
Helium	Ne	2	4.0
Hydrogen	H	1	1.01
Indium	In	49	114.8
Iodine	I	53	126.90
Iron	Fe	26	55.85
Krypton	Kr	36	83.8
Lanthanum	La	57	138.90
Lead	Pb	82	207.2
Lithium	Li	3	6.94
Magnesium	Mg	12	24.32
Manganese	Mn	25	54.94
Mercury	Hg	80	200.60
Molybdenum	Mo	42	95.95
Neon	Ne	10	20.18
Nickel	Ni	28	58.70
Nitrogen	N	7	14.01
Osmium	Os	76	190.2
Oxygen	O	8	16.00
Palladium	Pd	46	106.5
Phosphorus	P	15	30.98
Platinum	Pt	78	195.1
Plutonium	Pu	94	242.06
Potassium	K	19	39.10
Radium	Ra	88	226.05
Radon	Rn	86	222.0
Rhenium	Re	75	186.3
Rubidium	Rb	45	102.90
Samarium	Sm	62	150.4
Selinium	Se	34	78.96
Silicon	Si	14	28.08
Silver	Ag	47	107.88
Sodium	Na	11	22.99
Strontium	Sr	38	87.63
Sulphur	S	16	32.06
Tantalum	Ta	73	180.9
Technitium	Tc	52	127.60
Thallium	Tl	81	204.38
Thorium	Th	90	232.1
Tin	Sn	50	118.7
Titanium	Ti	22	47.90
Tungsten	W	74	183.87
Uranium	U	92	238.05
Vanadium	V	23	50.95
Ytterbium	Yb	70	173.04
Yttrium	Y	39	88.9
Zinc	Zn	30	65.38
Zirconium	Zr	40	91.22

Glossary

A-band : A region of muscle sarcomere that corresponds to the myosin (thick) filaments.

Absolute configuration : Correct stereochemistry (handedness) of a chiral molecule.

Absorbance (A) : A measure of the decrease in the intensity of the transmitted light. ($A = \log I/I_0$)

Absorption edges : Discontinuities in the absorption profile of X-rays by an atom.

Actin : Thin filamental coiled-coil protein in actomyosin complex; a polymer of G-actin monomers.

Action potential (AP) : A nerve impulse potential triggered when the membrane is depolarised beyond a "threshold" value.

Action spectrum : Photo-response spectrum of a pigment molecule as a function of wavelength.

Active site : Region of an enzyme where the interaction with the substrate takes place.

Active transport : Transport of matter against a concentration gradient. Requires expenditure of energy.

Actomyosin : Actin-myosin protein complex.

Adiabatic process : Process without exchange of heat with the surroundings (no change of entropy).

Adsorption : Retention of material on the surface of an adsorber.

Aerobic : Oxygen-present environment.

Agglutination : Aggregation of particles into groups.

Alpha helix : One of the ordered secondary structural elements found in macromolecules.

Alpha rays : High energy helium nucleus ($^4He^{2+}$).

Amino acid sequence : The sequential order of the amino acids, from N-terminal towards C-terminal, of a protein/enzyme.

Amino acids : Organic molecules which are the constituents of proteins (L-amino acids).

Amorphous state : Non-crystalline material with internal random arrangement of matter.
Amphoteric : Acidic and basic nature of a molecule.
Anaerobic : Oxygen-absent environment.
Anisotropy : Non-uniformity of a physical property of a substance in different directions.
Antibody : A protein synthesised by an organism against a substance, an antigen
Antibonding : See orbitals.
Antigen : A substance that induces a specific immune response.
Antiport : Carrier-mediated transport whereby transport of one solute is linked to transport of different solutes in the opposite direction.
Antiserum : Serum containing antibodies.
Apoprotein : A protein without the prosthetic group.
Asymmetric unit : A part of a unit cell containing part of the molecule or molecules (not related by symmetry) which by symmetry operations would generate the structure in the unit cell.
Atomic mass number (Z) : Number of protons in an atomic nucleus.
Atomic scattering factor : The scattering strength of an atom, related to its atomic number (number of electrons).
Atomisation : Conversion of sample to atomic vapour.
ATPase : Adenosine triphosphatase (an enzyme).
Atrium : Upper chamber of the heart.
Aufbau principle : **Pauli**'s principle of conditions governing the electronic configuration of elements in the periodic table.
Axon : Long, thin, cable-like nerve cell structure.
Bacteriophage : A virus that multiplies in bacteria and lyses.
Bandwidth : Wavelength or frequency region.
Base-stacking : Arrangement of planar aromatic moieties at 3.4 Å apart in nucleic acids (and similar environment).
Beta rays : Electrons and positrons.
Beta sheet : β-pleated sheet; secondary structural element found in macromolecules.
Bioluminescence : See chemiluminescence.
Birefringence : Double refraction in anisotropic crystals with two rays of light with different velocities and orthogonally polarised.
Biuret : A urea derivative.
Blackbody : An object that absorbs and emits all wavelengths of radiation uniformly.
Bohr magneton : Moment of a single electron spin.
Bonding : See orbitals.
Bosons : Particles with zero or integral spin obeying **Bose-Einstein** statistics (e.g. photons).
Bremsstrahlung : Continuous X-ray spectrum produced by decelerating electrons.
Brownian motion : Random motion of particles in a solution (at molecular level).
Buffer : Medium incorporating resistance to change (e.g. **pH** buffer).

Bundle of His : A small band of cardiac muscle fibres transmitting the waves of depolarisation from atria to the ventricles during cardiac muscle contraction.

Carcinogen : Cancer causing substance.

Cell constants : See unit cell dimensions.

Centrosymmetric structure : Structure having centre of symmetry. The phase angle of a reflection is either 0° or 180°.

Cerebroside : Glycolipid found mostly in brain.

Channel-linked proteins : Membrane spanning proteins which are highly selective in allowing passive diffusion of only specific ions.

Chemical bond : Combination of atomic orbitals to form molecules.

Chemical potential : Molecular Gibbs free energy.

Chemical shift : Effective magnetic field experienced by a given nucleus depending on its chemical environment.

Chemiluminescence : Internal reactions in a solution leading to products in excited states.

Chirality : Handedness. A chiral molecule and its mirror image are not superimposable.

Chlorophyll : Principal photoreceptor in green plants.

Chloroplast : A specialised photosynthesis organelle in plants, containing chlorophyll.

Chromatography : Technique of molecular separation of compounds in a mixture.

Chromophore : A molecule with π- or **n**-electrons.

Circular dichroism : Difference in the absorbances of left and right circularly polarised light rays.

Clone : Genetically identical cells or organisms.

Closed system : A thermodynamic system where only energy, but not matter, is transported in and out of the system.

Cochlea : The "snail-like" coiled structure in the inner ear.

Codon : A triplet sequence of nucleotides in m-RNA that codes for an amino acid synthesis.

Coenzyme : A cofactor molecule (a small molecule) of an enzyme.

Coenzyme Q : An organic molecule which acts as electron carrier shuttle in the main pathway of the respiratory chain.

Cofactor : A small molecule essential for an enzyme activity (a coenzyme).

Coherence : Highly oriented. In optics, radiation that gives rise to interference with maximum fringe contrast.

Collagen : A structural protein found in bone, skin and tendon.

Colorimetry : Spectroscopic technique that is based on the colour change for the identification of compounds.

Compton scattering : Inelastic scattering of γ- and X-rays.

Concentration gradient : Change in concentration as a function of distance.

Configuration : Spatial arrangement of the constituents of a molecule (e.g. *cis* and *trans*).

Conformation : Spatial orientations of a molecule and its constituents arising out of rotations about single bonds.

Convergent evolution : Different genes leading to proteins with common functional features.

Copolymer : A polymer with two or more number of types of monomers.

Corpuscles : Particles. Ray nature of light expounded by **Newton**.

Cosmic rays : Extraterrestrial high energy radiations.

Coulombic : Force field where potential is inversely proportional to distance ($V \propto 1/r$).

Covalent bond : Chemical bond formed by sharing of electrons between atoms in a molecule.

Critical velocity : Velocity above which fluid flow is turbulent. Velocity above which the motion of a particle is not easily determinable.

Crystal lattice : Representing of repeating units (groups of atoms) by points (Bravais lattice). There are fourteen Bravais lattices—primitive(P), face-centred(C), all face-centred(F) and body-centred(I).

Crystal structure : The arrangement of atoms/molecules/groups on lattice points, three-dimensional translation of which would form a crystal.

Crystalline state : See single crystal.

Cytoskeleton : Network of protein complexes involved in cell contraction, and motility.

Dalton : A unit of molecular mass generally equivalent to the mass of hydrogen atom.

Dark reactions : Photosynthetic reactions (**Hill** reactions) to "fix" CO_2 to carbohydrates, which do not require sunlight.

Debye-Hückel sea : Environment of charged particles amidst other ionic particles.

Degeneracy : Different orbitals occupying the same energy state: Degeneracy is lifted under the influence of magnetic or electric field.

Dendrite : A neural process that receives synaptic input.

Density gradient : Spatial distribution of concentration, in a gradient.

Diastole : The dilation of the cavities of the heart as they fill with blood.

Dichroic ratio : Ratio of the difference between the absorbances of the parallel and perpendicular light beams.

Difference Fourier : A Fourier synthesis where the coefficients are the differences between the observed and calculated structure factors ($\Delta F = |F_o - F_c|$).

Diffraction grating : A polished surface with equidistant lines etched on it to produce diffraction.

Diffractometer : Electronic counting device to measure intensity data of the X-ray diffraction reflections (h k I).

Direct methods : Statistical methods of obtaining relative phases of a set of reflections, h, k, l, on the basis of the relationships among the indices and their structure factors.

Dispersion : Divergence. Variation in the velocity of propagation of a wave with wavelength.

Dispersion elements : Prism and grating elements used in spectroscopies for wavelength-dispersion and selection.

Divergent evolution : Production of different genes by mutation from a common ancestral gene.

DNA-polymerases : Enzymes responsible for the correct duplication of DNA.

Domain structure : Regions of a macromolecule that can be structurally separable into different moieties.

Doppler broadening : Broadening of spectral lines due to molecular motions.
Dosimetry : Measurement of high energy radiation.
Echocardiogram : Sonographic record of the size and movement of the heart and its valves.
Eddy current : Turbulence in fluid motion.
Electrocardiogram (ECG) : Biopotential record of the heart function.
Electrochemical gradient : Combination of solute concentration gradient as well as electrical potential gradient.
Electroencephalogram (EEG) : Biopotential record of the brain function.
Electrogenic : Process involving transfer of a net charge across a cell membrane.
Electrolyte : An ionisable substance in solution that can conduct electricity.
Electron density map : A plot of electron density (of atoms) at various points in an asymmetric unit (or unit) cell.
Electronic transitions : Transitions of molecules and atoms from one electronic state to another.
Electron-transport chain : Enzymatic machinery in the inner membrane to transport electrons down the redox potential.
Electrophoresis : Movement of charged groups under the influence of an electric field.
Elution : Separation process of adsorbed molecules in a chromatography system.
E-map : A Fourier map with phases of a set of reflections obtained from "direct methods".
Endocrine signalling : Transmission of information *via* hormones.
Endocytosis : The process of import of molecules to the interior of the cell.
Endonuclease : An enzyme that cleaves the phosphodiester bonds in the sugar-phosphate backbone of a nucleic acid.
Enthalpy (H) : Heat content of a system at constant pressure.
Entropy (S) : A thermodynamical measure of molecular disorder.
Equilibrium potential : The membrane potential at which there is an electrochemical potential equilibrium for a particular ion species.
Eucaryotes : Cell or organism with membrane-bound discrete nuclei.
Excimer : Combination of atoms that survive in the excited state only.
Exocytosis : The process of molecular transport by secretory vesicles.
Exon : That part of the gene sequence that is expressed in m-RNA.
Facilitated diffusion : Passive transport, mediated by specific carrier molecules across a lipid bilayer.
Feedback : Self-amplification process.
Fermion : Particle with non-integral spin, obeying **Fermi-Dirac** statistics (e.g. electron).
Fibrous molecule : A macromolecule with its length far greater than its width (e.g. DNA).
Fluorescence quenching : Attenuation of fluorescence due to molecular collisions.
Fourier map : A Fourier synthesis computed with diffraction amplitudes as coefficients.
Fourier transform (T) : Transform of time-domain data (signal data) to frequency-domain (spectrum) and *vice versa*.
Free energy : A thermodynamic property of a system defining the amount of energy

available for work.

Free induction decay : Time-domain decay profile (FID) of a pulse.

Friedel's law : The condition that satisfies the diffracted intensities of a set of reflections, $(I_{hkl} = I_{\overline{hkl}})$.

Friedel's pair : A pair of symmetrically-related reflections whose intensities are the same (e.g. $(I_{hkl} = I_{\overline{hkl}})$.

FT-Spectroscopy : Data acquisition in frequency-domain (spectrum) from the time-domain (signal) measurements by Fourier transformation procedures.

G-protein : GTP-binding regulatory protein involved in metabotropic signal transduction.

Gamma ray : High energy photon produced in nuclear reactions.

Gap junction : A channel that provides direct electrical coupling between cells.

Geiger counter : Geiger-Müller instrument used in high energy radiation measurements.

Gene : Ordered sequence of nucleotide sequence of a DNA.

Gene duplication : Evolution of new proteins by gene fusion and gene multiplication.

Genetic code : Assignment of codons to the twenty amino acids.

Gibbs free energy : Energy of a system at whose expense work is done at constant pressure.

Globular macromolecule : Macromolecule with compact globular shape (e.g. myoglobin).

Glycans : Polypeptides with glycosamine constituents.

Glycolipids : Lipids containing carbohydrate moieties.

Glycoproteins : Proteins with carbohydrate moieties.

G-protein : GTP-binding regulatory protein, involved in metabotropic signal transduction.

H^+-ATPase : Proton-translocating enzyme involved in the oxidative phosphorylation of ATP from ADP and P_i.

Half-life : The period during which the radioactive decay of a radionuclide is half of its initial value.

Hamiltonian : Mathematical operation to describe total energy of a system.

Helix-coil transition : See order-disorder phenomena.

Helmoltz free energy : Energy available at constant volume for work.

Heme : Prosthetic group of a macromolecule.

Histones : Basic proteins found in the chromosome of eucaryotic cells (except in sperm cells).

Hologram : A record of the interference pattern between the reference and object waves.

Hybridisation : Combination of more than one configurations leading to a new configuration.

Hydrogen bond : A polar non-covalent interaction involving a hydrogen atom and two polar atoms.

Hydrolysis : Cleavage of a molecule by the addition of water molecule(s) in that process.

Hydrophilic : Water-loving molecules and reactions.

Hydrophobic : Water-repelling molecules and reactions.

Hyperpolarisation : An increase in the membrane potential toward a more negative value.

H-zone : The light zone in the muscle sarcomere (the region between actin filaments).

I-band : The region between A-band and Z-disk of a sarcomere.

Immuno assay : Identification and quantification procedure employing antibodies.

Immunogenic : Capable of causing immune response.

Immunoglobulin : See antibody.

Incus : The middle bone of the ossicles in the mammalian middle ear.

Intercalation : Stacking interactions of planar aromatic moieties in macromolecules and their molecular complexes.

Internal conversion : Radiationless transition between two electron states.

Internal energy : Heat content of a system at constant volume.

Intron : Part of the gene that is not expressed in m-RNA.

Ion pumps : Internal membrane proteins that use cellular energy to pump ions across the cell membrane against their concentration gradients (active transport).

Ion-channel proteins : See channel-linked proteins.

Ionic bond : Attractive interaction between two oppositely charged ions.

Ionisation : Removal of electron(s) to render atom(s)/molecule(s) in ionic state.

Ionisation chambers : High energy radiation-measuring devices based on their ionisation characteristics.

Ionophore : Molecule that facilitates the transport of ions across membrane bilayer.

Ionotropic : See ligand-gated channels.

Ischemia : Deficiency of blood in a part.

Isoelectric point : pH of an ionisable species at zero net charge.

Isoelectric : State of zero electric potential.

Isotopes : Nuclides having same number of protons (Z = constant) but different number of neutrons.

Isotropic : Exhibiting the same physical property in all directions.

Laminar flow : Steady flow (non-turbulent flow) of fluids.

Ligand : A molecule that forms a non-covalent complex with another molecule (generally a macromolecule).

Ligand-gated channel : Gated ion channel that functions with response to the binding of a ligand.

Light : Electromagnetic radiation exhibiting both wave and corpuscular natures photons.

Light reactions : Photosynthetic reactions requiring light energy to produce NADH and ATP.

Line spectrum : Discrete spectral lines characteristic of the atom/molecule due to transitions between discrete energy levels.

Lipoprotein : A protein with covalently-bound lipid molecule(s).

Lysis : Rupture of cellular components.

Malleus (Hammer)	: Outermost bone of the ossicles in the mammalian middle ear.
Matrix	: Region enclosed by the inner membrane in mitochrondrion where many soluble proteins are found.
Matrix mechanics	: See quantum mechanics.
Melting temperature	: The temperature corresponding to half-complete melting transition.
Membrane potential	: Electrochemical potential that exists across a plasma membrane.
Microtubule	: One of the protein complexes of cytoskeleton.
Molecular engineering	: See protein engineering.
Monochromator	: A device to select radiation of a particular wavelength.
Monoclonal	: From a single cell.
Mössbauer spectroscopy	: Nuclear gamma-ray spectroscopy.
Motif	: Aggregate of secondary structural elements (helix, sheet and loop).
mRNA	: RNA that serves as template for protein synthesis.
Mutant	: Altercd molecule.
Mutation	: Alteration of a molecule by natural or by other means.
Myelin sheath	: Lipid sheath wound around an axon fibre providing it electrical insulation.
Myeloma	: Antibody producing tumour cells.
Myocardium	: A muscle layer of the heart.
Myosin	: Thick filamental protein complex in actomyosin protein assembly.
Na^+/K^+ pump	: Na^+/K^+ transport enzyme, an Na^+/K^+ ATPase involved in the transport of Na^+/K^+ ions across the plasma cell membrane.
Nerve impulse	: Electrical signa! along a nerve fibre.
Neuron	: Nerve cell which transmits nerve impulses.
Neurotransmitter	: Molecule which acts as a carrier/conduit of nerve impulse transmission.
Neutron	: One of the fundamental constituents of atomic nucleus with no charge.
Nodes of Ranvier	: Exposed regions in myelinated segments of axons.
Non-bonded interactions	: Non-covalent interactions, such as hydrogen bond, ionic, dipole-dipole, van der waals and dispersion interactions.
Nuclear mass number	: Number of protons and neutrons in a nucleus.
Nucleases	: Enzymes that hydrolyse nucleic acids by cleaving phosphodiester bonds.
Nucleic acid bases	: Purine and pyrimidine constituents of nucleic acids (A, G, T, U and C).
Nucleic acids	: Genetic materials. Linear polymers of nucleotides.
Nucleoside	: A nucleic acid base and a ribose (oxy or deoxy) sugar unit.
Nucleotide	: The basic unit of a nucleic acid, consisting of a base + a ribose sugar + a phosphate group.
Oligomer	: A linear aggregate of several monomers (≤ 10).
Oligonucleotide	: A small (≤ 10 monomers) nucleic acid fragment.
Oligopeptide	: A small polypeptide fragment (≤ 10 amino acids).
Oncology	: Study of tumours.
Open system	: A thermodynamic system where both energy and matter are transported **in** and **out** of the system.

Operator : In genetics, an operator is a sequence of bases in DNA to which a repressor binds. In physics, an operator is a mathematical operation performed on a function to obtain information.

Opsin : A visual pigment protein.

Optical activity : The ability of a substance to rotate the plane of polarisation of a plane polarised beam of light.

Orbital : Most probable spatial position of a particle of particular energy. Bonding orbitals lower the energy states. Antibonding orbitals raise the energy states.

Order/disorder phenomena : Change in molecular conformation/configuration due to substrate binding or other types of interactions.

Organ of Corti : The portion of the cochlea in which sound energy is transduced into neural activity.

Orthogonal system : Cartesian coordinate system.

Osmosis : Diffusion of solute across a semi-permeable membrane.

Ossicles : Tiny bones (malleus, incus and stapes) of the mammalian middle ear.

Oxidative phosphorylation : Aerobic respiration in mitochondrion to systhesise ATP, from ADP and P_i, coupled to the energy released from the electron-transport reactions.

Pacemaker : An electrically excitable heart muscle cell that fires spontaneously and rhythmically.

Pair-production : Conversion of γ-ray photon energy into matter ($e^- + e^+$).

Particle annihilation : Reverse of pair-production. Conversion of matter (particle-antiparticle pair) into energy (γ-ray).

Passive diffusion : Diffusion solely due to concentration gradient.

Pathogen : Disease producing agent (bacterium and virus).

Patterson method : 'Heavy atom' method of resolving the 'phase problem' in X-ray structure determination. Isomorphous replacement methods are also based on 'heavy atom' method.

Peptide bond : Covalent C-N linkage in a polypeptide backbone.

Permeability : Ability to diffuse.

pH : Measure of $[H^+]$ concentration in a solution.

'Phase' problem : The problem of determining the phase angle of each reflection, *hkl*, to compute an electron density map from a Fourier series with Fo's and calculated phases, ϕ_c's.

Photon : A packet (quantum) of light energy.

Photo-receptors : See receptor-proteins.

Photosynthesis : Process by which chlorophyll molecules absorb light energy to "fix" atmospheric CO_2 to produce carbohydrates in plants.

Photosystem I : Photosynthetic reaction centre, P_{700}, with absorption maximum at 700 nm.

Photosystem II : Photosynthetic reaction centre, P_{680}, with light absorption maximum at 680 nm.

Pixel : Area element in a raster for scanning.

Planck's constant : Fundamental constant, **h**, relating energy to frequency, as integral multiples of **h**.

Plasma	: Liquid fraction of blood (in biology). State of ionised matter (in physics).
Polarisation	: Orientation or orientated state.
Polymer	: A macromolecule composed of large number of covalently-linked monomers.
Polymerase	: An enzyme that polymerises nucleotides via phosphodiester linkages.
Population inversion	: Creation of situation where higher energy levels are more populated (contrary to the normal situation) than lower energy levels. Necessary condition for lasing action.
Postsynaptic	: Downstream process at a synapse.
Presynaptic	: Upstream process at a synapse.
Primary structure	: Sequential positions of amino acids of a protein.
Procaryote	: Cell or organism lacking membrane-bound, and structurally-discrete nuclei.
Promoter	: A region of DNA molecule at which RNA polymerase binds and initiates transcription.
Prosthetic group	: A non-peptide molecular unit required for the biological activity of a protein or enzyme.
Protein	: A linear polymer (other than S-S linkage) of amino acids.
Protein engineering	: Methodologies of protein modification to obtain proteins with altered functions.
Protein folding	: The three-dimensional architecture of a polypeptide.
Proton	: One of the fundamental constituents of a nucleus with positive charge ($^{1}H^{+}$).
Proton gradient	: The difference between H^{+} ion concentrations inside and outside of a membrane bilayer.
Pulse-height analyser	: Device to limit the recording pulses of preset amplitudes only.
P-wave	: The portion of the ECG associated with depolarisation of the atria.
Quantum mechanics	: Methodologies of explaining the interactions between sub-microscopic particles (matrix and wave matrixness).
QRS-wave	: The portion of the ECG related to depolarisation of the ventricles.
Quantum numbers	: Integral numbers representing the wave function according to quantum mechanics.
Quaternary structure	: The spatial arrangement of subunit polypeptides in a multi-subunit protein (e.g. hemoglobin).
Racemic mixer	: A mixture composed of equal amounts of L- and D-enantiomers.
Radiation dose	: Radiation absorbed by unit mass of the material.
Radioactive labelling	: Incorporation of a radionuclide as a marker in biological assays and functions (e.g. radio-immunoassay, autoradiography).
Radioactive tracers	: See radio (active) labelling.
Radioactivity	: Spontaneous disintegration of nuclei of higher atomic number.
Radioimmunoassay (RIA)	: A radio-immunological technique for quantitative measurement of a substance.
Radionuclide	: Radioactive isotope.

Radius of gyration	: Square root of the average of the squares of the mutual distances between all particles in an ensemble.
Raman scattering	: Inelastic scattering of light due to molecular vibrations.
Random coil	: Non-helical and non-sheet type secondary and tertiary organisation of macromolecules.
Rayleigh scattering	: Elastic scattering of light by molecules.
Receptor-proteins	: Protein complexes involved in the processes of signal transduction (photo-, mechano- and chemo-receptors etc.).
Refractive index	: The ratio of velocities of light in vacuum and in the medium.
Relaxation process	: Process by which a perturbed system tries to regain its equilibrium state.
Reliability index	: A goodness of fit between experimentally determined and calculated structure factors.
Repolarisation	: The return to resting potential of a cell that is not (electrically) active.
Repressor protein	: A protein capable of binding to an operator and preventing the initiation of transcription.
Resolution	: Degree of detail obtainable.
Resolving power	: Ability of a system/instrument to distinguish two nearest components as separate entitites.
Respiratory chain	: See electron transport train.
Restriction nuclease	: Enzyme that cleaves duplex DNA at specific nucleotide sites.
Reverse turn	: Sudden reversal of the direction of the polypeptide backbone (e.g. β-turn. hairpin turn).
Reynolds number	: The ratio of inertial to viscous force. High Reynolds numbers indicate turbulent flow of liquids.
Rhodopsin	: Enzyme complex involved in the retinal photo-reception.
Rotatory dispersion	: Variation of optical rotation with wavelength.
Salt bridge	: Polar non-bonded interaction.
Salting out	: Separation of a substance by adding salt.
Sarcoma	: Malignant tumour.
Sarcomere	: The contractile unit of a striated muscle fibre.
Scattering factor	: See atomic scattering factor.
Scintillation counter	: An instrument to measure fluorescence pulses as a measure of the intensity of incident radiation.
SDS	: Sodium dodecyl sulphate. An ionic detergent used in SDS-gel electrophoresis.
SDS-electrophoresis	: An electrophoretic technique of molecular separation based on size.
Secondary active transport	: Passive transport of a solute coupled to a primary active transport mechanism.
Secondary structure	: Ordered and coil segments of a protein.
Signal transduction	: Conversion of a signal from one form to another (e.g. mechano to electrical).
Single crystalline	: A solid with periodic arrangement of matter in all three dimensions.
Single-point mutation	: Alteration of a protein by a single amino acid.

Singlet/triplet states : Electronic states corresponding to the number of degenerate states in the absence of an external magnetic (or electric) field.

Site-directed mutagenesis : Specific mutation strategy employed in protein engineering to modify a protein structure.

Smooth muscle : Muscle without sarcomere and striated organisation.

Soma : A nerve cell body.

Somatic : Referring to body tissue.

Space group : Group of operations to convert the contents of an asymmetric unit cell into an (infinitely) extending pattern.

Space lattice : Three-dimensional arrangement of points on a lattice.

Spectral lines : Discrete energy lines arising out of transitions between allowed states.

Spin-lattice relaxation : Process of thermal equilibrium between spin system and the lattice (surrounding space).

Spin-spin relaxation : Process of equilibrium between randomly-distributed spin orientations.

Stapes (stirrup) : The innermost auditory bone of the ossicles in the mammalian ear.

Structure factor : The amplitude and phase information of reflections, *hkl*. It is the Fourier transform of the contents in a unit cell, sampled at reciprocal lattice points, *hkl*.

Substrate : A molecule that is acted on by an enzyme.

Svedberg : A unit of sedimentation.

Synapse : Transmission process between two neuronal junctions.

Synaptic cleft : Space between pre- and postsynaptic neurons.

Synaptic signalling : Transmission of information by the nervous system from one cell to the other.

Synchrotron radiation : Radiation emitted by high-energy electrons. It has high intensity and continuous spectral range.

Syncitium : A group of cells forming a functional continuum.

Systole : A portion of the cardiac sequence when the heart muscle is contracting.

Target theory : Empirical methods to correlate dose-response relationship in radiation therapy.

Tautomers : Isomers with different chemical properties.

Template : A master mold molecule for synthesis of a macromolecule (one strand of a nucleic acid).

Tertiary structure : The three-dimensional structure of a polypeptide.

Tetanus : Uninterrupted muscle contraction.

Thermistor : Thermal resistor.

Thermogram : A plot of variation ofa physical parameter as a function of temperature.

Thermophilic : Heat-loving or heat-tolerant.

Thick filament : Myosin filament.

Thin filament : An F-actin filament.

Thylakoid : Membrane disc containing photosynthetic organelles.

Torsion angle : Rolation about a single bond.

Trajectory : Path of a particle in motion (according to classical physics).

Transcytosis : The process of transporting macromolecules from one cell to another.

Transducin : A GTP-binding protein, $G_{t\alpha\beta\gamma}$, involved in the photo-signal amplification.

Transduction : Process of converting one form of energy to another.

Tropomyosin : Coiled coil fibrous protein; a component of thin filaments.

Troponin complex : Three-subunit protein complex that regulates the muscle contraction.

Turbulence : Fluid flow with chaotic variation in time.

Tyndall scattering : Scattering of light by colloidal particles.

T-wave : The portion of the ECG associated with repolarisation (relaxation) of the ventricles.

Ubiquinone : See Coenzymc Q.

Uncertainty principle : Heisenberg's hypothesis elucidating the indeterminant nature of observables at submicroscopic levels.

Unit cell : Imaginary parallelopiped that contains the basic unit of a crystal structure.

Unit cell content : The number of formula units in a unit cell.

Unit cell dimensions : Unit cell parameters, a, b, c (in Å) and angles, α, β, γ (in degrees).

Van der Waals interactions : Non-covalent interactions between atoms and molecules (see also non-bonded interactions).

Vibrational transitions : Molecular transitions due to bond stretching and bending (studied by Raman and IR spectroscopies).

Voltage-gated : Membrane spanning ion channels whose functions are sensitive to membrane potential.

Volterra system : Non-equilibrium, open systems in a state of perpetual change.

Voxel : Volume element in raster scan.

Wave mechanics : See quantum mechanics.

Waveguide : Channelled-wave propagation (due to total internal reflection).

Wobble hypothesis : The ability of non-Watson–Crick type base-pairing of nucleic acid bases in codon–anticodon interactions.

X-rays : High-energy radiation (0.1 to 10 Å) produced by bombardment of heavy element targets with fast electrons.

Z-band : A narrow band at either side of the muscle sarcomere.

Zeemann effect : Splitting of degenerate energy levels under the influence of an external magnetic field.

Zwitter ion : Molecule/compound of dipolar nature, which can be acidic or basic depending on the pH.

Index

D

E